Wheat

Wheat

Environment, Food and Health

Mike J. Gooding
Aberystwyth University,
Aberystwyth, UK

Peter R. Shewry
Rothamsted Research,
Harpenden, UK
&
University of Reading,
Reading, UK

WILEY

This edition first published 2022

Registered Offices
John Wiley & Sons, Inc., 111 River Street, Hoboken, NJ 07030, USA
John Wiley & Sons Ltd, The Atrium, Southern Gate, Chichester, West Sussex, PO19 8SQ, UK

Editorial Office
The Atrium, Southern Gate, Chichester, West Sussex, PO19 8SQ, UK

For details of our global editorial offices, customer services, and more information about Wiley products visit us at www.wiley.com.

Wiley also publishes its books in a variety of electronic formats and by print-on-demand. Some content that appears in standard print versions of this book may not be available in other formats.

Library of Congress Cataloging-in-Publication Data

Names: Gooding, Mike J., author. | Shewry, Peter R., author.
Title: Wheat : environment, food and health / Mike J. Gooding, Aberystwyth University, Aberystwyth, UK, Peter R. Shewry, Rothamsted Research, Harpenden, UK & University of Reading, Reading, UK.
Description: Hoboken, NJ : Wiley, 2022. | Includes bibliographical references and index.
Identifiers: LCCN 2021046095 (print) | LCCN 2021046096 (ebook) | ISBN 9781119652557 (hardback) | ISBN 9781119652588 (adobe pdf) | ISBN 9781119652595 (epub)
Subjects: LCSH: Wheat. | Wheat–History. | Wheat trade–History.
Classification: LCC SB191.W5 G665 2021 (print) | LCC SB191.W5 (ebook) | DDC 633.1/1–dc23
LC record available at https://lccn.loc.gov/2021046095
LC ebook record available at https://lccn.loc.gov/2021046096

Cover Design: Wiley
Cover Images: Courtesy of Richard Casebow, Crops Research Unit, University of Reading, UK; Courtesy of Mike Gooding; Courtesy of Rothamsted Research Ltd

Set in 9.5/12.5pt STIXTwoText by Straive, Pondicherry, India
Printed and bound by CPI Group (UK) Ltd, Croydon, CR0 4YY

C9781119652557_190122

We would like to dedicate this volume to our friends and colleagues Prof. W. Paul Davies (1949–2016) and Dr. Donald D. Kasarda (1933–2021).

Contents

Foreword

Wheat was one of the first cereals to be domesticated and it is now produced on over 700 million hectares globally, making it the most widely grown crop worldwide. This is attributable to its wide adaptability across climate and elevation ranges, as well as its central contribution to human diets. The energy-rich grain is also the major vegetable protein source in human diets and, alongside maize and rice, the cereal staples underpin food and nutrition security. Importantly, wheat grain can be stored and processed into a diverse range of end products.

The versatility of wheat as both a field crop and as a grain commodity make it unique. There are many strands of knowledge, research, and breeding that have contributed to the wheat that is grown around the world today. Many features are covered in this book, including the major facets of wheat production (soils, water, temperature, sunshine), the optimal characteristics and management of the canopy to deliver harvestable outputs, and the wheat grain itself. A wealth of information describes the structure, development, and composition and quality for processing and health of wheat grain, all vital for its versatility as a food and feedstuff. Finally, the genetic improvement of wheat continues apace. Much recent innovation has uncovered previously intractable aspects of the wheat genome and we see accelerated progress in understanding genetic improvement. This expertly assembled book brings together a wealth of knowledge on wheat production and processing along with the exciting prospects of modern wheat breeding methods. The increasing global population and climate change threats necessitate our use of the synthesis of this information in continuing to breed and produce the wheat of the future.

Dr. Alison Bentley
Director, Global Wheat Program, CIMMYT

Acknowledgements

We have studied wheat for many years and benefitted greatly from discussions and collaborations with colleagues in our UK institutions (University of Reading, Harper Adams University, Royal Agricultural University, IBERS/University of Aberystwyth, Rothamsted Research, and Long Ashton Research Station/University of Bristol), in other UK bodies (universities, research institutes, and industry), and internationally. These people are too many to list individually!

We are grateful to Dr. Keith Goulding (Rothamsted Research) and Dr. Simon Griffiths (John Innes Centre, Norwich, UK) for their constructive comments on the text relating to soils and the environment and to genetic improvement, respectively. We thank VSN International Ltd. for the use of Genstat for data analysis throughout this book. We are also grateful to a number of colleagues who have provided advice, figures or datasets: Dr. Frederic Beaudoin, Dr. Yongfang Wan, Dr. Alison Lovegrove, Dr. Cristina Sanchis-Gritsch, Dr. Andy Neal, Margaret Glendining (all Rothamsted Research), Dr. Brittany Hazard (Quadram Institute Bioscience, Norwich), Prof. Filiz Koksel (University of Manitoba, Canada), Dr. Paola Tosi (University of Reading, UK), Dr. Katy Moore (University of Manchester, UK), Dr. Simon Berry (Limagrain UK Ltd.), Dr. Louise Wells (ATC, Maidenhead, UK), Professor Marie Sofie Møller (DTU, Lyngby, Denmark), and Dr. Ellen F. Mosleth (Nofima, Ås, Norway).

Finally, would like to thank Sue Steele (Rothamsted Research) and Ruth Gooding for their help throughout the project.

1

Wheat and Humans

An Introduction to the Development and Utilisation of the Wheat Crop

Nobody is qualified to become a statesman who is entirely ignorant of the problem of wheat.

Socrates

1.1 Wheat Production in the Past and Present

1.1.1 Co-Evolution of Wheat Production and Human Societies

The wheat story is one of growing mutual reliance between a remarkable mammal, us (*Homo sapiens sapiens*), and a remarkable group of grain-bearing grasses, the wheats (*Triticum* spp.). This relationship has affected our societal development and the way we spend our lives. At the same time, we have manipulated wheat plants to our own designs: the domesticated types of wheat grown by farmers fail to establish and flourish in the wild. The most widely grown wheat species, *Triticum aestivum*, appears never to have existed outside of cultivation.

We are increasingly aware that the interdependency between wheat and humans has implications for our health and environment. Our understanding of this association informs how we plan to maintain food supplies, how we attempt to sustain the quality of our nutrition, and how we strive to protect the ecosystems on which we depend: for ourselves and future generations.

It has not always been like this. For at least 95% of their time on earth, modern humans existed as sparse groups of mobile hunter-gatherers. From around 12 000 years ago (before present, BP), however, there is scattered evidence of sedentism. It is not certain that this change from nomadic to sedentary behaviour resulted from the need to tend plants such as wheat, but from 10 500 to 8000 BP there is evidence of increased planting. Over the following millennia, the settlements became organised into states which were largely founded, funded, and organised around the production of grain from grasses such as wheat (Scott 2017). Increases in production have been associated with political and military strength (Perkins 1997) while interruptions in the supply of wheat and its products continue to contribute to famine, civil unrest, and conflicts (Bellemare 2015).

Wheat: Environment, Food and Health, First Edition. Mike J. Gooding and Peter R. Shewry.

If the mobility of the hunter-gatherers vs. sedentism, and freedom from states vs. state control are articulated as conflicts (e.g. Scott 2017), then we should also recognise a further battle, one in which humans have only recently joined: that between the trees and the grasses (the Poaceae). This *war* has been playing out since the origins of the grasses, perhaps 70 million years ago (MYA) (Thomas 2017). The grasses are flowering plants (angiosperms) that have one seed leaf (monocotyledon). They produce a fruit of one seed (a caryopsis) with a mature embryo to facilitate rapid establishment. Compared with most trees, grasses are drought tolerant (Chapter 3). They can either recover or establish rapidly after dry spells and fires (Chapman 1996). Many of them, including wheat (Gooding et al. 1998), can survive being grazed for a significant part of their life cycle; again, this is an advantage over the trees. Humans have strongly backed the grasses: through clearance of forests by fires; through grazing with livestock; and most recently through soil cultivation and planting. Today the cultivated grasses are commonly grown on soils that developed under trees (Chapter 2). Grasses, including those grown for grain (717 million ha [Mha]) and for sugar cane (27 Mha), with the land covered in permanent meadows and pastures (3400 Mha; O'Mara 2012), together account for 85% of the world's agricultural land.

Wheat is a cereal. The cereals are grass plants grown primarily for their seeds to be used as a foodstuff. We name them after the Roman god of food plants, Ceres. For early gatherers and then farmers, cereal grains provided a concentrated source of nutrients. Wheat, like other cereals, is multifunctional, providing carbohydrates (mostly starch), protein, fat, minerals, vitamins, and fibre. Growing wheat would have been comparatively easy; wild wheats still grow in dense stands today without assistance (Evans 1993; Fu et al. 2019). Although the natural dispersal mechanisms of wheat were compromised during domestication, a seed merely falling on good soil has always been capable 'of multiplying thirty, some sixty, some a hundred times' (Mark's Gospel c. 65 Christian Era [CE]). Some early forms of agriculture just involved broadcasting seeds onto the fertile and damp silts left by a retreating annual river flood (Scott 2017). In a modern-day equivalent, in some rice-wheat systems, the wheat is still established by broadcasting seed onto wet soils before or after the rice harvest (Erenstein and Laxmi 2008). Cereal grains are dry, easy to reap, simple to store, and readily transported (Evans 1993). In comparison, roots and tubers are moist, bulky, perishable, and more labour intensive to harvest and move. Compared to the early legume grain crops such as peas, beans, chickpeas, and lentils, cereal growth is more *determinate*. This meant that the cereal grains tended to ripen together rather than over an extended duration, depending on their position within an inflorescence and on the maturity of different stems and branches. The cereals could, therefore, be harvested efficiently over a shorter period. In terms of importance for state building, Scott (2017) emphasises the suitability of cereals for paying tribute and taxes, and for providing rations and payment. The cereal crops were 'visible, divisible, assessable, transportable' (Scott 2017). Additionally, cereal straw was valued for multiple uses, including roofing, livestock fodder, and animal bedding (Sinclair 1998).

The early adoption of the cereals, and their subsequent importance for the growth and functioning of states, had consequences that increased their dominance to current times. Economically and politically, it is easier to justify investment in research, plant breeding, fertilizer and irrigation technology, agrochemical discovery, machinery improvement, advisory and education services, processing technology, and trading efficiency for the most important crops such as wheat (e.g. Perkins 1997). Investment results in greater competitive advantage and further increased status. This is a continuing process that contributes to other pressures for agricultural systems to become simpler. That is, there is a tendency to focus on fewer crops, which are grown and processed with increasing efficiency.

World crop production is now dominated by the cereals (Figure 1.1). Nine of the twenty-two most important grain and oilseed crops are cereals, and they account for c. 85% of all grain and oilseed production. In comparison with the roots and tubers, the cereal area of over 700 Mha dwarfs that of cassava (27 Mha), potatoes (17 Mha), yams (9 Mha), and sweet potatoes (8 Mha). Similarly, the major perennial plantation crops of sugar, tea, bananas, plantains, coffee, cocoa, coconuts, rubber, and oil palm *only* account for c. 125 Mha in total. Wheat is one of the few major crops: one of the *big three* with maize and rice (Figure 1.1c).

The major factors in the widespread adoption of wheat, leading to its current pre-eminence, have been its perceived superiority for human consumption, its unique processing properties, and its wide adaptation to different climatic conditions. Consequently, while barley, oats, and rye have been the dominant cereals in different times and places, even in temperate areas, none have matched the reputation of wheat as a human food. Indeed, higher proportions of barley, oats, and rye were traditionally relegated to the feed of livestock and draft animals (Wrigley 2009). Wheat is now produced on a greater area and grown over a wider geographic range than any other arable crop. It is also the arable crop most traded internationally as a commodity (Figures 1.1 and 1.2). It accounts for over a quarter of all cereal production. Wheat alone provides 20% of the calories and protein in

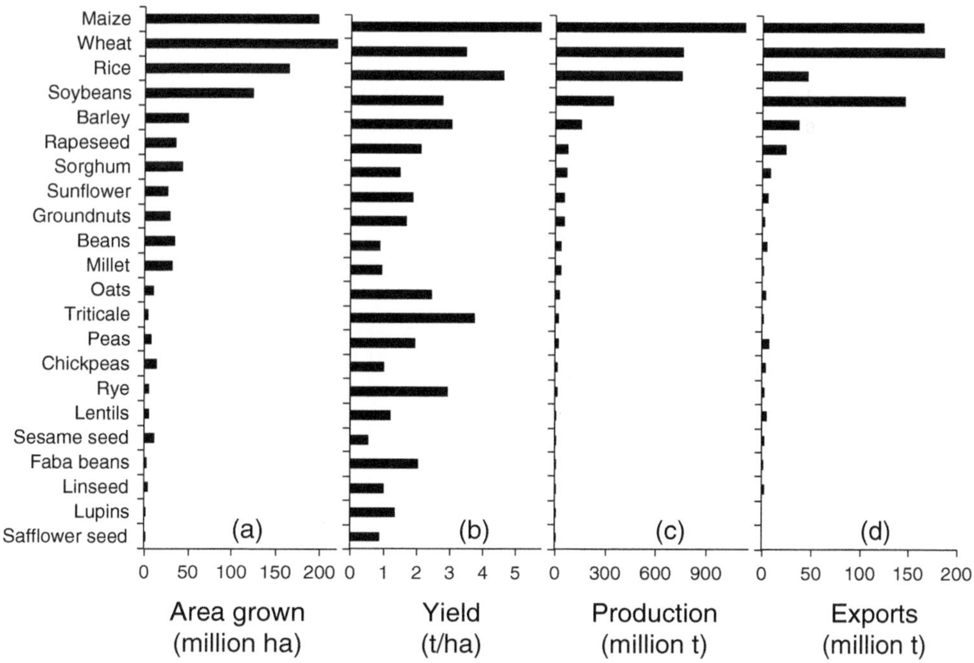

Figure 1.1 The world production and exports of grain and oilseeds. *Source:* Data are means of 2015–2019 from FAOSTAT (FAO 2021).

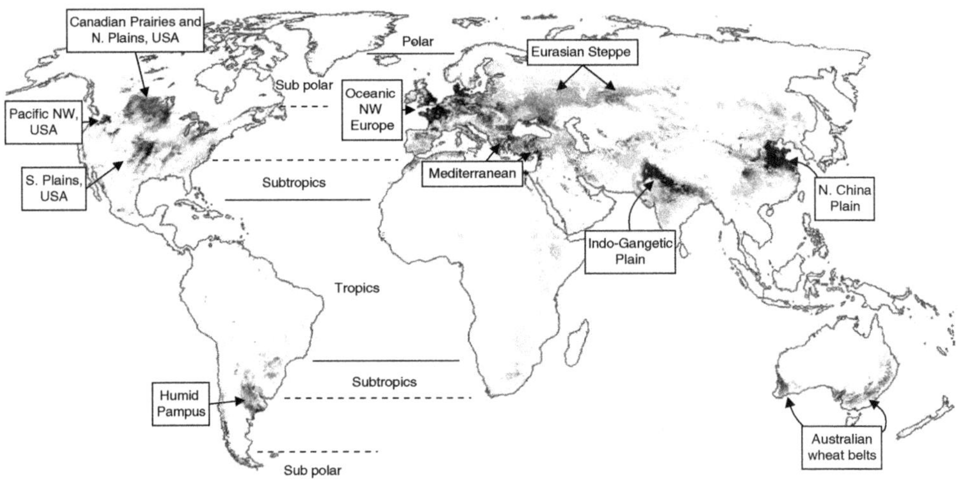

Figure 1.2 The global distribution of wheat production. Shading intensity represents estimated yield × area per pixel. Further details on the climate of the labelled major areas of production given in Figure 1.6. *Source:* Data from Monfreda et al. (2008). Mapping modified from Andrew MT, CC BY-SA 3.0, https://commons.wikimedia.org/w/index.php?curid=9643590

the global human diet, a greater contribution than from either maize or rice (Shiferaw et al. 2013). Dependency on wheat is particularly high in areas of North Africa, the Middle East, and Central Asia, where wheat often contributes more than 50% of daily calorific intake (e.g. Figure 1.3). The top 10 wheat-producing countries are China, India, Russian Federation, USA, France, Canada, Pakistan, Ukraine, Australia, and Germany (Table 1.1). These account for 70% of global production, with Asia and Europe accounting for 44% and 34%, respectively.

1.1.2 Wheat Supply and Demand

As stated previously, there have been significant pressures to invest in and therefore further increase the dominance of the major cereals. There is undoubtedly a trade-off between increasing the efficiency of production through specialization and increasing the resilience of systems through exploiting diversity (Urruty et al. 2016; Dardonville et al. 2020). Nonetheless, investment and focus on wheat have enabled production to be increased by more than threefold over the past 60 years (Figure 1.4). Indeed, the rate of growth has significantly exceeded that of the number of humans on the planet, and has met further demands arising from changing eating habits. Of note is the increase in the per capita availability of wheat between 1960 and 1990 in populous countries such as India and China, despite rapid population growth during the same period (Figure 1.3). Furthermore, the increase in global wheat production since 1960 occurred without using more land (unlike rice, maize, and soybean). Global production of wheat, therefore, closely follows the increases in yield per unit area (Figure 1.4). In other words, we have become much more efficient in the use of land for wheat production (Dibb 2000). There are clear threats to being able to sustain or increase land-use efficiency to produce wheat, but these challenges

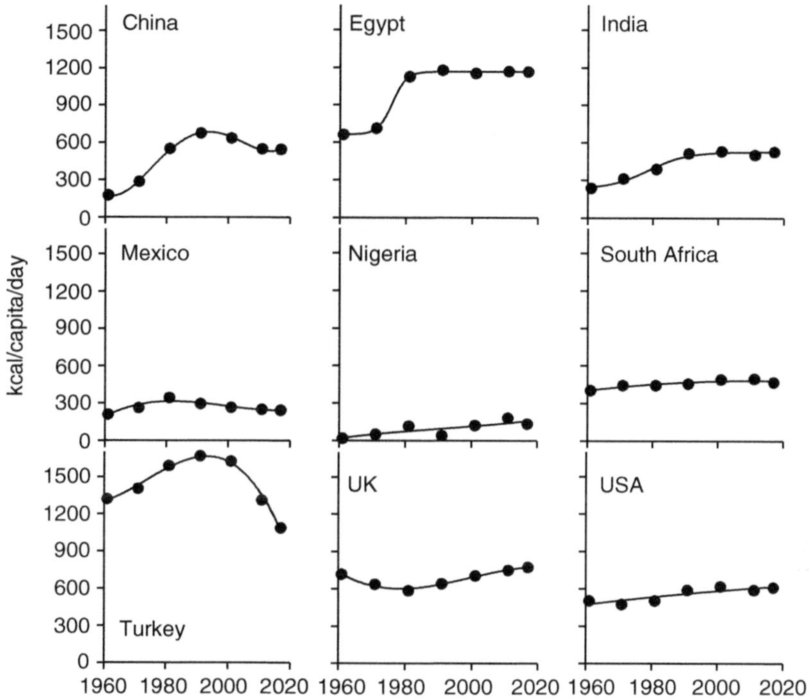

Figure 1.3 Changes in the availability of wheat (kcal/capita/day) in the diets of nine countries. *Source:* Data from FAOSTAT (FAO 2021).

need to be set against the environmental impacts of using more land or the difficulties of achieving significant changes to human diets (e.g. Smith et al. 2018; Cassman and Grassini 2020). It is salutary to reflect that many past civilisations have failed to meet the challenge of satisfying short-term demands for food, without severely impairing the ability of future generations to do likewise (Hillel 1991; Montgomery 2007; Mithen 2012). However, given current trends in population growth and diet, which also drive increased use of grain for livestock production, it is predicted that demand for grain could increase by around 50% between 2020 and 2050. It is highly unlikely that new crops and/or novel cropping systems will be able to provide the intensification of production that will be required to meet this increased demand. This argument will hold even if the increase in demand is limited to 25% by dramatic changes to the relationships between income, urbanization, and diet seen over recent decades (Cassman and Grassini 2020). There is, therefore, continuing pressure to increase the efficiency with which we use land to produce wheat and other grains.

Despite the past success in increasing grain production (Figure 1.4), the last 30 years have seen food price spikes (Figure 1.5) associated with perturbations in the supply and demand for wheat. These *shocks* have resulted from many factors, including extreme weather events in major wheat-growing areas, changes in land-use policy (particularly regarding biofuel production) affecting the area of food grains, the oil price, the behaviour of financial markets, and the buying and selling strategies of major wheat exporters and importers seeking

Table 1.1 Production, yield, and area of wheat in the world and selected regions and countries.

Region/Country	Production (10^6 t)	Yield (t/ha)	Area (10^6 ha)
World	752	3.5	217.3
Regions			
Asia	331	3.3	99.3
Europe	256	4.1	62.7
Americas	114	3.3	34.7
Africa	26	2.6	10.1
Australia/NZ	25	2.2	11.5
Top 10 countries			
China	133	5.4	24.5
India	97	3.2	30.3
Russian Federation	77	2.9	27.1
USA	54	3.3	16.3
France	35	6.5	5.4
Canada	31	3.3	9.3
Pakistan	26	2.9	9.0
Ukraine	26	4.0	6.4
Australia	25	2.3	11.5
Germany	23	7.3	3.2
Other countries			
UK	14	8.0	1.8
Kazakhstan	15	1.3	11.9

Source: Data are means of 2016 to 2018, from FAOSTAT (FAO 2021).

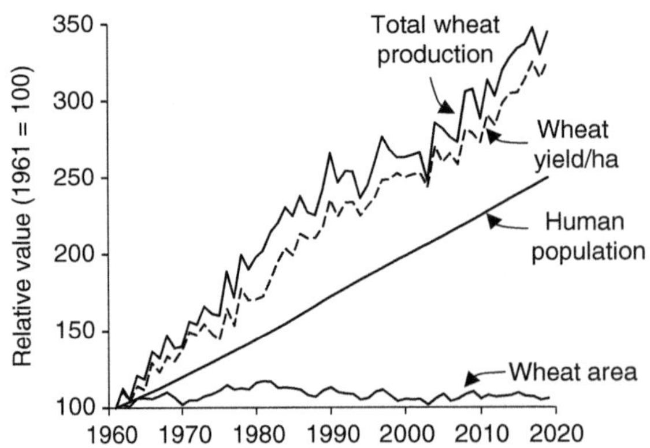

Figure 1.4 Growth in wheat production since 1961 compared with the wheat area, the wheat yield per unit area, and human population growth. *Source:* Data from FAOSTAT (FAO 2021).

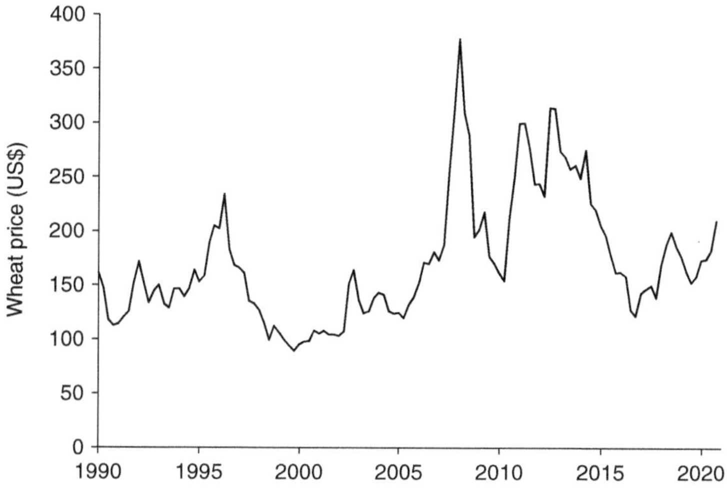

Figure 1.5 The world primary commodity price of wheat (IMF 2021).

to secure domestic supplies (Tadesse et al. 2016). The most dramatic spike in recent decades was in 2008; this represented a fourfold increase in the wheat price compared to that of eight years earlier (Figure 1.5). The origins of this spike have been attributed to drought, crop disease, grain use for biofuels (and therefore the oil price), and increasing demand resulting from the economic growth of China and India. Wheat price spikes result in increased poverty and have a disproportionately harmful effect on the livelihoods and well-being of the poor (Anderson et al. 2013).

1.1.3 Wheat Adaptation

Much of the current success of wheat, as well as the hopes for its future production, relies on the adaptability of different cultivated varieties (*cultivars*) of wheat to a wide range of growing environments and production systems. Wheat production stretches through temperate areas, extending into the sub-tropics and the sub-polar regions, but also includes the tropical highlands of East Africa. Major areas of production (Figure 1.2) include river plains such as the Indo-Gangetic plain (Figure 1.6h) and the North China plain (Figure 1.6i), where wheat is often irrigated and grown in the dry winters while rice and maize are common summer crops. By contrast, wheat is also grown over large mid-continental expanses that would have originally supported extensive grasslands and/or forests such as the Eurasian Steppe (Figure 1.6e), the Canadian Prairies (Figure 1.6a), and the plains of the USA (Figure 1.6c). At high latitudes, wheat is frequently grown in the summer months to avoid the harsh winters (Figure 1.6a). In semi-arid regions, such as some areas in the Pacific Northwest (Figure 1.6b), rainfed wheat is often grown every other year whilst water is replenished in soils left to fallow in intervening years. In contrast, intensive production in Oceanic NW Europe (Figure 1.6e) is made possible by evenly distributed rainfall and moderate temperatures. A wet, mild environment is also experienced in the Humid Pampas of

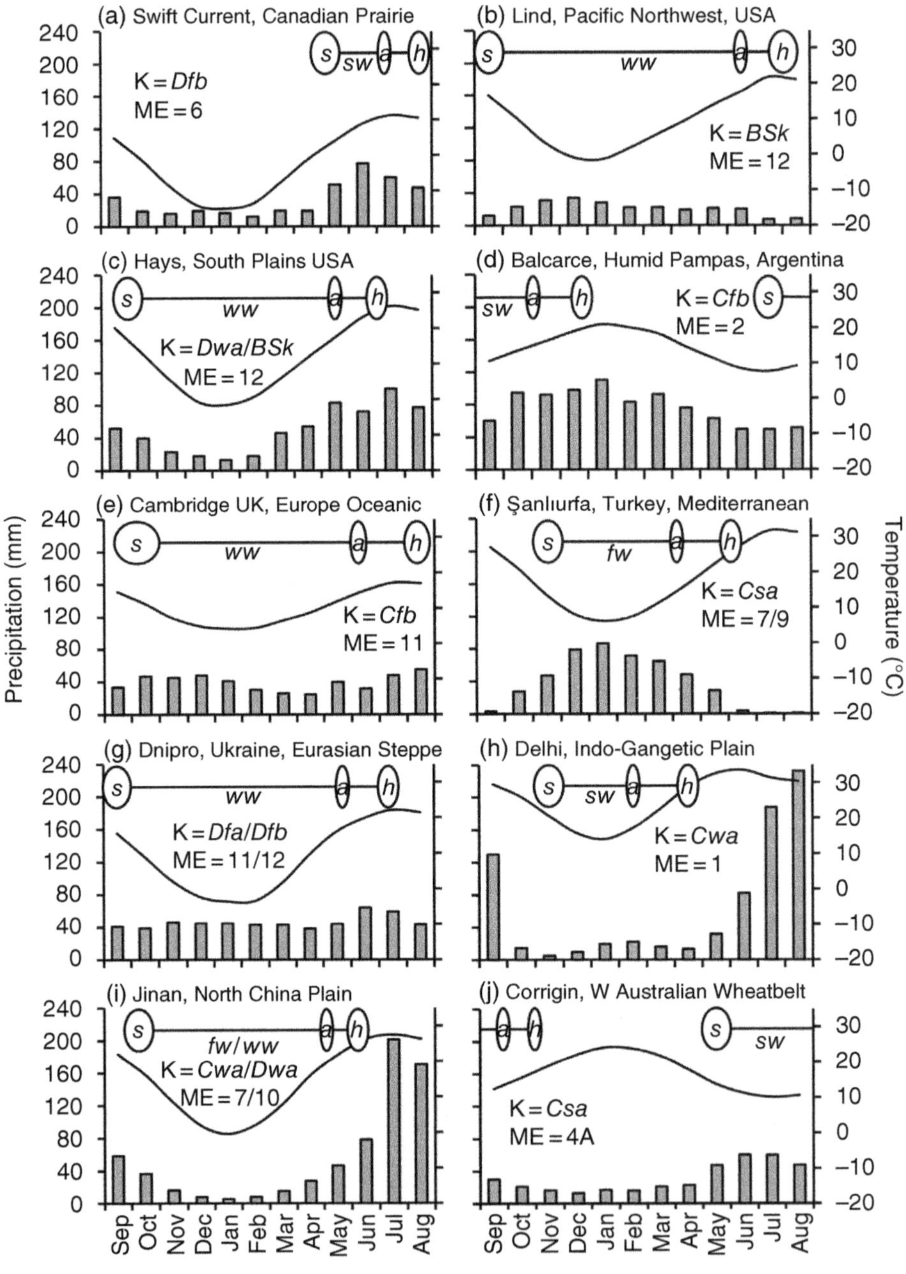

Figure 1.6 Mean temperature (line) and precipitation (bars) for locations in the 10 wheat growing regions labelled in Figure 1.2. Typical timings for sowing, flowering (anthesis), and harvest at location are denoted by *s*, *a*, and *h*, respectively. Wheat types (Section 1.2) are spring wheat (*sw*), facultative wheat (*fw*), and winter wheat (*ww*). K is the Köppen climate classification (Table 1.2), and M is the mega-environment (Table 1.3) for the location. Note: both climate and mega-environment can vary significantly within the growing regions. Mega-environments and wheat types (and hence different cropping seasons) can also overlap at the same location (Gbegbelegbe et al. 2017).

Argentina (Figure 1.6d), where three crops are often grown (e.g. wheat, soybean, maize) in two-year rotations. The wild wheats still growing in and around the eastern Mediterranean (Zohary et al. 1969), like their modern crop counterparts, are adapted to grow in the mild, wet winters, maturing before the hot dry summers (Figure 1.6f). The Australian wheat belts span numerous climatic zones, including the Mediterranean pattern (Figure 1.6j).

In the Köppen climate classification system, wheat production spans the semi-arid, temperate, and continental zones (Table 1.2). For breeding purposes, particularly for the developing world, the International Maize and Wheat Improvement Centre (i.e. CIMMYT: Centro Internacional de Mejoramiento de Maíz y Trigo) has divided wheat-growing areas into mega-environments, which again emphasise the wide adaption of the crop (Table 1.3).

It is evident that the different growing environments contribute to a diversity of cropping systems, varying in sowing and harvest dates, rotational contexts, and the use of irrigation. The environment also plays a large role in determining the justification of inputs such as mineral fertilizers and crop protection agrochemicals. For example, although national production levels of wheat in the UK and Kazakhstan are similar, the two countries contrast in how these levels are achieved (Table 1.1). In Kazakhstan, low yields per hectare are produced over a large area where wheat growth is mostly limited by low water availability (Chapter 3) and high temperatures (Chapter 4). Much of the wheat is planted as late as May, giving a short growing season, but still exposed to abiotic stresses. By contrast, in the UK high land-use efficiency is achieved on a much smaller area. Growth proceeds in clement conditions where yield is, therefore, mostly limited by light interception (Chapter 5). Most of the wheat grown in the UK is planted early in the autumn and proceeds to harvest over a period of 10–11 months. If, as in the UK, abiotic stresses are less frequently severe, wheat crops are particularly responsive to inputs, such as fertilizer and agrochemical applications. Together, these inputs serve to maximise light capture and therefore photosynthesis and yield (Chapter 6).

The environment clearly has an impact on wheat supply and, conversely, wheat influences the environment. If we are to address the sustainability of wheat production, both factors need to be understood. Our current reliance on cereal grain, and on wheat in particular, has significant implications for human health and well-being. The relationships between wheat production, the environment, and human diet and health are therefore the overarching themes of this book.

For the remainder of this chapter, we will introduce three of the main factors that account for the global success of the wheat crop. Firstly, the wheat plant is described in terms of its growth and development as a grass. Secondly, we introduce the evolution of wheat and explain its wide adaptability. Finally, we introduce the unique processing properties of wheat and the diversity of wheat grain types and wheat-based products.

1.2 The Wheat Plant

As already said, wheat, like other cereals, is a grass plant grown primarily for its edible seed. The lineages of wheat, rice, and maize may have diverged from a common ancestor about 40 MYA (Huang et al. 2002; Gill et al. 2004). The wheats (*Triticum* spp.) form part of the Triticeae tribe, which also includes closely related cereals, notably barley (*Hordeum vulgare*) and rye (*Secale*

Table 1.2 Köppen climate classifications used in Figure 1.6.

Grouping	Water distribution	Temperature description	Composite classification
B Desert/ semi-arid			
	S Semi-arid[a]		
		k cold[b]	*BSk*
C Temperate/ mesothermal[c]			
	s Dry summer[d] (Mediterranean)		
		a Hot summer[e]	*Csa*
	w Dry winter[f] (e.g. subtropical)		
		a Hot summer[e]	*Cwa*
	f No dry season (e.g. oceanic)		
		b Warm summer[g]	*Cfb*
D Continental/ microthermal[h]			
	w Dry winter[f]		
		a Hot summer[e]	*Dwa*
	f No dry season		
		a Hot summer[e]	*Dfa*
		b Warm summer[g]	*Dfb*

Definitions:
[a] 'Semi-arid' is where the mean annual precipitation is > 5× but < 10 × a threshold precipitation. When precipitation is relatively evenly distributed, the threshold precipitation is 2 × mean annual temperature (°C) +14.
[b] 'Cold' is where mean annual temperature is < 18 °C.
[c] 'Temperate' is where the mean temperature of the hottest month is > 10 °C and the coldest month is between 0 and 18 °C.
[d] A 'dry summer' is where the precipitation in the driest summer month is < 40 mm and also less than a third of the wettest month in winter.
[e] A 'hot summer' is where the mean temperature of the hottest month is ≥ 22 °C.
[f] A 'dry winter' is where the wettest winter month has less than a tenth of the precipitation than the wettest summer month.
[g] A 'warm summer' is where the hottest month is < 22 °C but there are at least four months where the mean temperature is > 10 °C.
[h] A continental climate is where the hottest month is > 10 °C but the coldest month is ≤ 0 °C.
Source: adapted from Peel et al. (2007).

cereale), as well as many wild grasses, such as the weed common couch grass (*Elymus repens*). The lineages of wheat and barley appear to have separated 10–14 MYA, and of wheat and rye 7 MYA.

The Triticeae have common structures and patterns of development typical of many grasses. It is normal to define phases of development in terms of growth stage scores

Table 1.3 Mega-Environments (ME) for wheat.

ME	Wheat type	Agro-climatic conditions
1	Spring	Irrigated. Temperate in winter to late heat stress.
2	Spring	High rainfall (> 500 mm during cropping season). Temperate.
3	Spring	High rainfall. Temperate. Acid soils.
4A	Spring	Low rainfall. Winter rain followed by drought. Temperate.
4B	Spring	Low rainfall. Early drought preceding summer rain. Temperate.
4C	Spring	Low rainfall. Soil water stored from monsoon. Temperate.
5A	Spring	High rainfall. Hot (coolest month > 17.5 °C).
5B	Spring	Irrigated. Hot.
6	Spring	Moderate rainfall. Winter would kill wheat. Coolest month < −10 °C.
7	Facultative	Irrigated. Cool.
8A	Facultative, daylength sensitive	High rainfall. Cool.
8B	Facultative, daylength insensitive	High rainfall. Cool.
9	Facultative	Low rainfall (< 500 mm available to crop). Cool.
10A	Winter	Irrigated. Cold winter.
10B	Winter	Rainfall + irrigation. Cold winter.
11A	Winter, daylength sensitive	High rainfall. Cold winter.
11B	Winter, daylength insensitive	High rainfall. Cold winter.
12	Winter	Low rainfall. Cold winter, hot summer.

Source: Adapted from Rajaram et al. (1993) and Gbegbelegbe et al. (2017).

(Table 1.4; Figure 1.7). Growth stage scores are used extensively to interpret the effects of the environment on crop growth and yield, and to optimize the timings of agronomic treatments including irrigation and the application of fertilizers, growth regulators, and crop protection agrochemicals (Barber et al. 2015). They can also be used to define safe grazing periods for dual-purpose crops (Gooding et al. 1998). The most widely used growth stage scoring method for wheat, and that used throughout this book, is the decimal growth stage (DGS) system of Zadoks et al. (1974). Equivalents between the Feekes and Zadoks scales are given in Table 1.4. Equivalents with lesser-used scoring systems are provided by Landes and Porter (1989) and Harrell et al. (1993, 1998).

The growth of wheat can be broadly divided into two phases: the vegetative phase and the reproductive phase. The distinction between the two phases depends on events in the bud tip, or stem apex (i.e. the apical meristem if referring to the main stem) of the plant.

Table 1.4 Growth stage scores of wheat plants.

	Decimal growth stages (DGS) according to Zadoks et al. (1974)		The scale of Feekes (1941)		
Major growth phases	**Major division**	**Minor divisions described by *n***	**Score**	**Description**	**Equivalent DGS score**
Germination	0*n*	*n* indicates development before leaf emergence i.e. from dry seed ($n = 0$) to first leaf at coleoptile tip ($n = 9$)			
Leaf production	1*n*	*n* is the number of unfolded leaves on the main stem to a max. $n = 9$	1	Seedling emergence to three leaf stage	10–13
Tiller production	2*n*	*n* is the number of tillers with their own unfolded leaf to a max. $n = 9$	2	Start of tillering	21–25
			3	End of tillering	26–29
Stem extension	3*n*	*n* is the number of nodes to a max. of 6, then $n = 7$ for flag leaf just visible, and $n = 9$ flag leaf unfolded	4	Leaf sheaths lengthen	30
			5	Leaf sheaths strongly erect	30
			6	One node detectable	31
			7	Two nodes detectable	32
			8	Flag leaf just visible	37
			9	Flag leaf emerged	39
Booting	4*n*	*n* indicates degree of swelling of the boot	10	Booting	45
Ear emergence	5*n*	*n* indicates proportion of ear or ears emerged	10.5	Full ear emergence	59
Anthesis	6*n*	*n* indicates degree of completion	10.51	Start of anthesis	60
Grain expansion	7*n*	i.e. droplet exuded when caryopsis squeezed changes from watery ($n = 1$) to milky ($n = 7$)	10.54	Caryopsis water ripe	71
			11.1	Caryopsis milky ripe	75
Dough development	8*n*	i.e. no droplet exuded from squeezed caryopsis, thumbnail imprint not retained ($n = 3$) to thumbnail imprint retained ($n = 5+$)	11.2	Soft dough	85
Ripening	9*n*	Grain dry enough to harvest ($n = 2$) to seed with no primary dormancy ($n = 7$)	11.4	Harvest ripe	92

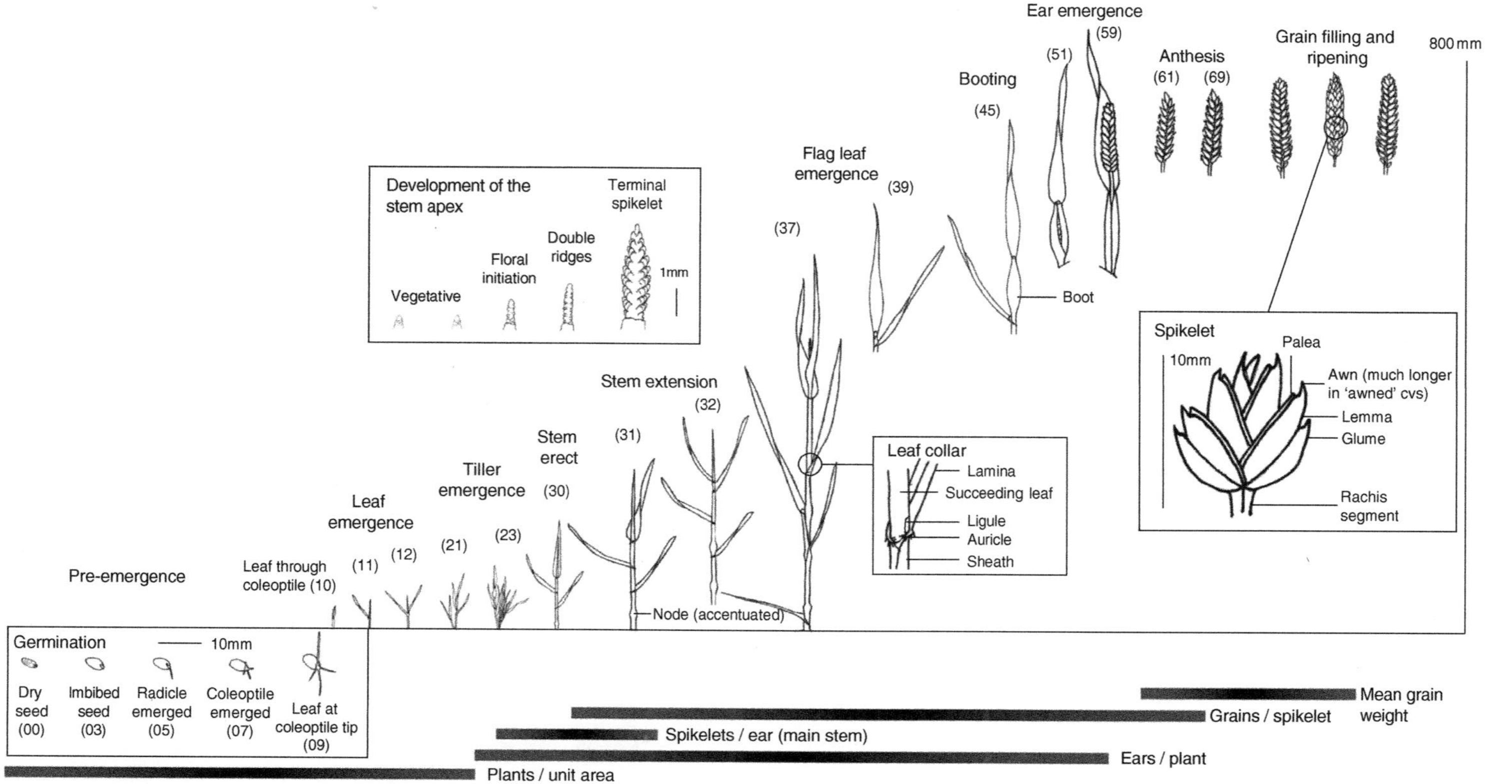

Figure 1.7 Wheat growth and development. Only main stem shown from stem erect growth stage. Horizontal bars indicate when yield components are principally determined on the main stem. Numerals in brackets are decimal growth stage scores (Table 1.4). *Source:* Reproduced from Gooding (2009) with permission.

The vegetative phase encompasses germination, and then a period during which growth from the stem apex gives rise only to vegetative tissues: leaves, tillers (i.e. the *branches* or side stems of grass plants), and adventitious roots.

1.2.1 Vegetative Phase

1.2.1.1 Germination

Starting with the germination stages, the dried grain (DGS 00) is markedly asymmetric (Figure 1.8) with the ventral side divided by a crease parallel to the long axis. The embryo is present at the germ end but typically 80% of the grain weight is allocated to the nutrient store of the seed, i.e. the starchy endosperm. The outer parts of the grain consist of compressed cell layers, notably the true seed coat (the testa) and the fruit coat (the pericarp), which, together with the outermost layer of endosperm cells (the aleurone), collectively form the bran on milling (Evers et al. 2006). Germination (DGS = 0*n*) starts in conditions that break or circumvent dormancy, and with adequate moisture, temperature, and aeration (Chapters 3 and 4). Water is imbibed and the aleurone cells are stimulated to produce hydrolytic enzymes that break down the endosperm to release simple sugars and amino acids for the growth of the seedling. Externally, the radicle (root) and coleoptile emerge (Figure 1.8). The coleoptile is a pale tube-like structure that protects and facilitates the emergence of the first true leaf. The coleoptile elongates to the soil surface. At least three seedling (seminal) roots will usually have been produced before the plants emerge above the soil surface. The germination phase is complete when the first true leaf reaches the tip of the coleoptile which encases it. The apical meristem is raised towards the soil surface by expansion of an underground stem.

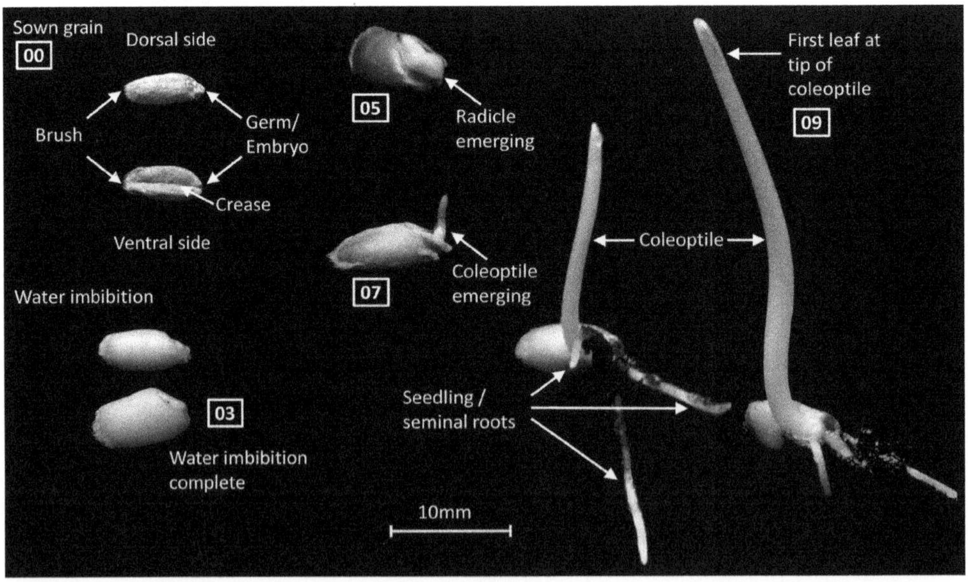

Figure 1.8 The germination phases of wheat. Boxed numerals are the decimal growth stage scores from Table 1.4.

1.2.1.2 Leaves

The stem apex usually remains just below ground level during the further vegetative phases, where it is partially protected from herbivory and extremes of temperature. The apex itself may only be 0.1 mm in length (Figure 1.7) and produces primordia, from which leaves expand and are pushed upwards. The leaf production growth stages (DGS = 1*n*) start with the emergence of the first leaf through the coleoptile (Figure 1.9). A leaf is counted as being fully emerged or unfolded based on the emergence of the collar, which is the junction between the leaf lamina (or blade) and the leaf sheath. The leaf sheaths are usually hairy (non-glabrous). The leaf laminas are narrow with about 12 veins. Small additional structures on the collar become visible, particularly on the larger leaves; these are the auricles and the ligule (Figure 1.9). The auricles are useful for distinguishing wheat from other cereals before the ears emerge. Whereas the wheat auricles are commonly 1–3 mm long and are usually covered in hairs, barley auricles are hairless, oats have no auricles, and rye auricles are much shorter. The ligule is blunt and between 2 and 4 mm long. The sheath and lamina are contiguous. The sheath connects the leaf to its node on the true stem. While the plant is in the vegetative stage, the nodes remain compressed on a short stem ($<$ 5 mm) behind the stem apex at the base of the plant. New leaves become externally visible as they emerge from within the leaf sheath of the previous leaf. The DGS progresses as each leaf becomes fully unfolded (Figure 1.9).

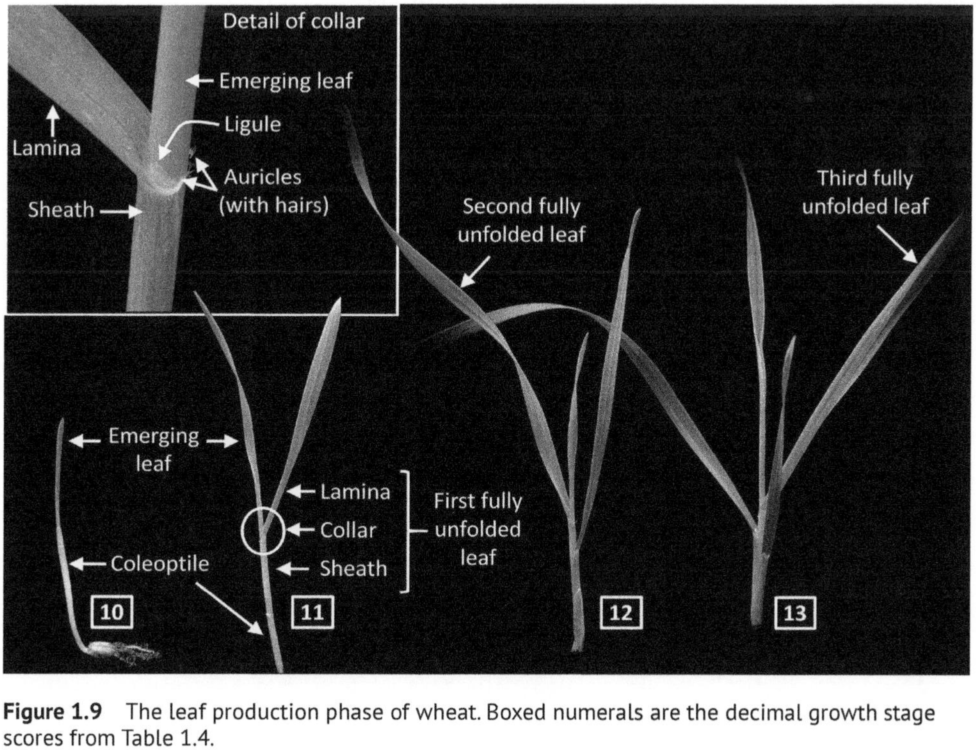

Figure 1.9 The leaf production phase of wheat. Boxed numerals are the decimal growth stage scores from Table 1.4.

1.2.1.3 Tillers

The production of tillers (DGS = $2n$) allows the plant to spread and increase its canopy size more than would be possible through just increasing the number of leaves on the main stem. Tillers originate in the axils of the leaf nodes and rise between the leaf sheaths. Tillers are counted as such when they have their own fully unfolded leaf (Figure 1.10). Each tiller has the potential to produce an ear, but the plant will often produce more tillers than survive to maturity. Tillering is particularly important when inter-plant competition is low. This can arise, for example, when seedling establishment has been poor due to difficult seed bed conditions and/or when seed of low vigour has been used. Additionally, tillering can help compensate when neighbouring stems or plants have been lost through frost or hail damage, or through herbivory by insects and mammals.

1.2.1.4 Roots

The seminal roots continue to develop during the vegetative phase, but further adventitious (or nodal) roots start to develop about one to two months after germination. The adventitious roots derive from the coleoptile node or lower nodes of the main stem and tillers. They generally overtake the seminal roots and can occupy over 90–95% of the total root volume of a fully grown crop.

1.2.2 Reproductive Phase

The reproductive stages commence when the stem apex or growing point starts producing the structures of the ear (or *spike*). The timing of this transition is strongly dependent on wheat genotype and the environment. Of particular relevance are the interactions between genetic factors, temperature, and day length (or photoperiod), i.e. cultivars vary greatly in

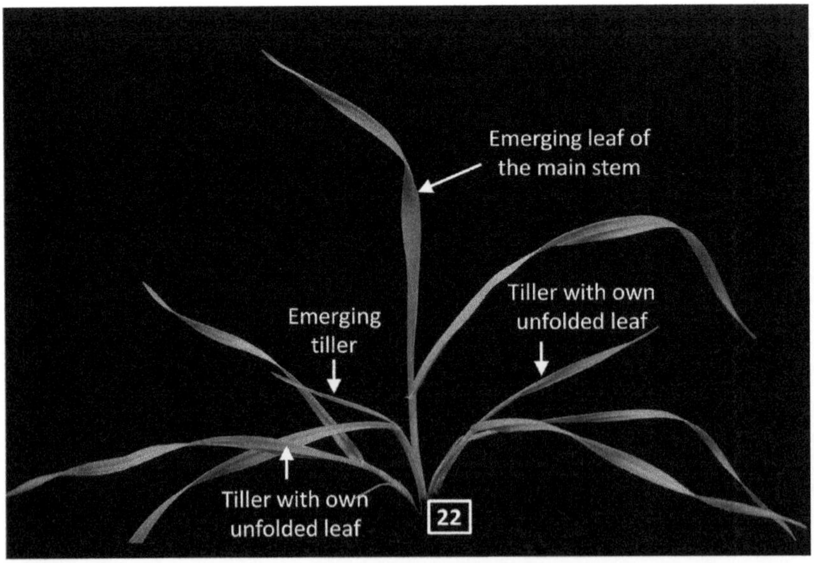

Figure 1.10 The tiller production phase of wheat. Boxed numeral is the decimal growth stage score from Table 1.4.

their need to pass through a cold period (i.e. to become *vernalised*), and for days to lengthen. These interactions are detailed in Chapter 4, but for now it suffices to recognise that those cultivars that have a significant requirement to pass through a cold period for timely transition to the reproductive phase and hence to maturation are classed as *winter* wheats (as in Figure 1.6, Table 1.3). Those that have little such requirement are classed as *spring* wheats. *Facultative* wheats are intermediate between winter and spring wheats in their requirement for vernalization (Braun and Săulescu 2002). Those cultivars that do or do not have a strong requirement for days to lengthen for timely development are respectively classed as *day length sensitive* and *day length insensitive* cultivars (Table 1.3).

1.2.2.1 Stem Extension

The first signs of the transition from the vegetative to the reproductive phase are that the stem apex elongates to 0.3 mm with the appearance of primordia as single ridges. At this stage, the stem apex is still close to ground level. The buds in the axils of the apex ridges are spikelet primordia and, with their leaf initials form double ridges as the developing spike elongates to between 0.8 and 1 mm (Figure 1.7). In broad terms, double ridges occur as the plants take on a more upright appearance as leaf sheaths elongate and pass through the pseudostem erect growth stage (Figure 1.11) (Barber et al. 2015).

The structures of the ear develop as it is simultaneously elevated through the leaf sheaths of the canopy by the lengthening stem. Stem elongation is evidenced by the detection of nodes behind the rising apex (Figure 1.11). As the spike continues to elongate, the central spikelets swell while additional double ridges are formed acropetally (upwards) until the terminal spikelet is formed at the apex. By this time, the developing ear is usually 10–30 mm above the base (or crown) of the plant, and between one or two nodes are detectable (DGS31–32; Figure 1.7; Barber et al. 2015). The spikelets continue to differentiate and

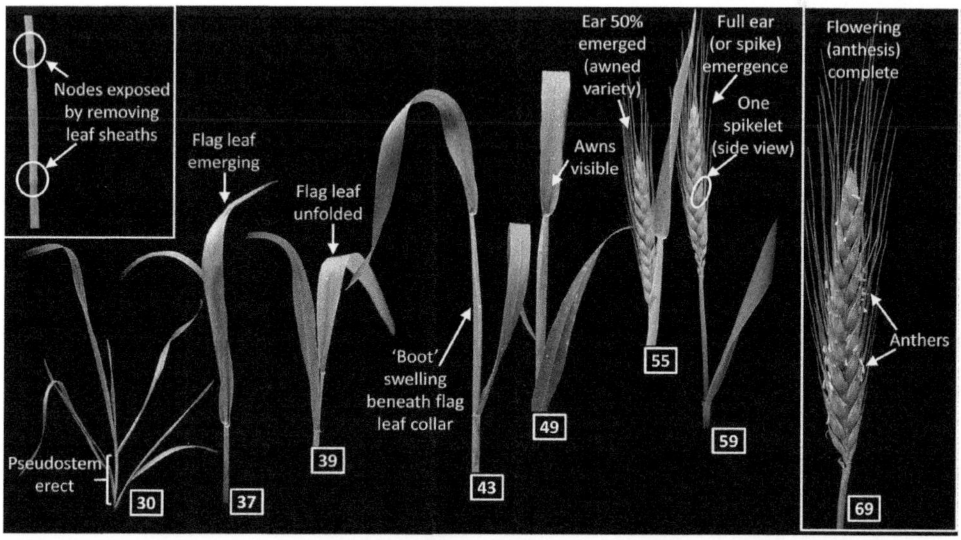

Figure 1.11 Stem extension, booting, ear emergence, and anthesis of wheat. Boxed numerals are the decimal growth stage scores from Table 1.4.

develop scaly tissues (glume, lemma, palea) and potential sites for grain formation (florets) (Figure 1.12). Floret initiation continues until about the time that the last leaf (the flag leaf) emerges and unfolds (DGS37–39; Figure 1.11). Between 7 and 11 florets may be initiated in each spikelet but, as with tillers, many more florets are initiated than develop to maturation with a grain.

1.2.2.2 Booting and Ear Emergence

Once the flag leaf has emerged, the swelling of the developing ears within the flag leaf sheath becomes visible as a *boot* (DGS 40–49; Figure 1.11). It is approximately during this booting phase that cell division by meiosis occurs to produce the male (pollen) and female (ovule) gametes of the flowers. Booting is soon followed by the emergence of the ear above the flag leaf (DGS 50–59; Figure 1.11). At ear emergence, the structure and colour of the ears can be used to distinguish wheat cultivars. For example, the length of the awns extending from the lemmas (Figure 1.12) can either be short (awnless cultivars) or long (awned cultivars), as compared in Figure 1.13. Awns are a wild-type adaptation that would have aided seed dispersal. However, there appears to be a small yield advantage for awned cultivars in warmer areas, and particularly for preventing excessive grain shrivelling in dry conditions (Rebetzke et al. 2016).

1.2.2.3 Anthesis

Anthesis (flowering; DGS 60–69; Figure 1.11) usually occurs between three and eight days after ear emergence, starting in the lower florets of the central spikelets. Most flowering throughout the spike is often completed within two to four days, although late flowering in distal spikelets and florets can extend this period to seven days (Percival 1921). Wheat is mostly self-pollinated, but some outcrossing is possible, particularly in genotypes with more open florets and when plants are grown close together. However, cross-pollination is limited because wheat pollen is heavy and short lived.

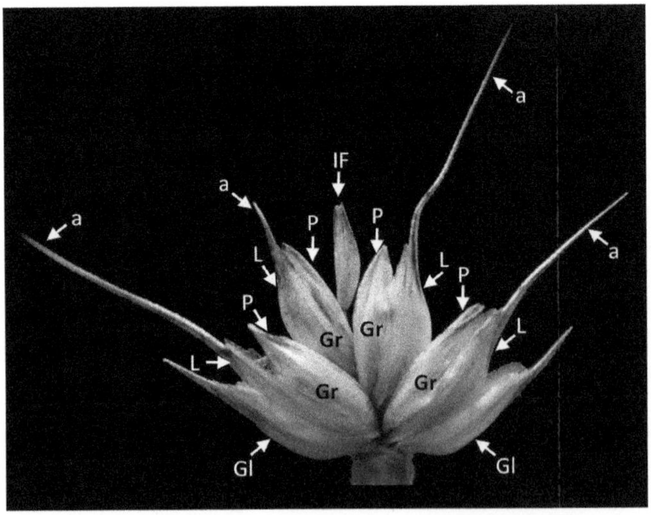

Figure 1.12 Structures of a single mature spikelet. Gl = Glume; L = Lemma; P = Palea; Gr = Site of grain, obscured by lemma and palea; a = awn; IF = Infertile floret.

Figure 1.13 Ears (spikes) of wheats and their relatives.

1.2.2.4 Grain Growth

After fertilization, the developing caryopses (*grain*) expand rapidly with net ingress of water. When squeezed, the immature grains first exude a clear watery droplet, i.e. signifying the watery-ripe growth stage (DGS = 71). Later, the liquid becomes milky as starch is accumulated (DGS = 73). A wet mass, surrounded by milky liquid, denotes *medium milk* (DGS = 75), whereas *late milk* occurs when the grain contents are wet and sticky (Tottman 1987). The growth stage scores move from milk development to dough development when no liquid droplet can be squeezed from the grain. At *early dough* (DGS = 83) the grain contents are soft and cheesy. At *soft dough* the grain contents cannot be easily squeezed out, but a fingernail impression into the dorsal side quickly disappears (DGS = 85). This stage is broadly coincident with the senescence of the flag leaf and with the maximum accumulation of dry matter in the grain (*physiological maturity*; Barber et al. 2015). Grain moisture content (w/w) is often between 50 and 40% at this stage. Later stages describe the net loss of water from the grain as harvest is approached (*harvest maturity*). At the *hard dough* stage (DGS = 87) a fingernail impression is retained, whereas by DGS = 92 the caryopsis is hard and cannot be dented. Further details of grain development, particularly as they pertain to grain quality, are given in Chapter 7.

Harvest can proceed once grain moisture content drops below 18–20% but at this level further drying before or during storage will be necessary. In the UK, the target moisture content for safe temporary storage at ambient conditions going into autumn after harvest is 14.5% or below (AHDB 2011). The equivalent for Australia, with its warmer conditions, is 12%, and grain loads with higher than 12.5% moisture are prevented from entering the marketing system (Newman 2008). In Canada, wheat is graded as *Straight*, *Tough*, and *Damp* for moisture contents of < 14.6, 14.6–17.0, and > 17% respectively (CGC 2020). In the US moisture does not affect wheat grading, but the price of wheat will depend on contract specifications: various discounts can be applied as moisture contents rise above 12% (Wilson and Dahl 2002).

1.3 Wheat Evolution and Migration

1.3.1 Origin in the Fertile Crescent

Wheat was first used, domesticated, and farmed in and around what has become known as the Fertile Crescent. This term was first coined by J.H. Breasted (1916), when he described an area covering a narrow arc from the southeast corner of the Mediterranean Sea to the northern tip of the Persian Gulf (Figure 1.14). For him, the Fertile Crescent comprised hilly flanks limited by mountains to the north and east, by plains and a drying desert to the south, and by the sea to the west. The Fertile Crescent has been redrawn many times since Breasted (1916), reflecting new archaeological discoveries and the expansion of agriculture from its origins. Hence, over time the area scholars referred to as the Fertile Crescent became broader and extended down to include more of Mesopotamia between the Tigris and Euphrates rivers, and the Nile delta of Egypt. As well as wheat, other plants used and domesticated in the Fertile Crescent included barley, lentil (*Len culinaris*), pea (*Pisum sativum*), chickpea (*Cicer arietinum*), broad bean (*Vicia faba*), and flax (*Linum usitatissimum*). Rye and oats (*Avena* spp.) may well initially have been identified by early farmers as weeds in their wheat fields; later they became domesticated crop species, possibly as agriculture moved into cooler areas (Evans 1993; Kilian et al. 2009).

Although the climate of the Fertile Crescent has changed since the earliest cultivation, wild wheats persist in the area today. Notably, they still occur in regions where early farming settlements have been identified (Evans 1993; Nesbitt 2001). It seems likely, therefore, that early cultivated wheats were adapted to growing in the cooler and wetter winter and spring, maturing before the excessive heat and drought of summer.

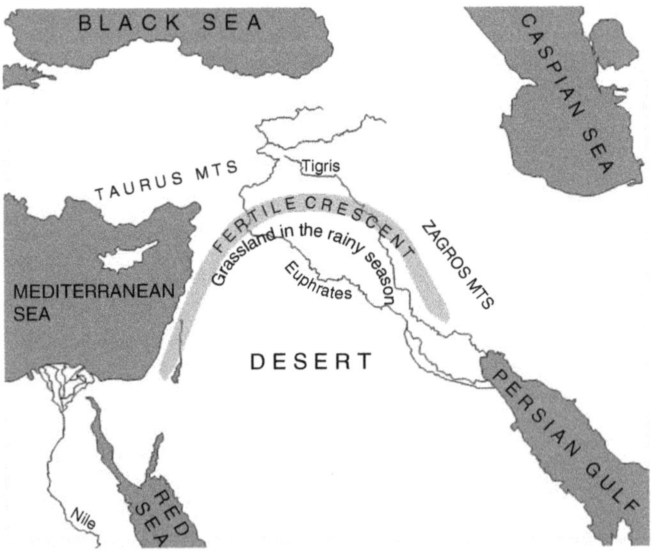

Figure 1.14 The Fertile Crescent as originally described by J. H. Breasted. *Source:* Redrawn from Breasted (1916).

1.3.2 Wild Wheats

Chromosome and genetic analysis have facilitated our understanding of the relationships between the wild wheats of prehistory and the cultivated wheats of today. All members of the Triticeae have multiples of a related set of seven chromosomes (genomes), numbered 1–7. Diploid (literally *two-fold likeness*) species have two copies of each chromosome ($2x = 2n = 14$). Some sets of chromosomes are more similar than others and they have thus been grouped accordingly. Those sets of chromosomes relevant to wheat evolution are labelled the A, B, D, G, and S genomes (Figures 1.13 and 1.15; Kihara 1929, 1954; Dewey 1984).

The diploid species carrying the different, but related, sets of chromosomes appear to have diverged between 2.5–4.5 MYA (Huang et al. 2002). Two species of wild diploid wheat (wild einkorn) are found today: *Triticum uratu* and *Triticum monococcum* var. *boeoticum* (Figure 1.15). Both species have the genome formula AA, i.e. they have a diploid complement of the A set.

Although the diploid species are biologically isolated, *hybrids* have been formed in the Triticeae through allopolyploidy. This is a type of hybridization where the offspring inherit the complete chromosome complement from each of the parents. For example, wild emmer contains four sets of chromosomes, the diploid complement from *T. uratu* (AA) and the

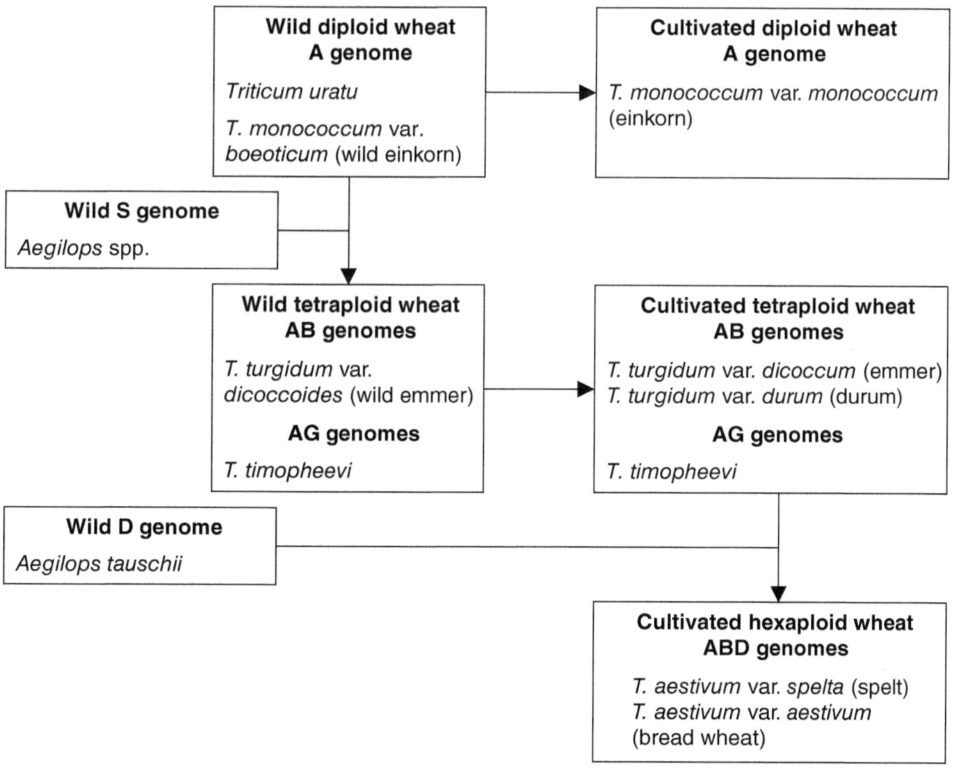

Figure 1.15 Summary of the genome constitutions and evolutionary relationships of the major wild and cultivated forms of wheat (*Triticum* spp.).

diploid (BB) complement from a type of goat grass (*Aegilops* spp.). Wild emmer (*Triticum turgidum* var. *dicoccoides*) is therefore an allotetraploid (4x = 28) with genome formula AABB. This allopolyploidy is made possible because the different sets of chromosomes from the diploid species do not pair during meiosis in the polyploids (Chapter 10). Such sets of chromosomes, which are homologous but non-pairing, are termed homoeologous (Glover et al. 2016), for example, chromosomes 1A and 1B in emmer are homoeologues.

The hybridization to produce wild emmer occurred naturally, possibly between 0.2 and 0.5 MYA (Huang et al. 2002; Gill et al. 2004). A second wild tetraploid species also occurs (*Triticum timopheevi* var. *araraticum*) with the genome formula AAGG. Although the B and G genomes are distinct, they are known to be related and both may be derived from the S genome present in the Sitopsis section of *Aegilops*, with *Ae. speltoides* being the closest currently existing species. The two tetraploid wheat species may, therefore, have diverged after the formation of an ancestral AASS tetraploid. In fact, Dvořák et al. (2006) considered that the B, G, and S genomes are sufficiently alike to be classed as the same group.

1.3.3 Domestication

Genetic changes occurred as gatherers and early farmers selected traits in the wild diploid and tetraploid species that facilitated their cultivation, giving rise to distinct cultivated forms that are regarded here as variety groups (var.). Variety groups within the *Triticum* species are sometimes referred to as subspecies (subsp.) in other classification systems. The cultivated forms of wheat have traits collectively known as the *domestication syndrome*. Two traits have been particularly important in the domestication of wheat. The first is the loss of shattering of the spike at maturity through having a *semi-tough rachis*. It is important for the rachis of mature ears of wild plants to be brittle, such that the ear shatters or disarticulates to aid seed dispersal. By contrast, mutations at the *Br* (brittle rachis) loci on the short arms of group 3 chromosomes (Nalam et al. 2006) confer a semi-tough rachis that keeps the ear intact. This prevents seed loss before manual harvesting; although this is highly advantageous to gatherers and farmers, it represents a significant loss of fitness for the wild plant. The appearance of the non-shattering phenotype in the archaeological record is, therefore, often assumed to be confirmation of wheat cultivation (Allaby et al. 2017). However, it is also likely that cultivation of wild types predated that of domesticated forms. Both shattering and non-shattering forms are found within the Fertile Crescent dated to the same time, and often within the same site (Tanno and Willcox 2012). There are very early dates for the first possible appearance of non-shattering einkorn (> 11 000 BP) and emmer (> 10 000 BP), at camps or settlements where shattering types dominated (Allaby et al. 2017). Thereafter the shift from non-shattering to shattering forms was a slow evolutionary process. The greatest selection pressures from the presumed adoption of cultivation are estimated to be from 10 500 to 9500 BP (Allaby et al. 2017), but the complete transition to domesticated forms probably took three to four millennia (Fuller et al. 2012). Today, all cultivated forms of wheat have a semi-tough rachis.

The second important change in domestication has been the transition from hulled forms in which the grain is harvested within glumes and other spikelet structures (Figure 1.12), to free-threshing forms in which grains are released *naked* during harvesting processes. The free-threshing trait of wheat arose by a dominant mutant at the *Q* locus which modified the effects of recessive mutations at, e.g. the *Tg* (tenacious glume) locus on the group 2 chromosomes (Dubcovsky and Dvorak 2007; Kilian et al. 2009). Not all cultivated forms of wheat yield naked grain at harvest. For example, most cultivated einkorns are hulled (*T. monococcum* var. *monococcum*). However, the var. *sinskajae* has free-threshing grain and can be used as a source of this trait in einkorn breeding programmes.

Wild emmer (*T. turgidum* var. *dicoccoides*) has given rise to several cultivated tetraploid variety groups. Cultivated emmer (var. *dicoccum*) is hulled but the most widely grown form is free-threshing durum wheat (var. *durum*). Durum wheat accounts for about 5% of global wheat production. It is particularly adapted to Mediterranean climates and is mostly used for making pasta and regional foods such as couscous and bulgar. Other free-threshing variety groups are only cultivated to a limited extent. These minor groups include rivet wheat (var. *turgidum*), Polish wheat (var. *polonicum*), Persian wheat (var. *carthalicum*), and Khorasan wheat (var. *turanicum*). *T. timopheevi* is represented in both hulled (var. *timopheevi*) and naked forms (var. *militinae*) and is still cultivated in small areas in the Caucasus.

Cultivated einkorn and emmer clearly developed from the domestication of natural populations. By contrast, hexaploid wheat (*T. aestivum*, genome formula AABBDD) has never existed as a wild species. It originated through hybridisation of cultivated emmer with the wild goat grass *Aegilops tauschii* (also called *Triticum tauschii* and *Ae. squarrosa*), which contributed the D genome (Figures 1.13 and 1.15). This hybridisation probably occurred several times, with farmers selecting hexaploid lines for their larger grains and higher yields. Modern wild populations of *Ae. taushcii* are found growing in the north and north-east of the Fertile Crescent and populations carrying genomes closely related to the D genome of hexaploid wheat occur in Transcaucasia, between the Black Sea and the Caspian Sea (particularly in modern-day Armenia), and around the southwest of the Caspian Sea (modern-day northern Iran) (Dvořák et al. 1998). The first hexaploid wheats may, therefore, have occurred around the south or west of the Caspian Sea about 8000 BP (Kilian et al. 2009).

1.3.4 The Spread of Wheat Cultivation

The cultivation of bread wheat spread via Anatolia to Greece, and then both northwards through the Balkans to the Danube (7000 BP), and westwards to Italy, France, and Spain (7000 BP), reaching the UK and Scandinavia by about 5000 BP. Similarly, wheat spread via Iran into central Asia, reaching China by about 3000 BP, and to Africa, initially via Egypt. It was introduced by the Spanish to Mexico in 1529 CE and by the British to Australia in 1788 CE. These migration routes have been described in detail by Feldman (2001).

This wide geographic expansion of wheat was undoubtedly facilitated by the allohexaploidy of *T. aestivum*. The environmental and geographic distributions of allopolyploids are often wider than for their diploid parents (Miller 1987). Furthermore, *T. aestivum* has adapted to a wider range of environments than the tetraploid *T. turgidum* (Dubcovsky and Dvorak 2007). Feldman and Levy (2005) list several mechanisms by which allopolyploidy has led to the rapid evolution of wheat genomes and gene expression. The multiple genomes would be expected to confer *fixed* hybrid vigour, genetic buffering, and evolutionary adaptability (Simmonds and Smartt 1999). Having multiple copies of similar genes means that mutations in one or two copies can occur without catastrophic effects. This can, for example, allow the development of intermediate types and better *fine tuning* of growth and development patterns to the environment. The addition of the D genome appears to have been particularly important given the different and wider geographic distribution and environmental adaptation of *Ae. taushcii* compared with the wild donors of the A and B genomes. Modern wild stands of *Ae. taushcii* extend beyond the Mediterranean zones into continental climates (Zohary et al. 1969).

Although *T. aestivum* is one of the *youngest* major crop species, it is now also one of the most diverse. Feldman (2001) estimated that about 25 000 genotypes were available globally at the turn of the millennium, but this is clearly an underestimate as material present in countries such as China was not readily available at the time of publication. Although several genotypes of *Ae. tauschii* contributed to the formation of hexaploid bread wheat, most of the diversity in bread wheat has clearly arisen since the initial hydridisation occurred. Dubcovsky and Dvorak (2007) suggest that this has several origins. Firstly, extensive gene flow has occurred between bread wheat and cultivated and wild forms of tetraploid wheat, particularly wild emmer (He et al. 2019; Allen et al. 2021), capturing diversity from these species. However, gene flow into the D genome has been more limited; it remains less diverse than the A and B genomes (Allen et al. 2021). Secondly, bread wheat has an unusually large genome, comprising about 17Gb of DNA, (about 40 times the size of the rice genome). About 85% of the genome is non-coding repetitive DNA including mobile elements, which facilitates the generation of variation by gene duplication and deletion and by insertion into regulatory and coding sequences. These events are more likely to be tolerated due to the buffering capacity of the three genomes. Hence, the combination of gene flow from related wild species, polyploidy, and high genome plasticity have contributed to the variation which has developed over the past 10 000 years.

The dominant grouping of *T. aestivum* is modern, free-threshing, bread wheat (var. *aestivum*). This form accounts for about 95% of current global wheat production and is widely used to make bread, other baked goods including cakes and biscuits, Asian noodles (but not pasta), and as an ingredient in food processing. Two additional naked forms are club wheat (var. *compactum*), which has highly compact heads and is grown principally in the USA (Pacific Northwest and California), and Indian shot wheat (var. *sphaerococcum*), which is characterized by rounded grains and is from northwest India and Iran. *T. aestivum* also occurs in three hulled forms, the most widely cultivated of which is spelt (var. *spelta*). The other recognised hulled variety groups are var. *vavilovi* from Armenia and var. *macha*, from Georgia.

1.3.5 Increases in Harvest Index

In addition to the characteristics of the classic *domestication syndrome*, modern cultivated wheats now differ in many ways from their wild progenitors. The increase in harvest index for grain has been of great significance, i.e. the increase in the ratio of the grain mass to the total crop biomass (but often neglecting the roots and fallen leaves) (Donald 1962). Austin et al. (1982) reported harvest indices of only 0.15, 0.05, and 0.16 for *T. uratu*, *Ae. speltoides*, and *T. tauschii*, respectively, while the harvest index of tetraploid wild emmer was a little higher at 0.28. Even at this level, however, it would be necessary to produce more than 3.5 t of crop biomass to yield each tonne of grain. Although farmers have long recognized that shorter wheats have higher harvest indices (Roberts 1847; Garnett 1883), cultivars released before 1900 still had indices below 0.35 (Austin et al. 1989). During the twentieth century, significant increases in harvest index were achieved, particularly through breeding programmes that incorporated dwarfing alleles from Japanese wheats, reducing crop heights by 10–20% (Borojevic and Borojevic 2005). By the mid-1980s, the harvest index of UK wheats approached 0.55 (Austin et al. 1989), i.e. at harvest most of the crop biomass in the field was present in the grain. This progression was seen in wheats throughout the world and across ploidy levels (Evans 1993). The considerable implications of the adoption of high harvest index, shorter wheats on wheat agronomy, quality, and sustainability are covered in Chapter 6.

1.4 Wheat as Food

Although wheat still accounts for half of the total intake of calories in some countries, the contribution to nutrition in Europe in historical times was even greater. For example, the cost of bread was estimated to account for between a third and two thirds of the total budgets of working families (as opposed to landowners and gentry) in the UK in the period between 1760 and 1836, with the highest estimate being almost 90% for one labourer's family (Peterson 1995). Apparently, therefore, the coarse bread that was consumed at the time was able to meet most of the nutritional requirements of adults engaged in physical work.

The wider contributions of wheat and bread to health are often ignored, with both being widely regarded as little more than a source of energy. In fact, even in modern Europe wheat, and particularly bread, still provide surprisingly high contributions of a range of essential nutrients. For example, although bread only contributes 10–13% (depending on age and gender) of the daily intake of energy in the UK, it also contributes 10–12% of the daily intake of protein and between 10 and 20% of the daily intakes of minerals (iron, zinc, copper, magnesium, selenium, and calcium) and B vitamins (B1 thiamine, B3 niacin, B9 folate). More importantly, it is also a major source of dietary fibre, contributing about 20% of the daily intake (Bates et al. 2014a, b). These contributions are higher in Poland, with bread providing 21.9% of energy, 16.5% of protein, 35.4% of fibre, 24.9% of iron, and 20.7% of folate (vitamin B9) (Laskowski et al. 2019). The wide consumption and low price of bread therefore make it an excellent vehicle for delivering improved nutrition to large populations at low cost.

The nutritional and health impacts of wheat are discussed in detail in Chapter 9, with the following sections focusing on more practical aspects of the processing of wheat for food, particularly bread.

1.4.1 The Development of Milling and Baking

The remains of grains of wild cereals and goat grasses from c. 50 000 to 60 000 BP have been found in cave dwellings at Kebrara in modern-day Israel (Lev et al. 2005), while remains of c. 23 000 years old wild emmer have been identified from the Ohalo II hunter-gatherer camp on the shores of the Sea of Galilee (Weiss et al. 2004). The seed collections from both sites are diverse. Kebrara is dominated by the remains of legume seeds; at Ohalo II, the emmer is a rarity amongst greater amounts of wild barley (*Hordeum spontaneum*) and an extensive range of smaller seeded grasses. Nonetheless, there is good evidence that grains of wild wheats were eaten at least 13 000 years before domestication, and a much earlier date is probable.

Raw wheat seeds are tough, unpalatable, and relatively indigestible unless milled or crushed. At Ohalo II, there is evidence that seeds were crushed between grindstones, long before the advent of agriculture. Crushing would have exposed the starchy endosperm and dramatically increased the bioavailability of nutrients, particularly that of carbohydrate. It would also have speeded the rate with which blood sugar was increased, i.e. the glycaemic index (Rubel 2011).

The ancient process of milling involved manually grinding the grain between stones; a lower, usually concave, static quern stone (called a *saddle quern*) and an upper hemispherical or cylindrical (like a rolling pin) handstone pushed over the grain. A fine example from Norway is shown in Figure 1.16a.

Of particular importance for the adoption of wheat as a food crop were the properties of the dough formed by mixing the flour released from crushed or milled grain with water. Wheat doughs have unique visco-elastic properties that confer significant functional and organoleptic qualities on wheat products, particularly bread and other baked goods. Baking dates back at least 14 000 years (Henry et al. 2011, 2014; Arranz-Otaegui et al. 2018), and possibly much earlier (Rubel 2011). In particular, doughs made from wheat, far more than from other cereals, are able to trap carbon dioxide released during fermentation by yeast, and hence can produce baked foods with low density. This is the basis of the diverse forms of leavened bread. Yeast spores may occur naturally on the surface of cereal grains and some fermentation occurs readily in wheat dough left to rest. However, it is probable that early bakers used sourdough systems (mixtures of lactobacilli and yeasts), with starters being carried over from batch to batch as in artisan bakeries today. Bread production was clearly important in the Uruk culture (6000–5100 BP) of lower Mesopotamia, and cuneiform writings from around 4000 BP include Sumerian poems and myths about the invention of bread, along with recipes. The Egyptians appear to have perfected leavened bread production by 4000 BP.

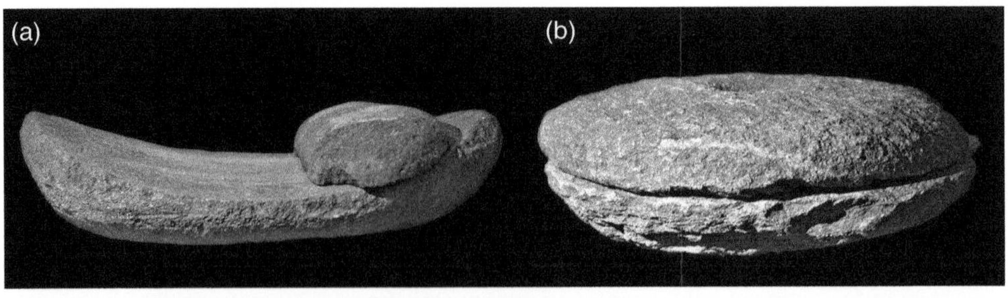

Figure 1.16 Stone querns on display at the Archaeological Museum, University of Stavanger, Norway. a, neolithic (possibly bronze age) saddle quern; b, pair of rotary quern stones (46–47 cm diameter) from about 550 CE. *Source:* Photographs kindly provided by Per Storemyr (Per Storemyr Archaeology and Conservation).

From around 2500 BP the laborious use of saddle querns had been replaced by rotary querns. These comprised two circular stones with the upper being rotated by hand and the grain fed through a hole in the upper stone and the flour emerging from the edge (Figure 1.16b). Further refinement of this basic system led to mechanized stone milling. In Roman times, grain was ground as it passed down between a lower *meta* stone and the upper, rotated, *catillus* (Figure 1.17). The Romans prized wheat as the empire's principal staple food. In Rome, at any one time, up to 320 000 of the poorer population were supported by free wheat grain and bread (Rickman 1980).

By the 1800s CE stone milling had evolved into a highly sophisticated process, with the mill stones being carved (*dressed*) to improve efficiency. Dressing divides the surface of the stone into flat areas, called lands, which are separated by furrows. Further small *feathering* grooves are then carved from the furrows onto the lands. These provide cutting edges, acting like scissors, and channel the meal from the centre to the outside of the stones. A gap is also maintained between the stones to prevent wear and overheating. Dressing mill stones is a highly skilled art and needs to be carried out regularly (see Cookson 2019). Stone milling produces wholemeal flour which can be *dressed* (sieved through cloth filters) to separate white flour. However, the separation of the flour and bran is poor, and the flour produced is expensive. Hence, before the introduction of roller mills, white flour was a luxury, associated with purity and the social elite due to its appearance and cost. This situation was transformed with the advent of roller milling (Figure 1.18).

The first successful roller mill was developed in Switzerland about 1830 CE and the first complete roller mill (Radford's Albert Mill, Liverpool) established in the UK in 1870 CE (Cookson 2019). Roller milling was widely adopted by the end of the nineteenth century. Roller mills use a series of rollers; these initially break the grain open and then scrape the starchy endosperm away from the bran (comprising the pericarp, testa, aleurone layer, and usually also the germ) to give *break fractions*. Further rollers reduce the size of the flour particles in these fractions to give *reduction fractions*. When combined with sieving, these

Figure 1.17 The milling room and detail of a mill in a bakery complex at the port of Rome, Ostia; constructed c. 200 CE (*Source:* described by Bakker 1999).

Figure 1.18 A modern roller mill and miller. *Source:* Photograph kindly provided by Dr. Mervin Poole, Heygates Ltd., UK.

two types of roller generate over 20 white flour fractions in modern mills; when recombined the flour may account for up to 80% of the grain dry weight, essentially the whole of the starchy endosperm. The white flour produced by roller milling is very fine and pure, and the process is highly efficient and economical to operate. It therefore led to the current global dominance of white bread.

1.4.2 The Cultural Significance of Bread

In view of the *life or death* importance of wheat, it is not surprising that it has become firmly entrenched in both culture and religion. One of the most culturally embedded types of bread is *non* (Figure 1.19g), a leavened bread which is consumed throughout Central Asia. It is cooked on the wall of a clay oven (tandoor) and stamped with a wooden tool studded with pins (a chekich) to raise a pattern. It has a special status in Central Asian culture; it must be kept upright and cannot be placed on the ground or thrown away. It must be torn by hand rather than cut. At a wedding ceremony, the bride and groom both take a bite from a non and finish it the next day as part of their first meal as man and wife. Similarly, when a soldier goes to war, he takes a bite of non as he leaves. His family will then hang the bread up and only take it down once he is safely home. Uzbek-style non is old enough to have been mentioned in the *Epic of Gilgamesh* (tales of the ruler of the Sumerian Kingdom of Uruk), dating from over 4000 BP.

Many readers will be aware of the significance of wheat and bread in Judaism and Christianity (Jacob 1977); both are mentioned in the Old and New Testaments of the Christian Bible. The Jews probably learned to make leavened bread during their time in Egypt; unleavened bread (matzos) is used at Passover to celebrate their escape, as a symbol that they would not have had time to make leavened bread before their flight. Unleavened

Figure 1.19 Examples of breads consumed around the world. (a) Turkish frankala hearth bread; (b) Yemeni maloung flat bread; (c) Chinese steamed bread; (d) Indian chapati; (e) Arabic pocket bread; (f) Iranian sangak; (g) Central European non; (h), Iranian lavash. *Source:* All images from the author's collection except (d), which was provided by Dr. Anil Gaddameddi (Hyderabad, India).

bread was also used for sacrifices in the Temple; this may represent the nomadic origin of the Jews and/or the fact that early sourdough bread would deteriorate rapidly if not eaten. The major symbolic role of bread in Christianity arises from the Last Supper; it represents (in Protestant churches) or becomes (in Catholic and Orthodox churches) the body of Christ during the Eucharist/Holy Communion.

1.4.3 Bread Today

Breads are produced in a vast array of forms, many of them restricted to specific geographical regions. They can be broadly divided into two types: high specific volume breads such as Western pan breads (although some such as French baguettes, have lower volumes) and low specific volume flat breads, which are dominant in the Middle East and Asia. Flat breads include both leavened and unleavened types, while both Western and flat breads can be leavened with yeast or sourdough systems. They also vary in how they are cooked. Although baking is the most common method, they may also be steamed or fried.

High-volume Western breads are baked on the bottom or shelves of an oven, either unsupported (hearth breads) or in a pan which supports the loaf leading to a higher volume (pan breads). Most Western breads are produced using bakers' yeast (*Saccharomyces cerevisiae*) for fermentation, but sourdough systems are becoming increasingly popular, usually as artisanal hearth breads. Yeast-based systems use well-defined genetic strains of yeast that are purchased from commercial producers and, in most cases, used fresh for each batch of bread. Hence, their effects are well understood and reproducible across processes. By contrast, sourdough systems use combinations of natural lactobacilli and yeasts. In artisan systems, the sourdough cultures are perpetuated by carrying starters over between batches of dough, while for large-scale production they may be purchased from ingredient supplies. In both cases there is wide variation in the compositions of microorganisms and their corresponding effects on the dough and bread. This poses a major challenge for researchers and food manufacturers as research carried out on one sourdough system may not apply to other systems. Nonetheless sourdough systems have become more widely used in Western-style wheat breads over the past two decades, as part of the trend towards increased consumption of traditional foods (with perceived health benefits). Hence, there is a flourishing sourdough research community with a rapidly increasing number of publications. However, because of the issues discussed above, the generic significance of much of the work is difficult to assess.

The effects of fermentation, whether using yeast or soughdough systems, will depend on the conditions employed. In traditional breadmaking systems, the dough may be fermented overnight, or even longer. By contrast, most of the high-volume *factory* bread produced in the UK and many other Commonwealth countries is produced using the Chorleywood Breadmaking Process (CBP), or processes related to this. This is a very rapid method, with the mixing and dough development being carried out in a single operation using a large amount of yeast, high energy, and control of the headspace pressure during mixing. After this, only short intermediate and final proving periods are required before baking. The effects of fermentation on the health benefits of breads are discussed in Chapter 9.

Although much of the high-volume white bread produced globally is Western pan bread, highly attractive regional types also exist. For example, Figure 1.19a shows the most popular bread consumed in Turkey, an airy white hearth bread with pointed ends called *somum ekmek* or *frankala*. Steamed breads (Figure 1.19c) are popular in China and Southeast Asia. Although wholemeal products have been introduced, they are traditionally made from white flour, and may be fermented using yeast or sourdough systems. Quality requirements in China range from light and open to dense texture, and from firm and cohesive to soft structure, depending on the region (Huang and Miskelly 2019).

Faridi (1988) lists over 50 major types of flat bread; only a few of these can be mentioned here. Arabic bread (often called pita) is a yeast-leavened flat bread from the Middle East and North Africa, which can be produced from white or high extraction (bran-enriched) flours. It is baked for a short time on the bottom of an oven at high temperature; this turns the water into steam, causing the bread to puff up, separating the two sides to form a pocket in the baked loaf (Figure 1.19d).

Other flat breads in the region are Yemeni maloung, which is cooked on the inner wall of a tandoor (Figure 1.19b), and sangak, which is a large thin sourdough white bread baked on a bed of small stones in Iran (Figure 1.19f). One of the most popular breads in Iran is lavash, a white sourdough bread that exists in a range of forms including in thin crispy sheets which can be dried and stored in plastic bags for weeks (Figure 1.19h).

About 80% of the bread consumed in the Indian sub-continent is chapati (Figure 1.19d), an unleavened bread which is made from atta flour (about 95% flour extraction) and water and is cooked on a hotplate. Puri is made from a similar recipe to chapati but is fried, whereas paratha is cooked on a hotplate but contains salt and about 40% fat. Unlike these breads, naan is made from white flour, is yeast fermented, and then cooked on the inside of a tandoor.

1.4.4 The Fall and Rise of Whole Grain Foods

The adoption of roller milling made white flour available to all the population. This led to massive changes in the diet of working people with coarse wholemeal or brown breads being almost completely replaced by white products in the UK by 1880 CE. Although the science of nutrition was then in its infancy, concerns were nevertheless expressed about the impact of this change in diet on the health of the population.

An early supporter of wholemeal bread was Sylvester Graham (1794–1851), an American Presbyterian minister, who proposed a lifestyle that combined temperance, vegetarianism, and the consumption of whole-grain bread. His views were expounded in his preaching and in publications, notably *Lectures on the Science of Human Life,* published in 1849. They led to a movement referred to as *Grahamism* and inspired the development and marketing of Graham flour, Graham bread, and Graham crackers. The last of these are still widely consumed in the USA. However, Graham did not endorse any products or benefit financially from their sale.

Foremost among the critics of white bread in the UK were May Yates (née Mary Corkling) (1850–1938) and Thomas Allinson (1858–1918). Allinson qualified as a doctor in 1879 and established a practice in London. He believed that diet was crucial for health, and particularly advocated the consumption of stone ground wholemeal wheat. He was frequently in dispute with conventional medicine and was *struck off* (disqualified from practicing) in 1892, having been found guilty of *infamous conduct* (self-promotion). In the same year he purchased a stone mill and established a milling and baking company that continues to produce wholemeal bread to the present day. By contrast, May Yates was not trained as a scientist but became convinced of the benefits of wholemeal bread during a visit to Sicily. She founded the Bread Reform League in 1880 and spent 40 years campaigning for the use of high extraction (about 85%) flours. The late nineteenth century also saw the introduction of improved *patent* breads. The most well-known of these, and the only one still produced today in the UK, is *Hovis*, which is enriched in wheat germ.

Despite the compulsory production of high extraction and wholemeal breads in the UK during the two World Wars, white bread remained the favourite for much of the British population, and in many other countries. It is not difficult to understand why this is the case; factory-produced white bread stores well (with a shelf life of up to a week), while its mild flavour and soft texture means that it has high consumer acceptability and combines well with other foods. By contrast, wholemeal products are more expensive, strongly flavoured, may have a coarse texture, and store less well.

Nevertheless, it is now recognized that the bran fractions are rich in fibre, minerals, vitamins, and phytochemicals and, therefore, that white flour is depleted in beneficial components compared to whole grain (e.g. Wang et al. 2013). As its health benefits have been promoted, the consumption of wholemeal has increased in some countries over the last decade. The catalyst for this increase was the approval in 1999 by the US Food and Drug Administration (FDA) of an application to 'use the following claim on the label and in labelling of any product that meets the eligibility criteria described in the notification: Diets rich in whole grain foods and other plant foods and low in total fat, saturated fat, and cholesterol, may help reduce the risk of heart disease and certain cancers' (USFDA 1999). The FDA further defined whole grain foods as containing 51% or more whole grain ingredient(s) by weight, meaning that the amount of fibre required to qualify could be below 6% (based on whole grain containing 11% fibre). This was followed by the approval of several health claims for whole grain fibre by the European Food Safety Authority (EU 2006, 2012) (see Chapter 9). These approvals have provided an economic stimulus to the production and marketing of wholegrain products.

The marketing of wholegrain products has, however, been confusing for the consumer because of the use of different definitions. Although the FDA minimum of 51% is clearly not equivalent to the whole grain, the omission of only a small proportion of the bran fraction can have a significant impact on palatability and consumer acceptability without seriously compromising the health benefits. The consortium of the HEALTHGRAIN EU project therefore launched a series of discussion meetings that resulted in an agreed definition (van der Kamp et al. 2014), which should facilitate the development and acceptance of whole grain products in the future. This states that:

1) Whole grains shall consist of the intact, ground, cracked, or flaked kernel after the removal of inedible parts such as the hull and husk. The principal anatomical components – the starchy endosperm, germ and bran – are present in the same relative proportions as they exist in the intact kernel.
2) Small losses of components, that is, less than 2% of the grain (< 10% of the bran), that occur through processing methods consistent with safety and quality are allowed.

1.4.5 Producing White and Wholemeal Flours by Roller Milling

Although there has been increased interest in recent years in the use of whole grain flour, most wheat-based foods, including breads, pastries, cakes, cookies, noodles, and pasta, are still made from white flour. The art of the miller is to maximize the recovery of white flour by increasing the yields of the purest fractions and combining them to achieve a high total flour yield from a given weight of grain (i.e. the flour extraction rate). However, the purity of the combined flour will inevitably decrease as the extraction rate increases, with a *trade off* between yield and purity. The purity of the flour is often monitored as ash content, as ash is derived from minerals which are concentrated in the bran fractions. In addition to varying in purity, the white flour fractions produced by milling will also vary in their contents and compositions of other components, resulting from gradients in composition within the starchy endosperm (Chapter 8). Thus, although the > 20 fractions produced by modern roller milling do not correspond to precise regions of the starchy endosperm they do differ in their compositions (e.g. Gonzalez-Thuillier et al. 2015). There is, therefore, an opportunity to combine selected mill streams to produce specialist flours differing in their compositions and properties for processing and health (Shewry et al. 2019).

Wholegrain flour is produced by combining all the milling fractions in the proportions that they are produced, rather than by grinding the whole grain as is often thought. However, it is also possible to combine fractions to give *high extraction* flours, in which a proportion of the bran, notably the aleurone, is included. These flours may correspond to an extraction rate of 85% or more and are enriched in the fibre, minerals, vitamins, and phytochemicals present in the bran. It is also possible to use milling to produce germ-rich fractions that can then be incorporated into flours to increase the contents of B vitamins, as in the Hovis patent of 1887. The relationship between extraction rate and the presence of other tissues in white flour is illustrated in Figure 1.20, based on the analysis of durum wheat milling fractions using biochemical markers.

1.5 Grain Quality

1.5.1 Grain Size, Shape, and Specific Weight

Plump grain is preferred for milling because it gives higher flour yields than thin grain. The reason for this is simple; plump grain has a higher proportion of starchy endosperm to bran. Conversely, well filled, heavier grain tends to yield less bran (Figure 1.21). The proportion of endosperm to bran is relevant to animal feed and alcohol production as well as to milling, because high starch contents are related to energy contents and alcohol yields, respectively.

Some of the scatter in Figure 1.21 may be due to cultivar differences in grain shape rather than purely grain size. More rounded grain with a less pronounced crease should give higher flour extraction rates (Evers et al. 1990). Genetic analyses and molecular studies of grain size and shape have been reported (Gegas et al. 2010; Brinton and Uauy 2019). These should allow selection for grain size and shape in breeding programmes but direct selection for these qualities, as opposed to yield, is not usually carried out.

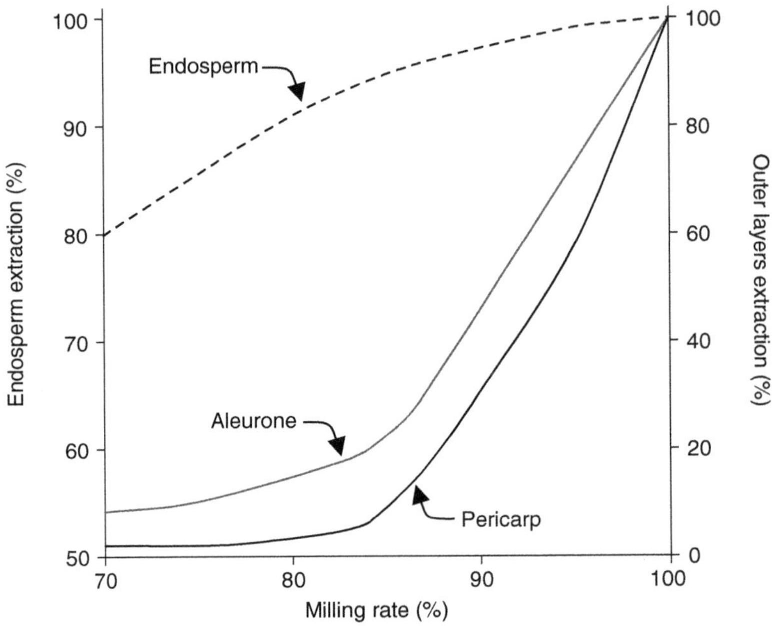

Figure 1.20 Determination, using phenolic acids as biochemical markers, of the grain tissues proportions (aleurone, pericarp, endosperm), as a function of total flour extraction rate (durum wheat grains, cv. Ardente). *Source:* Redrawn from Hemery et al. (2007) with permission.

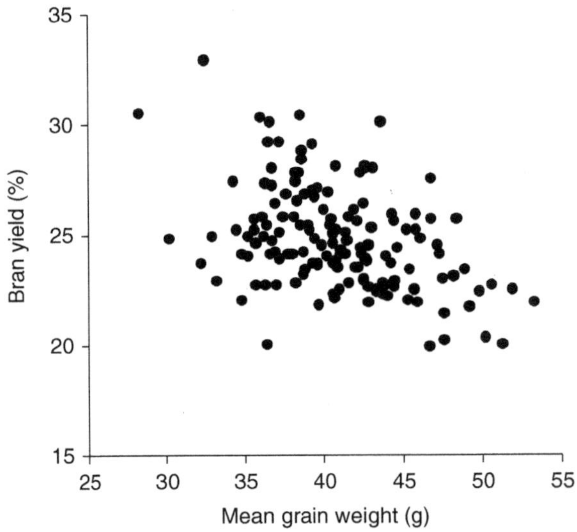

Figure 1.21 Relationship between grain weight and bran yield for 150 wheat genotypes from a single field experiment. Bran yields are determined using a small-scale laboratory mill and are therefore higher than those achieved by large-scale commercial mills. *Source:* Redrawn from Ward et al. (2008) with permission.

What has long been selected for, however, is grain-specific weight, also known as the hectolitre (hl)-, test-, or the bushel weight (e.g. Reeves and Roy 1974). Specific weight is the bulk packing density of grain; it was originally used over 150 years ago in wheat buying and selling to estimate the weight of wheat in a grain bin. It is one of the most widely used measures of grain quality for two main reasons. Firstly, it is easy and rapid to measure. Assessment requires the weighing of a known volume filled with grain in a controlled manner. Secondly, it is loosely related to grain size, shape, and shrivelling (e.g. Clarke et al. 2004). Grain samples with large numbers of shrivelled grain may have specific weights as low as 40 kg/hl. In the UK, the highest prices for breadmaking quality wheat commonly require specific weights of at least 76 kg/hl, while for animal feed a lower threshold is tolerated (often around 72 kg/hl). Unfortunately, the relationships between specific weight and flour extraction rates are not close (e.g. Hook 1984), particularly for specific weights above 70 kg/hl (e.g. Evers et al. 1990). This is perhaps unsurprising as significant reduction in endosperm contents only occurs in markedly shrivelled grain (Evers et al. 1990). As well as grain size and shape, specific weight is affected by the density of individual grains, preharvest weathering, moisture content, wetting, drying, grain handling, and other factors influencing surface properties. In addition to the genetic component, specific weight is highly dependent on environmental conditions and crop management during grain filling (Chapters 3, 4, and 6).

1.5.2 Endosperm Texture

The classification of wheat into hard and soft types (Chapter 8) is determined largely by the degree of cohesion between the protein and starch granules within the starchy endosperm, with hard wheats requiring greater force to break the grain and reduce the starchy endosperm to flour. Hardness is strongly but not entirely genetically determined, particularly by variation at the *Hardness* (*Ha*) locus on the short arm of chromosome 5D (Turnbull and Rahman 2002). When milled, hard wheats yield coarser but more evenly shaped flour particles. The flour is also more easily separated from the bran, and the particles flow easily over surfaces and through sieves (Kent and Evers 1994; Carson and Edwards 2009).

Endosperm texture can be determined by standardized grinding and sieving to weigh the proportion of material that will pass through a mesh. This is the basis of the Particle Size Index (PSI) determination with soft wheats having a greater proportion of *throughs* (and therefore greater PSI) than hard wheats (Carson and Edwards 2009). Near-infrared reflectance (NIR) calibrated to PSI is routinely used for rapid determination of grain hardness at mill intake. Automated determination of hardness can also be made on a single grain basis using the Perten Single Kernel Characterization System (SKCS) (Pearson et al. 2007). This measures the weight, moisture content, and crush profiles of individual seeds. The standard sample size is 300 kernels and the software provides data on the distributions of the parameters.

1.5.3 Water Absorption

Determining the amount of water absorbed by flour is a crucial test used by bakers to optimise the mixing conditions for baking. Bakers therefore specify the level of water absorption (WA) for the flours they purchase, and millers adjust their milling conditions to achieve this level.

Although WA is affected by starch, protein, and fibre (see Chapter 8), the major concern of millers is the contribution of starch damage. This is because starch is the major component in the grain and damaged starch absorbs 3–4 times its own weight of water, compared with 0.5 times for undamaged starch. Starch damage is easier to achieve when milling hard wheats because they respond to higher grinding pressures during roller milling (Ford 1987). It is difficult to obtain high starch damage in soft wheats without excessive pressures leading to overheating and bran contamination (Ford 1987). Damaged starch leads to greater WA in dough formation, and more rapid fermentation rates. Both processes are desirable for modern breadmaking processes (Carson and Edwards 2009). The large amounts of water absorbed by hard flours also contribute to the longer shelf life of bread (shelf life being determined by drying and starch retrogradation). Hence, hard wheats are preferred for breadmaking (Figure 1.21). By contrast, flours from soft wheats absorb little water and are preferred for the manufacture of most types of biscuit (cookies) where the low water content reduces the required baking time, cost, and the risk of cracking during cooling after baking. Soft wheats with low WA are also favoured for making cakes and pastries (Figure 1.21).

WA is usually determined using the Brabender Farinograph. This automated mixer measures the resistance of flour and water while they are being mixed to form a dough. The level of maximum resistance can be adjusted to a predetermined optimum by altering the amount of water added, the ideal amount of water being a measure of WA (Kent and Evers 1994).

1.5.4 Gluten

In Section 1.4.1 we introduced the unique properties of wheat doughs, pointing out that they have been significant factors behind the spread and adoption of wheat. Those unique properties are, in large part, due to the gluten proteins of wheat flour. There is a conundrum that the very component of wheat flour that contributes to its pre-eminence as a foodstuff is also the proposed source of several adverse reactions to wheat consumption (Sapone et al. 2012). We therefore provide a brief introduction here to the gluten fraction and its role in processing and health. These properties, impacts, and potential mitigations are discussed in greater detail in Chapters 7–10.

1.5.4.1 The Origin and Properties of Gluten

Gluten is a continuous proteinaceous network that is formed when wheat flour is mixed with water to form dough. It can be prepared at home by gently washing a ball of dough under running tap water or with dilute salt solution. This removes most of the soluble components, including the albumin and globulin proteins, plus the starch and other particulate matter, leaving a cohesive mass that can be extended but also shows elastic recoil (Figure 1.23). Gluten consists predominantly (75–80%) of proteins, most of which can be classified into two related groups called gliadins and glutenins. These two groups form the major storage protein fraction of the grain. However, they differ in that whereas the gliadins are present as monomers, the glutenins form high molecular mass polymers stabilised by inter-chain disulphide bonds. Both groups can also be defined as *prolamins* in that the individual subunits (but not necessarily the polymers) are insoluble in water but soluble in alcohol-water mixtures.

As storage proteins, the biological role of the gliadins and glutenins is to provide nitrogen (N), sulphur (S), and carbon skeletons to support seed germination and seedling growth.

It is therefore serendipitous that the interactions of gliadins and glutenins with each other and with other components in wheat doughs have such dramatic effects on wheat utilization. This is because they largely define the rheological properties of doughs, including the balance between dough viscosity and elasticity that determines the quality for different end uses.

1.5.4.2 Gluten and Health

The most widely known and the most well-characterised adverse reaction to wheat consumption is coeliac disease (CD). CD has been recognised since ancient Greek times; the first modern description was given by the British paediatrician Samuel Gee in 1887. Gee also adopted the classical Greek name from *koiliakós*, meaning abdominal. However, the link to wheat was not made until the 1940s by Willem Karel Dicke, a Dutch scientist. The link with gluten was established by 1952. CD is an autoimmune response which results in damage to the small intestine. This leads to several symptoms, notably malabsorption of nutrients and diarrhoea. It is triggered in genetically susceptible individuals by the ingestion of wheat gluten or related proteins from barley or rye. The aetiology of CD is well understood and there is no cure except avoidance of the proteins responsible for triggering the response. It is estimated to affect about 1% of the global population but may exceed this in some countries. Small proportions of coeliac patients may also suffer from dermatitis herpatiformis or neurological symptoms (including ataxia).

True allergy to the ingestion of wheat (or gluten proteins) is relatively rare. However, there is greater concern about a loosely defined group of symptoms known as *non-coeliac gluten sensitivity* or *non-coeliac wheat sensitivity*, which has been reported to affect between 0.5 and 10% of the population. This condition is still poorly understood but current work suggests that the prevalence may be higher than that of CD, with proteins other than gluten responsible for triggering the response. These conditions are discussed in detail in Chapter 9 (see also Brouns et al. 2019).

There is no doubt that concerns about adverse effects on health, propagated particularly in the social and popular media, have affected the consumption of wheat in some counties. However, they must be considered in perspective and not allowed to overshadow the health benefits from wheat consumption or the degree of global food security based on increasing wheat production (Lillywhite and Sarrouy 2014; Peña et al. 2017).

1.5.4.3 Dough Properties that Determine Processing Quality

As discussed, the gluten network is largely responsible for the unique biophysical properties of dough, a balance between extensibility and elasticity. Highly viscous doughs are readily extended when stretched whereas elastic doughs resist extension and exhibit elastic recoil when the stress is relaxed (Chapter 8). Highly elastic doughs are referred to as strong and are preferred for making leavened bread (Figure 1.22). This is because the gas released during fermentation (proofing) is captured in bubbles, allowing the dough to rise. The protein network is then denatured by baking, giving a light porous structure to the bread. By contrast, biscuit production requires weak but extensible doughs that can be spread out into an even layer without recoil after rolling. Intermediate strength doughs can be used to produce steamed bread, chapatis, and noodles (Figure 1.22), but puffed breakfast cereals require strong wheat to prevent disintegration when pressure is released after cooking.

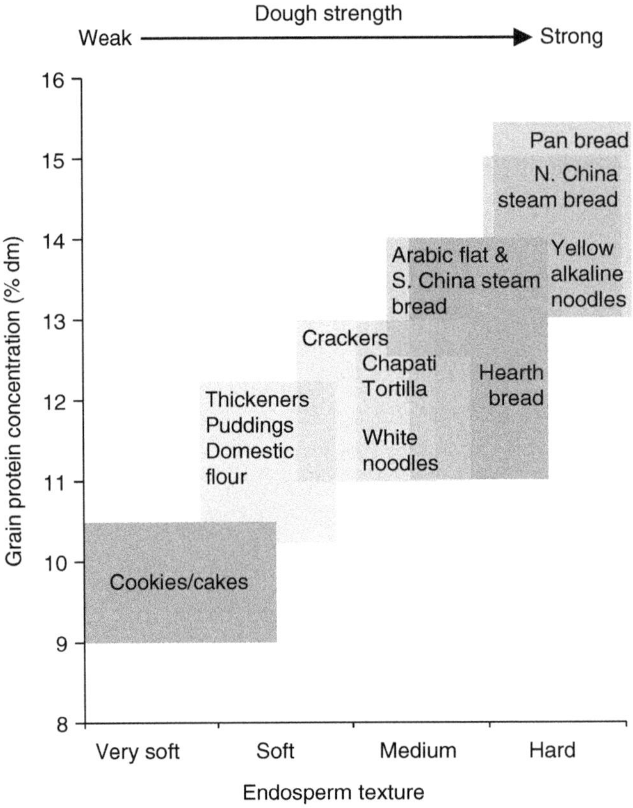

Figure 1.22 The relationships between grain protein content, dough strength, grain texture, and the quality of bread wheat for various products. *Source:* Adapted from Peña (2002), Uthayakumaran and Wrigley (2017), and Moss (1973).

Flaked and shredded products are made from wheats with weak gluten similar to those used for biscuits (Blackman and Payne 1987; Lin et al. 1990).

The cooking quality of pasta made from durum wheat is mainly determined by the ability to absorb water while retaining firmness and shape and without becoming sticky (Sarrafi et al. 1989). Both the processing and cooking of pasta are, therefore, dependent on strong gluten. Tenacious doughs have high initial resistance to extension but break after only a relatively small distance; they are often described as *short* and their suitability is limited to some domestic uses (Guzman et al. 2016). The role of gluten proteins in determining dough rheology and end-use functionality has led to the importance of grain protein concentration and protein quality in the marketing and classification of wheat (e.g. CGC 2020).

1.5.4.4 Importance of Total Protein Concentration

Gluten proteins account for up to 80% of the total grain protein, and their proportion becomes larger with increasing protein concentration. The quantity of wet gluten derived from a flour is, therefore, highly correlated with grain protein concentration. Grain protein concentration is partly determined by genetic factors, with cultivars bred for breadmaking

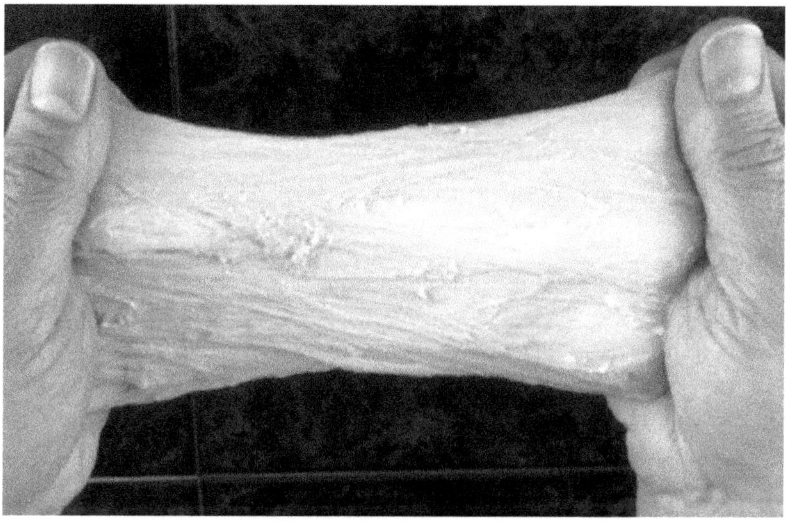

Figure 1.23 Gluten produced by hand washing, stretched to show the cohesive properties. *Source:* Taken from Shewry et al. (1995).

often containing about two percentage points more protein than cultivars bred for livestock feed when grown under the same conditions (Snape et al. 1993). However, there are also strong environmental effects, with protein content ranging between about 6 and 20% due to variation in, for example, temperature, and availabilities of water, N, and S for grain filling (Carson and Edwards 2009; Gooding 2017).

Grain protein concentration is not usually determined directly but calculated based on the N concentration determined by chemical analysis (Kjeldahl wet chemistry or Dumas oxidative combustion) or by near-infrared spectroscopy (NIRS) calibrated based on N analysis (Carson and Edwards 2009). It is often assumed that there is a constant relationship between the amount of N and the amount of protein in biological samples and that the crude protein concentration can be calculated by multiplying the N concentration by a constant factor, with N × 6.25 being most widely used. However, this is not the case because it wrongly assumes that all proteins have a similar N content. Gluten proteins, for example, have higher N contents than most other proteins due to the presence of between 30 and 50% of glutamine, an amino acid which contains two N atoms (as opposed to one in most other amino acids). Consequently, N × 5.7 is widely accepted as a conversion factor for wheat grain and flour (Draper and Stewart 1980). However, this factor is still imprecise as it will vary with the proportion of gluten proteins in the sample. Hence, it will be lower for high protein grain. Similarly, it will be lower for white flour than for wholemeal (with N × 5.83 having been suggested [Kent and Evers 1994]) and vary between mill streams. Nevertheless, it is clearly impractical to use a range of values and a conversion factor of N × 5.7 is almost universally used to calculate protein concentration for marketing and utilization of wheat grain.

Although protein concentration is not the same as protein quality, a minimum protein concentration is required for breadmaking and increasing protein concentration can, to some extent, compensate for lower quality. Hence, higher protein concentrations are required for pasta and breadmaking than for other uses (Figure 1.22). The minimum requirement for the CBP is around 13% DM. Products requiring a weak, extensible dough, such as biscuits, cakes, and pastries, usually require a protein content of less than 10.5% DM and levels down to 9% can be tolerated for some products (Carson and Edwards 2009). Low protein concentrations, and therefore high starch contents, are associated with higher alcohol yields in distilling and bioethanol industries (Taylor et al. 1993).

1.5.4.5 Importance of Protein Quality

In addition to protein concentration, the relative proportions and subunit compositions of the gliadin and glutenin fractions also have significant effects on dough rheology and functionality for different end uses. In broad terms, gliadins contribute to the viscosity and extensibility and glutenins to the elasticity and strength of doughs (Rustgi et al. 2019). The proportions of gliadins and glutenins may vary between genotypes (Wieser 2000) and there are also significant differences in quality associated with allelic variation in the protein compositions of the two fractions. In particular, one group of gluten proteins, the high molecular weight subunits of glutenin, are widely used as biochemical markers for dough strength (see Chapter 8 and Payne et al. 1987) based on analysis by sodium dodecylsulphate polyacrylamide gel electrophoresis or by capillary or reversed-phase high-performance liquid-chromatography (RP-HPLC) (Rustgi et al. 2019).

The strong genetic control of protein quality means that cultivars are often grouped based on their intended end use. For example, in the UK cultivars are grouped depending on whether they have potential for breadmaking (Groups 1 and 2), biscuit making (Group 3), or neither end use (Group 4).

The proportions and compositions of the gliadin and glutenin fractions are also affected by environmental factors, which can lead to effects on processing quality. For example, the availability of S is particularly important as it affects the ratio of N : S in the grain and, therefore the proportions of proteins which are rich in cysteine (a suphur-containing amino acid required for the formation of disulphide bonds in glutenin polymers) (Wieser et al. 2004). Indeed, several studies showed that the grain S concentration was a better predictor of breadmaking quality than N concentration (Zhao et al. 1999a, b; Ruske et al. 2004).

1.5.4.6 Measurement of Dough Rheology and Quality

The properties of dough and gluten can be measured using rheology, a branch of physics that deals with the deformation and flow of matter. Several mechanical systems have been developed to measure the rheological properties of doughs in research and grain utilization.

These fall broadly into two types. Recording dough mixers, such as the Farinograph, Mixograph, and Reomixer, measure the mechanical energy required to mix a dough to its maximum (peak) resistance. They thus provide data on the peak resistance, the time required to mix to the peak (both measures of dough strength), and the breakdown of the dough when overmixed (a measure of stability). The second type of system, such as the Extensograph, Alveograph, and Kieffer Rig, measures the properties (resistance and extensibility) of the dough after it has been mixed to peak resistance and allowed to rest, by extending the dough until it breaks. However, these mechanical systems are low throughput and clearly not appropriate for the analysis of large numbers of samples in breeding programmes.

Several simpler systems are therefore widely used, by breeders as well as in the food industry. For example, in the SDS-sedimentation volume test, wholemeal flour is mixed with weak lactic acid and sodium dodecyl sulphate (SDS) in a measuring cylinder. This solution dissolves much of the protein but not the high molecular weight proteins, particularly the glutenins, which are a major contributor to the volume of suspended sediment recorded after a period of settling (Fullington et al. 1987).

Solvent retention capacity (SRC) determines the ability of flour to retain a set of four solvents (water, 50% sucrose, 5% sodium carbonate, 5% lactic acid) (Slade and Levine 1994). Because these solvents are preferentially absorbed by one or more of the major grain components (glutenin [lactic acid], starch [sodium carbonate and sucrose], and pentosans [sucrose]), their relative retention can be used to predict aspects of grain composition and quality. The relative retention is expressed as the gluten performance index (GPI), which is calculated as:

$$GPI = \frac{lactic\ acid\ SRC}{sodium\ carbonate\ SRC + sucrose\ SRC} \quad (1.1)$$

The GPI has been reported to explain the performance of glutenin in dough development (Kweon et al. 2011).

1.5.5 Other Factors Affecting the Acceptability of Wheat for Different End-Uses

For millers sourcing wheat grain for breadmaking, protein quality is one of the primary concerns. However, several other factors are of crucial importance in determining whether grain is acceptable.

1.5.5.1 Alpha-Amylase Activity

Alpha-amylase is a naturally occurring enzyme that breaks down starch to release sugars during seed germination. Some α-amylase is beneficial for breadmaking because the sugars released are required for fermentation. In fact, commercially produced enzyme may be added as an improver in commercial breadmaking. However, excessive activity results in a dark crust (due to Maillard reactions) and a sticky crumb with poor resilience and texture, the latter preventing the use of mechanical loaf slicing (Chamberlain et al. 1982).

Excessive α-amylase can result from several syndromes, three of which come into play early in grain development (Lunn et al. 2001). These are the retention of high levels of α-amylase in the pericarp (retained pericarp α-amylase activity, RPAA), the deposition of specific α-amylase isoforms in the endosperm cavity during grain development (pre-maturity α-amylase activity, PMAA, or late maturity α-amylase, LMA) and, very rarely, the germination of the grain early in development (pre-maturity sprouting, PrMS). However, by far the most common syndrome is germination (sprouting) of the mature grain before harvest, i.e. *pre-harvest sprouting* (PHS). This occurs when the crop is subject to wet conditions just prior to harvest, and when the grain has low levels of dormancy.

Levels of α-amylase are assessed using the Hagberg Falling Number (HFN) test. This was initially developed to monitor the levels of α-amylase in wheat growing in the field in

Sweden to decide whether the crop was at risk of spoilage. It is now widely used by the grain industry as a measure of α-amylase activity in harvested grain. HFN is a measure of the viscosity of a mixture of water and milled wheat mixed in a tube and placed in a water bath at 100 °C. Alpha-amylase breaks down the starch and, therefore, reduces the viscosity of the water: wheat mixture. The falling number is the time in seconds required for stirring (60 seconds) plus the time taken for a stirrer to fall through the flour suspension while it is being liquefied by the enzyme (Vaidyanathan 1987).

1.5.5.2 Seed Coat Colour

Wheats can be described as red or white based on the intensity of red pigmentation in the seed coat (Lachman et al. 2017). Seed coat colour is largely determined by homologous genes at the *R-1* loci on the long arms of the group 3 chromosomes (i.e. *R-A1*, *R-B1*, and *R-D1*). Red cultivars carry one or more of the red (dominant) alleles and the intensity of pigmentation increases as *gene dosage* increases to three (Flintham 2000). Most American and European wheats are red. This is because the red seed coat is associated with greater resistance to PHS (Mares et al. 2009; Ji et al. 2018). By contrast, white wheats are more suited to areas that are dry during ripening and harvest and are favoured for the manufacture of certain types of flat bread, steamed breads, and noodles. White wheats can also be used to produce a higher extraction rate of flour during milling without products becoming discoloured with bran flecks (Hatcher et al. 2006).

1.6 Further Chapters

In this chapter, we have discussed the importance of wheat and briefly explained some of the reasons for that importance in terms of the diversity of wheat adaptation and of wheat use. In the next four chapters we will explore the environmental requirements and sustainability of wheat production. The chapter headings reflect what has always been known by wheat farmers: that for high yields you need a good soil, ample water, mild temperatures, and sunshine. Chapter 6 (Canopy management) evaluates farming practices associated with increasing land-use efficiency. In subsequent chapters we return to the wheat grain and its use, with respect to grain development and composition (Chapter 7); grain processing and flour functionality (Chapter 8); and the role of wheat in diet and health (Chapter 9). We conclude with a chapter on modern breeding methods and targets for further sustaining wheat production and its use for food security.

References

AHDB (2011). *HGCA Grain Storage Guide for Cereals and Oilseeds*, 3e. Agriculture and Horticulture Development Board.

Allaby, R.G., Stevens, C., Lucas, L. et al. (2017). Geographic mosaics and changing rates of cereal domestication. *Philosophical Transactions of the Royal Society B: Biological Sciences* 372: 20160429. https://doi.org/10.1098/rstb.2016.0429.

Allen, A.M., Wilkinson, P., Burridge, A. et al. (2021). The role of gene flow and chromosomal instability in shaping the bread wheat genome. *Nature Plants* 7: 172–183. https://doi.org/10.1038/s41477-020-00845-2.

Anderson, K., Ivanic, M., and Martin, W. (2013). Food price spikes, price insulation, and poverty. In: *The Economics of Food Price Volatility* (eds. J.-P. Chavas, J. Hummels and B.D. Wright), 311–339. National Bureau of Economic Research https://www.nber.org/books-and-chapters/economics-food-price-volatility/food-price-spikes-price-insulation-and-poverty.

Arranz-Otaegui, A., Carretero, L.G., Ramsey, M.N. et al. (2018). Archaebotanical evidence reveals the origins of bread 14400 years ago in northeastern Jordan. *Proceedings of the National Academy of Science, USA* 115: 7925–7930. https://doi.org/10.1073/pnas.1801071115.

Austin, R.B., Morgan, C.L., Ford, M.A. et al. (1982). Flag leaf photosynthesis of *Triticum aestivum* and related diploid and tetraploid species. *Annals of Botany* 49 (2): 177–189. https://doi.org/10.1093/oxfordjournals.aob.a086238.

Austin, R.B., Ford, M.A., and Morgan, C.L. (1989). Genetic improvement in the yield of winter-wheat – a further evaluation. *Journal of Agricultural Science* 112: 295–301. https://doi.org/10.1017/S0021859600085749.

Bakker, J.T. (1999). The mills-bakeries of Ostia: description and interpretation. *Dutch Monographs on Ancient History and Archaeology* 21: 131.

Barber, H.M., Carney, J., Alghabari, F. et al. (2015). Decimal growth stages for precision wheat production in changing environments? *Annals of Applied Biology* 166 (3): 355–371. https://doi.org/10.1111/aab.12207.

Bates, B., Lennox, A., Prentice, A. et al. (Eds.) (2014a). *National Diet and Nutrition Survey: Results from Years 1–4 (combined) of the Rolling Programme (2008/2009–2011/2012). Executive Summary*. Public Health England. https://assets.publishing.service.gov.uk/government/uploads/system/uploads/attachment_data/file/594360/NDNS_Y1_to_4_UK_report_executive_summary_revised_February_2017.pdf

Bates, B., Lennox, A., Prentice, A. et al. (Eds.) (2014b). *National Diet and Nutrition Survey: Results from Years 1-4 (combined) of the Rolling Programme (2008/2009–2011/2012).* Public Health England. https://assets.publishing.service.gov.uk/government/uploads/system/uploads/attachment_data/file/216484/dh_128550.pdf

Bellemare, M.F. (2015). Rising food prices, food price volatility, and social unrest. *American Journal of Agricultural Economics* 97 (1): 1–21. https://doi.org/10.1093/ajae/aau038.

Blackman, J.A. and Payne, P.I. (1987). Grain quality. In: *Wheat Breeding, Its Scientific Basis* (ed. F.G.H. Lupton), 455–485. Chapman and Hall.

Borojevic, K. and Borojevic, K. (2005). The transfer and history of "reduced height genes" (Rht) in wheat from Japan to Europe. *Journal of Heredity* 96 (4): 455–459. https://doi.org/10.1093/jhered/esi060.

Braun, H.J. and Săulescu, N.N. (2002). Breeding winter and facultative wheat. In: *Bread Wheat Improvement and Production. FAO Plant Production and Protection Series No. 30* (eds. B.C. Curtis, S. Rajaram and H. Gómez Macpherson). Food and Agriculture Organization of the United Nations http://www.fao.org/3/Y4011E/y4011e0f.htm#bm15.

Breasted, J.H. (1916). *Ancient Times: A History of the Early World. An Introduction to the Study of Ancient History and the Career of Early Man.* Ginn and Company.

Brinton, J. and Uauy, C. (2019). A reductionist approach to dissecting grain weight and yield in wheat. *Journal of Integrated Plant Biology* 61: 337–358. https://doi.org/10.1111/jipb.12741.

Brouns, F., van Rooy, G., Shewry, P. et al. (2019). Adverse reactions to wheat or wheat components. *Comprehensive Reviews in Food Science and Food Safety* 18: 1437–1452. https://doi.org/10.1111/1541-4337.12475.

Carson, G.R. and Edwards, N.M. (2009). Criteria of wheat and flour quality. In: *Wheat: Chemistry and Technology*, 4e (eds. K. Khan and P.R. Shewry), 97–118. AACC International.

Cassman, K.G. and Grassini, P. (2020). A global perspective on sustainable intensification research. *Nature Sustainability* 3 (4): 262–268. https://doi.org/10.1038/s41893-020-0507-8.

CGC (2020). *Official Grain Grading Guide*. Canadian Grain Commission.

Chamberlain, N., Collins, T.H., and McDermott, E.E. (1982). The influence of α-amylase on loaf properties in the UK. *Proceedings of the 7th World Cereal and Bread Congress*: 841–845.

Chapman, G.P. (1996). *The Biology of Grasses*. CAB International.

Clarke, M.P., Gooding, M.J., and Jones, S.A. (2004). The effects of irrigation, nitrogen fertilizer and grain size on Hagberg falling number, specific weight and blackpoint of winter wheat. *Journal of the Science of Food and Agriculture* 84 (3): 227–236. https://doi.org/10.1002/jsfa.1657.

Cookson, M. (2019). The rise and fall of the millstone. *Cereal Foods World* 64 (4) https://doi.org/10.1094/CFW-64-4-0038.

Dardonville, M., Urruty, N., Bockstaller, C. et al. (2020). Influence of diversity and intensification level on vulnerability, resilience and robustness of agricultural systems. *Agricultural Systems* 184: 102913. https://doi.org/10.1016/j.agsy.2020.102913.

Dewey, D.D. (1984). The genomic system of classification as a guide to intergenome hybridisation with the perennial Triticeae. In: *Gene Manipulation in Plant Improvement. 16th Stadler Genetics Symposium* (ed. J.P. Gustafson), 209–279. Plenum Press.

Dibb, D.W. (2000). The mysteries (myths) of nutrient use efficiency. *Better Crops with Plant Food* 84 (3): 3–5. http://www.ipni.net/publication/bettercrops.nsf/0/02DEA6DA6511C494852579800081FC4B/$FILE/Better%20Crops%202000-3%20p03.pdf.

Donald, C.M. (1962). In search of yield. *Journal of the Australian Institute of Agricultural Science* 28: 171–178.

Draper, S.R. and Stewart, B.A. (1980). Procedures for the comparative assessment of quality in crop varieties. III. Methods used in assessing grain protein content, Hagberg falling number, ease of milling and the baking quality of wheat varieties. *Journal of the National Institute of Agricultural Botany* 15: 194–197.

Dubcovsky, J. and Dvorak, J. (2007). Genome plasticity a key factor in the success of polyploidy wheat under domestication. *Science* 316: 1862–1866. https://doi.org/10.1126/science.1143986.

Dvořák, J., Luo, M.C., Yang, Z.L. et al. (1998). The structure of the *Aegilops tauschii* genepool and the evolution of hexaploid wheat. *Theoretical and Applied Genetics* 97 (4): 657–670. https://doi.org/10.1007/s001220050942.

Dvořák, J., Akhunov, E.D., Akhunov, A.R. et al. (2006). Molecular characterization of a diagnostic DNA marker for domesticated tetraploid wheat provides evidence for gene flow from wild tetraploid wheat to hexaploid wheat. *Molecular Biology and Evolution* 23 (7): 1386–1396. https://doi.org/10.1093/molbev/msl004.

Erenstein, O. and Laxmi, V. (2008). Zero tillage impacts in India's rice–wheat systems: a review. *Soil and Tillage Research* 100 (1–2): 1–14. https://doi.org/10.1016/j.still.2008.05.001.

EU (2006). Regulation (EC) No 1924/2006 of the European Parliament and of the Council of 20 December 2006 on Nutrition and Health Claims made on Foods. *Official Journal of the European Union* L404/9 https://eur-lex.europa.eu/eli/reg/2006/1924/oj.

EU (2012). Commission Regulation (EU) No 1047/2012 of 8 November 2012 amending Regulation (EC) No 1924/2006 with regard to the list of nutrition claims. *Official Journal of the European Union* L310/36 http://data.europa.eu/eli/reg/2012/1047/oj.

Evans, L.T. (1993). *Crop Evolution, Adaptation and Yield*. Cambridge University Press.

Evers, A.D., Cox, R.I., Shaheedullah, M.Z. et al. (1990). Predicting milling extraction rate by image analysis of wheat grains. In: *Aspects of Applied Biology, 25, Cereal Quality II* (eds. G.F.J. Milford, P.S. Kettlewell, J.H. Orson, et al.), 417–426. Association of Applied Biologists.

Evers, A., Nesbitt, M., and Gooding, M.J. (2006). Wheat. In: *The Encyclopedia of Seeds: Science, Technology and Uses* (eds. J.D. Bewley, M. Black and P. Halmer), 752–755. CAB International.

FAO (2021). *FAOSTAT Food and Agriculture Data*. Food and Agriculture Organization of the United Nations http://www.fao.org/faostat/en/#home.

Faridi, H. (1988). Flat breads. In: *Wheat: Chemistry and Technology Volume II*, 3e (ed. Y. Pomeranz), 457–505. American Association of Cereal Chemists.

Feekes, W. (1941). De Tarwe en haar milieu. In: *Verslag XVII. Technkche*, 560–561. Tarwe Commission.

Feldman, M. (2001). Origin of cultivated wheat. In: *The World Wheat Book: A History of Wheat Breeding* (eds. A.P. Bonjean and W.J. Angus), 3–56. Lavoisier Publishing.

Feldman, M. and Levy, A.A. (2005). Allopolyploidy – a shaping force in the evolution of wheat genomes. *Cytogenetic and Genome Research* 109 (1–3): 250–258. https://doi.org/10.1159/000082407.

Flintham, J.E. (2000). Different genetic components control coat-imposed and embryo-imposed dormancy in wheat. *Seed Science Research* 10: 43–50. https://doi.org/10.1017/S0960258500000052.

Ford, M. (1987). Quality requirements for milling and baking. In: *Aspects of Applied Biology 15, Cereal Quality*, 10–17. Association of Applied Biologists.

Fu, Y.-B., Peterson, G.W., Horbach, C. et al. (2019). Elevated mutation and selection in wild emmer wheat in response to 28 years of global warming. *Proceedings of the National Academy of Sciences* 116 (40): 20002–20008. https://doi.org/10.1073/pnas.1909564116.

Fuller, D.Q., Willcox, G., and Allaby, R.G. (2012). Early agricultural pathways: moving outside the 'core area' hypothesis in Southwest Asia. *Journal of Experimental Botany* 63 (2): 617–633. https://doi.org/10.1093/jxb/err307.

Fullington, J.G., Miskelly, D.M., Wrigley, C.W. et al. (1987). Quality related endosperm proteins in sulfur-deficient and normal wheat grain. *Journal of Cereal Science* 5: 233–246. https://doi.org/10.1016/S0733-5210(87)80025-5.

Garnett, T. (1883). The cultivation of wheat. In: *Essays in Natural History and Agriculture*. Chiswick Press.

Gbegbelegbe, S., Cammarano, D., Asseng, S. et al. (2017). Baseline simulation for global wheat production with CIMMYT mega-environment specific cultivars. *Field Crops Research* 202: 122–135. https://doi.org/10.1016/j.fcr.2016.06.010.

Gegas, V.C., Nazari, A., Griffiths, S. et al. (2010). A genetic framework for grain size and shape variation in wheat. *The Plant Cell* 22: 1046–1056. https://doi.org/10.1105/tpc.110.074153.

Gill, B.S., Appels, R., Botha-Oberholster, A.M. et al. (2004). A workshop report on wheat genome sequencing: International Genome Research on Wheat Consortium. *Genetics* 168 (2): 1087–1096. https://doi.org/10.1534/genetics.104.034769.

Glover, N.M., Redestig, H., and Dessimz, C. (2016). Homoeologs: what are they and how do we infer them? *Trends in Plant Science* 21: 609–621. https://doi.org/10.1016/j.tplants.2016.02.005.

Gonzalez-Thuillier, I., Salt, L., Chope, C. et al. (2015). Distribution of lipids in the grain of wheat (cv. Hereward) determined by lipidomic analysis of milling and pearling fractions.

Journal of Agricultural and Food Chemistry 63: 10705–10716. https://doi.org/10.1021/acs.jafc.5b05289.

Gooding, M.J. (2009). The wheat crop. In: *Wheat: Chemistry and Technology*, 4e (eds. K. Khan and P.R. Shewry), 35–70. AACC International.

Gooding, M.J. (2017). The effects of growth environment and agronomy on grain quality. In: *Cereal Grains*, 2e (eds. C. Wrigley, I. Batey and D. Miskelly), 493–512. Woodhead Publishing https://doi.org/10.1016/B978-0-08-100719-8.00018-8.

Gooding, M.J., Cosser, N.D., Thompson, A.J. et al. (1998). Sheep grazing and defoliation of contrasting varieties of organically grown winter wheat with and without undersowing. *Grass and Forage Science* 53 (1): 76–87. https://doi.org/10.1046/j.1365-2494.1998.00106.x.

Guzman, C., Peña, R.J., Singh, R. et al. (2016). Wheat quality improvement at CIMMYT and the use of genomic selection on it. *Applied & Translational Genomics* 11: 3–8. https://doi.org/10.1016/j.atg.2016.10.004.

Harrell, D.M., Wilhelm, W.W., and McMaster, G.S. (1993). Scales – a computer-program to convert among 3 developmental stage scales for wheat. *Agronomy Journal* 85: 758–763. https://doi.org/10.2134/agronj1993.00021962008500030043x.

Harrell, D.M., Wilhelm, W.W., and McMaster, G.S. (1998). Scales 2: computer program to convert among developmental stage scales for corn and small grains. *Agronomy Journal* 90: 235–238. https://doi.org/10.2134/agronj1998.00021962009000020021x.

Hatcher, D.W., Lukow, O.M., and Dexter, J.E. (2006). Influence of environment on Canadian hard white spring wheat noodle quality. *Cereal Foods World* 51 (4): 184–190.

He, F., Pasan, R., and Shi, F. (2019). Exome sequencing highlights the role of wild-relative introgression in shaping the adaptive landscape of the wheat genome. *Nature Genetics* 51: 896–904. https://doi.org/10.1038/s41588-019-0382-2.

Hemery, Y., Rouau, X., Lullien-Pellerin, V. et al. (2007). Dry processes to develop wheat fractions and products with enhanced nutritional quality. *Journal of Cereal Science* 46: 327–347. https://doi.org/10.1016/j.jcs.2007.09.008.

Henry, A.G., Brooks, A.S., and Piperno, D.R. (2011). Microfossils in calculus demonstrate consumption of plants and cooked foods in Neanderthal diets (Shanidar III, Iraq; Spy I and II, Belgium). *Proceedings of the National Academy of Science, USA* 108: 486–491. https://doi.org/10.1073/pnas.1016868108.

Henry, A.G., Brooks, A.S., and Piperno, D.R. (2014). Plant foods and the dietary ecology of Neanderthals and early modern humans. *Journal of Human Evolution* 69: 44–54. https://doi.org/10.1016/j.jhevol.2013.12.014.

Hillel, D. (1991). *Out of the Earth: Civilization and the Life of the Soil*. Aurum Press Ltd.

Hook, S.C. (1984). Specific weight and wheat quality. *Journal of the Science of Food and Agriculture* 35 (10): 1136–1141. https://doi.org/10.1002/jsfa.2740351013.

Huang, S. and Miskelly, D. (2019). Steamed bread – a review of manufacturing, flour quality requirements, and quality evaluation. *Cereal Chemistry* 96 (1): 8–22. https://doi.org/10.1002/cche.10096.

Huang, S., Sirikhachornkit, A., Su, X. et al. (2002). Genes encoding plastid acetyl-CoA carboxylase and 3-phosphoglycerate kinase of the Triticum/Aegilops complex and the evolutionary history of polyploid wheat. *Proceedings of the National Academy of Sciences* 99 (12): 8133–8138. https://doi.org/10.1073/pnas.072223799.

IMF (2021). *IMF Primary Commodity Prices*. International Monetary Fund https://www.imf.org/en/Research/%20commodity-prices.

Jacob, H.E. (1977). *Six Thousand Years of Bread. Its Holy and Unholy History*. Skyhorse Publishing.

Ji, T., Penning, B., and Baik, B.K. (2018). Pre-harvest sprouting resistance of soft winter wheat varieties and associated grain characteristics. *Journal of Cereal Science* 83: 110–115. https://doi.org/10.1016/j.jcs.2018.08.006.

Kent, N.L. and Evers, A.D. (1994). *Kent's Technology of Cereals*, 4e. Woodhead Publishing.

Kihara, H. (1929). Conjugation of homologous chromosomes in genus hybrids *Triticum* x *Aegilops* and species hybrids of *Aegilops*. *Cytologia* 1: 1–15.

Kihara, H. (1954). Considerations of the evolution and distribution of *Aegilops* species based on the Analyser method. *Cytologia* 19: 336–357.

Kilian, B., Özkan, H., Pozzi, C. et al. (2009). Domestication of the Triticeae in the Fertile Crescent. In: *Genetics and Genomics of the Triticeae. Plant Genetics and Genomics: Crops and Models*, vol. 7 (eds. G. Muehlbauer and C. Feuillet). Springer https://doi.org/10.1007/978-0-387-77489-3_3.

Kweon, M., Slade, L., and Levine, H. (2011). Solvent Retention Capacity (SRC) testing of wheat flour: principles and value in predicting flour functionality in different wheat-based food processes and in wheat breeding: a review. *Cereal Chemistry* 88: 537–552. https://doi.org/10.1094/CCHEM-07-11-0092.

Lachman, J., Martinek, P., Kotíková, Z. et al. (2017). Genetics and chemistry of pigments in wheat grain – a review. *Journal of Cereal Science* 74: 145–154. https://doi.org/10.1016/j.jcs.2017.02.007.

Landes, A. and Porter, J.R. (1989). Comparison of scales used for categorising the development of wheat, barley, rye and oats. *Annals of Applied Biology* 115: 343–360. https://doi.org/10.1111/j.1744-7348.1989.tb03393.x.

Laskowski, W., Górska-Warsewicz, H., Rejman, K. et al. (2019). How important are cereals and cereal products in the average polish diet? *Nutrients* 11: 679–692. https://doi.org/10.3390/nu11030679.

Lev, E., Kislev, M.E., and Bar-Yosef, O. (2005). Mousterian vegetal food in Kebara cave, Mt. Carmel. *Journal of Archaeological Science* 32 (3): 475–484. https://doi.org/10.1016/j.jas.2004.11.006.

Lillywhite, R.D. and Sarrouy, C. (2014). *A Review of the Dietary, Health and Environmental Status of Whole Grain Cereals*. University of Warwick.

Lin, Z.-J., Miskelly, D.M., and Moss, H.J. (1990). Suitability of various Australian wheats for chinese-style steamed bread. *Journal of the Science of Food and Agriculture* 53: 203–213. https://doi.org/10.1002/jsfa.2740530208.

Lunn, G.D., Major, B.J., Kettlewell, P.S. et al. (2001). Mechanisms leading to excess alpha-amylase activity in wheat (*Triticum aestivum*, L) grain in the UK. *Journal of Cereal Science* 33 (3): 313–329. https://doi.org/10.1006/jcrs.2001.0369.

Mares, D., Rathjen, J., Mrva, K. et al. (2009). Genetic and environmental control of dormancy in white-grained wheat (*Triticum aestivum* L.). *Euphytica* 168 (3): 311–318. https://doi.org/10.1007/s10681-009-9927-2.

Miller, T.E. (1987). Systematics and evolution. In: *Wheat Breeding: Its Scientific Basis* (ed. F.G.H. Lupton), 1–30. Chapman and Hall.

Mithen, S. (2012). *Thirst: Water & Power in the Ancient World*. Orion Publishing Group.
Monfreda, C., Ramankutty, N., and Foley, J.A. (2008). Farming the planet: 2. Geographic distribution of crop areas, yields, physiological types, and net primary production in the year 2000. *Global Biogeochemical Cycles* 22: GB1022. https://doi.org/10.1029/2007GB002947.
Montgomery, D.R. (2007). *Dirt: The Erosion of Civilisations*. University of California Press.
Moss, H.J. (1973). Quality standards for wheat varieties. *Journal of the Australian Institute of Agricultural Science* 39: 109–115.
Nalam, V.J., Vales, M.I., Watson, C.J.W. et al. (2006). Map-based analysis of genes affecting the brittle rachis character in tetraploid wheat (*Triticum turgidum* L.). *Theoretical and Applied Genetics* 112: 373–381. https://doi.org/10.1016/j.gene.2014.03.034.
Nesbitt, M. (2001). Wheat evolution: integrating archaeological and biological evidence. In: *Wheat Taxonomy: The Legacy of John Percival* (eds. P.D.S. Caligari and P.E. Brandham), 37–59. Academic Press.
Newman, C. (2008). *Grain Storage: Maintaining Grain Quality*. Western Australia, Department of Agriculture and Food.
O'Mara, F.P. (2012). The role of grasslands in food security and climate change. *Annals of Botany* 110 (6): 1263–1270. https://doi.org/10.1093/aob/mcs209.
Payne, P.I., Nightingale, M.A., Krattiger, A.F. et al. (1987). The relationship between HMW glutenin subunit composition and the bread-making quality of British-grown wheat varieties. *Journal of the Science of Food and Agriculture* 40 (1): 51–65. https://doi.org/10.1002/jsfa.2740400108.
Pearson, T., Wilson, J., Gwirtz, J. et al. (2007). Relationship between single wheat kernel particle-size distribution and Perten SKCS 4100 hardness index. *Cereal Chemistry* 84 (6): 567–575. https://doi.org/10.1094/CCHEM-84-6-0567.
Peel, M.C., Finlayson, B.L., and Mcmahon, T.A. (2007). Updated world map of the Köppen-Geiger climate classification. *Hydrology and Earth System Sciences Discussions, European Geosciences Union* 4 (2): 439–473. https://hal.archives-ouvertes.fr/hal-00298818.
Peña, R.J. (2002). Wheat for bread and other foods. In: *Bread Wheat Improvement and Production* (eds. B.C. Curtis, S. Rajaram and H. Gómez Macpherson), 483–542. Food and Agriculture Organization of the United Nations http://www.fao.org/3/Y4011E/y4011e0w.htm#bm32.
Peña, R.J., Braun, H.J., and Mollins, J. (2017). Anti-wheat fad diets undermine global food security efforts: wheat consumption healthy despite claims in self-help publications. In: *The Wheat and Nutrition Series: A Compilation of Studies on Wheat and Health*, 1–19. CIMMYT https://repository.cimmyt.org/xmlui/handle/10883/19130.
Percival, J. (1921). *The Wheat Plant: A Monograph*. Duckworth and Co.
Perkins, J.H. (1997). *Geopolitics and the Green Revolution: Wheat, Genes, and the Cold War*. Oxford University Press.
Peterson, C. (1995). *Bread and the British Economy c1770–1870*. New York: Scolar Press and Ashgate Publishing 246 pp.
Rajaram, S., Van Ginkel, M., and Fischer, R.A. (1993). CIMMYT's wheat breeding mega-environments (ME). In *Proceedings of the 8th International Wheat Genetics Symposium* (pp. 1101–1105).
Rebetzke, G.J., Bonnett, D.G., and Reynolds, M.P. (2016). Awns reduce grain number to increase grain size and harvestable yield in irrigated and rainfed spring wheat. *Journal of Experimental Botany* 67 (9): 2573–2586. https://doi.org/10.1093/jxb/erw081.

Reeves, J.T. and Roy, N. (1974). Wheat breeding. *Journal of the Department of Agriculture, Western Australia, Series 4* 15 (2): 39–43. https://researchlibrary.agric.wa.gov.au/journal_agriculture4/vol15/iss2/3.

Rickman, G. (1980). The grain trade under the Roman Empire. *Memoirs of the American Academy in Rome* 36: 261–275. https://doi.org/10.2307/4238709.

Roberts, E. (1847). On the management of wheat. *Journal of the Royal Agricultural Society of England* 8: 60–77.

Rubel, W. (2011). *Bread: A Global History*. Reaktion Books.

Ruske, R.E., Gooding, M.J., and Dobraszczyk, B.J. (2004). Effects of triazole and strobilurin fungicide programmes, with and without late-season nitrogen fertiliser, on the baking quality of Malacca winter wheat. *Journal of Cereal Science* 40 (1): 1–8. https://doi.org/10.1016/j.jcs.2004.03.003.

Rustgi, S., Shewry, P., Brouns, F. et al. (2019). Wheat seed proteins: factors influencing their content, composition, and technological properties, and strategies to reduce adverse reactions. *Comprehensive Reviews in Food Science and Food Safety* 18 (6): 1751–1769. https://doi.org/10.1111/1541-4337.12493.

Sapone, A., Bai, J.C., Ciacci, C. et al. (2012). Spectrum of gluten-related disorders: consensus on new nomenclature and classification. *BMC Medicine* 10 (1): 1–12. https://doi.org/10.1186/1741-7015-10-13.

Sarrafi, A., Ecochard, R., and Grignac, P. (1989). Genetic variability for some grain quality characters in tetraploid wheats. *Plant Varieties and Seeds* 2: 163–169.

Scott, J.C. (2017). *Against the Grain: A Deep History of the Earliest States*. Yale University Press.

Shewry, P.R., Tatham, A.S., Barro, F. et al. (1995). Biotechnology of breadmaking: unravelling and manipulating the multi-protein gluten complex. *Biotechnology* 13: 1185–1190. https://doi.org/10.1038/nbt1195-1185.

Shewry, P.R., Wan, Y., Hawkesford, M.J. et al. (2019). Spatial distribution of functional components in the starchy endosperm of wheat grains. *Journal of Cereal Science* 91: 102869. https://doi.org/10.1016/j.jcs.2019.102869.

Shiferaw, B., Smale, M., Braun, H.J. et al. (2013). Crops that feed the world 10. Past successes and future challenges to the role played by wheat in global food security. *Food Security* 5 (3): 291–317. https://doi.org/10.1007/s12571-013-0263-y.

Simmonds, N.W. and Smartt, J. (1999). *Principles of Crop Improvement*, 2e. Blackwell Science.

Sinclair, T.R. (1998). Historical changes in harvest index and crop nitrogen accumulation. *Crop Science* 38: 638–643. https://doi.org/10.2135/cropsci1998.0011183X003800030002x.

Slade, L. and Levine, H. (1994). Structure-function relationships of cookie and cracker ingredients. In: *The Science of Cookie and Cracker Production* (ed. H. Faridi), 23–141. New York: Chapman & Hall.

Smith, L.G., Jones, P.J., Kirk, G.J. et al. (2018). Modelling the production impacts of a widespread conversion to organic agriculture in England and Wales. *Land Use Policy* 76: 391–404. https://doi.org/10.1016/j.landusepol.2018.02.035.

Snape, J.W., Hyne, V., and Aitken, K. (1993). Targeting genes in wheat using marker mediated approaches. *Proceedings of the Eighth International Wheat Genetics Symposium* 2: 749–759.

Tadesse, G., Algieri, B., Kalkuhl, M. et al. (2016). Drivers and triggers of international food price spikes and volatility. In: *Food Price Volatility and its Implications for Food Security and Policy* (eds. M. Kalkuhl, J. von Braun and M. Torero), 59–82. Springer.

Tanno, K.I. and Willcox, G. (2012). Distinguishing wild and domestic wheat and barley spikelets from early Holocene sites in the near east. *Vegetation History and Archaeobotany* 21 (2): 107–115. https://doi.org/10.1007/s00334-011-0316-0.

Taylor, B.R., Cranstoun, D.A.S., and Roscrow, J.C. (1993). The quality of winter wheat varieties for distilling from Scottish sites. In: *Aspects of Applied Biology 36, Cereal Quality III* (eds. P.S. Kettlewell, J.R. Garstang, C.M. Duffus, et al.), 481–489. Association of Applied Biologists.

Thomas, H. (2017). *The War Between Trees and Grasses*. Cambrian Printers.

Tottman, D.R. (1987). The decimal code for the growth stages of cereals, with illustrations. *Annals of Applied Biology* 110 (2): 441–454. https://doi.org/10.1111/j.1744-7348.1987.tb03275.x.

Turnbull, K.M. and Rahman, S. (2002). Endosperm texture in wheat. *Journal of Cereal Science* 36 (3): 327–337. https://doi.org/10.1006/jcrs.2002.0468.

Urruty, N., Tailliez-Lefebvre, D., and Huyghe, C. (2016). Stability, robustness, vulnerability and resilience of agricultural systems. A review. *Agronomy for Sustainable Development* 36 (1): 15. https://doi.org/10.1007/s13593-015-0347-5.

USFDA (1999). *Health Claim Notification for Whole Grain Foods*. U.S. Food & Drug Administrationhttps://www.fda.gov/food/food-labeling-nutrition/health-claim-notification-whole-grain-foods.

Uthayakumaran, S. and Wrigley, C. (2017). Wheat: grain-quality characteristics and management of quality requirements. In: *Cereal Grains: Assessing and Managing Quality*, 2e (eds. C. Wrigley, I. Batey and D. Miskelly), 91–134. Woodhead Publishing https://doi.org/10.1016/B978-0-08-100719-8.00005-X.

Vaidyanathan, L.V. (1987). Precision and reliability of measuring Hagberg falling number of wheat including variability associated with crop husbandry and grain handling. In: *Aspects of Applied Biology 15, Cereal Quality*, 495–513. Association of Applied Biologists.

Van der Kamp, J.W., Poutanen, K., Seal, C.J. et al. (2014). The HEALTHGRAIN definition of "whole grain". *Food & Nutrition Research* 58: 22100. https://doi.org/10.3402/fnr.v58.22100.

Wang, L., Yao, Y., He, Z. et al. (2013). Determination of phenolic acid concentrations in wheat flours produced at different extraction rates. *Journal of Cereal Science* 57: 67–72. https://doi.org/10.1016/j.jcs.2012.09.013.

Ward, J.L., Poutanen, K., Gebruers, K. et al. (2008). The HEALTHGRAIN cereal diversity screen: concept, results, and prospects. *Journal of Agricultural and Food Chemistry* 56 (21): 9699–9709. https://doi.org/10.1021/jf8009574.

Weiss, E., Wetterstrom, W., Nadel, D. et al. (2004). The broad spectrum revisited: evidence from plant remains. *Proceedings of the National Academy of Sciences* 101 (26): 9551–9555. https://doi.org/10.1073/pnas.0402362101.

Wieser, H. (2000). Comparative investigations of gluten proteins from different wheat species. I. Qualitative and quantitative composition of gluten protein types. *European Food Research and Technology* 211: 262–268. https://doi.org/10.1007/s002170000165.

Wieser, H., Gutser, R., and Von Tucher, S. (2004). Influence of sulphur fertilisation on quantities and proportions of gluten protein types in wheat flour. *Journal of Cereal Science* 40 (3): 239–244. https://doi.org/10.1016/j.jcs.2004.05.005.

Wilson, W.W., and Dahl, B.L. (2002). Marketing of wheat on a constant and nil moisture basis. *Agribusiness & Applied Economics Report No. 496*. North Dakota State University. doi:10.22004/ag.econ.23638

Wrigley, C.W. (2009). Wheat: a unique grain for the world. In: *Wheat Chemistry and Technology*, 4e (eds. K. Khan and P.R. Shewry), 1–17. AACC International https://doi.org/10.1016/B978-1-891127-55-7.50008-2.

Zadoks, J.C., Chang, T.T., and Konzak, C.F. (1974). A decimal code for the growth stages of cereals. *Weed Research* 14: 415–421.

Zhao, F.J., Hawkesford, M.J., and McGrath, S.P. (1999a). Sulphur assimilation and effects on yield and quality of wheat. *Journal of Cereal Science* 30 (1): 1–17. https://doi.org/10.1006/jcrs.1998.0241.

Zhao, F.J., Salmon, S.E., Withers, P.J.A. et al. (1999b). Variation in the breadmaking quality and rheological properties of wheat in relation to sulphur nutrition under field conditions. *Journal of Cereal Science* 30 (1): 19–31. https://doi.org/10.1006/jcrs.1998.0244.

Zohary, D., Harlan, J.R., and Vardi, A. (1969). The wild diploid progenitors of wheat and their breeding value. *Euphytica* 18: 58–65. https://link.springer.com/content/pdf/10.1007/BF00021982.pdf.

2

A 'Good' Soil

What befalls the earth befalls all the sons of the earth

attrib. Chief Seattle 1854

Hudson (2020) group soil with fresh air and freshwater as the Earth's three most vital natural resources. As a resource for wheat production a good soil is a *fertile* soil: one that consistently supports the germination, growth, and development of a crop to greatly multiply the number of seeds produced compared with what is sown. The concept of *fertility* being associated with land used for crop production in Mesopotamia is at least 4500–4000 years old (Lal et al. 2007), while the Chinese had a system of taxation based on soil classification in the Yao dynasty around 4300 BP (Montgomery 2007a). Soil being the source of fertility and productivity is reflected in many of today's religions. In the Jewish scriptures man is derived from the soil both physically and linguistically: 'Adam' from the Hebrew *adama* meaning earth or soil. Our Latin name for man, *Homo*, derives from the Indo-European root for 'earth,' which is also a stem for *humus*.

Soil provides anchorage and the substrate in and through which nutrients and water are made available to crop plants; a substrate into which roots can penetrate and interact. Soil fertility has, thus, underpinned food production and the development of communities, sedentism, states, and civilisation. Conversely, those same communities and states can decline when soil and its fertility are denuded. Instability, collapse, or migration of developed societies have been at least partly associated with soil depletion and reduced fertility for the Sumerians, the Romans, the Greeks, the Mayans, and the ancient Chinese (Jacobsen and Adams 1958; Hillel 1991; Bakker et al. 2007; Montgomery 2007a). Depopulation of islands associated with land degradation has been particularly well documented for Easter Island and Iceland (Montgomery 2007a).

Understanding the links between maintaining soil fertility and productive agriculture is core to any desire to improve sustainability and food security. Many systems purporting to be *sustainable* highlight the need to protect and improve the functioning of the soil. For example, *Organic Farming* stresses, amongst several aspects, the importance of the soil organic matter. One of the early pioneers of organic farming, Lady Eve Balfour, emphasised that 'The health of man, beast, plant and soil is one indivisible whole' (Balfour 1943). More latterly *Conservation Agriculture* highlights the value of soil retention, structure, and function for improved and sustained food security (FAO 2021). In addition to agriculture,

Wheat: Environment, Food and Health, First Edition. Mike J. Gooding and Peter R. Shewry.

soils deliver wider *public* benefits. Soil properties determine whether rainfall is retained to replenish groundwater or runs off the land to contribute to flooding. The soil and its associated biology can help purify water by filtering and breaking down pollutants and pathogens. Increasingly, soil is seen as a potential site of carbon sequestration, as a pool for biodiversity and genes, and as an archive of geological and archaeological heritage (FAO and ITPS 2015).

2.1 Soils for Wheat Production

Soils vary depending on their origins, their subsequent movement and weathering, the vegetation they support, and their long-term management. Many of these effects are dependent on, and interact with, climate and topography. At the broadest scale, soils are classified with hierarchical schemes of *taxonomy*, based principally on outcomes of soil development processes. At the field scale, variation can be captured in a range of soil qualities that will define a soil's innate fertility and appropriate management for wheat production. These qualities include soil texture, organic matter, pH, sodicity, salinity, structure, and depth. Deficiencies in any one of these factors can significantly limit wheat growth, to the point of crop failure.

2.1.1 Soil Taxonomy

The two principal methods for classifying soils are those of the US Department of Agriculture (Soil Survey Staff 2014) and the World Reference Base (WRB) for Soil Resources (IUSS Working Group WRB 2015). Using the former, den Biggelaar et al. (2004a) estimated that around 80% of the wheat area is on three soil orders – the Alfisols, the Mollisols, and the Inceptisols – with most of the remainder on Vertisols, Entisols, and Ultisols (Figure 2.1).

Alfisols (mostly Luvisols or Lixisols in WRB) are typically formed in humid areas under broadleaf, deciduous forest. The subsoil is rich in clay (Section 2.1.2) so alfisols are comparatively moisture and nutrient retentive. For agriculture, these soils are therefore often

Figure 2.1 Areas of wheat according to soil order as estimated by den Biggelaar et al. (2004a).

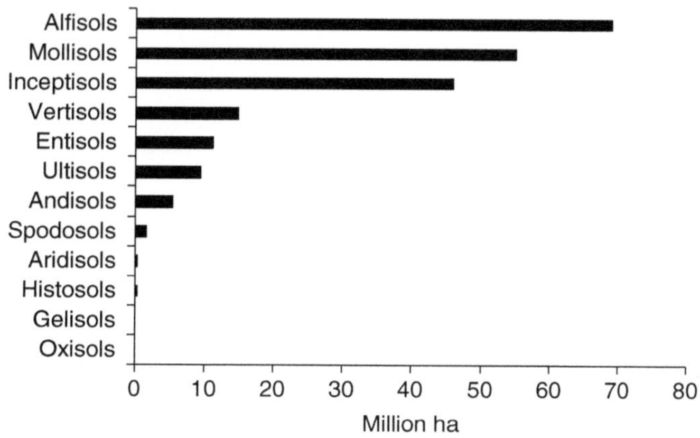

fertile and resilient, although significant erosion can occur under cultivation, particularly on slopes with high rainfall.

Mollisols (mostly Chernozems, Kastanozems, and Phaeozems in WRB) form under grassland and are therefore usually present in drier areas than alfisols. Typical regions are found in the Great Plains of North America, the Humid Pampas of Argentina, and the Eurasian Steppe (Figure 1.2). The parent material is often limestone and windblown fine particles (loess; aeolian deposits). The soils have a relatively high organic matter content typical of accumulations that develop under permanent grassland cover (Section 2.1.3). The organic matter darkens the soil and contributes to a soft, well-aggregated soil that can be highly fertile. Under annual cropping and cultivation, however, mollisols are at significant risk of erosion and loss of organic material (Fenton 2012; Lu et al. 2016).

Inceptisols (mostly Cambisols or Umbrisols in WRB) are comparatively young soils, but older than the entisols described later. Relative youth means that there has been less weathering and development, i.e. there has been insufficient time or, in the case of northern latitudes, also temperature, for much definition of distinct soil horizons or layers. This order includes a disparate range of soil types and origins. In areas of North America and northern Europe, many of the inceptisols have formed on and from late-Pleistocene glacial drift. Inceptisols can, however, also be underpinned by sediments transported by water (alluvial deposits) in river plains, basins, and deltas. Similarly, some of the soils formed on loess are classed as Inceptisols.

Vertisols have a high clay content, such that they are hard and form deep cracks when dry. Conversely, when wet vertisols are very sticky and swell. They are found under wheat crops in sub-tropical and semi-arid areas including around the Mediterranean, the northern wheat belt of Australia, the rice-wheat rotations in central India, the Ethiopian highlands, central Mexico, and the Sudan. The cracking and swelling cycles provide some *natural* cultivation (often called 'self-structuring'), which helps to incorporate organic matter from crop residues. If machinery and management is sufficient to cultivate the heavy soil despite the challenges, and when sufficient water is available for crop growth (often from irrigation), the land can be fertile.

Entisols are the youngest and least developed of the soil orders. Their composition closely resembles their parent material with little to no development of horizons. Entisols can be alluvial or aeolian deposits and are often sands and/or silts.

Ultisols (mostly Acrisols and Alisols in WRB), in contrast to the entisols, have undergone long-term and continuous weathering. They have no calcareous material and the surface layers have less than 10% weatherable minerals. The ultisols can occur in temperate and tropical regions but the long weathering is often associated with being heavily leached and acidic. Under wheat they are found as the red clay soils in the southeastern United States, in sub-tropical areas of southern China, and in southern Chile.

2.1.2 Soil Texture

The texture of a soil refers to the size distribution of particles from which it is composed (Figure 2.2).

Clays have the smallest particles (< 0.002 mm) of soil. They originate from the chemical weathering of rocks, and are minerals mostly of aluminium, silicon, and oxygen

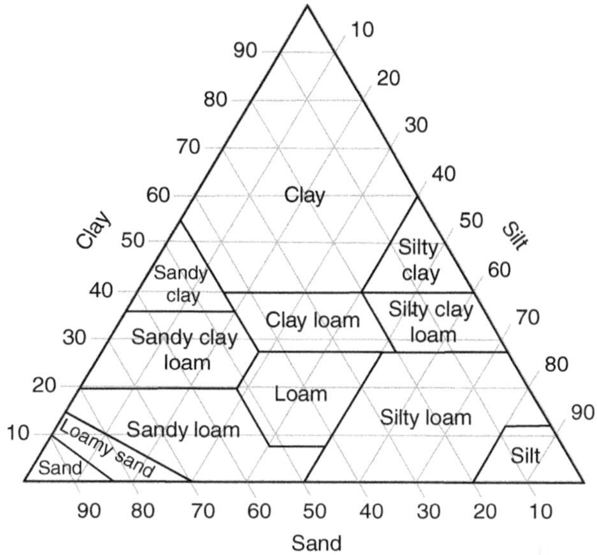

Figure 2.2 Soil texture classifications based on particle size distributions (% content, w/w).

(aluminosilicates) such as in kaolinite, illite, and smectite. The layers of silicate and the small particle size provide a remarkably high surface area (up to 80 ha/kg) on which water and solutes can be absorbed. Clayey soils, therefore, provide a reservoir of nutrients that become slowly available for plant growth. Clay particles, as well as oxides and organic matter, carry both positive and negative charges (Matocha 2017). A permanent negative charge is caused by substitution of aluminium in place of silicon or of magnesium in place of aluminium in the clay structure. A pH-dependent charge can also occur on the edges of clays: positive at low/acid pH values and negative at high/alkaline pH values. Clays therefore usually retain positive ions (cations) of, for example, hydrogen (H^+), sodium (Na^+), potassium (K^+), calcium (Ca^{2+}), and magnesium (Mg^{2+}). Conversely, soluble, negatively charged ions (anions) such as nitrate (NO_3^-) and sulphate (SO_4^{2-}) are generally highly mobile within soils. Despite their nutrient and water retaining properties, clayey soils can be difficult to manage. Colloquially, these soils are termed as *heavy* and may require two to four times as much tractor power to cultivate compared with *light* soils. If significant damage to the soil structure is to be avoided, wet conditions restrict the use of cultivation and machinery traffic for establishment of winter and spring wheats. Low soil temperatures may result in a further restriction to growth, as wet clayey soils take longer to warm up in the spring. Minimum cultivation techniques (Section 2.4.1), which are not reliant on inverting the soil by ploughing, can be appropriate, particularly because the cycles of swelling and cracking in wet and dry weather respectively can help structure the soil.

Sands, including gravel, comprise the largest soil particles and can be subdivided into fine sand (0.06–0.20 mm), medium sand (0.20–0.60 mm), coarse sand (0.60–2.00 mm), and gravel (> 2.00 mm). Sands derive from the physical weathering of rocks and are mostly grains of quartz. Sandy soils are *light* and easily worked, and more often accessible in autumn and spring than heavier soils. They are, however, also comparatively nutrient poor. The soils are readily leached with rainfall and irrigation, but droughts develop rapidly in dry periods. Furthermore, there is little natural structuring and therefore ploughing is often required to alleviate compaction in the upper layers. The sandy soils can also have poor stability and are easily eroded by water on slopes and by wind on the flat (Finch et al. 2014).

Silts are intermediate between the clays and the sands (0.002–0.06 mm). Formed from glacial, river, marine, and wind-blown deposits, their composition is like sand. Unlike sand, however, silt particles have an affinity with clay particles with which they are often coated. The medium or *loamy* soils can be made up with various combinations of clay, sand, and silt particles (Figure 2.2).

Wheat can be grown on most soil textures, with farmers adjusting the intensity and timing of tillage, and of nutrient management to partially account for the qualities of the soil. Modern wheat cultivars, however, are responsive to high levels of fertility and thrive particularly on well-drained, medium to heavy textured soils, such as silt and clay loams with high nutrient status (Wibberley 1989).

2.1.3 Soil Organic Matter

Farmers have long known of the benefits of organic matter for crop fertility (Russell 1977; Johnston et al. 2009). Our connection with the soil organic matter is direct. All terrestrial organisms deposit organic matter from which they are composed onto the soil, a process fully appreciated by our ancestors: 'All come from dust, and to dust all return' (Ecclesiastes, possibly 2450–2200 BP).

As well as excrement and decaying fauna, the organic matter of the soil includes the growing roots and root exudates of crop plants and weeds. When the above-ground crop senesces, the roots die and decay, as will any residue from the above-ground crop left on the surface or incorporated into the soil after harvest. Much of the initial processing and transport of this new organic matter will be by invertebrates. Earthworms have significant value to soil fertility in trash burial, nutrient cycling, soil aeration, and drainage (Bailey et al. 1999). Once the particle size of the organic matter has been reduced, the microorganisms of the soil take over. It is the products of this microbial activity, and the decay of microbes, that form the bulk of the true humus fraction in the soil.

The living microbes will themselves account for 2–4% of the soil organic matter and often at least 1 t/ha of organic carbon. The microbial population is dominated by bacteria and fungi, contributing 100–10 000 times more biomass than the other major components of what is called the *Soil Microbiome*, such as protozoa, single-celled algae, archaea, and viruses. The composition, activity, and interaction with plant roots of the soil microbiome do much to define soil health and plant productivity. Soil microorganisms mediate the cycling of nutrients, influence soil acidity, regulate soil carbon dynamics, can improve plant growth and abiotic stress resistance, and can suppress plant pathogens. Microbes can also alter water availability by influencing soil hydraulic conductivity and hydrophobicity, degrade synthetic compounds, chelate metals, and detoxify poisons (Chaparro et al. 2012; Fierer 2017). The function and structure of the microbiome depends particularly on soil pH, and the availability of oxygen, carbon, nitrogen, phosphorus, and moisture, and the plants growing in the soil (Fierer 2017).

Soil organic matter is usually assumed to contain 58% organic carbon. It is also a significant reserve of crop nutrients. These can be divided between those required by the crop in quantities of kilogrammes per tonne of crop (the *macronutrients*: nitrogen [N], potassium [K], phosphorus [P], sulphur [S], magnesium [Mg], and calcium [Ca]), and those required in quantities of grammes per tonne of crop (the *micronutrients*: boron [B], chlorine [Cl], copper [Cu], iron [Fe], manganese [Mn], molybdenum [Mo], nickel [Ni], and zinc [Zn]). The crop nutrients can be released as organic matter is oxidised and broken down by microbes. Additionally, organic matter has both anion and cation exchange sites and hence retains nutrients and helps reduce nutrient leaching, particularly on sandy soils (Russell 1977; Johnston et al. 2009). Soil organic matter improves soil friability, soil aggregation, aggregate stability, water infiltration, water retention, and resilience against erosion, and can reduce the power required for cultivation (Johnston et al. 2009; Lal 2009). Organic matter can increase the potential water holding capacity of soils (NRCS 2018) and increase the amount of grain produced per unit of available water (Mbava et al. 2020). Given these multiple roles, it appears to be impossible to fully compensate for lower organic matter in soil by increasing fertilizer inputs and irrigation with regards to crop yield and yield stability. Wheat productivity can be significantly limited on soils with less than 1.7% organic matter (1% carbon) in the top 30 cm of soil (Figure 2.3; Holland 2004; Ghimire et al. 2018).

In addition to its direct importance for agricultural productivity, soil organic matter delivers several public goods. As noted above, organic matter is composed mostly of carbon. Soil is, therefore, a significant reservoir of sequestered carbon, concentrated mostly in the upper layers (Figure 2.4). It is estimated that three times as much carbon is stored in the top metre of soil than is in the atmosphere (Scharlemann et al. 2014). Even under continuous annual cultivation, a sandy loam soil with low organic matter (c. 2%) will have 30–40 t C/ha in just the top 30 cm. The same depth of a mineral soil under continuous pasture with, say, 6% organic matter may have around 100 t C/ha. Much higher levels are present in soils classed as organic soils (> 10% organic matter) and peat (> 50% organic matter) (MAFF 1988). Increasing soil organic matter is, therefore, seen as a potential method of offsetting CO_2 emissions from the burning of fossil fuels (Poulton et al. 2018). Conversely, however, mismanagement of agricultural soils with high organic matter can release large quantities of CO_2 into the atmosphere.

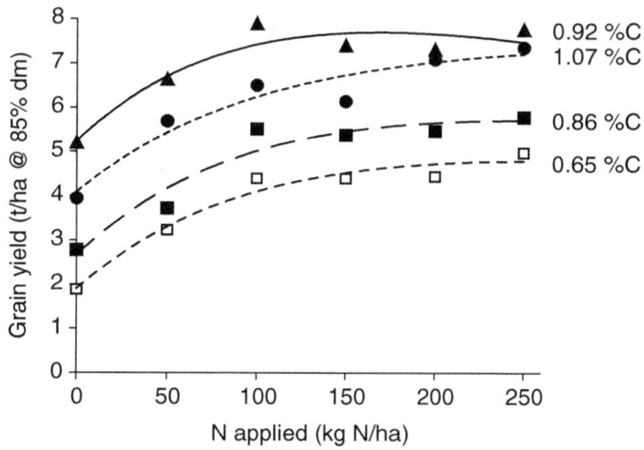

Figure 2.3 The response of winter wheat to nitrogen fertilizer after previous long-term treatments to manipulate soil organic carbon contents. *Source:* Data from Johnston et al. (2009).

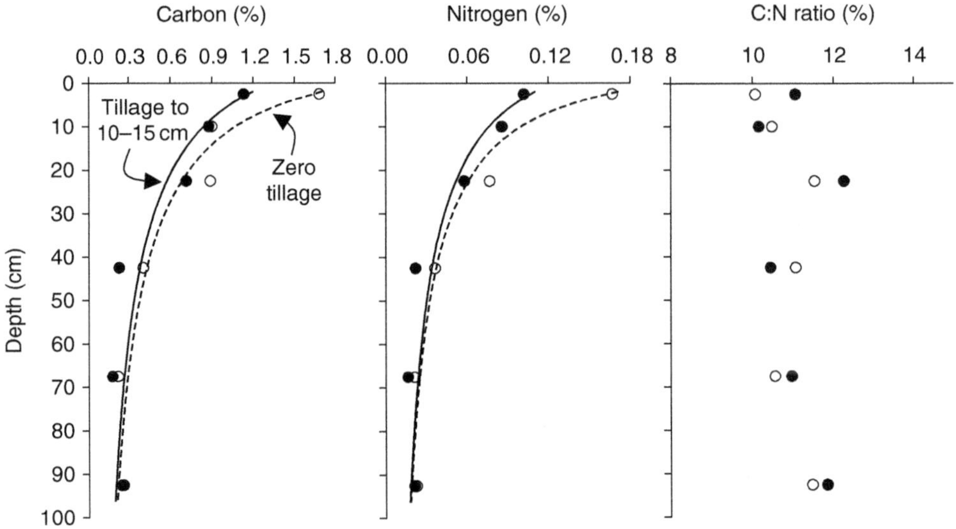

Figure 2.4 The effects of depth and either cultivating soil with discs or zero tillage for 20 years on the soil carbon and nitrogen distributions. Mean of continuous wheat, soyabean-wheat rotation, and sorghum-wheat-soyabean rotation. *Source:* From data presented in Wright et al. (2007).

2.1.4 Soil pH and Sodicity

The pH is the log of the inverse of the effective concentration of hydrogen ions, and hence an assessment of the soil alkalinity or acidity. Acidic, neutral, or alkaline soils have pH values of < 7, 7, and > 7, respectively. The pH of the soil affects the relative solubility and mobility of ions in the soil. As soils become increasingly acidic many of the nutrients required by wheat become less available to the crop (Figure 2.5). Conversely the availabilities of other ions such as manganese and aluminium increase to toxic concentrations. When taken up, aluminium inhibits the growth of roots (Figure 2.6), such that wheat grown on acid soil becomes stunted and yellow (Baquy et al. 2017). In long-term experiments in the UK, the yield and the yield stability of wheat both declined as the pH dropped below 6 (Figure 2.7). Wheat is thus more susceptible to acidity than several other crops including oats, rye, and potatoes (Wibberley 1989; Holland et al. 2019). Although maintaining a pH in the region of 6.5–7.5 is seen as being ideal (Wibberley 1989), and some yield reduction is possible between pH 5.5 and 6.0 (Figure 2.7; see also Cook and Veseth 1991), the target minimum pH in regions of soil acidity is often 5.5 (Schroder et al. 2011; Orton et al. 2018). A pH of less than 5.5 has been associated with reduced productivity and crop failures in major wheat-growing areas (Schroder et al. 2011). There have, however, been significant advances in breeding wheat lines that are more tolerant of acidity, mostly due to increased tolerance to Al^{3+} ions. For example, tolerance can result from wheat roots exuding malate into the rhizosphere, where it chelates Al^{3+} ions, forming compounds that are not taken up by the root (Kochian et al. 2015).

Although acidity is the most common problem associated with pH, values for soil pH above 7.5, for example on calcareous soils, are associated with reduced availability of some

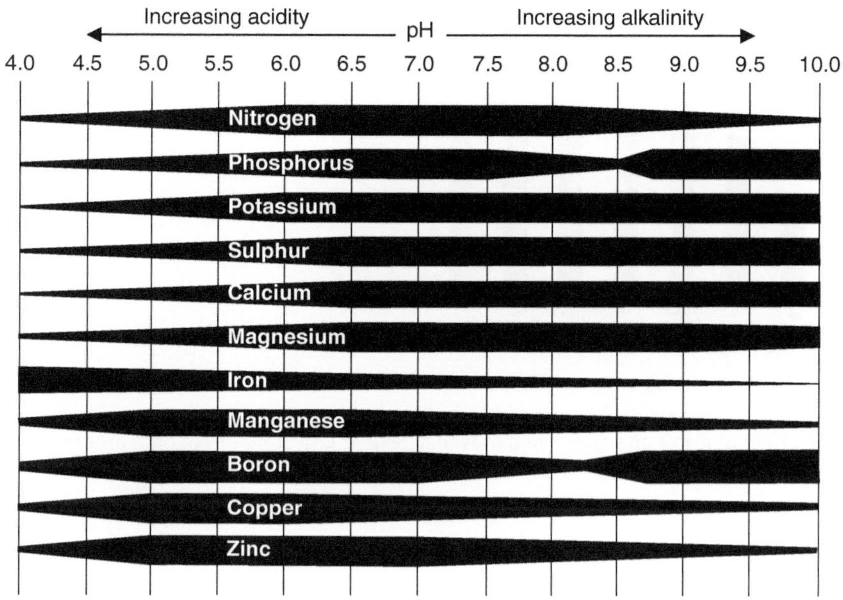

Figure 2.5 The effect of soil pH on the availability of some wheat nutrients in mineral soils. *Source:* Adapted from Truog (1947).

Figure 2.6 The effect of soil pH on the seedling root growth of wheat, and the amount of exchangeable aluminium in the soil. *Source:* Calculated from data from two soil types in Baquy et al. (2017).

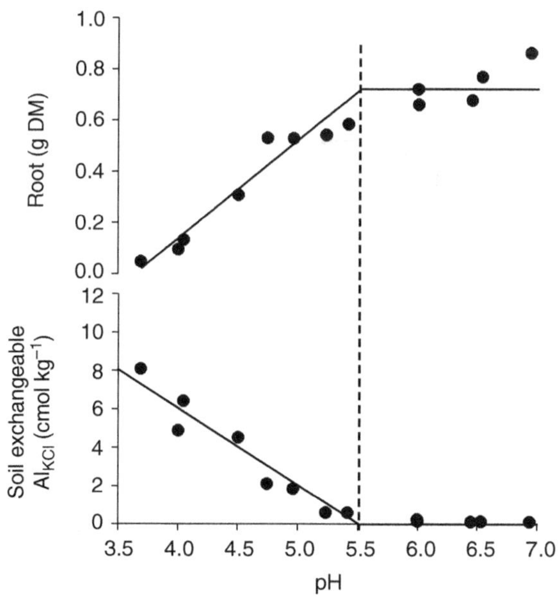

essential nutrients, most notably P but also several micronutrients including B, Cu, Fe, Mn, and Zn (Figure 2.5). This can lead to lower wheat yields and/or poor response to P and micronutrient containing fertilizers.

Soils with a pH above 8.2 may also be sodic soils (FAO and ITPS 2015). Sodic soils are those with a high sodium ion content (Na^+) relative to the amounts of other cations

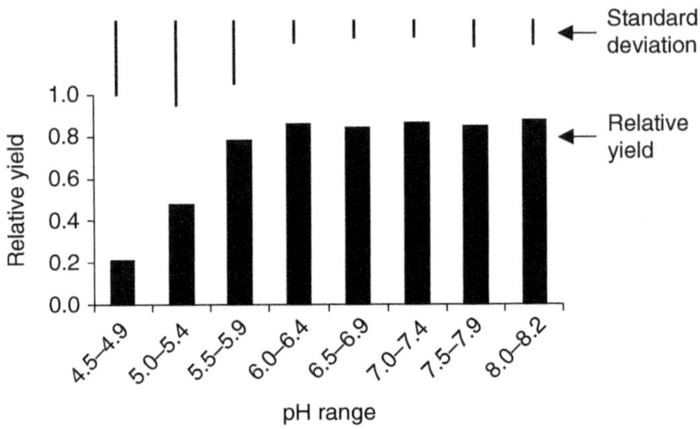

Figure 2.7 The effect of pH on the grain yield of wheat. *Source:* Calculated from data presented in Holland et al. (2019).

(particularly calcium, Ca^{2+}), expressed as an exchangeable sodium percentage (ESP). These soils can be found in arid and semi-arid areas where the high levels of exchangeable sodium cause ion imbalances and dispersion of clay particles. The resultant poor soil structure is often associated with severe capping of the soil surface (surface crusts), a high bulk density, poor water infiltration, and waterlogging (Gill et al. 2008). Most of the world's sodic soils are in Australia (FAO and ITPS 2015). Orton et al. (2018) estimated that the production on two-thirds of the Australian wheat lands was affected by sodicity. Estimating the effect of sodicity per se on wheat yields is difficult because many factors are involved, including effects on emergence through surface crusts, poor root growth through compacted and/or waterlogged soils, and poor nutrient availability. However, an ESP of 6% has been used to define areas in which sodicity will have a significant effect on wheat yield (Dang et al. 2006; Orton et al. 2018). Methods to increase wheat yields on sodic soils include the application of gypsum, and/or the deep incorporation of organic matter (Gill et al. 2008; Rasouli et al. 2013). Gypsum is composed of calcium sulphate dihydrate ($CaSO_4{\cdot}2H_2O$) and derived from quarries and mines, particularly in China, Iran, Spain, and North America (Escavy et al. 2012; Fixen and Johnston 2012). Application of gypsum supplies calcium and sulphur to the crop, but additionally reduces the Na:Ca ratio and the ESP. Gypsum, therefore, improves clay flocculation and structure stability and has been used as a soil conditioner for over 250 years (Rasouli et al. 2013).

2.1.5 Salinity

The salinity of a soil relates to its salt concentration, principally that of sodium chloride (NaCl), but sodium sulphate (Na_2SO_4) and other neutral salts may contribute. As salt concentration increases so do osmotic pressure, ion toxicity, and nutritional imbalances (Rengasamy 2006; Singh et al. 2018a). The increased osmotic pressure reduces water uptake, and excessive salinity in wheat fields results in reduced germination, poorer plant growth, and premature ripening (Grieve et al. 1994). Salinity is usually measured by determining the electrical conductivity of an extracted soil paste (EC_e) in units of deci Siemens

Figure 2.8 The effect of salinity on the grain yield of wheat. *Source:* Fitted to data from Francois et al. (1986).

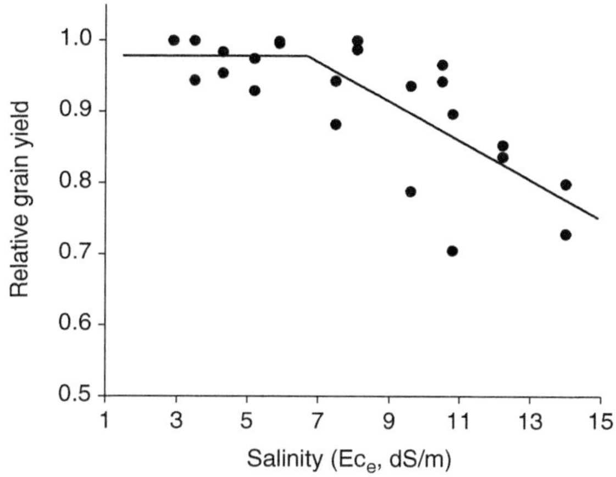

per metre (dS/m). As not all salts are equally harmful, the concentration of chloride can provide additional qualification, e.g. concentrations over 300 mg/kg in the topsoil or 600 mg/kg in the subsoil can be indicative of saline soils (Orton et al. 2018). The salinity levels at which wheat yields will be reduced depend on climate, soil type, crop management, and the depth of salt layers (Orton et al. 2018). However, critical EC_e values for topsoil generally vary between about 4 and 8 dS/m (e.g. Figure 2.8; Lobell et al. 2007). This range, and the observation that appreciable (but not maximal) yields are possible on much more saline soils is indicative of wheat being moderately tolerant to saline conditions: significantly more tolerant than maize and rice, but less so than rye and barley (U.S. Salinity Laboratory 2020).

2.1.6 Soil Structure

Soil structure encompasses a number of properties, including its structural form, i.e. the size and arrangement of aggregates and pores; the ability to retain its form when under stress; and the ability to recover structural form through natural processes (Ball et al. 2007). Structural form is determined by the way clay, sand, and silt particles, with organic matter and oxides, are grouped or aggregated. The size and shape of these aggregates will depend on the soil constituents, but their stability will also be influenced by wetting, drying, traffic, and cultivations. Optimum soil aggregate sizes for wheat establishment range between 1 and 3 mm. More generally, however, the spaces or pores between the aggregates are important. It is the pore diameter and volume that determine many qualities of soil including aeration, water retention and conductivity, bulk density and compaction, optimal soil water content for cultivation, penetration resistance and root growth, and soil structural stability. Taking a range of factors into account, Reynolds et al. (2009) suggest that ideally most soil pores should be in the range of 0.06–0.14 mm.

In practice, it is important to be able to determine the quality of soil structure in the field. Table 2.1 summarizes the scheme presented by Ball et al. (2007) whereby a soil block is extracted with a spade to 15 cm and is then broken up and inspected. The scheme emphasises friability, aggregate size, porosity, and rooting.

Table 2.1 Description of soil structure quality.

Structure quality (Sq)	Ease of break up of moist soil	Size and appearance of aggregates	Visible porosity	Roots
Sq1 Friable	Aggregates crumble readily	Mostly < 6 mm	High	Throughout the soil
Sq2 Intact	Aggregates easy to break	2–70 mm, no angular non-porous aggregates (clods)	Most aggregates are porous	Throughout the soil
Sq3 Firm	Not difficult	2 mm–100 mm; most > than 10 mm. May be some clods	Macropores and cracks present	Most are around aggregates
Sq4 Compact	Quite difficult	Some > 100 mm aggregates and clods, most are > 70 mm	A few macropores and cracks	In macropores and around aggregates
Sq5 Very compact	Difficult	Mostly > 100 m, very few < 70 mm; angular and non-porous	Very low; may have anaerobic zones	Few, if any, restricted to cracks

Source: Based on Ball et al. (2007), where further description and illustrations are available.

Poor structure is often associated with compaction. This can occur naturally on some soils if particle shape and size distribution lead to close packing. In wheat fields, however, compaction is most often caused when compressive forces derived from wheels and tillage machinery act on a compressible soil. The degree of compaction will depend on the weight of machinery, tyre pressure, soil moisture content, soil type, and frequency of traffic. Tractor traffic on, and cultivation of, wet soil is particularly problematic in reducing pore volume, increasing the proportion of large (> 20 mm) aggregates, increasing erosion by run-off, and increasing nitrogen losses through denitrification (Chapter 6) (Bakken et al. 1987). One form of compaction is the formation of plough pans, as it is almost inevitable that ploughing will force some soil downwards, producing a compacted layer just below the plough depth.

Compaction can be assessed visually (e.g. Sq4 and Sq5 in Table 2.1) but is often inferred from quantifiable effects on bulk density, and/or resistance to penetration (*soil strength*). Bulk density is the packing density of the soil in the field, i.e. the dry mass of the soil divided by the volume of the undisturbed soil (g/cm^3). Correlations between crop yield and bulk density are not particularly close over different soil types because of modifying effects on porosity and penetration resistance (Håkansson and Lipiec 2000). However, bulk densities of 0.9–1.3 g/cm^3 have been considered optimal for crop production, at least in the upper layers (e.g. the plough layer) of fine to medium textured soils (Reynolds et al. 2009). Håkansson and Lipiec (2000) consider the optimal compactness for wheat (and barley) to be higher than those for peas, oats, oilseed rape, field beans (*Vicia faba*), and potatoes. At bulk densities of less than 0.9 g/cm^3 the soil provides poor anchorage and retains less water. Bulk densities above 1.3 g/cm^3 can be associated with yield loss due to a reduction in porosity and therefore significantly less permeability to air, water, and roots (Batey 2009). In compacted soils, wheat roots can be limited to passing through any larger macropores or biopores (e.g. as formed by earthworms or channels left by decayed roots) that may be present rather than ramifying throughout the whole soil (Atkinson et al. 2020).

In addition to bulk density, compaction can increase soil strength as measured with a penetrometer (Gao et al. 2016). A penetration resistance of > 2.5 MPa is assumed to impede the ability of roots to deform the soil and thereby grow through it. As penetration resistance increases with soil depth, deep roots may be confined to exploiting existing pores (Gao et al. 2016). Compaction will exacerbate other stresses such as from soilborne diseases and deficits of moisture and nutrients (Allmaras et al. 1988; Barraclough and Weir 1988).

2.1.7 Soil Depth

Wheat plants can produce roots down to at least 2 m and exploit the resources of deep soils (Zhang et al. 2004). Wheat is, however, also grown on much shallower soils, for example over limestone and chalk. Soil depth is often considered as a factor in the evaluation of agricultural land, e.g. in Table 2.2, Grades 1 and 2 require soil depths of at least 60 and 45 cm, respectively, but much wheat is also grown on Grades 3a and 3b, which only require 30 and 20 cm of depth (MAFF 1988). The influence of soil depth on wheat yield is difficult to quantify and will depend on the extent to which nutrient and water deficiencies in shallower rooting zones can be offset by the application of fertilisers and, more rarely, irrigation. In a review including wheat, Bakker et al. (2004) showed that the strategy used to compare the effects of soil depth affected the results obtained. If the depth was manipulated by removing or adding topsoil, then estimates of above 20% yield loss for every 10 cm of soil depth reduction were common. By contrast, if the effect of soil depth was ascertained by transects across soils of various depths, or by comparing matched plots on mature soil profiles, then estimates of the effect of depth were much lower (often < 5% yield loss for every 10 cm). Unsurprisingly, the effects of soil loss were greater when the soil used as a starting point for the experiments was shallow (Webb et al. 2001).

2.1.8 Land Classification

Given that wheat is grown on such a large and varied area, it is implicit that production is possible on a wide range of soil types. In terms of the crops chosen by farmers, the economic margin from wheat is balanced against those for other land-uses for a given site. Consequently, wheat may not be grown on the best land if this is suitable for more valuable, but more demanding crops. As an example, Table 2.2 gives the agriculture land classification system of England and Wales (MAFF 1988). For this, land quality is based on climate, gradient, flooding risk, soil texture, structure and organic matter, depth, stoniness, water retention, drought risk, and erosion. Land quality is then related to the types of crop that can be accommodated, their potential yields, and the stability of yields.

To summarise, wheat is adaptable to growing well in many situations but the soil qualities that will pose the least restriction on growth, i.e. those of a *good* soil, are shown in Table 2.3.

Table 2.2 The agricultural land classification system of England and Wales (MAFF 1988).

Grade	Quality of agricultural land	Cropping potential
1	Excellent	Wide range of crops can be grown, which may commonly include top fruit, soft fruit, salad crops, and winter harvested vegetables. Yields are high and relatively stable.
2	Very good	Wide range of crops can be grown but some possible constraints with regards to winter harvested vegetable crops and arable root crops. High and stable yields but less so compared with Grade 1.
3a	Good	Moderate to high yields of a narrow range of arable crops, especially cereals, or moderate yields of a wide range of crops including cereals, grass, oilseed rape, potatoes, sugar beet, and the less demanding horticultural crops.
3b	Moderate	Moderate yields of a narrow range of crops, principally cereals and grass or lower yields of a wider range of crops or high yields of grass, which can be grazed or harvested over most of the year.
4	Poor	Mainly grass with occasional arable crops (e.g. cereals and forage crops), the yields of which are variable.
5	Very poor	Permanent pasture or rough grazing, except for occasional pioneer forage crops.

Table 2.3 Soil qualities associated with little or no constraint to wheat growth.

Quality	Unit	Level
Soil depth	cm	45
Soil texture	w/w	< 50% sand, < 50% clay
Conditions in the top 25 cm		
Bulk density	g/cm^3	< 1.3
Organic matter	%C	> 1
Acidity, Alkalinity	pH	> 6, < 7.5
Salinity	EC_e dS/m	< 4
Sodicity	ESP	< 6

2.2 The Rise of the Plough

Mechanical manipulation of the soil to aid the sowing and growing of crops (i.e. *tillage*) has been associated with traditional agriculture for millennia. Tillage has been imbued with near spiritual properties of 'nourishing the earth' and 'breaking the drought' (Lal et al. 2007). Various digging sticks (or *ards*) and paddles designed to be pulled through the soil were evident in Mesopotamia from 6000 to 8000 BP. These acted as primitive chisel ploughs and could be pulled by oxen in Sumerian and Egyptian agriculture. By the time of the Romans, wooden structures were strengthened with an iron share and were shaped with *ears* to widen the furrow. By 1000 BP, ploughs were heavier and fitted with a cutter (or *coulter*) to penetrate the stubble, which was then sliced with a share and passed over a mouldboard to invert and partially break up the soil. In the 1830s a cast iron plough was marketed in the US by John Deere. With the advent of tractors, and their increasing size and power, the area that could be ploughed in a sowing season has greatly facilitated the increase in the number of people that can be fed from the produce from one farmer (Lal et al. 2007).

Nowadays, before ploughing a field it may first be subsoiled to relieve compaction, particularly that associated with plough pans (Zhang et al. 2004). Subsoilers are varied in design but essentially a robust tine, often with a blade designed to lift the soil from below the plough layer, is drawn through the soil. Mouldboard ploughing to 20–30 cm is then often used as a *primary* cultivation to relieve compaction in the plough layer (Table 2.4) and level a rutted surface. Primary cultivation can also be achieved with rotary cultivators, heavy disc harrows, and chisel ploughing (Spedding 1983).

Inverting the soil with mouldboard ploughing is particularly effective at burying surface residues and thereby killing weeds that have already established in the stubble of previous crops or fallows. The seeds of weeds that have fallen on the stubble are also buried beyond the depth from which many of them can emerge. Deep burial is particularly effective at suppressing small-seeded weed species with relatively low levels of dormancy (Lutman et al. 2013). Inversion tillage also buries the structures and sporulating bodies of numerous wheat diseases and hence contributes to practices to enhance stubble hygiene (Smiley 1996). Furthermore, burial reduces the chance of green material from late tillers and volunteers persisting in the field that could act as a *green bridge* for diseases to pass from one crop to the next. For a spring sown crop, it is possible to plough in the preceding autumn, leaving the clods exposed to wetting–drying and freeze–thawing cycles, making it easier to produce a seed bed after winter. As will be discussed, ploughing can also be used to incorporate and mix lime into the soil.

After ploughing, a seed bed is prepared with *secondary* cultivation. Discs, tines, harrows, and similar equipment can act on a ploughed surface to prepare a suitably fine surface tilth to facilitate accurate depth placement of seed, good seed to soil contact, improved seedling

Table 2.4 Effect of cultivation method on bulk density, available N after sowing, and grain yield of winter wheat. Mean of first two years of experiment.

System	Cultivated depth (cm)	Bulk density (kg/l)	Available N (kg N/ha)		Grain yield (t/ha)
			Ammon.-N	Nitrate-N	
Plough+harrow+drill	30	1.49	7.0	48.2	7.97
Harrow+drill	3	1.58	6.4 (Harrow+drill and Drill combined)	39.5 (Harrow+drill and Drill combined)	5.67
Drill	0	1.59			5.95
S.E.D.		0.013	1.22	4.01	0.659

Source: From Uppal and Gooding (2013).

rooting and growth, and improved efficacy of soil-acting residual herbicides (see Section 2.3.6). Primary and secondary cultivations aerate the soil and facilitate microbial activity and the oxidation of organic matter. This breakdown of organic matter produces a flush of nutrients including nitrate, phosphate, and sulphate (Table 2.4; Powlson 1980); usually more than enough to suffice for seedling emergence and establishment.

2.3 Soil Change and Land Degradation

Against the backdrop of appreciation and dependence on soil and its fertility, the record of agricultural practice to maintain the quantity and quality of soil over decades and longer time spans is poor. The development of soil under natural processes is slow: perhaps a mean rate in the order of 0.03–0.04 mm/yr (Montgomery 2007b). When soils and the ecosystems they support are undisturbed they can maintain depth, biodiversity, organic matter, and nutrient content over the timescales of human generations (Montgomery 2007b; Amundson et al. 2015). By disturbing this equilibrium, agriculture and the cultivation of land for crops will inevitably be associated with changes in soil quality. The changes introduced by agriculture are rapid compared with the timescales needed to develop soil, so bringing new land into cultivation risks a degree of degradation, particularly in comparison to the inherent fertility and relative stability of many natural states (Amundson et al. 2015). Components of land degradation include loss of topsoil by erosion by wind and water; loss of nutrients and organic matter; salinisation; acidification; pollution; compaction, capping and physical degradation; waterlogging; and subsidence of organic soils (Lal 2009; FAO and ITPS 2015). The rate at which this degradation from the natural state occurs depends greatly on crop and land management, soil type, topography, and climate. On fragile soils, where little attempt is made to replenish nutrients, soil fertility may be lost rapidly following clearance for crop land such that communities of people have traditionally moved on once land has become *exhausted*. On less fragile soils, where there has been greater opportunity for more sedentism and nutrients to be returned, the degradation process may take centuries (Jacobsen and Adams 1958; Colledge et al. 2019). Estimates of the land and crop yield lost to agriculture due to degradation vary widely. It has been estimated that the Earth has already lost a third of its soil, that we lose 10Mha of agricultural land every year due to degradation (Pimentel 2006), and that most agriculture is now on degraded soil (Hudson 2020). Alternatively, a less pessimistic analysis estimates that the global crop land lost to degradation, together with the land equivalent of yield reductions where crops continue to be grown, does not exceed 3 Mha per year out of a total of about 1500 Mha of crop land (Lambin and Meyfroidt 2011).

2.3.1 Loss of Soil

The most visible and widespread form of land degradation is soil erosion by water and wind (Lal 2009). The level of risk depends particularly on the slope, soil type, crop and soil management, and weather. Water erosion is more likely to occur on sloping land with sparse vegetation cover, on light textured soils with fine sand or coarse silt, and where the soil is weakly structured. In many wheat-growing areas, these conditions are particularly prevalent shortly after soil cultivation and during crop establishment. Heavy rains can saturate the soil surface and break up the aggregates. Water and suspended soil particles then run down the slope as *run-off*, forming rills and gullies. This immediately reduces the yield potential of the growing crop, either because the seeds or plants are destabilised, uprooted, and washed away, or because plants are buried down slope. Gullies and rills may become wide and deep enough to impede traffic of machinery (MAFF 1988). This also results in off-site consequences such as streams and rivers being polluted with excess nutrients (eutrophication) and silt, and increased risk of floods downstream. In contrast to water erosion, wind erosion (or soil *blows*) is mostly seen on dry, flat or gently sloping land. Like water erosion, blows are most common where light soils are poorly covered with vegetation. Seeds and seedlings can be blown away, and more established plants killed by abrasion. In the longer term, erosion by either water or wind reduces the fertility of the soil. The upper soil layers and the finest soil particles are the richest in organic matter and nutrients (Figure 2.4), which can be preferentially lost as the surface is denuded. As soil depth declines over time, yield reductions can be expected due to impedance of root growth, and increased deficits of water and nutrients. The effects of soil loss are long lasting. Larney and Olson (2018) found the effects of losing 10–20 cm topsoil on spring wheat yields to last for at least 54 years.

Globally, erosion rates under wheat production are estimated to be around 12 t soil/ha/year (den Biggelaar et al. 2004a, b). This compares with the estimate of an average of 6 t/ha/year from all the world's farmland (Wilkinson and McElroy 2007). As emphasised above, however, erosion rates will depend greatly on local conditions and management. Scherr (1999) concluded that the soils of c. 75% of the agricultural area had been broadly stable since the middle of the twentieth century but that erosion was widespread and accelerating on the remaining 25%. The distribution of erosion rates is also highly skewed, i.e. the mean rate of erosion is higher than the median such that most farmland will experience less erosion than the global mean rate (Wilkinson and McElroy 2007).

Den Biggelaar et al. (2004a, b) attempted to account for different erosion rates and yield potentials for different soil orders across the major wheat growing areas. They estimated that global wheat yields would be progressively reduced by 0.2% each year due to ongoing soil loss. There is, however, significant variation in the estimates for different continents: smaller losses for Europe and North America and larger losses in Asia, Australia, and South and Central America. In a modelling exercise for spring wheat, Bakker et al. (2007) also predicted low yield reductions associated with soil erosion in Europe. In fact, the modelled effects were small enough to be overcome with technological innovation. They also, however, highlight the variation within Europe and importance of original soil depth. Loss of the upper soil horizons is particularly damaging where subsoils limit growth, for example, if the soil is shallow over bedrock, or if the subsoil is compacted, acid or saline. Bakker et al. (2007) predicted, therefore, reductions in wheat yields of 6–12% over the next 100 years due to erosion in areas of southern Europe. These areas were typified as having already low yields; a Mediterranean climate associated with low rainfall and loss of soil carbon; severe erosion in ancient times (Greek and Roman); and therefore, shallow, and stony soils. Although anthropogenic climate change may lead to enhanced levels of wind and water erosion in the future the magnitude of these will differ greatly among regions.

The most severe soil loss in wheat growing areas is probably in the Loess Plateau in China (Wen and Zhen 2020; Figure 2.9). This was a primary, fertile region for ancient agriculture. Currently, over half of the arable area is cropped by wheat (Sun et al. 2019). The term 'loess' describes sediments that are derived from wind-blown particles, mostly fine grains of silt with a diameter of 0.02–0.05 mm. In the case of the Loess Plateau, the sediments have formed over millennia as continental winds transported the dust from the deserts of central Asia. The plateau covers over 60 Mha and the loess areas have an average depth of 50–80 m. The soil is homogeneous with no stones and little organic matter. When dry the soil is stable, to the extent that in some areas vertical walls can be cut without collapsing. When saturated with water, however, the loess loses its cohesion. Large areas receive intense rainfall during the summer and the Yellow River gets its name from the sediment it derives as it and its tributaries pass through the Loess Plateau. Water erosion of the loess will always have been significant, and much of the North China Plain to the east of the plateau (Figure 2.9) has been greatly modified by the sediment from the Yellow River system (Hillel 1991). However, erosion rates have increased with population growth, over-grazing, and removal of tree and grassland cover to make way for arable farming. Over 14 Mha are losing more than 3 mm of soil per year while 5 Mha are losing twice this amount (Qiangguo 2002). Some regions are reported to be losing over 20 mm of soil per year (Shi and Shao 2000). Protecting the soil from erosion is a constant battle and methods include the re-establishment of trees and grassland, particularly on the slopes; terracing slopes; using contour levels, ditches and dams; establishing crops without ploughing,

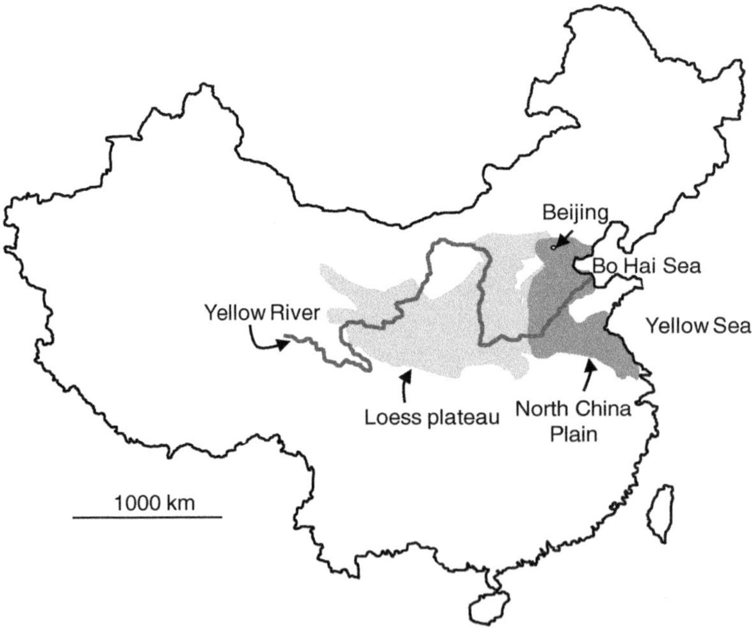

Figure 2.9 China: the relationship of the Loess Plateau to the North China Plain and the Yellow River. *Source:* Redrawn from Kidder and Zhuang (2015).

i.e. conservation tillage (Section 2.4.1); mulching; and retaining vegetation and crop residues on the soil surface (Fu et al. 2011; Wen and Zhen 2020).

Between the Loess Plateau and the Bo Hai Sea, the Yellow River passes through the North China Plain (Figure 2.9). The plain has good fertility and extends for 40.4 Mha. The area is hugely important for food production and security. The predominant crop is wheat and the area is responsible for two-thirds of China's, and therefore more than a tenth of global wheat production (Zhang et al. 2004; Lu and Fan 2013). The great load of silt carried by the Yellow River causes the riverbed to rise by 8–10 cm per year. On the plain, therefore, the river runs above the plain itself, sometimes 10 m above ground level. Levees contain the river and need to be continually maintained and raised to prevent the catastrophic flooding of the past.

As the Loess Plateau provides a powerful exemplar of erosion by water, the Dust Bowl disaster in USA in the 1930s amply demonstrates the damage that can be inflicted by the wind. As on the Loess Plateau, the semi-arid grasslands of the Southern Great Plains (Figure 2.10) were already fragile and at risk of erosion before an increase in extent and intensity of agriculture exacerbated the problem (Cunfer 2011). From around 1900 there was a significant increase in the number of farmers and in the use of tractors. Land was increasingly ploughed, principally to plant wheat. A severe drought started in the area in the early 1930s. The destabilised cultivated soil, without perennial vegetative protection was whipped up by winds to form vast clouds of dark dust that drifted across the continent (Lal et al. 2007). The drought and ensuing dust storms denuded the land of its top soil, and in conjunction with prevailing economic depression destroyed agriculturally based livelihoods, caused sickness, led to depopulation, and left long-term legacies of reduced soil fertility and poor human health (Cunfer 2011; Alexander et al. 2018; Arthi 2018; Long and Siu 2018). As in the Loess Plateau, the restoration and protection of Dust Bowl areas have involved re-establishment of grassland areas, and changes to cropping and soil management practices. There has been a marked increase in the use of conservation tillage, rather than of the plough.

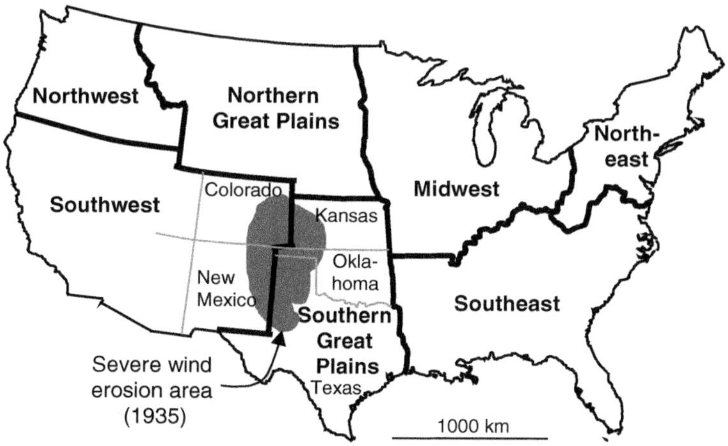

Figure 2.10 The classically recognized area of severe wind erosion in the USA at its maximum extent in 1935. *Source:* Redrawn from Bolles and Forman (2018) and USGCRP (2017).

Wind erosion and dust storms in wheat growing areas still threaten agricultural productivity and livelihoods, particularly when semi-arid areas have perennial vegetation removed and soil destabilized (e.g. Pimentel 2006; Wang et al. 2006; Abraham et al. 2016). From a review, Wang et al. (2006) estimated that wind erosion could be responsible for the loss of 53–1044 kg/ha of carbon and 5–90 kg/ha of nitrogen per year from the topsoil in Northern China. In relatively hot and dry areas wind erosion from cultivated land can contribute to desertification, i.e. previously fertile land transforms into desert. Wheat production areas particularly at risk of desertification extend through, e.g. South America (Abraham et al. 2016); Southern Europe, the Middle East and North Africa, (Shams and Kamel 2014; Martínez-Valderrama et al. 2016; Rinaldi et al. 2017; Akbari et al. 2020); and South Asia (Khan et al. 2017).

2.3.2 Organic Matter Loss and Amendment

As already noted, reducing the levels of soil organic matter is a threat to crop productivity and food security (Lal 2009). Of some minor comfort, are reports that the soil organic matter content often reaches a relatively stable, equilibrium level indicative of the soil type, management, and climate of the area (Johnston et al. 2009; Figure 2.11). Reaching this stable level may, however, take several decades. Indeed, no equilibrium is in sight even after 80 years of an experiment with continuous 2-year wheat: fallow rotation in Oregon, US, where the wheat has been established following ploughing (Figure 2.12).

Even if soil organic matter levels do tend to reach an equilibrium after an extended period with constant conditions, management practices change on shorter timescales so organic matter levels are often changing over time, and usually downwards in cropped agricultural soils. The rate of change will depend on the climate, the starting point, and the degree of change imposed by land management. In temperate areas, changes in soil organic matter resulting from variation in conventional agricultural practice are nearly always slow.

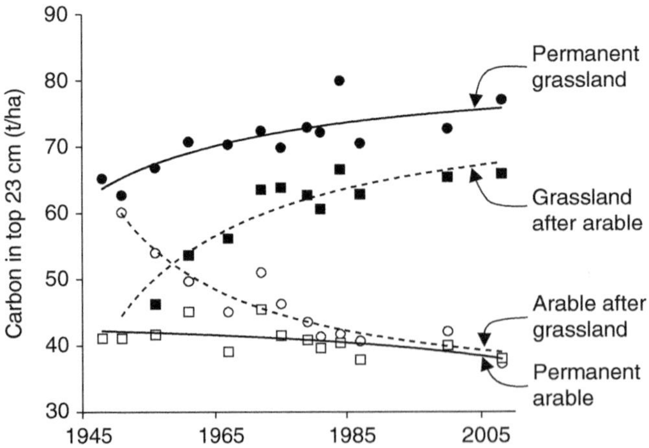

Figure 2.11 Effect of permanent pasture, arable rotation, and their conversion on soil organic matter contents. See text for details. *Source:* Data from Rothamsted Research (2018).

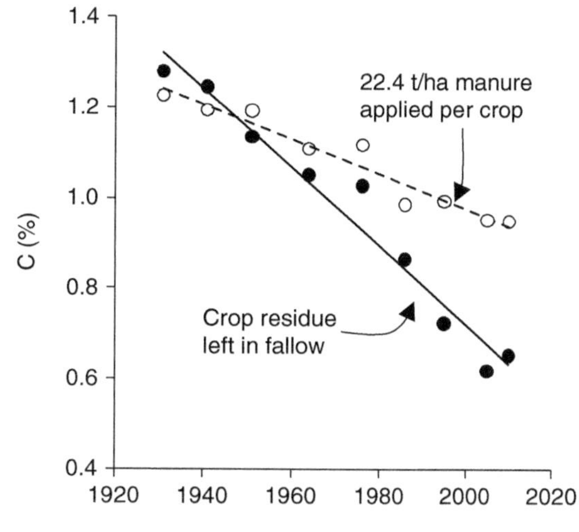

Figure 2.12 The effect of duration of a continuous wheat: fallow rotation and manure application on the soil carbon content in the top 60 cm of soil. *Source:* Calculated from Ghimire et al. (2018).

Allowing for errors in sampling and analysis, it may take more than a decade before effects reach statistical significance. Of particular concern, however, is when soil that has developed a high organic matter content under perennial systems such as grassland, woodlands, and forest is disturbed with cultivation to make way for arable crops (Gregory et al. 2016). Organic matter decomposition progresses faster as microbial activity is encouraged by aeration and adequate moisture and temperature. In the UK, for example, these three requirements are met by cultivating soils before crop establishment in autumn and spring. In temperate systems, yearly losses of soil organic matter, and release of carbon to the atmosphere, can last for decades after a perennial system is disturbed, until the organic matter content reaches the lower equilibrium of an arable soil (Johnston et al. 2009). Figure 2.11 shows the results from experiments on two fields on Rothamsted Farm (Johnston 1973). Highfield had been in long-term pasture, but an annual arable rotation was introduced. Fosters field had been in long-term arable, but a long-term pasture was also introduced. The fields are about 1.6 km apart, and both on a silt loam or loam over clay luvisol. Maintaining management in either permanent pasture or arable systems achieved average organic carbon levels at about 70 t C/ha (with a slight increase between 1948 and 2008) and 40 t C/ha (with a slight reduction during the experiment) respectively, in the top 23 cm (maximum plough depth). However, converting a permanent pasture to arable led to an average annual loss of 0.6 t C/ha for each of the 30 years following conversion. Further significant losses would also be evident in the subsoil (Gregory et al. 2016). Conversely, converting arable to permanent pasture led to an average increase of 0.7 t C/ha/year during the same period. Converting arable to woodland can also increase soil organic carbon (Poulton et al. 2018). Despite the hazards of cultivating previously undisturbed soils, Figure 2.13 shows that when a field has reached an equilibrium for a system on a relatively stable soil type, year-to-year changes in soil organic matter can be exceptionally low. In the Broadbalk long-term experiment at Rothamsted, there has been only a very small net loss of soil organic carbon in wheat grown continuously with plough-based cultivation on a silty clay loam with inorganic fertilizers for 150 years (Figure 2.13), in contrast to the results from Oregon (Figure 2.12). This is almost certainly because the Broadbalk field had been in an arable rotation for several centuries before the formal start of the experiment in 1843 (Johnston et al. 2009).

There are numerous ways to increase the organic matter content in impoverished soils to the extent that soil functioning is significantly improved. In mixed, animal plus crop livestock areas it is possible to increase organic matter from a low base in arable fields by repeatedly adding large quantities of livestock manures and slurries (e.g. Figure 2.13). In the same systems, replacing some of the arable crops with temporary grass leys of three to eight years' duration can also increase organic matter in the arable fields (Johnston et al. 2017). However, these strategies are not necessarily beneficial for the farming system. For example, increasing leys in the arable fields will reduce the area of the organic matter-retentive permanent pasture if the area of grass and arable is to be maintained. Applying large quantities of livestock manures and slurries to wheat fields diverts their use, and the nutrients they contain, from grassland and forage crop production, and may also contribute to increased emissions of nitrous oxide (N_2O) and pollution with nitrates and phosphates (e.g. Goulding et al. 2000; Chapter 6). Nonetheless, the application of manures will have a larger effect on soil organic matter where levels are lower to start with, which will usually be arable fields rather than grasslands. Other additions that can increase the organic matter content of soils include the long-term use of sewage sludge (biosolids) (Cogger et al. 2013) and vegetable and food waste composts (Poulton et al. 2018). These products have various effects on wheat related to nutrient content and availability as well as effects on soil organic matter per se. Their use on arable fields will depend on demand from competing uses such as for energy production in anaerobic digestion. Additionally, if applications of organic materials or ley farming stops, the organic matter content of the soils can be expected to revert, over time, to the low-level organic matter equilibrium reached

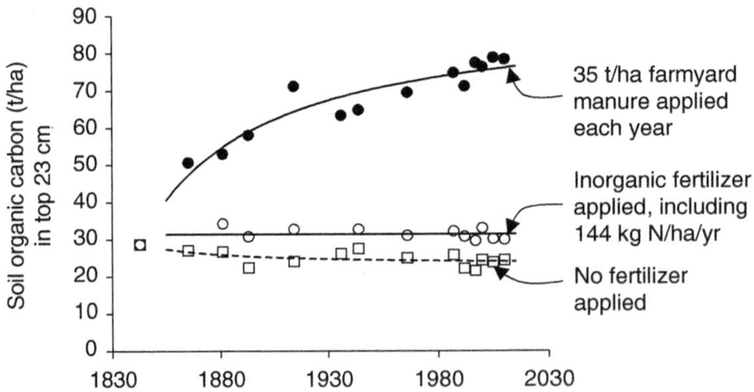

Figure 2.13 Effect of farmyard manure and inorganic fertilizers on the soil organic carbon under continuous wheat established with plough-based inversion tillage. *Source:* Data from Rothamsted Research (2021).

under continuous arable systems (Johnston et al. 2009), releasing carbon to the atmosphere as it does so. A more stable form of carbon may therefore provide a longer-term benefit to carbon sequestration if incorporated in soil. One such form is biochar which is produced by high temperature pyrolysis of biomass carbon. Despite its recalcitrance, depending on the source of the biomass, biochar incorporation into soil has been associated with increased water retention, cation exchange capacity, nutrient availability, pH, and wheat yields (Alburquerque et al. 2013; Farrell et al. 2014). However, widespread use on wheat fields is currently limited by cost of production and availability (Maroušek 2014).

Of more value to arable farmers may be the ability to retain, or hopefully reach an equilibrium with higher amounts of soil organic matter, by changing the practices within the arable rotation itself. Principal methods have included incorporating or retaining wheat straw on the soil surface after harvest; increasing biomass input by ensuring that the crop yield potential of the soil is met with appropriate fertilizer applications (Poulton et al. 2018); increasing cropping intensity by using cover crops and avoiding extended fallow periods (Cogger et al. 2013); and reducing the severity of soil disturbance by moving away from plough-based inversion tillage to less invasive minimum tillage, or even to no-till systems (Section 2.4.1). Figure 2.4 shows significantly more organic matter in the surface layers under wheat-based systems after 20 years without tillage in Texas. Similar results have been reported in long term wheat-based experiments on the Loess Plateau in China (Gao et al. 2019). In other experiments, however, particularly in temperate regions and/or on relatively stable soils, the effects of reduced or no-till systems on organic carbon have been small to nil (e.g. Bono et al. 2008; Powlson et al. 2011). Powlson et al. (2014) point out that much of the change in organic carbon with tillage practice is distributional rather than total. Table 2.5 shows that, in the UK, using tines to 10 cm rather than ploughing to 20 cm for 17 years retained significantly more soil organic carbon in the upper layers, but not when all of the plough depth was considered. The increase in organic carbon in the surface layers can benefit crop growth, but Powlson et al. (2014) argue that content can decrease in the subsoil, and so increases over the whole soil profile are often slight at best and represent little potential

Table 2.5 The effect of either ploughing or using tines for 17 years on the soil carbon and nitrogen. Mean of two sites in the UK (Johnston et al. 2009).

Cultivation	Depth	Carbon (%)	Nitrogen (%)	C:N
Plough to 20 cm	0–10 cm	1.52	0.128	11.9
	0–20 cm	1.53	0.128	11.9
Tine to 10 cm	0–10 cm	1.95	0.158	12.4
	0–20 cm	1.57	0.134	11.7
S.E. (5 d.f.)		0.014	0.0021	0.22

for large-scale carbon sequestration. It is also evident that the beneficial effects on soil organic carbon reported for some reduced tillage systems resulted from a reduced rate of loss, rather than an actual increase in organic carbon.

2.3.3 Acidification and Liming

The slow acidification of soils occurs naturally through precipitation. Rain is slightly acidic due to the dissolution of CO_2 to produce carbonic acid. This causes the leaching of basic (i.e. alkaline) cations such as potassium (K^+), calcium (Ca^{2+}), and magnesium (Mg^{2+}) with water passing through the soil. Increasing acidity will result in an initial build-up of hydrogen (H^+) ions, which rapidly attack mineral surfaces to release aluminium (Al^{3+}). Many soils have experienced increased rates of acidification due to deposition of atmospheric sulphur dioxide (SO_2) and nitrogen dioxide (NO_2), which when interacting with water form sulphuric acid (H_2SO_4) and nitric acid (HNO_3), respectively. Some atmospheric SO_2 and NO_2 have natural origins, from volcanic emissions and lightning in electric storms, respectively. Most, however, is anthropogenic, particularly due to the burning of fossil fuels such as sulphur-rich coal (*acid rain*). Some partial mitigation of this latter effect by governmental intervention has been achieved since the widespread recognition of the threat in the 1970s and 1980s. However, the effect of acid deposition in high rainfall areas, particularly around and downwind of industrial areas, can still be significant.

Several agricultural practices also lead to acidification. The leaching of calcium carbonate (40% Ca) from agricultural soils can range from 125 to 2000 kg/ha/year with higher values from soils that are well drained, rich in calcium, and receive high rainfall (Finch et al. 2014). Loss of calcium and magnesium under long-term continuous wheat on free-draining clay loam (Bolton 1972) averaged between 600 and 1000 kg/ha/yr calcium carbonate ($CaCO_3$) equivalents, mostly through leaching (Gasser 1985). Acid rain will have contributed to this, but the application of ammonium-based nitrogen fertilizers has had a major impact. The fertilizers are not acidic but when ammonium ions (NH_4^+) are oxidized to produce nitrate (NO_3^-), H^+ is produced thus lowering the pH. If no attempt is made to counteract this effect acidity can build up with rate and duration of nitrogen application to limit yields (Figure 2.14; Schroder et al. 2011). The soils that are most prone to increases in acidity are the lighter sandy soils, which lack the buffering capacity of heavier soils or of organic matter. Increases in acidity can be particularly damaging in soils that are already acidic, e.g. those from highly weathered soils in high rainfall areas.

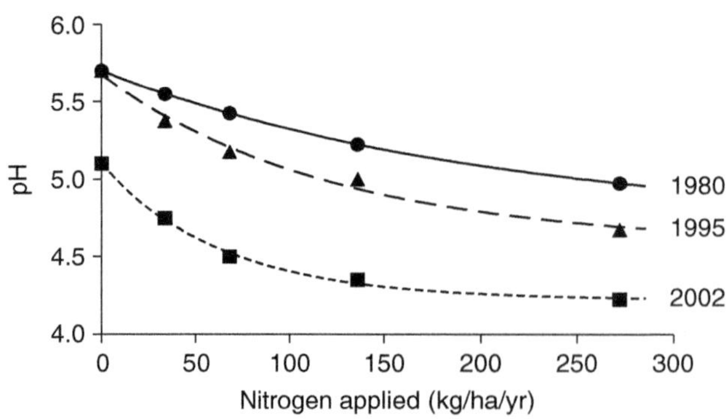

Figure 2.14 The effect of long-term, repeated application of nitrogen fertilizer (mean of anhydrous ammonia, ammonium nitrate, urea and sulphur-coated urea) to continuous wheat crops without liming since 1971. *Source:* From data in Schroder et al. (2011).

Acidity can be corrected by applying lime, usually calcium carbonate from quarried and pulverised limestone and chalk. Application rates depend on soil type and soil acidity (Table 2.6) but the amounts may be large and require specialist equipment and contractors to incorporate. Expense increases with transport costs and therefore distance from lime works. Local supplies and equipment may not, therefore, be available at economic rates. Alternative liming materials include *waste limes* from industries that use lime as a purifying material, such as in sugar beet processing; magnesium or dolomite limestone ($MgCO_3$ + $CaCO_3$); and shell rich sand (Finch et al. 2014). Liming abilities of different materials can be compared with that of calcium oxide, i.e. a neutralising value (NV) of 50% (w/w) means the product has half of the liming strength of calcium oxide.

It is important to ensure even applications of lime, not only to avoid missing acidic areas, but also to prevent localised areas with excessive alkalinity. Pelletised or *granular* lime is available and precision liming is being practiced, with soil pH recorded using global positioning systems (GPS) and the data transferred to an onboard spreader computer, but its use is limited due to cost (Goulding 2016). Where large applications are needed, e.g. for correcting pH levels of 5.5 and lower (Table 2.6), the application may be split, e.g. before and after ploughing. Incorporating and mixing lime with the soil is important as its movement down through the soil is generally slow (Tang et al. 2011). This slow movement presents a particular challenge for correcting sub-surface acidity in no-till systems, where surface applied lime has reduced efficacy. In such situations mixing lime with organic materials, and liquids, or through micronising calcite or dolomite has shown promise (Tao et al. 2019; Nogaroli and da Fonseca 2020). Even in ploughed systems, liming is inefficient at correcting sub-soil acidity. A further approach, therefore, is to apply less acidifying fertilizers, particularly avoiding those that involve the release of ammonium. Calcium nitrate ($Ca[NO_3]_2$), formed by treating lime with nitric acid, has thus been proposed as a nitrogen fertilizer for alleviating acidity problems in both plough-based and no-till situations (Tang et al. 2011; Galieni et al. 2016).

2.3.3.1 Calcium as a Nutrient

Although calcium is a macronutrient, it is not often identified as such during production as the amounts made available through pH adjustment usually significantly exceed crop requirements. Nonetheless wheat crops remove about 2 and 6 kg of calcium carbonate

Table 2.6 An example of lime application rates (t/ha, neutralizing value of 50%) for arable crops (AHDB 2020).

Soil type	Target pH	Original soil pH			
		6.2	6.0	5.5	5.0
Light (Sands and loamy sands)	6.7	3	4	7	10
Medium (Sandy loams and silt loams)	6.7	4	5	8	12
Heavy (Clay loams and clays)	6.7	4	6	10	14
Organic (5.8–14.5%C)	6.7	4	6	10	14
Peat (> 14.5%C)	6.0	0	0	8	16

equivalents per tonne of wheat grain and straw, respectively. In the plant, calcium is required for cell wall and membrane stabilization, osmoregulation, and in signalling in response to environmental stimuli such as drought (Hawkesford et al. 2012).

2.3.4 Depletion of Micronutrients

The crop and environmental aspects related to macronutrients are dealt with elsewhere, in respect to water (Chapter 3) and light interception (Chapter 6). The micronutrient requirements are discussed here because producers often rely entirely on their provision from the soil rather than supplementing soil availability with micronutrient fertilizers. Micronutrients are, however, essential for wheat production and deficiency of any one of them can have catastrophic effects on wheat yields. Micronutrient deficiency may arise on soils where the parent material is deficient, and on extensively leached and light soils. The availability of most of them tends to be limited at high soil pH (Figure 2.5), and therefore deficiency is more likely after liming, and localised effects within a field may appear if lime has not been distributed evenly. In particular, liming previously acidic, uncultivated soil can lead to significant reductions in micronutrient availability compared to soils left under natural vegetation. In addition, there is growing likelihood of deficiency as soil is eroded, and as crop growth, yields and cropping intensity are increased, thus placing greater demands on soil reserves (Alloway 2008a; Fageria and Stone 2008). Reducing organic matter contents will also reduce availability of some micronutrients whilst others (e.g. Cu) form complexes which reduce their availability in organic soils. Taking all the changes in intensity of production, liming, and soil degradation into account, it is unsurprising that micronutrient deficiency is increasingly recognised as limiting wheat yields in major production areas (Alloway 2008b).

2.3.4.1 Boron

Boron (B) is required for the metabolism of carbohydrates, phenols, and growth regulators; for the synthesis of DNA and cell walls; and for the integrity of cell membranes (Hänsch and Mendel 2009). Vegetative development of wheat is relatively insensitive to B deficiency but low availability from flag leaf emergence (DGS 37) to the start of anthesis (DGS 60) can result in poor development of anthers and pollen, pollen germination failure, and significant yield loss through sterility (Subedi et al. 1998, 1999, 2000). Deficiency has been recorded in many wheat-producing areas, but the most extensive region stretches through eastern Nepal, northeastern India, northwestern Bangladesh, and southwestern China (Rerkasem and Jamjod 2004). There is wide variation in the susceptibility to B deficiency amongst wheat cultivars (Subedi et al. 1999), with some high yield potential semi-dwarf wheats being particularly prone to sterility associated with low B availability (Rerkasem and Jamjod 2004). There is, however, significant potential for improving wheat yields on B-deficient land through plant breeding. Deficiency can also be mitigated by the application of B-containing fertilizers (e.g. Subedi et al. 1997).

2.3.4.2 Chlorine

Chlorine (as Cl^-) is involved in photosynthesis, enzyme activities, and in the control of water relations of the plant through charge compensation, osmoregultation, and the functioning of stomata (Chapter 3). In addition to its importance for water regulation the application of chloride salts has been associated with the suppression of many fungal pathogens attacking the roots and leaves of wheat (Heckman 2007; Deliopoulos et al. 2010).

Deficiency in field conditions is rare, with excesses associated with salinity being more commonly problematic. Indeed, coastal rainfall often supplies sufficient Cl for crop growth. In inland, continental areas, however, rainfall and its Cl content is often much less and deficiency can sometimes develop. Inadequate Cl has, for example, been identified in the plains of the USA (Fixen 1993). Engel et al. (1997) reported a leaf spotting syndrome on some wheat cultivars grown in fields in and around Montana that could not be controlled with fungicides but was alleviated with the application of chloride.

Deficiency is indicated when chloride concentration in wheat plants is below 1.0–1.5 g/kg DM at ear emergence (Engel et al. 1997; Heckman 2007). Plant concentration is significantly affected by soil availability, and levels below 43 kg Cl/ha in the top 60 cm have been associated with yield losses in wheat (Fixen 1993). As with boron, cultivars differ greatly in their response to Cl deficiency and application.

2.3.4.3 Copper

Copper (Cu) is a constituent of several enzymes involved in photosynthesis, carbohydrate, and protein metabolism, and in pollen formation. Wheat is amongst the most sensitive crops to Cu deficiency (Kopsell and Kopsell 2007; Alloway 2008a). In severe cases, young wheat leaves become pale with yellow veins, twisted, flaccid, and die back from the leaf tip (Brennan and Bolland 2003; Alloway 2008a). Copper deficiency is also associated with weakened stems, and tiller death. Even when deficiency symptoms have not been obvious in the foliage, reduced Cu availability can have significant effects on reproductive development: Cu deficiency disrupts pollen formation, delays flowering, and increases sterility (Dell 1981). The ear is often deformed, and the lack of grain contributes to a *rat's tail* appearance. Those grain that are set are often shrivelled.

Copper deficiency symptoms are most likely to occur on high organic matter soils because Cu forms strong and stable complexes with phenolic and carboxyl groups in the organic material (Kopsell and Kopsell 2007). Like other micronutrients, however, sandy, calcareous, limed, and shallow soils are prone to being deficient. Large areas susceptible to Cu deficiency for wheat occur in Australia, Brazil, and Europe (Alloway 2008a). Copper is comparatively immobile in soil but single applications of Cu containing fertilizers, such as copper sulphate, can often meet the requirements of a number of successive crops (e.g. Brennan 2006).

2.3.4.4 Iron

Iron (Fe) is bound to many proteins. It has a particular role as a redox-active metal, i.e. it can facilitate the transfer of electrons. Iron-containing enzymes are important in mitochondrial respiration, nitrogen assimilation, and hormone biosynthesis. Iron also has roles in the production and scavenging of reactive oxygen species, and in disease suppression (Hänsch and Mendel 2009). Most Fe is found in the chloroplasts where it has important functions in photosynthesis (e.g. in ferredoxin; Chapter 5). Deficiency in wheat, therefore, causes a reduction in chlorophyll content and interveinal chlorosis (yellowing), particularly on the younger leaves (Berg et al. 1993; Wang et al. 2019). Inadequate Fe is more likely in calcareous and alkaline soils but can also be associated with soil compaction or excessive applications of P and Cu (Romheld and Nikolic 2007; Alloway 2008a). Deficiency is widespread on calcareous soils in the Middle East (Alloway 2008a), where it may be indicated when extractable (with diethylenetriaminepentaacetic acid [DTPA]) levels of Fe drop below 2.5 mg/kg soil (Cakmak et al. 1996). Deficiency can be rectified by applying Fe-containing fertilizers and mitigated by selecting cultivars that are better able to maintain yields on deficient soils (Berg et al. 1993).

2.3.4.5 Manganese

Manganese (Mn) affects the activity of several enzymes. This can be through either the activation of enzymes, or as a catalytic metal in their active sites (Hänsch and Mendel 2009). Mn is involved the electron transfer and O_2 evolution systems of photosynthesis (Chapter 5). In wheat, it has also been reported that Mn can help suppress take-all disease of the roots caused by the fungus *Gaeumannomyces graminis* var. *tritici* (Humphries et al. 2007). Deficiency can occur on heavily weathered, sandy, high organic matter, and/or calcareous soils, or following heavy lime applications (Alloway 2008a; Humphries et al. 2007). Wheat is comparatively susceptible to Mn deficiency (Alloway 2008a), which can occur as tissue concentrations at the start of ear emergence drop below 25 mg/kg (Humphries et al. 2007). Deficiency reduces leaf chlorophyll concentrations and photosynthetic rates (Ohki 1984). There is little remobilisation of Mn in the plant so symptoms are frequently observed on the younger tissues, which may exhibit grey spotting. Deficiency can be rectified with application of Mn-containing fertilizers and/or mitigated by selection of appropriate cultivars (Bansal et al. 1992).

2.3.4.6 Molybdenum

Molybdenum (Mo) is contained in enzymes involved in nitrogen assimilation, sulphur metabolism, hormone biosynthesis, and stress reactions (Hänsch and Mendel 2009). Gupta (1971) suggested that 0.09–0.18 mg/kg was a sufficient concentration of Mo in wheat at booting to prevent constraints on yield. Due to the importance for nitrogen assimilation, symptoms of Mo shortage can resemble that of nitrogen deficiency, i.e. yellowing of leaves including of the older laminas, stunting, and poor tillering. Deficiency is most likely on acid mineral soils (Hänsch and Mendel 2009). Alloway (2008a) identifies large areas of Australia and China where Mo deficiency occurs on acid soils under wheat, together with reports from areas of India and the USA. Deficiency can sometimes be rectified by liming. Otherwise, Mo-containing fertilizers can be applied, together with the selection of appropriate cultivars.

2.3.4.7 Nickel

Nickel (Ni) is required in smaller amounts than any other nutrient (< 0.05 mg/kg dry weight; Brown 2007). It is a constituent of urease and therefore particularly important when urea is the major source of nitrogen for wheat (Singh et al. 1990). Smaller benefits of Ni supply can still occur when this is not the case (Gheibi et al. 2009). Although Brown (2007) identifies situations that might lead to Ni deficiency in the field, he did not consider there to be evidence of truly Ni-deficient soils for wheat production. More latterly, however, there have been reports of yield responses to Ni-containing fertilizers in field grown wheat on alkaline soils (e.g. Singh et al. 2020).

2.3.4.8 Zinc

Zinc (Zn) is a constituent of several enzymes involved in photosynthesis, protein synthesis, regulation of auxin production, and energy release. It is also important for maintaining the integrity of membranes. Zinc is required for pollen and seed development, and Zn-deficient plants take longer to mature (Alloway 2008a; Hänsch and Mendel 2009). For wheat, deficiency may be indicated by soils with less than 0.5 mg of DTPA extractable Zn/kg of soil. Deficient wheat flag leaves contain < 10 mg Zn/kg DM (Cakmak et al. 1996). Symptoms on wheat include severe stunting, with whitish-brown areas developing on the laminas of the older leaves first (Cakmak et al. 1996). Soil deficiency is indicated by soil testing in large areas of India, China, and Turkey. Deficiency is also common in Australia on the sandy and calcareous soils. In South America deficiency occurs on highly weathered and leached soils (Alloway 2008a). Deficiency can be rectified by application of Zn-containing fertilizers and there is large variation in response of different cultivars (Cakmak et al. 1996).

2.3.5 Salinisation

Primary salinisation refers to processes that lead to naturally saline soils. Saline soils may derive from parental rock and sediments with a high salt content. Alternatively, soils may have been previously inundated with seawater or been reclaimed from the sea. Salt derived from the sea may also fall on land in rain, or as solid deposits on the wind. Similarly, salt may be blown long distances from dried lake beds (FAO and ITPS 2015).

Secondary salinisation is associated with human activities, mostly with irrigation in semi-arid areas. Salinisation of irrigated wheat lands has been an issue for least 4400 years (Jacobsen and Adams 1958). Irrigation waters from the Euphrates and Tigris were used by the Sumerians in Mesopotamia for wheat, until increasing salinity led first to a switch to barley, and then to decline and breakup of the Sumerian civilisation (Jacobsen and Adams 1958; Hillel 1991; Mithen 2012). As a Sumerian document dated c. 3600 BP laments, '*The black fields became white. . .the broad plain was choked with salt*' (Mithen 2012). The legacy of long-term irrigation effects on salinisation remain a significant factor for crop production in areas of the Middle East today.

All irrigation water contains some salts. If the total water from irrigation and rainfall during the year is insufficient to exceed evapotranspiration from the soil and crop, the percolation of water and solutes down through the profile, away from the root zone, does not occur. Over time the salt concentration builds up, a process that is exacerbated and more rapid if drainage is poor (Singh et al. 2018b) and/or the irrigation water is particularly brackish (Farooq et al. 2008). Poor drainage on irrigated lands contributes to significant salinisation in areas of India, Pakistan, and China, while the use of brackish groundwater is a contributory factor in areas of Asia and Europe (FAO and ITPS 2015).

Conversely, too much irrigation may cause salinisation if a rising water table brings brackish groundwater towards the crop. Once within the root zone, groundwater may rise to the soil surface through capillary action and evaporate from the soil surface, precipitating the salts on and in the topsoil. Irrigation is not the only way in which farming practices can cause water tables to rise. In areas of southern Australia and North America, replacement of trees and deep rooting grasslands, respectively, with wheat has been associated with an increase in percolation beyond the root zone. The water table has, therefore, risen and brought ancient salt deposits to the surface in saline seeps (Conacher 1990; Salama et al. 1999; Ganf and Morris 2016).

Strategies for delaying or correcting for salinisation differ radically depending on the principal causes. If salt has built up because there is no net loss of water through the soil profile, then problems can be alleviated by improving drainage and increasing irrigation to increase leaching. Conversely, if salinisation is due to a rising water table bringing ancient salts up to the root zone, alleviation is sometimes possible by increasing the cropping intensity and reducing fallow periods.

2.3.6 The Weed Seedbank

The weed seedbank encompasses all the viable weed seeds on and within the soil. The seedbank is a reservoir of individuals that can potentially establish as weeds. Alternatively, the seedbank can be viewed as a reservoir of plant biodiversity, important in the conservation of rare species, even within agricultural fields. The number of viable seeds in arable soils is often large. From a survey of 140 arable fields in Germany, Albrecht and Auerswald (2003) reported an average of c. 9000 viable seeds/m^2, but with samples ranging from < 1000 to > 40 000. In a wheat field Sousa et al. (2003) reported > 12 000 viable seeds/m^2 from 52 identified species.

The composition of the seedbank, and also the proportion of viable seeds that may emerge and establish in any one season, depends greatly on soil conditions (e.g. soil texture; water availability; available N, P, K; C : N ratio), historic and current timing and intensity of tillage and herbicide use, climate and weather, and the species present (Moss 1980; Cosser et al. 1996a, 1997; Albrecht and Auerswald 2003; Tørresen 2003; Colbach et al. 2006). In cereals, typically 3–7% of the seeds present in the seedbank will germinate each year (e.g. Tørresen et al. 2003), mostly from the top 30 mm.

The most challenging weeds of wheat are other grasses. This is not surprising as weeds from the same family as the crop are more likely to exploit the same ecological niches as have been prepared for the crop. For similar reasons, grass weeds can be competitive against wheat for the same resources and at the same times. Those grass weeds that can establish at or shortly after crop emergence yet shed their seeds before or at harvest time are particularly problematic (Froud-Williams 2002). Weeds closely related to the crop are also more likely to act as alternative hosts for wheat pests and diseases. Large-seeded grass weeds, such as wild oats (*Avena* spp.), can be difficult to separate from wheat grain and, therefore, spread with the crop and limit the certification of wheat to be sold as seed. Grass weeds are also more likely to be difficult to control with herbicides, i.e. herbicides that control grasses in wheat crops need to be highly selective. This need for selectivity reduces the number of active ingredients available, increases reliance on just a few chemistries, and increases the chances of the development of herbicide resistance in the weed populations.

An exemplar grass weed species is blackgrass (*Alopecurus myosuroides*), a major weed of wheat in Western Europe and increasing in other areas of the continent. It is a small-seeded grass that is favoured by early sowing of autumn-sown cereals, reduced tillage depths particularly on heavier soils, and by the growing of less competitive cultivars of wheat and other crops in rotation with wheat (Cosser et al. 1996b; Sim et al. 2007a, b; Moss 2017). Blackgrass establishes in autumn with early-sown winter crops, and if not controlled can produce large numbers of seed before and at harvest. In uncontrolled conditions blackgrass numbers in the seedbank can rise rapidly and have been recorded at up to 56 000 seeds/m^2 (Moss 1985). Under winter wheat, seed viability declines to about 20% after two to three years. Nonetheless due to high seed returns year on year, blackgrass in the seedbanks have been modelled to reach 36 000 seeds/m^2 when not controlled and the established plants from such a population to be capable of causing 40% yield loss (Gonzalez-Andujar et al. 1999). In addition to yield loss, blackgrass is susceptible to the ergot fungus (Chapter 6), which can contribute to grain contamination (Chapter 9), either directly or through increasing the infection of wheat (Mantle et al. 1977).

Due to the threat of blackgrass and other weeds, UK wheat farmers have increased their reliance on herbicide applications for weed control (e.g. Figure 2.15). By 2018 they were, on average, spraying their wheat crops three times, with each spray pass including two active ingredients. One of the most common timings was pre-emergence, i.e. the application of a soil-acting residual herbicide before the crops (and weeds) emerge. This leaves a herbicidal layer in the soil which is taken up by weeds via their roots as they germinate and try to establish. For the farmers, soil-acting residual herbicides have the advantage that control can last for an extended period, although the activity and duration of control depend greatly on soil temperature, moisture, rainfall events, and particularly the amount of soil organic matter present. After the crop emerges selective foliar herbicides are applied in the autumn and spring. When surveyed, blackgrass was the single-most important species that UK farmers mentioned when justifying their use of herbicides (Garthwaite et al. 2019). The mostly commonly used active ingredients for blackgrass in UK wheat crops included flufenacet (a pre-emergence residual herbicide that inhibits plant cell division), and mesosulfuron plus iodosulfuron (post-emergence foliar herbicides targeting acetolactate synthase [ALS]). However, herbicide resistance is widespread and increasing and blackgrass is the most important herbicide-resistant weed in Europe (Moss et al. 2007).

The current trends in herbicide use do not appear to be sustainable. In a recent study Varah et al. (2020) estimated current wheat production in England to be reduced by > 5% because of herbicide resistance to blackgrass. Further financial losses accrue from the cost

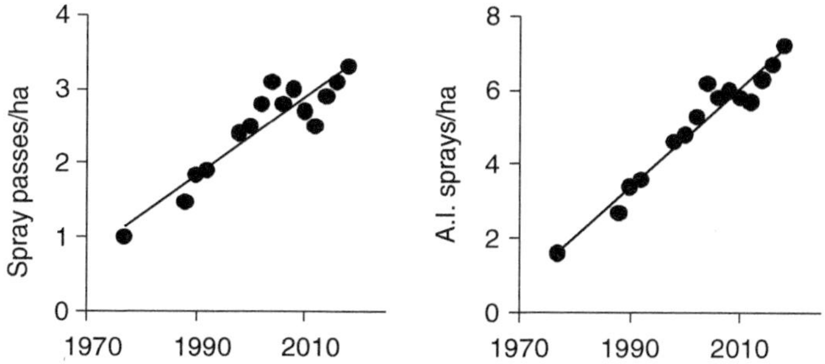

Figure 2.15 Applications of herbicides to UK wheat in terms of the number of spray passes per crop, and the number active ingredient (A.I.) sprays (sum of A.I. in each pass). *Source:* Based on data from Garthwaite et al. (2019) and previous editions.

of the extra herbicides applied. It is estimated that national wheat production could be reduced by 25% if there was total failure of herbicides to control blackgrass.

Blackgrass can be suppressed by non-chemical methods, particularly by later sowing in the autumn, including more spring-sown crops in the rotation, and by increased use of ploughing. Modifications to sowing date and rotations, however, also reduce the yield potential of wheat, and as discussed above, ploughing can be associated with land degradation. In short, grass weeds and the development of herbicide resistance are significant challenges to maintaining and increasing wheat production. Furthermore, climate change scenarios may increase weed problems and interact with herbicide efficacy (Varanasi et al. 2016). Irrespective of efficacy for weed control, the reported effects of herbicides on the environment are many and varied. With respect to soil quality and function, different effects are associated with the different herbicidal chemistries involved, but include reduced earthworm activity, reduced N cycling, and increased prevalence of certain crop diseases (Rose et al. 2016).

2.4 Systems for Protecting the Soil

2.4.1 Conservation Tillage

Despite the traditional and long-term use of ploughing, there are several drawbacks to its use. In addition to erosion and structural risks, the process of deep primary cultivation followed by secondary cultivation to produce the seed bed is consuming of time, labour, and energy (Table 2.7). Cultivation also dries out the soil, which can impede seed germination and crop growth. If nutrients released from the breakdown of organic matter are not taken up by the crop, they are at risk of being lost to the environment as pollutants. Additionally, although ploughing can suppress small-seeded weeds, it can favour the survival of larger seeded plants with long-term dormancy, such as wild oats. The drying of the soil, together with the disturbance and removal of groundcover, significantly reduces earthworm populations and activity (Edwards and Lofty 1982; Schmidt and Curry 2001; Stroud 2019). Ploughing is also a significant threat to the preservation of archaeological heritage beneath the soil surface (Wilkinson et al. 2006).

Table 2.7 Labour, fuel energy, and area capability assessments for cultivation systems in the UK, relative to ploughing with secondary cultivations.

Cultivation system	Depth of cultivation	Labour	Fuel energy	Area that can be covered by one man in a season
Plough, cultivator, drill	1.0	1.0	1.0	1.0
Shallow plough, drill	0.6	0.5	0.6	2.0
Spring tine x 3, drill	0.6	0.5	0.4	2.0
Sprayer, direct drill	0.0	0.3	0.1	4.0

Source: From Davies (1988).

After the Second World War the advent of herbicides for use on wheat pre- and post- emergence reduced the reliance on inversion tillage for weed control. Since then, there has been increasing adoption of reduced-, minimum-, or conservation tillage systems. Here, the term *conservation tillage* is used to encompass several methods that reduce the incursion into the soil compared with mouldboard ploughing to 20–30 cm. Methods can range, for example, from shallow or chisel ploughing to c. 13 cm, through various use of tines and harrows to c. 10 cm, to no-till systems where seed is delivered into a slot cut into otherwise undisturbed stubble. The motivations for adopting conservation tillage vary. Adoption has been most widespread in North and South America, and increasingly in semi-arid areas such as in Australia and China where the aim is to protect soils from erosion, and to conserve soil organic matter and soil moisture (Tebrügge and Böhrnsen 2001). Field studies suggest that soil losses by wind erosion can be reduced by up to 79% when farmers change from conventional soil tillage methods to no-till (Wang et al. 2006). Reducing tillage intensity and leaving more crop residues on the soil surface are suggested to increase water infiltration, reduce evaporation losses, and to reduce surface temperatures (Roberts and Johnston 2007).

In parts of North and South America the change from plough-based agriculture has been enshrined in legislation and/or requirements for economic support from governmental sources. In addition to sustaining food production capability, the political motivation for encouraging a change from ploughing is to improve the quality of surface waters. Where legislation, advisory services, and payment schemes are involved, there is a need for precise definition. In the US for example, conservation- and reduced- tillage are defined as systems that leave at least 30 and 15%, respectively, of the land covered with crop residue after sowing. These systems emphasise the role of surface crop residues and mulches to protect the soil from water erosion (Lal 2009). Water retention and/or the plant-available water holding capacity of the soil (PAWC, Chapter 3) have been increased by reduced- or no-till systems, mulching, and cover cropping. By 2008 nearly 60% of US wheat was established with either conservation- or reduced- tillage methods (Figure 2.16) and by 2011 nearly 80% of the grain cropped area in Argentina, including 4 Mha of wheat, was established with no-till (Peiretti and Dumanski 2014). The change from the plough to reduced- and no-till systems has been somewhat slower in European wheat production, with reduced tillage being significantly more acceptable than no-till (Meyer-Aurich et al. 2009; Lahmar 2010). There have been some governmental subsidies for conservation tillage in areas of high erosion risk but the prime motivation for reducing tillage intensity in Europe has been to reduce production costs and to be able sow large areas in a timely fashion (Table 2.7; Tebrügge and Böhrnsen 2001; Holland 2004; Lahmar 2010), factors that are particularly relevant to heavier soils. Drilling wheat in optimum sowing periods will increase yields, but there can be yield be penalties for moving away from the plough for wheat crops sown on relatively stable soils in humid and temperate conditions (Meyer-Aurich et al. 2009; Rial-Lovera et al. 2016). In a meta-analysis, Pittelkow et al. (2015) found that the yield of wheat was reduced by 2.6% in no-till compared with conventional systems, which was a significantly lower penalty than for rice (7.5%) and maize (7.6%). The yield penalty for wheat disappeared in rainfed dry areas where no-till outyielded conventional tillage by 2.6%, but no-till yields were significantly lower for temperate rainfed wheat (−7.5%). Yield reductions can be particularly marked on poorly structured soils, which are prone to compaction in the

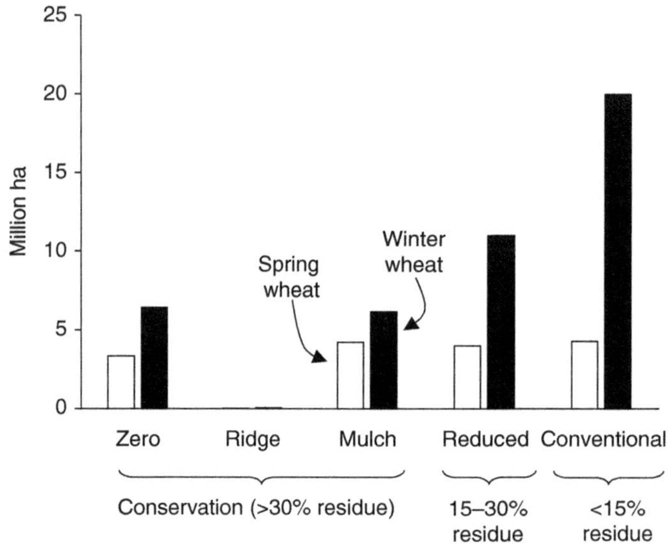

Figure 2.16 Estimated area of US wheat by method of tillage in 2008. *Source:* Data from CTIC (2020).

surface layers, and lack of ploughing may exacerbate problems (Table 2.4). Additionally, increased nitrogen fertiliser application rates may be required with reduced tillage, which could negate savings made on establishment costs (Meyer-Aurich et al. 2009). Similarly, diseases and small-seeded weeds can be more prevalent, or more costly to control in reduced- and conservation-tillage systems (Moss 1979; Smiley 1996; Camara et al. 2003; Lutman et al. 2013).

2.4.2 Organic Farming

Organic farming (also variously referred to as *ecological* or *biological* farming) has its origins with pioneers concerned for soil health and, as the name suggests, soil organic matter (Howard 1940; Balfour 1943). Organic farming can be defined in different ways but its guiding themes are well captured by the USDA (1980):

> Organic farming is a production system which avoids or largely excludes the use of synthetically compounded fertilisers, pesticides, growth regulators and livestock feed additives. To the maximum extent feasible, organic farming systems rely on crop rotations, crop residues, animal manures, legumes, green manures, off farm organic wastes, and aspects of biological pest control to maintain soil productivity and tilth, to supply plant nutrients and to control insects, weeds, and other pests.

Many farming systems can qualify as *organic* and to be successful these will be adapted to local physical and market conditions. Nevertheless, the over-arching *brand* of organic farming has advantages in being defined in regulation and legislation, is controlled through inspection and certification, and has significant consumer recognition (Seufert et al. 2017).

Although rates of adoption vary with country, the area of certified organic land continues to increase in many regions and currently stands at around 70 Mha, i.e. about 1.5% of all agricultural land. With respect to wheat production, there are currently about 1.0, 0.29, and 0.17 Mha of organic wheat grown in Europe, North America, and China, respectively (Willer et al. 2020). Organic bread is recognised as a premium product by some consumers and organic wheat can, therefore, achieve significantly higher ex-farm prices than conventional wheat (David et al. 2012). The added value can offset reduced yields of wheat, which are 30–70% of those achieved in systems relying on synthetic inputs (David et al. 2012). There are challenges for achieving breadmaking quality criteria in organic wheat, particularly with respect to grain protein concentration in high yielding climates (Gooding et al. 1993, 2012). However, some flexibility in minimum protein requirements is possible with modifications to the baking system and consumer expectations of the nature of organic bread. In environments with lower yield potential, achieving sufficient protein concentration is less of a challenge (Mason et al. 2007). The growth of specific cultivars may also reduce the requirement for high protein in organically grown wheat for breadmaking (Hussain et al. 2012).

Several studies have reported that organic farming results in higher soil organic matter contents than those found under conventional systems (e.g. Pulleman et al. 2000; Marriot and Wander 2006). This outcome is consistent with common practices of organic farming, which include more mixed livestock plus arable farming and the use of fertility building leys in rotation with grain and root crops; increased use of organic manures and composts; increased return of crop residues; and greater ground cover throughout the year with, e.g. undersowing of arable crops to establish following leys, other relay cropping systems, and cover and catch cropping before sowing spring cash crops. However, because of the lower yields achieved, the environmental effects of organic farming depend greatly on whether benefit is expressed on a per unit land area basis or on a per unit crop yield basis. For example, in a life cycle analysis Smith et al. (2019) modelled the hypothetical consequences of the whole of the UK agricultural area being converted to organic farming. They predicted that soil organic matter and carbon sequestration in the UK would increase. However, yields would decline, which if diets remained unchanged, would lead to an increase in imported foods. When increased overseas land use to compensate for shortfalls in domestic supply was factored in, large net emissions of carbon were predicted.

2.4.3 Conservation Agriculture

Conservation agriculture combines some of the principles of conservation tillage, some of the practices commonly used in organic farming, and recent technological advances in crop nutrition and crop protection. Defining features noted by Dumanski et al. (2006) and Kassam et al. (2019) are:

- Minimum soil disturbance, using zero tillage (or broadcasting) whenever possible, with the retention of crop residues on the soil surface.
- Maintaining a cover of the soil with crops for as long as possible. This can be achieved, for example, with cover crops between cash crops, and the use of relay cropping, where a following crop is established before harvest of the succeeding crop (as in undersowing a cereal to establish a following ley).

- Diversified crop rotations (at least three crop species) and cropping associations (for example with a balance of legume and non-legume crops; and a balance of perennial and annual crops).

These three points are not inconsistent with organic farming, although the first may pose a challenge as ploughing can make a significant contribution to weed control. Most of the practices in the second and third points are common organic farming practices but, in contrast to organic farming, conservation agriculture supports the use of synthetic fertilizers and agrochemicals when these are in balance with crop requirements. Finding this balance encourages the adoption of precision agriculture techniques for rate, spatial and temporal optimization of inputs to reduce costs, improve efficiency of operations, and reduce environmental damage (Dumanski et al. 2006). As with organic farming, the adoption of conservation agriculture has sometimes been reported to increase soil organic matter (e.g. Piccoli et al. 2016). In contrast with organic farming, however, wheat yields are more comparable to, and in the medium to long term (> 6 years), may exceed those from conventional systems in some production areas (Hobbs et al. 2008).

Conservation agriculture has significantly increased over the past three decades and currently accounts for around 200 Mha (12% of crop lands, 4% of the total agricultural area) (Kassam et al. 2019). Some of the areas with greatest adoption, such as North and South America, are also growing genetically modified (GM) crops. These do not currently include GM wheat, but the growth of other herbicide-tolerant GM crops in rotation with wheat reduces the challenges of weed control when ploughing is not practiced. The availability of herbicide-tolerant GM crops may be a factor in the adoption of conservation agriculture, but it is notable that the adoption is also high in areas where soils are fragile and where soil moisture retention is particularly important.

2.5 Land-Use Efficiency and Soils

Understanding the effects of wheat production on the soil environment provides a powerful argument against further expansion of wheat into areas currently covered with either perennial cropping or natural ecosystems and this argument is strengthened by considering the effects of wheat cultivation on biodiversity. There is ample evidence that annual cropping can cause damage by soil erosion and loss of carbon when perennial grasslands, woodlands, and forests are removed. This is particularly the case as many of these perennial systems have survived because they are on soils that are fragile or of low fertility, and/or are difficult to cultivate due to topography, soil type, or are poorly drained or susceptible to flooding. Some of the biggest erosion problems in wheat lands have been, and are, due to cultivation of land that was previously already fragile. In some cases, this erosion has necessitated reversion to grassland and woodland. Where wheat production persists on fragile soils, and particularly on those in semi-arid environments, conserving the soil and soil water with appropriate tillage and cropping is imperative, as are integrated methods for controlling grass weeds and diseases.

In addition to the need to protect the soil, the extent of wheat cultivation is constrained by factors that affect all arable crops. These competing factors include the requirement of land for growing biofuels, urban expansion, increased industrial forestry, and the expansion of protected areas. Each of these four additional factors could reduce the amount of arable area by at least as much as land degradation due to cropping (Lambin and Meyfroidt 2011). If the area is not to be increased, the growing demand for wheat must be satisfied by more efficient use of existing, relatively stable soils, i.e. grain yield per hectare needs to be increased. Increasing the intensity of wheat production on suitable land may also provide benefits for the soil in some situations, for example, for lowering the water table when high water tables are associated with salinisation, and in increasing the organic matter that is returned to the soil from the crop roots and residue. If good wheat land is to be irrigated, salinisation must be minimized with good drainage and efficient water delivery systems (Chapter 3). Finally, increasing acidity needs to be countered with appropriate liming and/or the use of non-acidifying fertilizers.

References

Abraham, E.M., Guevara, J.C., Candia, R.J. et al. (2016). Dust storms, drought and desertification in the Southwest of Buenos Aires Province, Argentina. *Revista de la Facultad de Ciencias Agrarias* 48 (2): 221–241. http://revistas.uncuyo.edu.ar/ojs3/index.php/RFCA/article/view/3169.

AHDB (2020). *Nutrient Management Guide (RB209) Section 1 Principle of Nutrient Management and Fertiliser Use*. Agriculture and Horticulture Development Board https://ahdb.org.uk/knowledge-library/rb209-section-1-principles-of-nutrient-management-and-fertiliser-use.

Akbari, M., Modarres, R., and Noughani, M.A. (2020). Assessing early warning for desertification hazard based on E-SMART indicators in arid regions of northeastern Iran. *Journal of Arid Environments* 174: 104086. https://doi.org/10.1016/j.jaridenv.2019.104086.

Albrecht, H. and Auerswald, K. (2003). Arable weed seedbanks and their relation to soil properties. In: *Aspects of Applied Biology 69, Seedbanks: Determination, Dynamics & Management* (eds. R.M. Bekker, F. Forcela, A.C. Grundy, et al.), 11–28. Association of Applied Biologists.

Alburquerque, J.A., Salazar, P., Barrón, V. et al. (2013). Enhanced wheat yield by biochar addition under different mineral fertilization levels. *Agronomy for Sustainable Development* 33: 475–484. https://doi.org/10.1007/s13593-012-0128-3.

Alexander, R., Nugent, C., and Nugent, K. (2018). The dust bowl in the US: an analysis based on current environmental and clinical studies. *American Journal of the Medical Sciences* 356 (2): 90–96. https://doi.org/10.1016/j.amjms.2018.03.015.

Allmaras, R., Kraft, J.M., and Miller, D.E. (1988). Effects of soil compaction and incorporated crop residue on root health. *Annual Review of Phytopathology* 26: 219–243. https://doi.org/10.1146/annurev.py.26.090188.001251.

Alloway, B.J. (2008a). Micronutrients and crop production: an introduction. In: *Micronutrient Deficiencies in Global Crop Production* (ed. B.J. Alloway), 1–39. Springer https://doi.org/10.1007/978-1-4020-6860-7_1.

Alloway, B.J. (ed.) (2008b). *Micronutrient Deficiencies in Global Crop Production*. Springer.

Amundson, R., Berhe, A.A., Hopmans, J.W. et al. (2015). Soil and human security in the 21st century. *Science* 348 (6235): 1261071. https://doi.org/10.1126/science.1261071.

Arthi, V. (2018). "The Dust Was Long in Settling": human capital and the lasting impact of the American Dust Bowl. *Journal of Economic History* 78 (1): 196–230. https://doi.org/10.1017/S0022050718000074.

Atkinson, J.A., Hawkesford, M.J., Whalley, W.R. et al. (2020). Soil strength influences wheat root interactions with soil macropores. *Plant Cell and Environment* 43: 235–245. https://doi.org/10.1111/pce.13659.

Bailey, A.P., Rehman, T., Park, J. et al. (1999). Towards a method for the economic evaluation of environmental indicators for UK integrated arable farming systems. *Agriculture, Ecosystems & Environment* 72 (2): 145–158. https://doi.org/10.1016/S0167-8809(98)00171-6.

Bakken, L.R., Børresen, T., and Njøs, A. (1987). Effect of soil compaction by tractor traffic on soil structure, denitrification, and yield of wheat (*Triticum aestivum* L.). *Journal of Soil Science* 38: 541–552. https://doi.org/10.1111/j.1365-2389.1987.tb02289.x.

Bakker, M.M., Govers, G., and Rounsevell, M.D.A. (2004). The crop productivity-erosion relationship: an analysis based on experimental work. *Catena* 57 (1): 55–76. https://doi.org/10.1016/j.catena.2003.07.002.

Bakker, M.M., Govers, G., Jones, R.A. et al. (2007). The effect of soil erosion on Europe's crop yields. *Ecosystems* 10 (7): 1209–1219. https://doi.org/10.1007/s10021-007-9090-3.

Balfour, E.B. (1943). *The Living Soil*. Faver & Faber.

Ball, B.C., Batey, T., and Munkholm, L.J. (2007). Field assessment of soil structural quality – a development of the Peerlkamp test. *Soil Use and Management* 23: 329–337. https://doi.org/10.1111/j.1475-2743.2007.00102.x.

Bansal, R.L., Nayyar, V.K., and Takkar, P.N. (1992). Field screening of wheat cultivars for manganese efficiency. *Field Crops Research* 29 (2): 107–112. https://doi.org/10.1016/0378-4290(92)90081-J.

Baquy, M.A.A., Li, J.Y., Xu, C.Y. et al. (2017). Determination of critical pH and Al concentration of acidic Ultisols for wheat and canola crops. *Solid Earth* 8: 149–159. https://doi.org/10.5194/se-8-149-2017.

Barraclough, P. and Weir, A. (1988). Effects of a compacted subsoil layer on root and shoot growth, water use and nutrient uptake of winter wheat. *Journal of Agricultural Science* 110 (2): 207–216. https://doi.org/10.1017/S0021859600081235.

Batey, T. (2009). Soil compaction and soil management – a review. *Soil Use and Management* 25: 335–345. https://doi.org/10.1111/j.1475-2743.2009.00236.x.

Berg, W.A., Hodges, M.E., and Krenzer, E.G. (1993). Iron deficiency in wheat grown on the Southern Plains. *Journal of Plant Nutrition* 16 (7): 1241–1248. https://doi.org/10.1080/01904169309364609.

den Biggelaar, C., Lal, R., Wiebe, K. et al. (2004a). The global impact of soil erosion on productivity II: effects on crop yields and production over time. *Advances in Agronomy* 81: 49–95. https://doi.org/10.1016/S0065-2113(03)81002-7.

den Biggelaar, C., Lal, R., Wiebe, K. et al. (2004b). The global impact of soil erosion on productivity I: absolute and relative erosion-induced yield losses. *Advances in Agronomy* 81: 1–48. https://doi.org/10.1016/S0065-2113(03)81001-5.

Bolles, K.C. and Forman, S.L. (2018). Evaluating landscape degradation along climatic gradients during the 1930s dust bowl drought from panchromatic historical aerial photographs, United States Great Plains. *Frontiers in Earth Science* 6: 153. https://doi.org/10.3389/feart.2018.00153.

Bolton, J. (1972). Changes in magnesium and calcium in soils of the Broadbalk wheat experiment at Rothamsted from 1865 to 1966. *Journal of Agricultural Science* 79 (2): 217–223. https://doi.org/10.1017/S0021859600032184.

Bono, A., Alvarez, R., Buschiazzo, D.E. et al. (2008). Tillage effects on soil carbon balance in a semiarid agroecosystem. *Soil Science Society of America Journal* 72: 1140–1149. https://doi.org/10.2136/sssaj2007.0250.

Brennan, R.F. (2006). Long-term residual value of copper fertiliser for production of wheat grain. *Australian Journal of Experimental Agriculture* 46 (1): 77–83. https://doi.org/10.1071/EA04271.

Brennan, R.F. and Bolland, M.D.A. (2003). Comparing copper requirements of faba bean, chickpea, and lentil with spring wheat. *Journal of Plant Nutrition* 26 (4): 883–899. https://doi.org/10.1081/PLN-120018572.

Brown, P.H. (2007). Nickel. In: *Handbook of Plant Nutrition* (eds. A.V. Barker and D.J. Pilbeam), 395–409. CRC Press.

Cakmak, I., Yilmaz, A., Kalayci, M. et al. (1996). Zinc deficiency as a critical problem in wheat production in Central Anatolia. *Plant and Soil* 180 (2): 165–172. https://doi.org/10.1007/BF00015299.

Camara, K.M., Payne, W.A., and Rasmussen, P.E. (2003). Long-term effects of tillage, nitrogen, and rainfall on winter wheat yields in the Pacific Northwest. *Agronomy Journal* 95: 828–835. https://doi.org/10.2134/agronj2003.8280.

Chaparro, J.M., Sheflin, A.M., Manter, D.K. et al. (2012). Manipulating the soil microbiome to increase soil health and plant fertility. *Biology and Fertility of Soils* 48: 489–499. https://doi.org/10.1007/s00374-012-0691-4.

Cogger, C.G., Bary, A.I., Kennedy, A.C. et al. (2013). Long-term crop and soil response to biosolids applications in dryland wheat. *Journal of Environmental Quality* 42: 1872–1880. https://doi.org/10.2134/jeq2013.05.0109.

Colbach, N., Dürr, C., Roger-Estrade, J. et al. (2006). AlomySys: odelling black-grass (*Alopecurus myosuroides* Huds.) germination and emergence, in interaction with seed characteristics, tillage and soil climate: I. Construction. *European Journal of Agronomy* 24 (2): 95–112. https://doi.org/10.1016/j.eja.2005.07.001.

Colledge, S., Conolly, J., Crema, E. et al. (2019). Neolithic population crash in northwest Europe associated with agricultural crisis. *Quaternary Research* 92 (3): 686–707. https://doi.org/10.1017/qua.2019.42.

Conacher, A.J. (1990). Salt of the Earth: secondary soil salinization in the Australian wheat belt. *Environment: Science and Policy for Sustainable Development* 32 (6): 4–42. https://doi.org/10.1080/00139157.1990.9931055.

Cook, R.J. and Veseth, R.J. (1991). *Wheat Health Management*. The American Phytopathological Society.

Cosser, N.D., Gooding, M.J., and Froud-Williams, R.J. (1996a). The impact of wheat cultivar, sowing date and grazing on the weed seedbank of an organic farming system. In: *Aspects of Applied Biology 47, Rotations and Cropping Systems*, 429–432. Association of Applied Biologists.

Cosser, N.D., Gooding, M.J., Davies, W.P. et al. (1996b). Effects of wheat dwarfing genes on *Alopecurus myosuroides*, wheat yield and grain quality. In: *Second International Weed Control Congress, Copenhagen*, 1089–1094. Flakkebjerg, Denmark: Department of Weed Control and Pesticide Ecology.

Cosser, N.D., Gooding, M.J., Thompson, A.J. et al. (1997). Competitive ability and tolerance of organically grown wheat cultivars to natural weed infestations. *Annals of Applied Biology* 130 (3): 523–535. https://doi.org/10.1111/j.1744-7348.1997.tb07679.x.

CTIC (2020). *National Crop Residue Management Survey*. Conservation Technology Information Center https://www.ctic.org/crm.

Cunfer, G. (2011). The Southern Great Plains wind erosion maps of 1936–1937. *Agricultural History* 85 (4): 540–559. https://doi.org/10.3098/ah.2011.85.4.540.

Dang, Y.P., Dalal, R.C., Routley, R. et al. (2006). Subsoil constraints to grain production in the cropping soils of the north-eastern region of Australia: an overview. *Australian Journal of Experimental Agriculture* 46: 19–35.

David, C., Abecassis, J., Carcea, M. et al. (2012). Organic bread wheat production and market in Europe. In: *Sustainable Agriculture Reviews*, vol. 11 (ed. E. Lichtfouse), 43–62. Springer https://doi.org/10.1007/978-94-007-5449-2_3.

Davies, B. (1988). *Reduced Cultivation for Cereals. HGCA Research Review No. 5*. Home-Grown Cereals Authority.

Deliopoulos, T., Kettlewell, P.S., and Hare, M.C. (2010). Fungal disease suppression by inorganic salts: a review. *Crop Protection* 29 (10): 1059–1075. https://doi.org/10.1016/j.cropro.2010.05.011.

Dell, B. (1981). Male sterility and anther wall structure in copper-deficient plants. *Annals of Botany* 48 (5): 599–608. https://doi.org/10.1093/oxfordjournals.aob.a086168.

Dumanski, J., Peiretti, R., Benites, J.R. et al. (2006). The paradigm of conservation agriculture. *Proceedings of World Association of Soil and Water Conservation* 1: 58–64.

Edwards, C. and Lofty, J. (1982). The effect of direct drilling and minimal cultivation on earthworm populations. *Journal of Applied Ecology* 19 (3): 723–734. https://doi.org/10.2307/2403277.

Engel, R.E., Bruckner, P.L., Mathre, D.E. et al. (1997). A chloride-deficient leaf spot syndrome of wheat. *Soil Science Society of America Journal* 61 (1): 176–184. https://doi.org/10.2136/sssaj1997.03615995006100010026x.

Escavy, J.I., Herrero, M.J., and Arribas, M.E. (2012). Gypsum resources of Spain: temporal and spatial distribution. *Ore Geology Reviews* 49: 72–84. https://doi.org/10.1016/j.oregeorev.2012.09.001.

Fageria, N.K. and Stone, L.F. (2008). Micronutrient deficiency problems in South America. In: *Micronutrient Deficiencies in Global Crop Production* (ed. B.J. Alloway), 245–266. Springer https://doi.org/10.1007/978-1-4020-6860-7_10.

FAO (2021). *Conservation Agriculture*. Food and Agriculture Organisation of the United Nations http://www.fao.org/conservation-agriculture/en.

FAO and ITPS (2015). *Status of the World's Soil Resources (SWSR) – Main Report*. Food and Agriculture Organization of the United Nations and Intergovernmental Technical Panel on Soils http://www.fao.org/3/i5199e/I5199E.pdf.

Farooq, M.R., Haq, M.A., Akhtar, J. et al. (2008). Comparative response of wheat (*Triticum aestivum* l.) genotypes to brackish water at seedling stage. *Pakistan Journal of Agricultural Science* 45 (4): 439–443.

Farrell, M., Macdonald, L.M., Butler, G. et al. (2014). Biochar and fertiliser applications influence phosphorus fractionation and wheat yield. *Biology and Fertility of Soils* 50: 169–178. https://doi.org/10.1007/s00374-013-0845-z.

Fenton, T.E. (2012). The impact of erosion on the classification of Mollisols in Iowa. *Canadian Journal of Soil Science* 92: 413–418. https://doi.org/10.4141/cjss2010-042.

Fierer, N. (2017). Embracing the unknown: disentangling the complexities of the soil microbiome. *Nature Reviews Microbiology* 15: 579–590. https://doi.org/10.1038/nrmicro.2017.87.

Finch, H.J.S., Samuel, A.M., and Lane, G.P.F. (2014). *Lockhart & Wiseman's Crop Husbandry Including Grassland*, 9e. Woodhead Publishing.

Fixen, P.E. (1993). Crop responses to chloride. *Advances in Agronomy* 50: 107–150.

Fixen, P.E. and Johnston, A.M. (2012). World fertilizer nutrient reserves: a view to the future. *Journal of the Science of Food and Agriculture* 92: 1001–1005. https://doi.org/10.1002/jsfa.4532.

Francois, L.E., Maas, E.V., Donovan, T.J. et al. (1986). Effect of salinity on grain yield and quality, vegetative growth, and germination of semi-dwarf and durum wheat. *Agronomy Journal* 78: 1053–1058. https://doi.org/10.2134/agronj1986.00021962007800060023x.

Froud-Williams, R.J. (2002). Weed competition. In: *Weed Management Handbook*, 9e (ed. R.E.L. Naylor), 16–38. Blackwell Science.

Fu, B., Liu, Y., Lü, Y. et al. (2011). Assessing the soil erosion control service of ecosystems change in the Loess Plateau of China. *Ecological Complexity* 8 (4): 284–293. https://doi.org/10.1016/j.ecocom.2011.07.003.

Galieni, A., Stagnari, F., Visioli, G. et al. (2016). Nitrogen fertilisation of durum wheat: a case study in Mediterranean area during transition to conservation agriculture. *Italian Journal of Agronomy* 11 (1): 12–23. https://doi.org/10.4081/ija.2016.662.

Ganf, G. and Morris, K. (2016). Salinisation and its impacts on Australia's riverine vegetation. In: *Vegetation of Australian Riverine Landscapes: Biology, Ecology and Management* (eds. S. Capon, C. James and M. Reid), 273–296. CSIRO Publishing.

Gao, W., Hodgkinson, L., Jin, K. et al. (2016). Deep roots and soil structure. *Plant, Cell and Environment* 39: 1662–1668. https://doi.org/10.1111/pce.12684.

Gao, L., Wang, B., Li, S. et al. (2019). Effects of different long-term tillage systems on the composition of organic matter by 13C CP/TOSS NMR in physical fractions in the Loess Plateau of China. *Soil and Tillage Research* 194: 104321. https://doi.org/10.1016/j.still.2019.104321.

Garthwaite, D., Ridley, L., Mace, A. et al. (2019). *Pesticide Usage Survey Report 284 Arable Crops in the United Kingdom 2018*. Fera https://secure.fera.defra.gov.uk/pusstats/surveys/documents/arable2018.pdf.

Gasser, J.K.R. (1985). Processes causing loss of calcium from agricultural soils. *Soil Use and Management* 1: 14–16. https://doi.org/10.1111/j.1475-2743.1985.tb00644.x.

Gheibi, M.N., Malakouti, M.J., Kholdebarin, B. et al. (2009). Significance of nickel supply for growth and chlorophyll content of wheat supplied with urea or ammonium nitrate. *Journal of Plant Nutrition* 32 (9): 1440–1450. https://doi.org/10.1080/01904160903092655.

Ghimire, R., Machado, S., and Bista, P. (2018). Decline in soil organic carbon and nitrogen limits yield in wheat-fallow systems. *Plant and Soil* 422: 423–435. https://doi.org/10.1007/s11104-017-3470-z.

Gill, J.S., Sale, P.W.G., and Tang, C. (2008). Amelioration of dense sodic subsoil using organic amendments increases wheat yield more than using gypsum in a high rainfall zone of southern Australia. *Field Crops Research* 107 (3): 265–275. https://doi.org/10.1016/j.fcr.2008.02.014.

Gonzalez-Andujar, J.L., Perry, J.N., and Moss, S.R. (1999). Modeling effects of spatial patterns on the seed bank dynamics of *Alopecurus myosuroides*. *Weed Science* 47 (6): 697–705. https://doi.org/10.1017/S0043174500091360.

Gooding, M.J., Davies, W.P., Thompson, A.J. et al. (1993). The challenge of achieving breadmaking quality in organic and low input wheat in the UK. A review. In: *Aspects of Applied Biology 36, Cereal Quality III* (eds. P.S. Kettlewell, J.R. Garstang, C.M. Duffus, et al.), 189–198. Association of Applied Biologists.

Gooding, M.J., Addisu, M., Uppal, R.K. et al. (2012). Effect of wheat dwarfing genes on nitrogen-use efficiency. *Journal of Agricultural Science* 150 (1): 3–22. https://doi.org/10.1017/S0021859611000414.

Goulding, K. (2016). Soil acidification and the importance of liming agricultural soils with particular reference to the United Kingdom. *Soil Use and Management* 32: 390–399. https://doi.org/10.1111/sum.12270.

Goulding, K., Poulton, P., Webster, C. et al. (2000). Nitrate leaching from the Broadbalk wheat experiment, Rothamsted, UK, as influenced by fertilizer and manure inputs and the weather. *Soil Use and Management* 16: 244–250. https://doi.org/10.1111/j.1475-2743.2000.tb00203.x.

Gregory, A.S., Dungait, J.A.J., Watts, C.W. et al. (2016). Long-term management changes topsoil and subsoil organic carbon and nitrogen dynamics in a temperate agricultural system. *European Journal of Soil Science* 67: 421–430. https://doi.org/10.1111/ejss.12359.

Grieve, C.M., Francois, L.E., and Maas, E.V. (1994). Salinity affects the timing of phasic development in spring wheat. *Crop Science* 34: 1544–1549. https://doi.org/10.2135/cropsci1994.0011183X003400060024x.

Gupta, U.C. (1971). Boron and molybdenum nutrition of wheat, barley and oats grown in Prince Edward Island soils. *Canadian Journal of Soil Science* 51 (3): 415–422. https://doi.org/10.4141/cjss71-056.

Håkansson, I. and Lipiec, J. (2000). A review of the usefulness of relative bulk density values in studies of soil structure and compaction. *Soil and Tillage Research* 53 (2): 71–85. https://doi.org/10.1016/S0167-1987(99)00095-1.

Hänsch, R. and Mendel, R.R. (2009). Physiological functions of mineral micronutrients (cu, Zn, Mn, Fe, Ni, Mo, B, cl). *Current Opinion in Plant Biology* 12 (3): 259–266. https://doi.org/10.1016/j.pbi.2009.05.006.

Hawkesford, M., Horst, W., Kichey, T. et al. (2012). Functions of macronutrients. In: *Marschner's Mineral Nutrition of Higher Plants*, 3e (ed. P. Marschner), 135–189. Academic Press https://doi.org/10.1016/B978-0-12-384905-2.00006-6.

Heckman, J.R. (2007). Chlorine. In: *Handbook of Plant Nutrition* (eds. A.V. Barker and D.J. Pilbeam), 279–291. CRC Press.

Hillel, D. (1991). *Out of the Earth: Civilization and the Life of the Soil*. Aurum Press Ltd.

Hobbs, P.R., Sayre, K., and Gupta, R. (2008). The role of conservation agriculture in sustainable agriculture. *Philosophical Transactions of the Royal Society B: Biological Sciences* 363 (1491): 543–555. https://doi.org/10.1098/rstb.2007.2169.

Holland, J.M. (2004). The environmental consequences of adopting conservation tillage in Europe: reviewing the evidence. *Agriculture, Ecosystems & Environment* 103 (1): 1–25. https://doi.org/10.1016/j.agee.2003.12.018.

Holland, J.E., White, P.J., Glendining, M.J. et al. (2019). Yield responses of arable crops to liming – an evaluation of relationships between yields and soil pH from a long-term liming experiment. *European Journal of Agronomy* 105: 176–188. https://doi.org/10.1016/j.eja.2019.02.016.

Howard, A. (1940). *The Agricultural Testament*. Oxford University Press.

Hudson, P.F. (2020). Water: how does agriculture impact freshwater resources? In: *Food and Sustainability* (eds. P. Behrens, T. Bosker and D. Ehrhardt), 71–101. Oxford University Press.

Humphries, J.M., Stangoulis, J.C.R., and Graham, R.D. (2007). Manganese. In: *Handbook of Plant Nutrition* (eds. A.V. Barker and D.J. Pilbeam), 351–374. CRC Press.

Hussain, A., Larsson, H., Kuktaite, R. et al. (2012). Towards the understanding of bread-making quality in organically grown wheat: dough mixing behaviour, protein polymerisation and structural properties. *Journal of Cereal Science* 56 (3): 659–666. https://doi.org/10.1016/j.jcs.2012.07.016.

IUSS Working Group WRB (2015). *World Reference Base for Soil Resources 2014, update 2015. International soil classification system for naming soils and creating legends for soil maps. World Soil Resources Reports No. 106*. FAO.

Jacobsen, T. and Adams, R.M. (1958). Salt and silt in ancient Mesopotamian agriculture. *Science* 128 (3334): 1251–1258.

Johnston, A.E. (1973). The effects of ley and arable cropping systems on the amounts of soil organic matter in the Rothamsted and Woburn ley-arable experiment, *Rothamsted Experimental Station, Report for 1972, Part 2*, pp. 131–159. https://doi.org/10.23637/ERADOC-1-7

Johnston, A.E., Poulton, P.R., and Coleman, K. (2009). Soil organic matter: its importance in sustainable agriculture and carbon dioxide fluxes. *Advances in Agronomy* 101: 1–57. https://doi.org/10.1016/S0065-2113(08)00801-8.

Johnston, A.E., Poulton, P.R., Coleman, K. et al. (2017). Changes in soil organic matter over 70 years in continuous arable and ley–arable rotations on a sandy loam soil in England. *European Journal of Soil Science* 68: 305–316. https://doi.org/10.1111/ejss.12415.

Kassam, A., Friedrich, T., and Derpsch, R. (2019). Global spread of conservation agriculture. *International Journal of Environmental Studies* 76 (1): 29–51. https://doi.org/10.1080/00207233.2018.1494927.

Khan, N., Bano, A., and Ali Babar, M.D. (2017). The root growth of wheat plants, the water conservation and fertility status of sandy soils influenced by plant growth promoting rhizobacteria. *Symbiosis* 72: 195–205. https://doi.org/10.1007/s13199-016-0457-0.

Kidder, T. and Zhuang, Y. (2015). Anthropocene archaeology of the Yellow River, China, 5000-2000 BP. *The Holocene* 25: 1627–1639. https://doi.org/10.1177/0959683615594469.

Kochian, L.V., Piñeros, M.A., Liu, J. et al. (2015). Plant adaptation to acid soils: the molecular basis for crop aluminium resistance. *Annual Review of Plant Biology* 66 (1): 571–598. https://doi.org/10.1146/annurev-arplant-043014-114822.

Kopsell, D.E. and Kopsell, D.A. (2007). Copper. In: *Handbook of Plant Nutrition* (eds. A.V. Barker and D.J. Pilbeam), 293–328. CRC Press.

Lahmar, R. (2010). Adoption of conservation agriculture in Europe: lessons of the KASSA project. *Land Use Policy* 27 (1): 4–10. https://doi.org/10.1016/j.landusepol.2008.02.001.

Lal, R. (2009). Challenges and opportunities in soil organic matter research. *European Journal of Soil Science* 60: 158–169. https://doi.org/10.1111/j.1365-2389.2008.01114.x.

Lal, R., Reicosky, D.C., and Hanson, J.D. (2007). Evolution of the plow over 10,000 years and the rationale for no-till farming. *Soil and Tillage Research* 93 (1): 1–12. https://doi.org/10.1016/j.still.2006.11.004.

Lambin, E.F. and Meyfroidt, P. (2011). Global land use change, economic globalization, and the looming land scarcity. *Proceedings of the National Academy of Sciences* 108 (9): 3465–3472. https://doi.org/10.1073/pnas.110048010.

Larney, F.J. and Olson, A.F. (2018). Wheat yield and soil properties reveal legacy effects of artificial erosion and amendments on a dryland Dark Brown Chernozem. *Canadian Journal of Soil Science* 98: 663–677. https://doi.org/10.1139/cjss-2018-0025.

Lobell, D.B., Ortiz-Monasterio, J.I., Gurrola, F.C. et al. (2007). Identification of saline soils with multiyear remote sensing of crop yields. *Soil Science Society of America Journal* 71: 777–783. https://doi.org/10.2136/sssaj2006.0306.

Long, J. and Siu, H. (2018). Refugees from dust and shrinking land: tracking the Dust Bowl migrants. *The Journal of Economic History* 78 (4): 1001–1033. https://doi.org/10.1017/S002205071800059.

Lu, H. and Fan, L. (2013). Winter wheat yield potentials and yield gaps in the North China Plain. *Field Crops Research* 143: 98–105. https://doi.org/10.1016/j.fcr.2012.09.015.

Lu, J., Zheng, F., Li, G. et al. (2016). The effects of raindrop impact and runoff detachment on hillslope soil erosion and soil aggregate loss in the Mollisol region of Northeast China. *Soil and Tillage Research* 161: 79–85. https://doi.org/10.1016/j.still.2016.04.002.

Lutman, P.J.W., Moss, S.R., Cook, S., and Welham, S.J. (2013). A review of the effects of crop agronomy on the management of Alopecurus myosuroides. *Weed Research* 53 (5): 299–313. https://doi.org/10.1111/wre.12024.

MAFF (1988). *Agricultural Land Classification of England and Wales*. Ministry of Agriculture, Fisheries and Food.

Mantle, P.G., Shaw, S., and Doling, D.A. (1977). Role of weed grasses in the etiology of ergot disease in wheat. *Annals of Applied Biology* 86 (3): 339–351. https://doi.org/10.1111/j.1744-7348.1977.tb01848.x.

Maroušek, J. (2014). Significant breakthrough in biochar cost reduction. *Clean Technologies and Environmental Policy* 16: 1821–1825. https://doi.org/10.1007/s10098-014-0730-y.

Marriott, E.E. and Wander, M.M. (2006). Total and labile soil organic matter in organic and conventional farming systems. *Soil Science Society of America Journal* 70 (3): 950–959. https://doi.org/10.2136/sssaj2005.0241.

Martínez-Valderrama, J., Ibáñez, J., Del Barrio, G. et al. (2016). Present and future of desertification in Spain: implementation of a surveillance system to prevent land degradation. *Science of the Total Environment* 563–564: 169–178. https://doi.org/10.1016/j.scitotenv.2016.04.065.

Mason, H., Navabi, A., Frick, B. et al. (2007). Does growing Canadian Western Hard Red Spring wheat under organic management alter its breadmaking quality? *Renewable Agriculture and Food Systems* 22: 157–167. https://doi.org/10.1017/S1742170507001688.

Matocha, C.J. (2017). Clay: charge properties. In: *Encyclopedia of Soil Science*, 3e (ed. R. Lal), 287–290. CRC Press.

Mbava, N., Mutema, M., Zengeni, R. et al. (2020). Factors affecting crop water use efficiency: a worldwide meta-analysis. *Agricultural Water Management* 228: 105878. https://doi.org/10.1016/j.agwat.2019.105878.

Meyer-Aurich, A., Gandorfer, M., Gerl, G. et al. (2009). Tillage and fertilizer effects on yield, profitability, and risk in a corn-wheat-potato-wheat rotation. *Agronomy Journal* 101: 1538–1547. https://doi.org/10.2134/agronj2008.0126x.

Mithen, S. (2012). *Thirst: Water & Power in the Ancient World*. Orion Publishing Group.

Montgomery, D.R. (2007a). *Dirt: The Erosion of Civilisations*. University of California Press.

Montgomery, D.R. (2007b). Soil erosion and agricultural sustainability. *Proceedings of the National Academy of Sciences* 104 (33): 13268–13272. https://doi.org/10.1073/pnas.0611508104.

Moss, S.R. (1979). The influence of tillage and method of straw disposal on the survival and growth of black-grass, Alopecurus myosuroides, and its control by chlortoluron and isoproturon. *Annals of Applied Biology* 91: 91–100. https://doi.org/10.1111/j.1744-7348.1979.tb07417.x.

Moss, S.R. (1980). A study of populations of black-grass (*Alopecurus myosuroides*) in winter wheat, as influenced by seed shed in the previous crop, cultivation system and straw disposal method. *Annals of Applied Biology* 94 (1): 121–126. https://doi.org/10.1111/j.1744-7348.1980.tb03903.x.

Moss, S.R. (1985). The survival of *Alopecurus myosuroides* Huds. Seeds in soil. *Weed Research* 25 (3): 201–211. https://doi.org/10.1111/j.1365-3180.1985.tb00636.x.

Moss, S.R. (2017). Black-grass (*Alopecurus myosuroides*): why has this weed become such a problem in Western Europe and what are the solutions? *Outlooks on Pest Management* 28 (5): 207–212. https://doi.org/10.1564/v28_oct_04.

Moss, S.R., Perryman, S.A., and Tatnell, L.V. (2007). Managing herbicide-resistant blackgrass (*Alopecurus myosuroides*): theory and practice. *Weed Technology* 21 (2): 300–309. https://doi.org/10.1614/WT-06-087.1.

Nogaroli, J.A. and da Fonseca, A.F. (2020). Yield index of crops grown under no-tillage after superficial application of micronized liming materials (MLM) on the soil. *Australian Journal of Crop Science* 14 (1): 187–195.

NRCS (2018). *Effects on Soil Water Holding Capacity Soil Water Retention Resulting from Soil Health Practices Implementation*. Natural Resources Conservation Service. U.S. Department of Agriculture https://www.nrcs.usda.gov/wps/PA_NRCSConsumption/download?cid=nrcseprd1392812&ext=pdf.

Ohki, K. (1984). Manganese deficiency and toxicity effects on growth, development, and nutrient composition in wheat. *Agronomy Journal* 76 (2): 213–218. https://doi.org/10.2134/agronj1984.00021962007600020011x.

Orton, T.G., Mallawaarachchi, T., Pringle, M.J. et al. (2018). Quantifying the economic impact of soil constraints on Australian agriculture: a case study of wheat. *Land Degradation and Development* 29: 3866–3875. https://doi.org/10.1002/ldr.3130.

Peiretti, R. and Dumanski, J. (2014). The transformation of agriculture in Argentina through soil conservation. *International Soil and Water Conservation Research* 2 (1): 14–20. https://doi.org/10.1016/S2095-6339(15)30010-1.

Piccoli, I., Chiarini, F., Carletti, P. et al. (2016). Disentangling the effects of conservation agriculture practices on the vertical distribution of soil organic carbon. Evidence of poor carbon sequestration in North-Eastern Italy. *Agriculture, Ecosystems & Environment* 230: 68–78. https://doi.org/10.1016/j.agee.2016.05.035.

Pimentel, D. (2006). Soil erosion: a food and environmental threat. *Environment, Development and Sustainability* 8: 119–137. https://doi.org/10.1007/s10668-005-1262-8.

Pittelkow, C.M., Linquist, B.A., Lundy, M.E. et al. (2015). When does no-till yield more? A global meta-analysis. *Field Crops Research* 183: 156–168. https://doi.org/10.1016/j.fcr.2015.07.020.

Poulton, P.R., Johnston, A.E., Macdonald, A. et al. (2018). Major limitations to achieving "4 per 1000" increases in soil organic carbon stock in temperate regions: evidence from long-term experiments at Rothamsted Research, United Kingdom. *Global Change Biology* 24: 2563–2584. https://doi.org/10.1111/gcb.14066.

Powlson, D.S. (1980). Effect of cultivation on the mineralization of nitrogen in soil. *Plant and Soil* 57: 151–153. https://doi.org/10.1007/BF02139653.

Powlson, D.S., Whitmore, A.P., and Goulding, K.W.T. (2011). Soil carbon sequestration to mitigate climate change: a critical re-examination to identify the true and the false. *European Journal of Soil Science* 62: 42–55. https://doi.org/j.1365-2389.2010.01342.x.

Powlson, D.S., Stirling, C.M., Jat, M.L. et al. (2014). Limited potential of no-till agriculture for climate change mitigation. *Nature Climate Change* 4 (8): 678–683.

Pulleman, M.M., Bouma, J., Van Essen, E.A. et al. (2000). Soil organic matter content as a function of different land use history. *Soil Science Society of America Journal* 64 (2): 689–693. https://doi.org/10.2136/sssaj2000.642689x.

Qiangguo, C. (2002). The relationships between soil erosion and human activities on the Loess Plateau. *Proceedings of the 12th ISCO Conference*, 1 (pp. 112–118). Tsinghua University Press.

Rasouli, F., Pouya, A.K., and Karimian, N. (2013). Wheat yield and physico-chemical properties of a sodic soil from semi-arid area of Iran as affected by applied gypsum. *Geoderma* 193–194: 246–255. https://doi.org/10.1016/j.geoderma.2012.10.001.

Rengasamy, P. (2006). World salinization with emphasis on Australia. *Journal of Experimental Botany* 57 (5): 1017–1023. https://doi.org/10.1093/jxb/erj108.

Rerkasem, B. and Jamjod, S. (2004). Boron deficiency in wheat: a review. *Field Crops Research* 89 (2–3): 173–186. https://doi.org/10.1016/j.fcr.2004.01.022.

Reynolds, W.D., Drury, C.F., Tan, C.S. et al. (2009). Use of indicators and pore volume-function characteristics to quantify soil physical quality. *Geoderma* 152 (3–4): 252–263. https://doi.org/10.1016/j.geoderma.2009.06.009.

Rial-Lovera, K., Davies, W., Cannon, N., and Conway, J. (2016). Influence of tillage systems and nitrogen management on grain yield, grain protein and nitrogen-use efficiency in UK spring wheat. *Journal of Agricultural Science* 154 (8): 1437–1452. https://doi.org/10.1017/S0021859616000058.

Rinaldi, M., Scalcione, E., Perniola, M. et al. (2017). Soil tillage and residues management in wheat continuous cropping in Southern Italy: a model application for agronomic and soil fertility assessment. *Computers and Electronics in Agriculture* 140: 77–87. https://doi.org/10.1016/j.compag.2017.05.037.

Roberts, T.L. and Johnston, A.M. (2007). Tillage intensity, crop rotation and fertilizer technology for sustainable wheat production. The North American Experience. In: *Wheat Production in Stressed Environments* (eds. H.T. Buck, J.E. Nisi and N. Salomon), 175–187. Springer.

Romheld, V. and Nikolic, M. (2007). Iron. In: *Handbook of Plant Nutrition* (eds. A.V. Barker and D.J. Pilbeam), 329–350. CRC Press.

Rose, M.T., Cavagnaro, T.R., Scanlan, C.A. et al. (2016). Impact of herbicides on soil biology and function. *Advances in Agronomy* 136: 133–220. https://doi.org/10.1016/bs.agron.2015.11.005.

Rothamsted Research (2018). *Rothamsted ley arable soil organic carbon content.* Electronic Rothamsted Archive. https://doi.org/10.23637/KeyRefOARLAsoc

Rothamsted Research (2021). *Broadbalk soil organic carbon content.* Electronic Rothamsted Archive. https://doi.org/10.23637/KeyRefOABKsoc-02

Russell, E.W. (1977). The role of organic matter in soil fertility. *Philosophical Transactions of the Royal Society London B.* 281: 209–219. https://doi.org/10.1098/rstb.1977.0134.

Salama, R., Otto, C., and Fitzpatrick, R. (1999). Contributions of groundwater conditions to soil and water salinization. *Hydrogeology Journal* 7: 46–64. https://doi.org/10.1007/s100400050179.

Scharlemann, J.P.W., Tanner, E.V.J., Hiederer, R. et al. (2014). Global soil carbon: understanding and managing the largest terrestrial carbon pool. *Carbon Management* 5: 81–91.

Scherr, S.J. (1999). *Soil Degradation: A Threat to Developing-Country Food Security by 2020? Food, Agriculture, and the Environment Discussion Paper 27.* International Food Policy Research Institute.

Schmidt, O. and Curry, J.P. (2001). Population dynamics of earthworms (Lumbricidae) and their role in nitrogen turnover in wheat and wheat clover cropping systems. *Pedobiologia* 45 (2): 174–187. https://doi.org/10.1078/0031-4056-00078.

Schroder, J.L., Zhang, H., Girma, K. et al. (2011). Soil acidification from long-term use of nitrogen fertilizers on winter wheat. *Soil Science Society of America Journal* 75: 957–964. https://doi.org/10.2136/sssaj2010.0187.

Seufert, V., Ramankutty, N., and Mayerhofer, T. (2017). What is this thing called organic? – how organic farming is codified in regulations. *Food Policy* 68: 10–20. https://doi.org/10.1016/j.foodpol.2016.12.009.

Shams, A.S. and Kamel, A.S. (2014). Rotations in coastal plains to combat desertification in Egypt. *International Journal of Water Resources and Arid Environments* 3 (2): 121–131.

Shi, H. and Shao, M. (2000). Soil and water loss from the Loess Plateau in China. *Journal of Arid Environments* 45: 9–20. https://doi.org/10.1006/jare.1999.0618.

Sim, L.C., Froud-Williams, R.J., and Gooding, M.J. (2007a). The influence of winter oilseed rape (*Brassica napus* ssp. *oleifera* var. *biennis*) cultivar on competition with grasses; crop yield, grass growth and grass seed return. *Journal of Agricultural Science* 145: 329–342. https://doi.org/10.1017/S0021859607006764.

Sim, L.C., Froud-Williams, R.J., and Gooding, M.J. (2007b). The influence of winter oilseed rape (*Brassica napus* ssp. *oleifera* var. *biennis*) canopy size on grass-weed growth and grass-weed seed return. *Journal of Agricultural Science* 145: 313–327. https://doi.org/10.1017/S0021859606006721.

Singh, B., Dang, Y.P., and Mehta, S.C. (1990). Influence of nitrogen on the behaviour of nickel in wheat. *Plant and Soil* 127 (2): 213–218. https://doi.org/10.1007/BF00014428.

Singh, P., Choudhary, O.P., and Singh, P. (2018a). Performance of some wheat cultivars under saline irrigation water in field conditions. *Communications in Soil Science and Plant Analysis* 49 (3): 334–343. https://doi.org/10.1080/00103624.2018.1427258.

Singh, P., Choudhary, O.P., and Mavi, M.S. (2018b). Irrigation-induced salinization effects on soil chemical and biological properties under cotton-wheat rotation on loamy sand soil in Northwest India. *Journal of the Indian Society of Soil Science* 66 (4): 386–391. https://doi.org/10.5958/0974-0228.2018.00048.8.

Singh, M.K., Singh, R.P., Yadav, P.K. et al. (2020). Effect of Ni and N sources on wheat (*Triticum aestivum* L.) yield and nutrient uptake. *The Pharma Innovation Journal* 9 (5): 390–395.

Smiley, R. (1996). Diseases of wheat and barley in conservation cropping systems of the semiarid Pacific Northwest. *American Journal of Alternative Agriculture* 11 (2–3): 95–103. https://doi.org/10.1017/S0889189300006858.

Smith, L.G., Kirk, G.J., Jones, P.J. et al. (2019). The greenhouse gas impacts of converting food production in England and Wales to organic methods. *Nature Communications* 10: 4641. https://doi.org/10.1038/s41467-019-12622-7.

Soil Survey Staff (2014). *Keys to Soil Taxonomy*, 12e. Natural Resources Conservation Service, U.S. Department of Agriculture https://www.nrcs.usda.gov/wps/PA_NRCSConsumption/download?cid=stelprdb1252094&ext=pdf.

Sousa, E., Caixinhas, L., and Rocha, F. (2003). Weed seedbank and surface flora on a wheat crop. In: *Aspects of Applied Biology 69, Seedbanks: Determination, Dynamics & Management* (eds. R.M. Bekker, F. Forcela, A.C. Grundy, et al.), 47–54. Association of Applied Biologists.

Spedding, C. (1983). *Fream's Agriculture*. Royal Agricultural Society of England.

Stroud, J. (2019). Soil health pilot study in England: outcomes from an on-farm earthworm survey. *PLoS One* 14 (2): e0203909. https://doi.org/10.1371/journal.pone.0203909.

Subedi, K.D., Budhathoki, C.B., Subedi, M. et al. (1997). Response of wheat genotypes to sowing date and boron fertilization aimed at controlling sterility in a rice-wheat rotation in Nepal. *Plant and Soil* 188 (2): 249–256. https://doi.org/10.1023/A:1004202725382.

Subedi, K.D., Gregory, P.J., Summerfield, R.J. et al. (1998). Cold temperatures and boron deficiency caused grain set failure in spring wheat (*Triticum aestivum* L.). *Field Crops Research* 57 (3): 277–288. https://doi.org/10.1016/S0378-4290(97)00148-2.

Subedi, K.D., Gregory, P.J., and Gooding, M.J. (1999). Boron accumulation and partitioning in wheat cultivars with contrasting tolerance to boron deficiency. *Plant and Soil* 214 (1): 141–152. https://doi.org/10.1023/A:1004792221715.

Subedi, K.D., Gregory, P.J., Summerfield, R.J. et al. (2000). Pattern of grain set in boron-deficient and cold-stressed wheat (*Triticum aestivum* L.). *Journal of Agricultural Science* 134 (1): 25–31. https://doi.org/10.1017/S0021859699007303.

Sun, L., Wang, R., Li, J. et al. (2019). Reasonable fertilization improves the conservation tillage benefit for soil water use and yield of rain-fed winter wheat: a case study from the Loess Plateau, China. *Field Crops Research* 242: 107589. https://doi.org/10.1016/j.fcr.2019.107589.

Tang, C., Conyers, M.K., Nuruzzaman, M. et al. (2011). Biological amelioration of subsoil acidity through managing nitrate uptake by wheat crops. *Plant and Soil* 338: 383–397. https://doi.org/10.1007/s11104-010-0552-6.

Tao, H., Pan, W.L., and Carter, P. (2019). *Wheat straw pulping by-product mixed with lime may address soil acidification in no-till fields*. Report 2019-11. Washington State University Extension.

Tebrügge, F. and Böhrnsen, A. (2001). Farmers' and experts' opinion on no-tillage in West-Europe and Nebraska. In: *World Congress on Conservation Agriculture*, vol. 1, 1–69. Food and Agriculture Organisation of the United Nations (FAO) and the European Conservation Agriculture Federation (ECAF).

Tørresen, K.S. (2003). Relationship between seedbanks and emerged weeds in long-term tillage experiments. In: *Aspects of Applied Biology 69, Seedbanks: Determination, Dynamics & Management* (eds. R.M. Bekker, F. Forcela, A.C. Grundy, et al.), 55–62. Association of Applied Biologists.

Tørresen, K.S., Skuterud, R., Tandsaether, H.J. et al. (2003). Long-term experiments with reduced tillage in spring cereals. I. Effects on weed flora, weed seedbank and grain yield. *Crop Protection* 22 (1): 185–200. https://doi.org/10.1016/S0261-2194(02)00145-X.

Truog, E. (1947). The liming of soils. In: *The Yearbook of Agriculture 1943-1947: Science in Farming*, 566–576. United States Department of Agriculture, U. S. Govt. Printing Office https://naldc.nal.usda.gov/download/IND43893966/PDF.

U.S. Salinity Laboratory (2020). *Databases: Salt Tolerance, Fiber, Grain, Special Crops*. U.S. Department of Agriculture https://www.ars.usda.gov/pacific-west-area/riverside-ca/agricultural-water-efficiency-and-salinity-research-unit/docs/databases/fgs-crops.

Uppal, R.K. and Gooding, M.J. (2013). Gibberellin-responsive and-insensitive dwarfing alleles on wheat performance in contrasting tillage systems. *Field Crops Research* 141: 55–62. https://doi.org/10.1016/j.fcr.2012.11.001.

USDA (1980). *Report and Recommendations on Organic Farming*. United States Department of Agriculture https://pubs.nal.usda.gov/report-and-recommendations-organic-farming-usda-1980.

USGCRP (2017). *Climate Science Special Report: Fourth National Climate Assessment*, vol. I. Washington: U.S. Global Change Research Program.

Varah, A., Ahodo, K., Coutts, S.R. et al. (2020). The costs of human-induced evolution in an agricultural system. *Nature Sustainability* 3 (1): 63–71. https://doi.org/10.1038/s41893-019-0450-8.

Varanasi, A., Prasad, P.V., and Jugulam, M. (2016). Impact of climate change factors on weeds and herbicide efficacy. *Advances in Agronomy* 135: 107–146. https://doi.org/10.1016/bs.agron.2015.09.002.

Wang, X., Oenema, O., Hoogmoed, W.B. et al. (2006). Dust storm erosion and its impact on soil carbon and nitrogen losses in northern China. *Catena* 66 (3): 221–227. https://doi.org/10.1016/j.catena.2006.02.006.

Wang, M., Kawakami, Y., and Bhullar, N.K. (2019). Molecular analysis of iron deficiency response in hexaploid wheat. *Frontiers in Sustainable Food Systems* 3: 67. https://doi.org/10.3389/fsufs.2019.00067.

Webb, J., Loveland, P., Chambers, B. et al. (2001). The impact of modern farming practices on soil fertility and quality in England and Wales. *Journal of Agricultural Science* 137 (2): 127–138. https://doi.org/10.1017/S0021859601001290.

Wen, X. and Zhen, L. (2020). Soil erosion control practices in the Chinese Loess Plateau: a systematic review. *Environmental Development* 34: 100493. https://doi.org/10.1016/j.envdev.2019.100493.

Wibberley, E.J. (1989). *Cereal Husbandry*. Farming Press.

Wilkinson, B.H. and McElroy, B.J. (2007). The impact of humans on continental erosion and sedimentation. *GSA Bulletin* 119 (1–2): 140–156. https://doi.org/10.1130/B25899.1.

Wilkinson, K., Tyler, A., Davidson, D. et al. (2006). Quantifying the threat to archaeological sites from the erosion of cultivated soil. *Antiquity* 80 (309): 658–670. https://doi.org/10.1017/S0003598X00094114.

Willer, H., Schlatter, B., Travnicek, J. et al. (eds.) (2020). *The World of Organic Agriculture. Statistics and Emerging Trends 2020*. Research Institute of Organic Agriculture (FiBL), Frick, and IFOAM – Organics International http://www.organic-world.net/yearbook/yearbook-2020.html.

Wright, A.L., Dou, F., and Hons, F.M. (2007). Soil organic C and N distribution for wheat cropping systems after 20 years of conservation tillage in central Texas. *Agriculture, Ecosystems and Environment* 121 (4): 376–382. https://doi.org/10.1016/j.agee.2006.11.011.

Zhang, X., Pei, D., and Chen, S. (2004). Root growth and soil water utilization of winter wheat in the North China Plain. *Hydrological Processes* 18: 2275–2287. https://doi.org/10.1002/hyp.5533.

3

Ample Water

We never know the worth of water, till the well is dry

Thomas Fuller

Farmers have always known that they depend on water for their crops and have been building structures to help water cereals for at least 9000 years (Mithen 2012; Fujii 2013). Knowledge of the role of water in plants is, however, much more recent. Jan Baptist van Helmont made an important deduction in the 1600s. Given the evident importance of soils, for many centuries it was assumed that plants consumed soil. To test this theory, van Helmont planted a willow in 90 kg of soil. He watered the willow for five years, after which the tree had increased in weight by about 75 kg. However, the soil weight had declined by just 0.06 kg. The willow had not *eaten* the soil. Van Helmont concluded that the willow gained its mass from the water. He was partly correct.

A clearer picture emerged in the following century. Stephen Hales (*Vegetable Staticks* 1727) found that water vapour was emitted from plant leaves and that this encouraged a flow of water and nutrients through the plants from the roots. He had described *transpiration*. Hales also found that air moves freely into leaves and surmised that plants probably drew part of their nourishment from the air. Jan Ingenhousz (*Experiments Upon Vegetables* 1779), and Jean Senebier (*Physiologie Vegetale* 1800) showed that light and green plant tissues were required to absorb this nourishment, i.e. carbon dioxide (CO_2). Also, when green plant tissues, light, and CO_2 were present, oxygen (O_2) was liberated. Nicolas-Théodore de Saussure (*Recherches chimiques sur la Végétation*, 1805) emphasised that the CO_2 came from the atmosphere, but also that some of the dry matter derived from the incorporation of water. From Hales, Ingenhousz, Senebier, and de Saussure we have the basic building blocks used to photosynthesise (literally 'join together with light') carbohydrate, i.e. water, carbon dioxide, light, and green leaves (Figure 3.1).

We now know that water has crucial roles in many of the photosynthetic reactions of plants (Chapter 5). Water is also the medium of nutrient transport from the soil to the roots, and for the movement of assimilates within the plant. It is also the solvent within which biochemical reactions occur. Water provides the necessary turgor pressure for the expansion of cells and the growth of tissues and organs (roots, stems, leaves, and grain). Water

Wheat: Environment, Food and Health, First Edition. Mike J. Gooding and Peter R. Shewry.

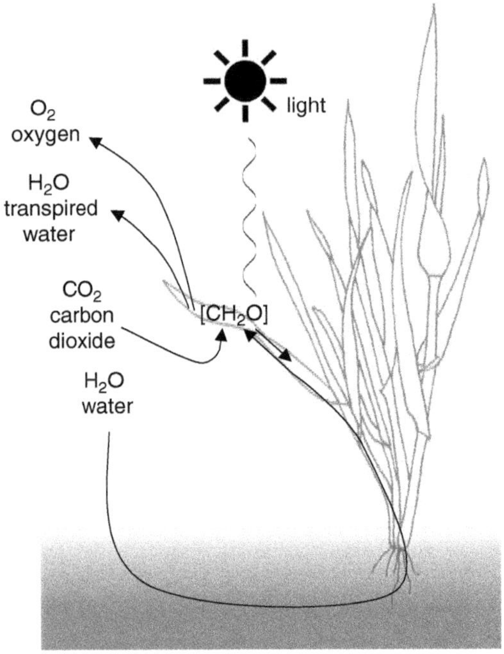

Figure 3.1 The inputs and outputs of photosynthesis.

dampens temperature fluctuations, and the loss of water from the leaf provides evaporative cooling and protection from heat stress. However, contrary to the inference of van Helmont, water is only a minor direct contributor to plant dry matter compared to the contribution from CO_2.

Most of the water taken up by wheat is lost from the plant through transpiration. Estimates of the water requirement of wheat vary greatly between studies, but reviews of the literature suggest that on average each tonne of wheat grain needs about 1000 t of water to be available (Zwart and Bastiaanssen 2004; Mbava et al. 2020), i.e. a *water use efficiency* (WUE) of 1 g/kg. The requirement is, indeed, for *ample* water. Wheat is not unusual in this respect. The WUE of wheat is generally better than that of oilseed, fibre, and legume crops. Amongst the cereals, the WUE of wheat is comparable to other C3 (Chapter 5) crops such as rice and barley, but poorer than the C4 maize and sorghum (Zwart and Bastiaanssen 2004; Mbava et al. 2020).

The large amount of water used by wheat and other crops is a significant challenge for food security. The water demand threatens the social, economic, and environmental sustainability of wheat production (e.g. Huang et al. 2020) and water scarcity in wheat-growing areas has been related to spikes in global food prices and political instability (Sternberg 2012). There are growing tensions between sectors, regions, and countries as to the governance, economics, and use of water supplies (e.g. Garrick et al. 2020). There is, therefore, an urgent need to understand the water requirements of wheat, the extent to which these may be reduced, and the sustainability or otherwise of water sources used for wheat production.

3.1 The Water Requirement of Wheat

The average WUE of 1 g/kg, which is stated in the previous section, relates water availability directly to grain yield. It is seductive in its simplicity. However, as already mentioned, there is wide variation around the value of 1 g/kg. On good soils, with high incident light, moderate temperatures, and with cultivars with high harvest indices, farmers in the east of the UK regularly harvest more than 10 tonnes of wheat grain from 600 to 700 mm (6000 to 7000 t/ha) of rainfall, i.e. 1.4 to 1.7 g/kg. In the inland Pacific Northwest of the US, Cook and Veseth (1991) present similar values for high yielding wheat crops. Conversely, there are many instances where WUE is less than 1 g/kg. Zwart and Bastiaanssen (2004) report some wheat systems producing only 0.6 g grain/kg water, while Mbava et al. (2020) include 2% of cases where the value was lower than 0.1 g/kg.

Clearly there are many processes that will influence the relationship between water supply in a field and the subsequent grain yield. An empirical relationship between water availability, uptake, use, and grain yield is shown in Figure 3.2. In water-limited environments there is often a series of approximately linear relationships amongst these components (Figure 3.3). Strategies to improve grain yield have, therefore, included increasing the amount of water available, the proportion of available water that is taken up, the amount of biomass produced per unit water transpired, and the proportion of crop biomass that is in the grain (i.e. the harvest index). To optimise water uptake and use, it is first necessary to understand the effects of water on wheat.

3.1.1 Germination and Seedling Emergence

The rates of seedling emergence and plant establishment can be key determinants of eventual crop yield, particularly in drier regions and when spring wheat is grown (Schillinger et al. 1998). Water must first be imbibed by the sown seeds. Wheat seeds need a water content of at least 30–35% to germinate, although at this level germination is slow and the developing seedlings have low vigour (Percival 1921). More rapid and complete germination occurs when 40–50% of the seed mass is water. In drier areas, farmers may drill wheat seeds as deep as 20 cm to access the necessary moisture (Schillinger et al. 1998). Deep sowing can also be justified to delay germination until significant amounts of rain have fallen, to ensure sufficient water to support crop establishment and early growth (Evans et al. 1975). After germination, seedling emergence and root growth need water for

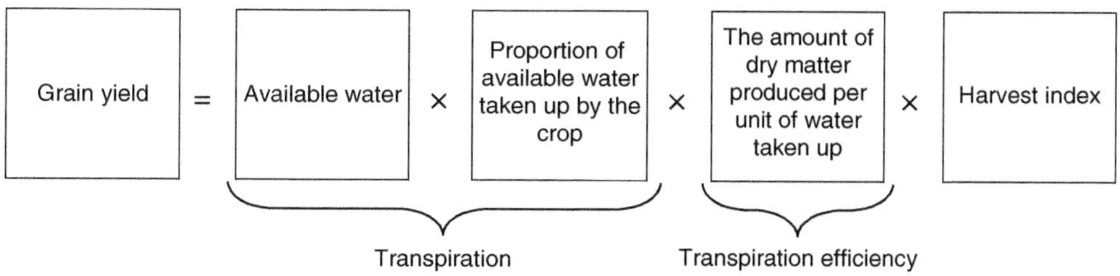

Figure 3.2 The components of water uptake and use related to grain yield. *Source:* After Hay and Porter 2006.

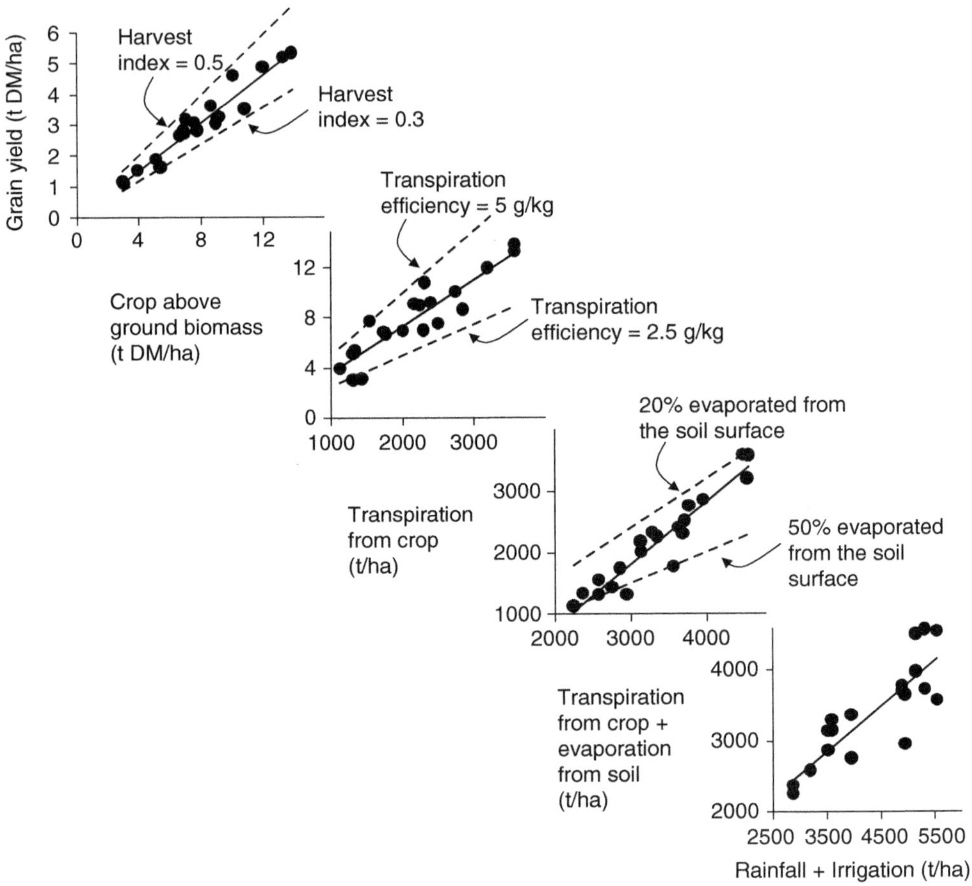

Figure 3.3 Relationships between water supply, evapotranspiration, transpiration, crop biomass, and grain yield. *Source:* Data is from Zhang et al. (1998) and are the treatment means for each of five years of a factorial combination of +/− irrigation and +/− nitrogen fertilizer application to wheat grown at ICARDA, Northern Syria.

numerous purposes, but particularly for cell expansion. Seed germination and seedling emergence are, therefore, inhibited when soil moisture concentration drops to levels that cannot be readily taken up by plants. The precise minimum moisture content will depend on soil type (Section 3.2.1). Seedling emergence can also be inhibited at high moisture contents when this is associated with saturation of the soil and declining oxygen availability. Between deficiency and excess, there is a broad range of soil moisture contents at which wheat seedlings can successfully establish (Figure 3.4). Within this range, the speed of emergence is positively associated with moisture concentration (Hanks and Thorp 1956).

3.1.2 Transpiration

Once the seedlings have emerged, the great majority of water is required to support transpiration. The transpiration stream is important for the movement of water and solutes

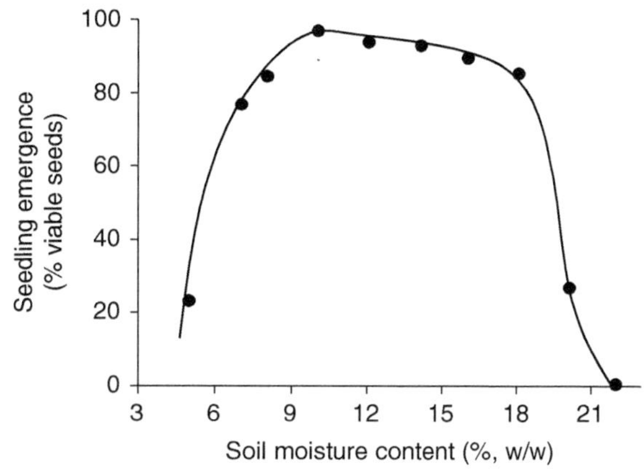

Figure 3.4 The relationship between moisture content of a sandy loam on seedling emergence of spring wheat. Mean of two soil temperatures from Khah et al. (1986).

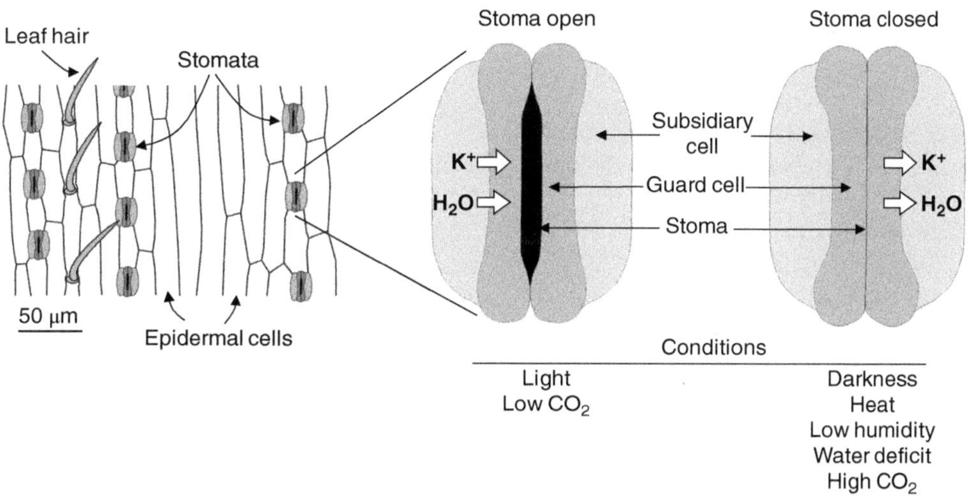

Figure 3.5 Diagrammatic view of the wheat leaf surface and functioning of the stomata. *Source:* After Dunn et al. (2019) and Nunes et al. (2020).

towards and through the roots, and into the above-ground biomass. However, much of the loss of transpired water is an apparently wasteful consequence of letting carbon dioxide enter the plant. The gatekeepers for the balance of *providing food* (ingress of CO_2) *while preventing thirst* are the stomata (Farquhar and Sharkey 1982; Moore et al. 1998).

The stomata (Figure 3.5) are small pores in the upper and lower leaf surfaces that can open and close depending on conditions. In wheat, the number and size of stomata are often negatively related to each other and depend on genotype and environment. Stomatal densities ranging from 40 to 100 mm^2, and mean apertures from 70 to 200 μm^2 are reported for the flag leaves of field-grown wheat genotypes by Shahinnia et al. (2016). On each side of the apertures are guard cells. In wheat these are elongated, dumbbell-shaped, and flanked by subsidiary cells (Figure 3.5). The elongated shape means that the aperture can reach a maximum area with relatively little lateral widening (Nunes et al. 2020). In wheat, as in other grasses, the speed of opening and closing of the stoma is helped by subsidiary cells working in tandem with the guard cells. There is a reciprocal ion (particularly potassium, K^+) shuffling between the guard cells and the subsidiary cells. To open the stomata, ions flow from the subsidiary cells into the guard cells. Water follows the ions by osmosis and the guard cells become turgid. When turgid, the guard cells bend due to their shape and because the cell walls adjacent to the aperture are thicker. To close the stomata the reverse happens. Ions and water move from the guard cells to the subsidiary cells and the guard cells become flaccid. In addition to supporting the ion transfer, the subsidiary cells also provide structural assistance to the guard cells as they also change shape during stomatal opening and closing. The stomata open in response to several stimuli including low CO_2 concentrations in the leaf and light. Conversely, closure occurs with high CO_2 in the leaf, darkness, low soil and air humidity, and high temperatures. Closure of stomata is particularly associated with the phytohormone abscisic acid, which is often induced in drought stress. The action of abscisic acid is modulated by the concentration of calcium ions (Ca^{2+}) (Mega et al. 2019; Nunes et al. 2020). Both potassium and calcium are, therefore, important in the control of the water relations of the plant.

The shape of the stomata and the combined movements and transfers of ions between subsidiary and guard cells mean that grasses, including wheat, can open and close their stomata relatively quickly compared to those in other crop families. This responsiveness may enable grasses to optimise the balance between letting CO_2 in, and water out, of the leaf when water is limiting. Nunes et al. (2020), for example, suggest that the shape and functioning of the grass stomata may have assisted the family in spreading and diversifying during a global dry period 30–40 MYA. With regards to wheat, the structure and functioning of the stomata may well contribute to its success in the drier areas that would otherwise support perennial grasslands.

If water availability is insufficient to meet the demands of transpiration the stomata close and photosynthesis is curtailed. Dry matter accumulation, cell expansion, and leaf turgor decline and the leaves can become flaccid or tightly rolled. If water stress continues, leaf size, tiller number, root growth, stem height, and stem thickness are all reduced, and the crop senesces prematurely (Campbell and Read 1968; Gallagher et al. 1979; Stark and Longley 1986; Assuero and Tognetti 2010).

The strength of the association between water uptake and yield in water-limited environments has supported the development of several measurement procedures that relate to crop transpiration. These include stomatal conductance, canopy temperature, and ^{13}C isotope discrimination (Brennan et al. 2007).

Stomatal conductance (g_s, $mmol/m^2/s$), describes the ease by which water exits, and CO_2 enters, the leaf. The conductance for water and CO_2 is linearly related, i.e. g_s for CO_2 is 0.625 of that of water (Fischer et al. 1998). At the leaf scale, conductance will be affected by the number of stomata, the maximum size of each aperture, and the extent to which the stomata are open. Stomatal conductance can be measured by porometers that assess the diffusion of water vapour from the leaf (Parkinson and Day 1980), or the resistance to air passing through the leaf (Rebetzke et al. 2003). There are linear relationships between stomatal conductance and photosynthetic rates, and sometimes also with wheat grain yield (e.g. Figure 3.6).

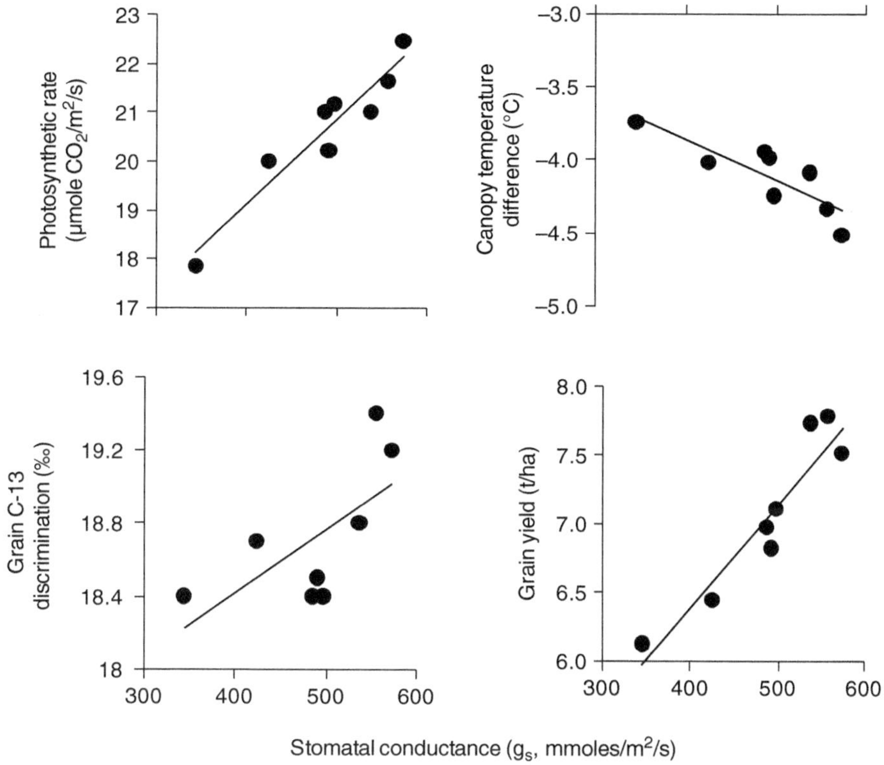

Figure 3.6 Relationships between stomatal conductance of the flag leaf, photosynthetic rate, grain ^{13}C isotope discrimination, and grain yield. Points are eight spring wheat cultivars. *Source:* Data from Fischer et al. (1998).

Canopy temperature (°C) is related to stomatal conductance because of the evaporative cooling imparted by transpiration. Canopy temperatures are readily recorded with infrared thermometers or thermal imaging cameras (Deery et al. 2016). The extent of evaporative cooling is often expressed as the disparity in temperature between the canopy and the air, i.e. the canopy temperature difference (e.g. Figure 3.6). In wheat, canopy temperature has been used to assess genotypic effects on root depth and function (Lopes and Reynolds 2010; Clarke et al. 2017a; Li et al. 2019). Canopy temperature can also be related to foliar disease severity (Dimmock and Gooding 2002).

^{13}C Isotope discrimination (Δ, ‰) is the extent to which the ratio of $^{13}C/^{12}C$ is less in the plant compared with that in the atmosphere, i.e.

$$\Delta = \left(R_a / R_p\right) - 1 \tag{3.1}$$

where R_a and R_p are the molar abundance ratios of $^{13}C/^{12}C$ in the air and plant, respectively (Hall et al. 1994). ^{13}C is a stable isotope of carbon but is 99 times rarer than ^{12}C. In wheat, as in other plants with the C3 pathway of photosynthesis, photosynthesis discriminates against ^{13}C. Wheat biomass, therefore, has a lower $^{13}C/^{12}C$ ratio than the atmosphere. The discrimination against ^{13}C increases as the partial pressure of CO_2 in the leaf (p_i) increases relative to that in the atmosphere (p_a). Δ can, therefore, be related to stomatal conductance (Figure 3.6b) although its use has been particularly advocated to identify wheats that can increase biomass for a given amount of transpired water when water is limiting (Hay and Porter 2006). The determination of the ratio of ^{13}C to ^{12}C requires mass spectrometry, but it has the advantage over temperature and porometer measurements in that analysis of mature material integrates effects over the lifetime of the crop (Hall et al. 1994).

Although the stomata are the gatekeepers of transpiration, the total demand for water to maintain the transpiration stream will depend on the size of the canopy and the evaporation potential of the atmosphere. In practice, measurements and calculations of the water lost through transpiration and evaporation are combined as evapotranspiration. The effects of canopy cover and meteorological conditions on crop evapotranspiration (E_t) are captured in the Penman-Monteith equation (Figure 3.7). This sets the upper limit for evapotranspiration subject particularly to roots accessing available water. Allan et al. (1998) provide details of how the Penman-Monteith equation is derived and how it can be used to predict evapotranspiration from solar radiation, air temperature, air humidity, and wind speed. These calculations allow us to compare water demands and soil moisture deficits over different times and environments, and to inform the timing and quantities needed for irrigation (Section 3.2.3).

3.1.3 Root Growth

As for the above-ground structures, water is required to maintain root cell turgor and expansion and wheat roots have mechanisms that enable them to grow towards water. Firstly, they are positively hydrotropic in response to a water potential gradient (Oyanagi et al. 1995), i.e. they grow towards higher water concentrations. Secondly, lateral root formation is increased in areas where water is available (and supressed otherwise). Thirdly, when water is scarce, wheat roots become more gravitropic, i.e. they have a greater

$$E_t = \frac{sR_n[1-e^{-kL}] + \rho\, C_p/r_a\,(e^* - e)}{[s + \gamma(1 + r_s/r_a)]}$$

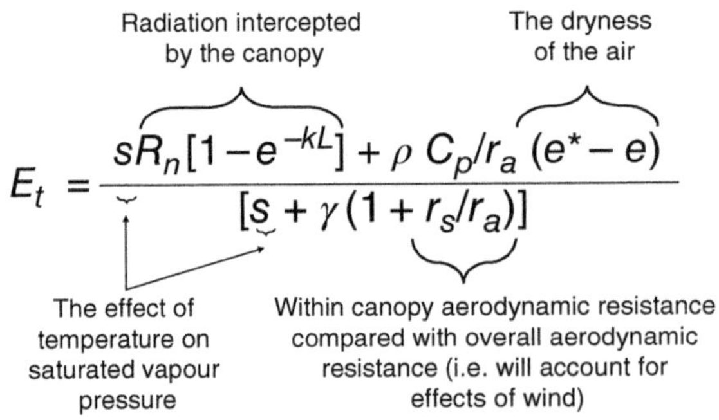

Figure 3.7 The relationship between crop evapotranspiration (E_t) and the size of the canopy and meteorological conditions as captured in the Penman-Monteith equation (Monteith 1965). s is the slope of the saturated vapour pressure temperature curve, R_n is the net irradiance, k and L are the light extinction coefficient and leaf area index of the crop (see Chapter 5), C_p is the air heat capacity, r_a is the aerodynamic resistance, e^* is the saturated vapour pressure, e is the actual vapour pressure, λ is the latent heat of vaporisation of water, γ is the psychometric constant, and r_s is the within canopy aerodynamic resistance (units and further explanation available in Hay and Porter 2006).

tendency to grow downwards to, hopefully, wetter soil (Hamada et al. 2012). Often, therefore, droughts can reduce root growth in the topsoil but increase root prevalence in deeper layers (Barraclough et al. 1991). The strength of the gravitropic and hydrotropic responses can vary among wheat genotypes and these varying and interacting effects can contribute to differences in wheat root architecture. Different root architectures and responses allow for adaptation to temporal and spatial variation in water availability. Furthermore, the roots have an important signalling role. Soil drying can increase the concentration of abscisic acid in the root such that growth and root hydraulic conductivity can continue thus delaying stress in the shoots. Abscisic acid is also transported in the xylem to the shoot, where it causes stomatal closure and reduces leaf expansion, thereby preventing the dehydration of leaf tissues (Barnabás et al. 2008).

3.1.4 Reproductive Growth and Grain Filling

The transition from the vegetative stage and the subsequent development of the ear structures can all be significantly damaged or delayed by drought (Saini and Westgate 1999). The growth stage most dependant on adequate moisture is meiosis. Drought can damage both male (pollen) and female (ovule) functioning, but most effects are on the pollen. Male meiosis occurs in the pollen mother cells of the anthers. This produces four haploid cells within a tetrad. After about 10 hours the tetrad breaks up to produce 4 cells that each develop further into pollen grains (Bennett et al. 1971). It is this period, from the start of meiosis to tetrad break up, that is particularly vulnerable to moisture stress (Saini and Westgate 1999). This leads to poor pollen fertility and therefore reduced grain set at anthesis. The anthers often become shrivelled, but even normal-looking anthers can contain sterile pollen that lack starch. Some of the effects on pollen are associated with them losing contact with the tapetum, a lining of cells on the inside of the anther walls that is important for the nutrition and development of pollen grains. This disruption to pollen development does not appear to be in response to lack of turgor as the water potential of the reproductive structures appears to be highly conserved during periods of stress. It is possible that the decline in fertility is associated with abscisic acid signalling and/or the ability of anthers to use sucrose (Morgan 1980; Westgate et al. 1996; Saini and Westgate 1999).

At the crop level, meiosis may occur over a period of up to five days in a wheat field. This reflects a range in timings between and within plants, tillers, spikes, and spikelets (Barber et al. 2015). Broadly, meiosis will be occurring during the booting growth stages, i.e. from the end of flag leaf emergence to the beginning of ear emergence (Fernández-Gómez et al. 2020). Grain number can be markedly reduced by water stress during this period (Figure 3.8). Grain yields are also reduced because any subsequent increases in mean grain weight are insufficient to fully compensate for the poor grain set (Figure 3.8). The effect of moisture stress at this stage is greatly exacerbated by heat stress. Conversely, well-watered crops are protected against hot conditions (Figure 3.9).

Water stress during anthesis can still reduce grain number but by this stage a reduction in yield would already be expected due to damaging effects on mean grain weight (Figures 3.8 and 3.9). Grain filling with dry matter appears to be highly dependent on water relations from shortly before anthesis until physiological maturity (Gooding et al. 2003).

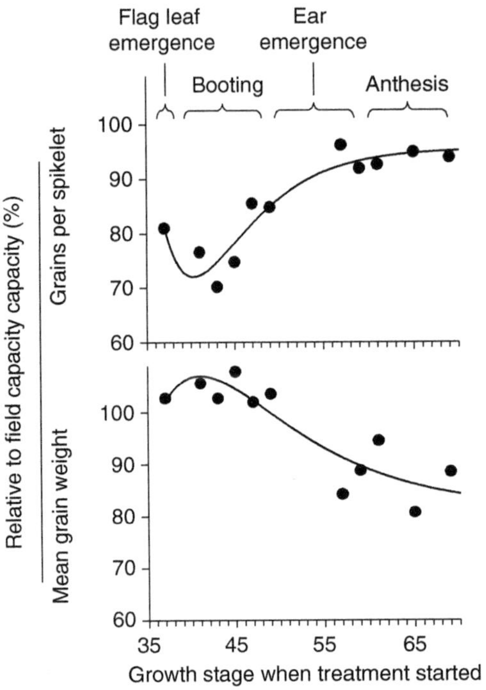

Figure 3.8 The effect of withholding water for three days starting at a range of growth stages (Zadoks et al. 1974) on grain number and mean grain weight of winter wheat. Further details in Barber et al. (2015).

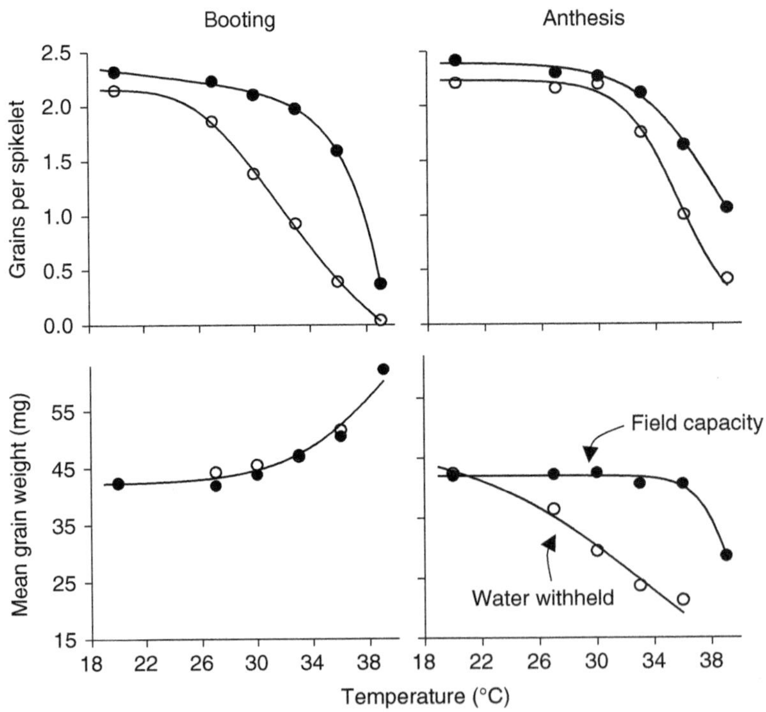

Figure 3.9 Effect of day temperature (16/8 h day/night cycles, night temperature 8 °C below day temperature) and water availability during three-day transfers to controlled environment cabinets during booting and anthesis on grain numbers and mean grain weight. *Source:* Based on Semenov et al. (2014).

Grain development can be divided in to three phases determined by the flux of water in and out of the grain (Figure 3.10). In Phase 1 the caryopsis expands with the rapid ingress of water. Water stress during this phase is particularly damaging to final mean grain weight and yield (Figure 3.11). In particular, there is an increase in shrivelled grain with a decrease in specific weight and an increased concentration of protein (i.e. nitrogen). The dramatic effect of drought early in grain filling, even when ample water may be available later, can be explained by effects on the maximum volume of water accumulated per grain by the end of Phase 1. Larger volumes of water per grain are associated with both longer durations and faster rates of grain filling with dry matter, and therefore with final mean grain weight (Figure 3.12).

In Phase 2 the volume of water in each grain changes little while there is an approximately linear accumulation of dry matter. Water stress during Phase 2 is not as damaging as it is in Phase 1 (Figure 3.11), but grain filling duration, final grain weight, and grain-specific weight can all still be reduced. These reductions are approximately linear; in proportion to water deficits below a threshold value (Figure 3.13). During this phase, however, Hagberg Falling Number (HFN) and grain protein concentration (Chapter 1) appear to benefit from drought compared to values obtained at field capacity (FC) (Figures 3.11 and 3.13). The lower HFN at FC does not result from wetting of the ear, because in these experiments the water was applied directly to the soil (Gooding et al. 2003). It should be remembered, however, that wheat will rarely be growing in soils at FC during grain filling in commercial production, where an increasing terminal drought is much more likely. It is

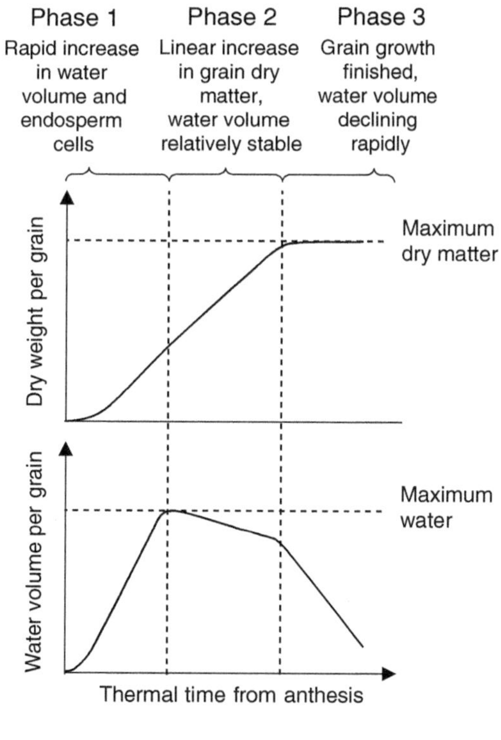

Figure 3.10 The phases of wheat grain growth as delineated by changes in water volumes. *Source:* Based on Pepler et al. (2006).

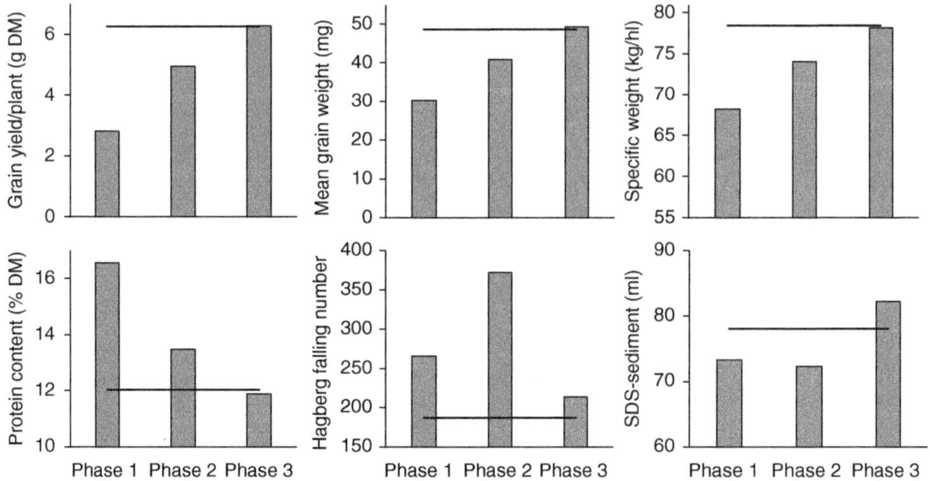

Figure 3.11 The effect of reducing water availability to 44% of field capacity during the different phases of grain growth (See Figure 3.10) on grain yield and quality. Solid horizontal bar is the result of maintaining field capacity throughout grain filling. *Source:* Data from Gooding et al. (2003).

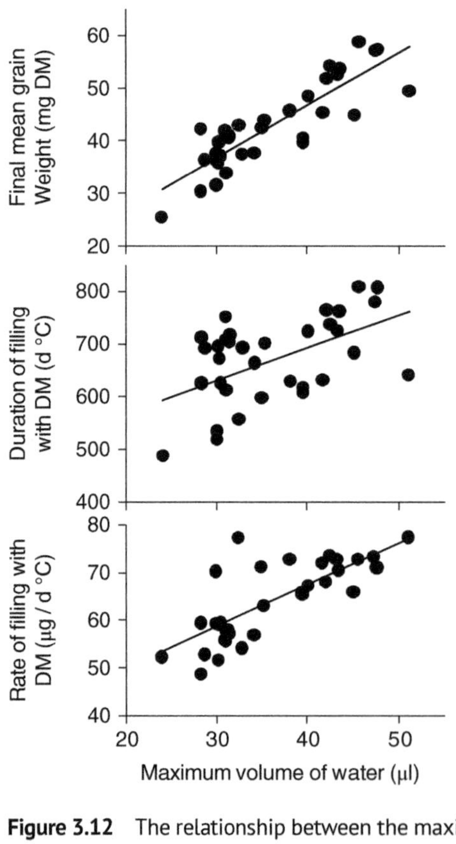

Figure 3.12 The relationship between the maximum volume of water attained in a grain and the filling of that grain with dry matter. Points are for winter wheat from a range of environments, genotypes, and positions within the ear. *Source:* Data from Pepler et al. (2006) and Gooding et al. (2003).

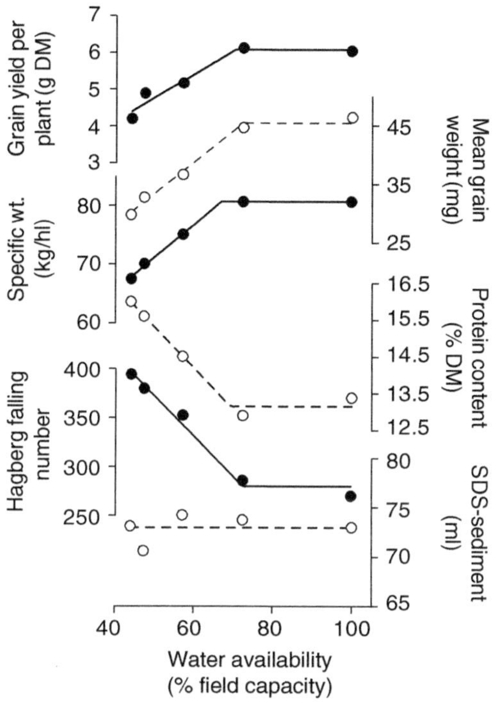

Figure 3.13 The effect of soil moisture deficits for 14 days during the linear phase of grain growth on grain yield and quality of winter wheat. *Source:* Data from Gooding et al. (2003).

possible that wet soils and associated signalling contributed to the production of prematurity α-amylase, which appears to be particularly sensitive to the environment in Phase 2 (e.g. Derkx and Mares 2020).

The beginning of Phase 3 is defined as the end of movement of water and dry matter into the grain. The grain moisture concentration of the grain at this stage can vary but it averages around 40% (Gooding et al. 2003). In Phase 3 the water volume declines rapidly as the grain dries and ripens before harvest.

3.2 Available Water

The water that is available to wheat crops includes all the water available from the soil at sowing to the maximum rooting depth, plus any water during the growing season from rainfall and/or irrigation.

3.2.1 Soil Water

Within limits, the soil acts as a repository of water from previous and concurrent rainfall and irrigation. The upper limit of this storage capacity is the maximum amount of water that a soil can retain against the pull of gravity. This is the FC of the soil and corresponds to a soil water potential of approximately zero (technically, 0.01–0.03 MPa is applied in its derivation).

Table 3.1 Approximate water holding capacities (mm) and equivalent moisture contents[a] (%, w/w) for different soil textures.

	Field capacity		Permanent wilting point		Plant available water
Soil texture	**mm/m**	**%**	**mm/m**	**%**	**mm**
Coarse sand	60	5	20	2	40
Fine sand	100	8	40	3	60
Loamy sand	140	10	60	5	80
Sandy loam	200	14	80	6	120
Light sandy clay loam	230	16	100	8	130
Loam	270	18	120	9	150
Sandy clay loam	280	19	130	10	150
Clay loam	320	21	140	10	180
Clay	400	25	250	17	150

[a] Calculated from the water holding capacities, assuming a bulk density of 1.2 t/m^3.

Rainfall or irrigation of a soil that is already at FC will cause leaching unless inadequate drainage leads to increasing pore saturation, waterlogging, and/or surface run-off.

Most crop plants, including most types of wheat, can continue to extract water from the soil until soil water potential falls to about −1.5 MPa. At this level plants have reached their permanent wilting point (PWP) and water has been extracted from soil pores larger than 2 μm in diameter (Hay and Porter 2006). The difference between FC and the PWP is the maximum amount of water that is available to plants that can be held by soil, i.e. the plant available water holding capacity (PAWC). Maximum PAWC will depend on the maximum depth of the rooting zone and soil texture, structure, and organic matter. Table 3.1 shows typical values of PAWC of different soil textures assuming adequate structure and average organic matter contents. The lighter soils hold less water at FC, but a large proportion of that water is available because there are few pores with diameters less than 2 μm. Clay soils have a larger FC but a smaller proportion of this is available because there are many pores less than 2 μm diameter.

Irrespective of the larger amount of water retained at PWP, the PAWC is still greater in medium to heavy soils. PAWC is particularly important for wheat yields in semi-arid areas when the soil has been used to *store* water from a preceding fallow (De Jong et al. 1984; Hunt and Kirkegaard 2011); when irrigation is mostly or only available at the start of the growing cycle; and/or when late-season rainfall is sparse (Lawes et al. 2009). In more humid areas higher PAWC can mitigate the effects of intermittent dry periods and delay the onset of late-season terminal droughts.

3.2.2 Rainfall

Of all arable crops wheat covers the greatest land area that is solely rainfed (1.45 million km^2). Wheat grown in this area depends entirely on the water stored in the soil, and on the timing and quantity of natural precipitation. Rainfall, therefore, has a major influence on wheat yields at the global, national, and sub-regional scales. This is not only due to effects of water shortage in droughts, but also the effects of waterlogging, floods, and erosion when water is in excess (Zampieri et al. 2017).

The projected impact of global warming on precipitation is complex and may lead to increases in both droughts and flooding (IPCC 2014). Warmer air can hold more water. Global precipitation is, therefore, likely to increase by 1–2% per degree of warming. Rainfall events are, however, also projected to become more intense but less frequent and more erratic. It is likely that drier areas, particularly in the low to mid latitudes, will become drier. Droughts already appear to be becoming more frequent (Mallya et al. 2016); a trend that is predicted to continue in climate change scenarios for many wheat-growing countries (e.g. Mpelasoka et al. 2008; Warren et al. 2009; Strzepek et al. 2010). By contrast, humid areas, particularly in higher latitudes, are projected to become wetter. Climate change scenarios are also predicted to exaggerate seasonal variation in rainfall distributions. Over much of Europe, for example, it is anticipated that wetter areas will become wetter, particularly in winter, while drier areas will become drier, particularly in summer (EEA 2017). Ruane et al. (2018) provide predictions of changes in the quantities of precipitation within wheat-growing seasons across the globe.

3.2.2.1 Rainfall Shortage

Insufficient rain in the major wheat-growing areas has for decades been associated with low yields and food insecurity (e.g. Zhang et al. 2017). Water shortage is, by definition, particularly likely in the semi-arid areas, i.e. areas where annual rainfall is insufficient to meet potential evapotranspiration. Clearly this definition depends on temperature, irradiation, and wind, as well as rainfall and its temporal distribution. Broadly, however, semi-arid areas commonly encompass regions that receive between 250 and 500 mm of rain per annum. This would include much of the world where vegetation would naturally be grassland and therefore may alternatively be described as steppe climate (Peel et al. 2007).

Even in relatively humid areas with well-watered crops (> 500 mm annual rainfall, mostly in the growing season), such as the UK, it is estimated that on average 10% of potential wheat production is lost because of insufficient water during one or more periods of development (Foulkes et al. 2007). For example, 20% of UK wheat crops are likely to have grain filling curtailed prematurely due the early development of a terminal drought.

As would be expected, in water limited environments rainfall is positively correlated with wheat yield. The size of the rainfall effect depends on timing, soil type, and quantity of stored water (e.g. Cornish 1950; Gangopadhyaya and Sarker 1965; Balaghi et al. 2008; Latiri et al. 2010; Basso et al. 2012). Rainfall just prior to, or shortly after sowing supports germination and crop establishment. In some semi-arid areas, the lack of early rain may cause delayed establishment or crop failure. If the soil is dry and no rain is forecast, some farmers may decide not to proceed with sowing. Rainfall during the growing season will support both vegetative and reproductive growth so that grain numbers and mean grain weight are increased. Conversely, deficits can significantly reduce responses to other inputs, particularly that of nitrogen fertilizer (Camara et al. 2003; Figure 3.14).

As well as total rainfall, the distribution of rainfall during the season is important. Limon-Ortega and Sayre (2012) ascribe poor yields of wheat to more erratic distribution of rainfall, as well as to less rainfall in total. Specifically, they concluded that low rainfall around the start of stem extension and/or during grain filling was particularly damaging.

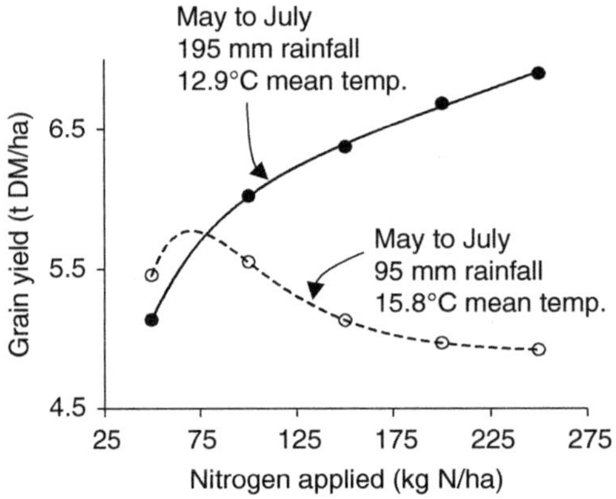

Figure 3.14 Effect of weather on response of winter wheat grown on a shallow brash over oolitic limestone to nitrogen fertilizer. *Source:* From data in Smith and Davies (1990).

3.2.2.2 Rainfall Excess

Excessive rain is frequent in more humid environments and is particularly problematic on medium to heavy soils, where correlations between annual rainfall and yield are often negative (Fisher and Russell 1997; Addy et al. 2020). Depending on timing, excessive rain can be associated with delayed sowing, nutrient leaching (Goulding et al. 2000), increased pressures from pests and diseases (Wiik and Ewaldz 2009), slower soil warming in spring, lodging (Easson et al. 1993), and delayed harvest (Bracken and Bailey 1928). Increased rainfall during the vegetative and stem extension stages can be associated with reduced grain protein concentration, presumably due to increased leaching of nitrogen and other forms of N loss, as well as dilution of protein with carbohydrate if canopy size and duration is increased (e.g. Smith and Gooding 1999). Excessive rainfall before harvest can reduce grain-specific weight, possibly due the weathering and cracking of grain (Bracken and Bailey 1928; Jošt et al. 1997; Atkinson et al. 2005). HFN is also reduced with wet weather close to harvest (Smith and Gooding 1996; Kettlewell et al. 1999; Yadav and Ellis 2017), as damp conditions around the ears in standing or lodged crops increase the risk of pre-harvest sprouting (Mares 1993).

Excessive rainfall is one of the factors, together with poor drainage and irrigation, that contributes to waterlogging. Waterlogging is estimated to affect between 15 and 20% of the global wheat area at some stage during the cropping season (Setter and Waters 2003). Waterlogging is particularly damaging to the early stages of emergence when seedlings can be killed. Later waterlogging reduces the growth of roots, which is associated with the development of anaerobic conditions. Brisson et al. (2002) determined a critical oxygen level of 0.12 mol/m^3 water in saturated soil. The rate of oxygen depletion, and therefore effects on root growth, will depend on many factors but will generally be more rapid at higher temperatures and for soils with a high biological activity (Setter and Waters 2003). The effects of waterlogging will also depend on its duration, and the proportion of the root system below the water table. In severely waterlogged wheat, the stomata close, photosynthesis declines, nutrient uptake is reduced, leaf lengths decline, fewer tillers are produced, and leaves become yellow and senesce prematurely (Belford 1981; Malik et al. 2002). Sometimes waterlogging in winter and spring of autumn sown wheat appears to exacerbate the effects of terminal droughts and curtail grain filling (Belford 1981). This interaction does not always occur (Dickin and Wright 2008), but when it does it may be due to the lasting effects of earlier waterlogging on reducing root growth and/or increasing stem base diseases.

3.2.3 Irrigation

In addition to the large area of rainfed wheat, the crop is also second after rice in the area that is irrigated (0.65 million km^2) (Portmann et al. 2010). Irrigation is often used in the semi-arid areas. For example, in the Great Plains of the US, groundwater from the Ogallala aquifer has been a major source of irrigation. The Ogallala stretches from northern Texas to South Dakota and has been used to help counter water deficits, droughts, and associated dust bowls since the disasters of the 1930s (Deines et al. 2020). Eighty-five percent of agriculture above the Ogallala now relies upon it for irrigation (Hudson 2020).

It is also common practice to irrigate wheat in more humid areas where annual rainfall may be high but where wheat is mainly grown in the dry season. In the North China Plain, for example, annual rainfall often exceeds 500 mm, but only 100 to 180 mm may fall in the wheat-growing period (Zhang et al. 1998). Similarly, in areas around and in the Indo-Gangetic Plain of India and Pakistan, paddy rice is grown in the summer with monsoon rains. The rice is then followed by a wheat crop, which needs irrigating in the dry winter. In India, long-term governmental policies have encouraged an expansion of the irrigated areas to increase wheat yields, increase food security, and reduce rural poverty. The country is responsible for 13% of the world's wheat supply and over 90% of that production is now irrigated (Zaveri and Lobell 2019).

As with rainfall, irrigation can increase crop establishment, ear numbers per plant, grain numbers per ear, and mean grain weight, depending on timing and prevailing water deficits (Gajri and Prihar 1983; Zhang et al. 1998; Clarke et al. 2004; Zaveri and Lobell 2019). Also, like rainfall, irrigation appears to be particularly important for reducing stress from the start of stem extension through to the early stages of grain filling (e.g. Zhang et al. 1998). In comparison to purely rainfed wheat, however, access to irrigation can lead to more stable yields, particularly in the face of increasingly erratic rainfall distributions. The stabilising effect of irrigation on yield was emphasised by Stanhill (1986). He compared the yields of wheat in Israel, which was mostly rainfed during the 1960s to 1980s, with those in Egypt, which was mostly irrigated. Not only were the yields in Egypt greater than in Israel, they were also much less variable. If the period is extended to 1990, as in Figure 3.15, the rate of yield increase was also greater in Egypt, signifying greater responses to improved cultivars and other inputs.

In addition to alleviating water stress, irrigation is also now recognised as having a significant role in alleviating heat stress across major wheat-growing areas (Tack et al. 2017; Zaveri and Lobell 2019).

Within field experiments, the grain yield response to increasing amounts of irrigation water is approximately quadratic (Figure 3.16). Wheat is more efficient at using irrigation water when water availability is low. In such circumstances, incremental increases in irrigation provide large grain yield responses, associated with increases in leaf area index and

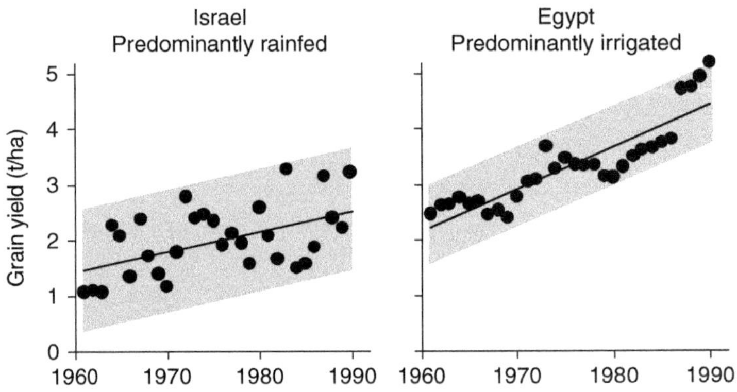

Figure 3.15 Progression in national wheat yields in neighbouring counties with contrasting systems of wheat production. Shaded areas are +/− two standard deviations from the trendline. *Source:* Data from FAOSTAT.

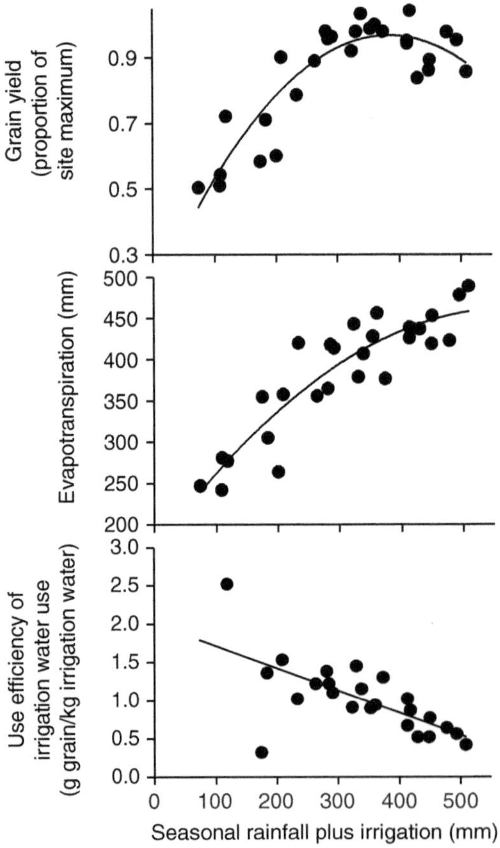

Figure 3.16 The response of wheat grain yield to increasing irrigation water. Values are from several sites and seasons, rescaled according to maximum yield for the site, and the season rainfall plus irrigation that gave the maximum yield. Mean seasonal rainfall for the sites was 123 mm. *Source:* From data in Zhang et al. (1998).

evapotranspiration. This rate of response declines as grain yield reaches a maximum with optimal levels of irrigation. Supra-optimal amounts of irrigation can lead to yield reductions, even though evapotranspiration is still increasing (Figure 3.16). These yield reductions are associated with increasing risks of several hazards associated with irrigation. Irrigating wheat can increase foliar and soilborne diseases (Olesen et al. 2000; Bithell et al. 2013). Saturating the soil, particularly after anthesis, renders the crop susceptible to lodging (Tripathi et al. 2005). High rates of irrigation, particularly when concentrated in a restricted number of doses, can lead to leaching of nutrients (Fang et al. 2006) and/or nitrous oxide (N_2O) emissions (Mahmood et al. 1998; Chapter 6). Waterlogging frequently impedes wheat growth in irrigated soils (Setter and Waters 2003), particularly in rice-wheat systems where paddies have been formed for the rice by compacting soil (Hossain and Uddin 2011).

The extent to which irrigation can be optimized depends on the method used and therefore the level of control possible; the cost, availability, and reliability of water supply; and interaction with many other management factors such as crop nutrition, crop protection, cultivar choice, soil type, and drainage. The methods of irrigation include surface, overhead, and subsurface techniques.

3.2.3.1 Surface Irrigation

Surface irrigation usually involves temporary flooding of crop areas with water supplied via field channels. Water can be fed from the field channels into the cropped areas by siphons, spiles (small pipes from the channel to the field), breaching the wall of the channel, or by pumping water from the field channel into pipes (Figure 3.17). *Basin* irrigation refers to where water in the crop area is constrained in a basin by bunds on all sides. It is unsuitable for sloping land, sandy soils, or soils that cap. *Border* irrigation is where the water flows down a gently slopped cropping area, constrained by bunds on either side of the direction of travel. The flow from the field channel is stopped when the water has reached, or is close to, the bottom of the cropped area. *Furrow* irrigation is where water is directed between rows or beds of wheat. At its simplest, wheat may be broadcast or drilled on the flat field, after which furrows are made in the soil at about 90 cm intervals. In this system, many plants will establish from within the furrows. A development of the furrow system is to drill 2–3 rows of wheat into raised beds (15 to 30 cm high, 70–90 cm wide), such that water is directed down the furrows between the beds (Sayre and Moreno Ramos 1997).

3.2.3.2 Overhead Irrigation

Overhead irrigation of wheat usually involves rain guns or sprinkler systems. Sprinklers can be attached to overhead booms that travel over the crop. For example, pivot systems irrigate in a circular pattern around a central point. Centre pivot irrigation is a common feature above the Ogallala aquifer in the US (Evett et al. 2020). Such systems usually provide more control and accuracy than surface flood irrigation methods. Given the required investment in infrastructure, reliability of water pressures, and maintenance, sprinkler systems can provide water more uniformly and more regularly in defined amounts to meet the demands of evapotranspiration. It is also possible to more precisely target water availability to responsive growth stages. Evapotranspiration can be estimated by calculation or observed as evaporation from uncovered pans (Liu et al. 2011). There are several examples,

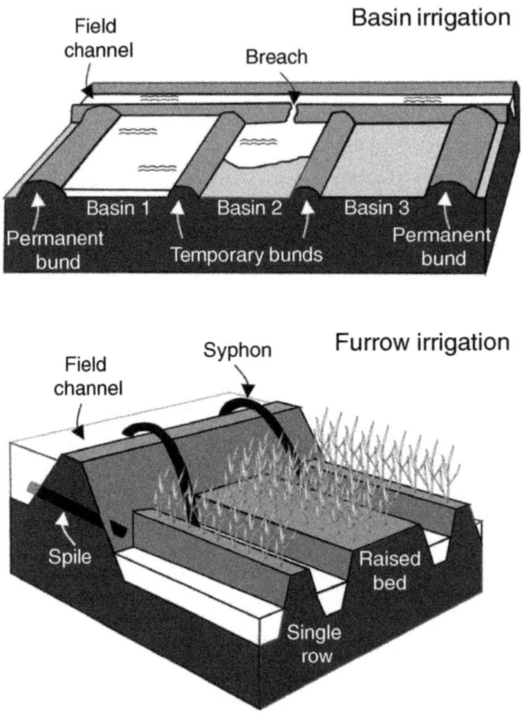

Figure 3.17 Components of surface irrigation systems. *Source:* Adapted from Brouwer et al. (1988).

therefore, where the adoption of sprinkler technology has increased wheat yields and/or the WUE compared with surface flood irrigation methods (Kahlown et al. 2007; Liu et al. 2011; Evett et al. 2020).

3.2.3.3 Sources of Irrigation Water

Despite the benefits of irrigation for higher and more stable yields, irrigated areas are heavily reliant on sustainability of supply and application, and subject to competing demand for water from other sectors. Evapotranspiration of irrigation water from crop lands represents around 90% of the water used by humans (Siebert et al. 2010). This irrigation can come from groundwaters or surface waters.

Groundwater accounts for 43% percent of water used in irrigation (Siebert et al. 2010). Water is stored as aquifers in porous rocks and sedimentary layers. These groundwaters are a vast store of freshwater, accounting for about 30% of the global freshwater resources. This compares with about 69% in glaciers and icecaps and 1% for all other sources (ground ice and permafrost, lakes, rivers, swamps, marshes, and the atmosphere). Water for irrigation schemes is pumped from the aquifers and directed through a series of channels or pipes to farmers fields. The largest areas equipped for groundwater irrigation are in India (39 million ha), China (19 million ha), and the US (17 million ha). Other important areas include the south of Spain, and western and southeastern Australia.

However, in most of these areas, water has been extracted at a significantly higher rate than the rate of groundwater recharge, leading inevitably to falling water tables

(Hu et al. 2005; Rodell et al. 2009; Hudson 2020). Increased extraction is commonly associated with an increased area of wheat production, which is often exacerbated by reduced rainfall. The unsustainability of water supply from groundwaters is, therefore, a major threat to continued intensive wheat production. In the North China Plain, for example, 70% of the groundwater irrigation is used for winter wheat. Since the 1970s, however, water tables have dropped by 20–40 m. In addition to the increasing difficulties of water extraction, the loss of water has been associated with significant land subsidence and proliferation of severe fissures (Wang et al. 2016). In the Punjab of India, under the rice-wheat system, the water table fell from 6 m in 1980 to 20 m in 2013. The rate of fall of the water table is also increasing, reaching 0.45 m/yr between 2008 and 2013 (Jalota et al. 2018). Similar rates of fall have been estimated for areas of the Ogallala aquifer in the US. It has been estimated that Ogallala aquifer has been depleted by about 33% over the last 50 years, with a further third projected to be lost by 2050 (Hudson 2020). When water tables decline, the continued use of groundwater is associated with significant increases in the energy and cost of water extraction and pumping (e.g. Jalota et al. 2018) and once aquifer saturated thickness is less than 9 m, irrigated crop production becomes impractical. Twenty-five percent of the Ogallala aquifer had reached this level in 2012 (Deines et al. 2020) and the irrigated area in the US Great Plains is likely to decline (Evett et al. 2020). Various schemes and policies have been proposed to reduce groundwater depletion by reducing cropping intensity, including cessation of wheat cropping in severe cases (e.g. Wang et al. 2016).

In addition to issues of depletion, there are also sustainability concerns over the agricultural pollution of groundwaters. This is particularly relevant when those waters are used directly for human drinking and cooking water (Skevas 2020). Increasing intensity of cropping, including cropping of wheat, has been associated with contamination of ground waters with, for example, nitrate (Chandna et al. 2011), herbicides (Battaglin et al. 2000), and insecticides (Kumari et al. 2008). Additionally, the agricultural depletion of aquifers near coasts can draw incursion from saltwater (Skevas 2020). Preventing agriculture from polluting groundwaters has, therefore, been addressed in several governmental policies and programmes. For example, in the US the Environmental Quality Incentives Program, and in the EU the Water Framework Directive (WFD) (Directive 2000/60/EC) and the Groundwater Directive (Directive 2006/118/EC) seek to prevent or limit inputs of pesticides and nitrates into groundwater.

Surface waters include rivers, springs, wetlands, and lakes, which form watersheds or drainage basins, usually discharging into the sea. Agricultural activity has often involved retaining, redirecting, and management of surface waters for crop production using hydraulic structures including dams, reservoirs, dykes, levees, canals, pumping systems, ditches, and drains. These structures are organised into irrigation schemes that require significant planning, investment, construction, and management. Additionally, they have major effects on landscape, environment, and the economies, health, and wellbeing of human societies within and beyond agriculture. Irrigation schemes, therefore, require considerable financial investment, governance, and political power (Hudson 2020).

The availability of surface waters is more immediately related to rainfall than are groundwaters, but dams and reservoirs buffer the effects of erratic rainfall, and the redirection of waters allow increased productivity on otherwise barren or marginal land. Dams, however, can also have significant environmental costs. The initial creation of reservoirs can flood areas important for biodiversity and local livelihoods and the effects on total flow and its temporal distribution impact upon the erosion and ecological functions of rivers. Silting behind dams and of beds beneath rivers with reduced flows have major impacts on downstream geomorphology. Damming rivers also hinders the downstream and upstream migration of aquatic organisms.

3.3 Water Use Efficiency

Water is a significant resource for wheat, but there are several threats to its continued availability to sustain production in many wheat growing areas. There is therefore a need to explore how we might reduce water losses and to maximise the amount of grain produced per kg of available water. For agronomic purposes, this is the broad definition of *water-use efficiency* used here (Mbava et al. 2020). In some texts, a narrower definition of WUE is used, i.e. the amount of biomass produced per kg of water transpired. Here we refer to this latter definition as *transpiration efficiency* (as per Figures 3.2 and 3.3).

3.3.1 Reducing Evaporation Losses

Early in crop growth, there is the possibility that water will be lost as evaporation from the soil. The extent of this loss depends greatly on location and prevailing conditions. For winter wheat production in northern latitudes, the evaporation of water from the soil may be less than one-tenth of that transpired by the crop. This is because evaporation will be low when the soil is exposed during the wet, cold, and dark conditions of winter. In the following spring and summer soil evaporation will be small because most of the radiant energy will be intercepted by the crop canopy. However, in the drier environments of lower latitudes evaporation from the soil can be significant, depending on when the water is available relative to crop growth. When canopy cover is low and/or uneven (Zhang et al. 1998; Zhang et al. 2004), the amount of water lost through soil evaporation can at least equal that which is transpired by the crop (Figure 3.3). Methods to reduce water loss through soil evaporation include those already discussed in Chapter 2, such as zero cultivation and the retention of residue mulches on the soil surface. Additionally, early, and complete canopy cover is beneficial. Early canopy cover can be assisted by ensuring adequate sowing rates into soil with sufficient soil moisture, the even establishment of vigorous seedlings, adequate nutrition and crop protection, and the use of appropriate genotypes (e.g. Zhang et al. 1998; Mullan and Reynolds 2010). Even where quick and complete canopy cover is achieved, however, evaporation of any rainfall intercepted by the canopy still represents a loss of available water (Leuning et al. 1994).

3.3.2 Increasing Rooting at Depth

The ability of wheat plants to access soil water will depend on rooting depth and distribution over time. In well-structured and deep soils, it is common for roots to reach 2 m deep by anthesis. Reports vary as to whether roots stop growing, degenerate, or continue to grow after anthesis (Evans et al. 1975). Much will also depend on the availability of moisture and nutrients. Gregory et al. (1978, 2005) found that after anthesis roots could senesce in the top 30 cm but continue to grow at deeper levels.

The ability of wheat to access water depends on the root length density (L_v), i.e. root length per unit volume of soil, commonly expressed as cm/cm^3. The proportion of available water that can be extracted (ϕ) is approximately related to L_v as in:

$$\phi = 1 - e^{-kL_v} \tag{3.2}$$

where k is the water capture coefficient such that larger values of k indicate more rapid uptake per unit L_v. A value of $k = 2$ appears reasonably appropriate for wheat, which implies that a root length density of less than 1 cm/cm^3 will significantly reduce water capture (King et al. 2003). It is often assumed that the root length density of wheat declines exponentially with depth, as observed in Figure 3.18. This pattern can be greatly modified by variation with depth in soil structure and strength, moisture content, and nutrient availability (Hoad et al. 2004; Gao et al. 2016). Nonetheless, the general pattern is for root length densities to decline with depth, associated with an increase in soil strength (Gao et al. 2016). Wheat root length densities are, therefore, often less than 1 cm/cm^3 below 40 cm depth (Hoad et al. 2004). Consequently, it is quite possible for available water to be exhausted in the upper 30–40 cm, whilst water from deeper down remains underutilized (Passioura 1983; Zhang et al. 2004). Inability to access deeper water efficiently has clear implications for stages of crop development (meiosis, anthesis, and grain filling), which are vulnerable to drought and heat stress. Improving water extraction with appropriate root architecture is one way to mitigate some of the effects of climate change (Senapati et al. 2020). Mechanisms to improve rooting at depth include: reducing impediments to rooting such as acidity, salinity, and early-season waterlogging; reducing seed rate (Hoad et al. 2004); reducing the frequency of irrigation (Li et al. 2010); and adoption of appropriate cultivars (Manschadi et al. 2006; Ford et al. 2006; Clarke et al. 2017b).

Figure 3.18 The distribution of wheat roots down a soil profile at anthesis. *Source:* Data from Gregory et al. (1978).

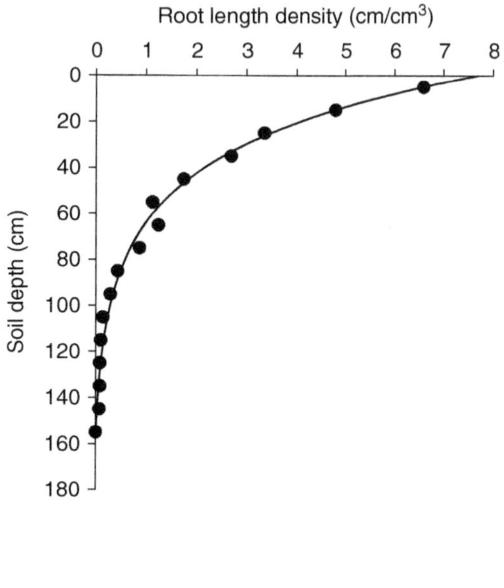

3.3.3 Deficit Irrigation

It is a common observation that as irrigation is increased to optimise grain yield, the efficiency with which the extra water increases grain yield declines (e.g. Figure 3.16). The grain yield per mm of available water will, therefore, be greatest at a lower level of water application than that required to maximise grain yield per unit area (Pereira et al. 2002). Consequently, there is a conflict between maximising WUE and maximising land-use efficiency and in areas where water availability is limited it may be beneficial to optimise the yield per unit water applied rather than maximise yield per unit area (Zwart and Bastiaanssen 2004; Alghory and Yazar 2019). These approaches are known as *deficit* irrigation strategies, i.e. the total amount of water applied is likely to be sub-optimal for grain yield. In such situations water is applied judiciously to protect the crop at critical growth stages (Alghory and Yazar 2019) and to allow for any rainfall that may accrue between irrigations. In this way, it is attempted to maximise WUE whilst minimising the yield penalties associated with water stress. Nonetheless, in a review of 41 studies on wheat, Yu et al. (2020) reported that deficit irrigation strategies had increased WUE by 6.6% but reduced grain yield 16.2%. However, the authors noted that yield penalties could be significantly reduced by further optimizing the systems. Deficit irrigation will be a rational consequence when the costs associated with irrigation are being borne by farmers: i.e. the economic margin over irrigation cost will be maximised below the irrigation rate that maximises yield. Irrigation costs can include energy to pump water, water rates, and depreciation on capital. The extent to which farmers are exposed to these costs, however, depends greatly on government policy in balancing competing demands for financial investment, energy, water, land, and food security.

3.3.4 Osmotic Adjustment

As stated in Section 3.2.1, most wheat cultivars will continue extracting water until the soil water potential drops to around −1.5 MPa, i.e. when the water in soil pores larger than 2 μm diameter has been removed. At this point the water potential in the leaf is −1.5 to −2.0 MPa, the stomata close in daylight, and photosynthesis is curtailed. There may be some small variation in timing of stomata closure associated with signalling responses mediated via plant hormones such as abscisic acid. However, there is also evidence that some wheat cultivars may be able to maintain leaf and guard cell turgor at water potentials less than −2.0 MPa. This is facilitated by raising osmotic potential in the leaf by increasing the concentration of solutes, principally potassium (Morgan 1992). Some relatively drought-tolerant wheats are, therefore, able to maintain some turgor down to leaf water potentials of −3.3 MPa. Such lines can keep their stomata open and photosynthesize at lower soil water potentials than lines with less osmotic adjustment (Hay and Porter 2006). Greater osmotic adjustment in the roots could also benefit continued root function and extension in drought conditions. In effect, osmotic adjustment will reduce the soil water potential at the PWP and thus increase the amount of available water (Hay and Porter 2006). In the field, therefore, osmotic adjustment of wheat lines has been associated with increased yield and water use under certain water stress treatments (Morgan and Condon 1986; Blum et al. 1999).

3.3.5 Transpiration Efficiency

Transpiration efficiency would be expected to be an important component of water-use efficiency in wheat (Figures 3.2 and 3.3). In typical field conditions, however, it varies substantially, from around 2.5 g (Figure 3.3) to 9 g of biomass growth per kg water transpired (Christy et al. 2018). Much of this variation is due to the environment. For example, transpiration efficiency declines with increased vapour pressure deficit of the air (Sinclair 2018). This is easy to understand if we accept that some transpiration is a *wasteful* consequence of letting CO_2 into the leaf. More water is thus wasted if the air is dry. Transpiration efficiency also declines with irrigation (Zhang et al. 1998; Zwart and Bastiaanssen 2004) and if the yield is limited by factors other than water availability such as insufficient nutrition and increased disease severity (French and Schultz 1984; Zhang et al. 1998; Angus and van Herwaarden 2001). Transpiration efficiency increases with increasing concentrations of atmospheric CO_2 (Christy et al. 2018), which is a positive effect associated with anthropogenic additions to the atmosphere.

Irrespective of the strong environmental component, improving transpiration efficiency may be a target for wheat breeding programmes for improving yield in water-limited environments. Selection has been assisted by exploiting the negative relationship between carbon isotope discrimination (Δ) and transpiration efficiency (e.g. Christy et al. 2018). Despite some success in these programmes, Sinclair (2018) argues that greater benefits will accrue from breeding for traits that match the availability of water to when it can be most useful for the crop. For example, if transpiration can be reduced early in the season, the conserved water will be more beneficial to grain yield by protecting reproductive growth and grain filling later in the season. Evans et al. (1975) suggest this might be achieved if plants had small, erect leaves with few tillers early in the season. This approach, however, could lead to greater loss of water through evaporation from the soil. Alternatively, it may be possible to reduce transpiration during less critical phases, for example, by increasing resistance to water uptake and transport via the seedling (seminal) roots (Passioura 1983). Furthermore, the effects of drought, particularly in the reproductive phases, may be reduced by the application of chemicals to the leaves to reduce transpiration rates, i.e. antitranspirants (Mphande et al. 2020).

3.3.6 Potassium

High levels of WUE depend on wheat growth not being limited by deficiency in any essential nutrient. However, the nutrient most directly implicated in the water relations of the plant is potassium (K^+). In addition to the role of K^+ in stomata, it is involved in maintaining the osmotic potential and turgor of cells throughout the plant. By influencing osmotic potential and water flow, K^+ is also crucial for cell and tissue expansion, and in the flow of assimilates in the phloem. As well as influencing the transport of amino-acids, potassium has roles in nitrate (NO_3^-) uptake, enzyme activation and protein synthesis (Pettigrew 2008). Potassium deficiency increases the negative effects of water stress on photosynthesis and reduces drought tolerance (Gupta et al. 1989; Wei et al. 2013). Deficient plants often become bleached around the margins and tips of older leaves, the stems become weak, and the crop therefore more likely to lodge. In addition to drought tolerance, increasing K^+ availability to deficient wheat crops can increase frost tolerance and reduce susceptibility to several diseases (e.g. Regmi et al. 2002).

Potassium is the most abundant cation in plants. It is, therefore, classed as a *macronutrient* because it is required at levels of kilogrammes per tonne of crop. By harvest, about 4.5 kg K/t is present in the grain and 8.0 kg K/t in the straw (Table 3.2). Although much K^+ is contained in soil, particularly most clay soils, it only becomes available to plants slowly. Being positively charged, K^+ is retained on the cation exchange sites (Chapter 2) of clay particles and organic matter. As K^+ does not move rapidly within most soils, it is less susceptible to leaching than some anions. However, the low mobility of K^+ has two further implications for managing crop supply. The first is that timing of application of any additional K^+ to the soil is not critical. Secondly, it is difficult to rectify deficiencies quickly, i.e. if symptoms of deficiency in the crop are visible, it is unlikely that any remedial action within the growing season will fully negate effects on yield. It is, therefore, common practice to try to maintain soil available K^+ levels at adequate levels over the long term.

Soil sampling and laboratory methods are often used to estimate how much of the soil reserves might become available to the crop, such as determining the amount of exchangeable K^+ from an ammonium nitrate soil extract (MAFF 1986). Using this approach, soil exchangeable K^+ should be maintained at between 120 and 180 mg/l. At these levels, applications of any fertilizers or manures should balance the amount of K^+ that is removed by the crop such that deficiency does not develop (Table 3.2). By convention, application rates are expressed in terms of K_2O equivalents (i.e. potash, which is 83% K). If the soil is deficient then the recommendation is to apply more K than is removed by the crop to build up soil levels in the long term. However, the amounts required to correct deficiencies are large: 300–500 kg K_2O equ./ha to increase availability by 50 mg/l. It is seldom practical or economic to do this in one season, so correcting for significant deficiencies may take 5 to 10 years (AHDB 2020).

The crop requirement for K^+ can be met through several sources. Some clay soils will release c. 50 mg/l of available K during the growing season. Even where this occurs, however, it is insufficient to match the demands of high-yielding wheat crops (Table 3.2). Long-term requirements will be less if straw is left as residue and/or incorporated into the soil. Organic materials such as farmyard manure and slurries can also provide significant quantities of K^+ (Table 3.3). Over 90% of the K^+ in these materials can be directly available for

Table 3.2 The effect of grain and straw yields on the removal rates of potassium by winter wheat, and the application rates required to balance that removal.

Grain		Straw		Application to balance offtake
Yield (t/ha)	K removed (kg)	Yield (t/ha)	K removed (kg)	K_2O equivalents (kg)
6	27	3	24	61
8	36	4	32	82
10	45	5	40	102
12	54	6	48	123

Source: Adapted from AHDB (2020).

Table 3.3 Typical potassium contents of livestock farmyard manure (FYM) and slurry.

	Dry matter content (%)	kg K_2O eqv./t fresh weight
Cattle slurry	6	2.5
Cattle FYM	25	9.4
Poultry FYM	30	1.2
Pig FYM	25	8
Sheep FYM	25	8

Source: Adapted from AHDB (2020).

crop uptake, whilst the remainder will add to long-term supply. However, additional sources are often required. The most widely used fertilizer is potassium chloride (KCl, muriate of potash, 60% K_2O), nearly all of which is obtained by mining and extraction of the mineral form of KCl, sylvite (63% K_2O), from deposits left by dried up oceans. Sylvite deposits are found and mined throughout the world, but particularly in Saskatchewan, Canada. Other minerals available include kainit ($MgSO_4.KCl.3H_2O$, 19.3% K_20) and sylvanite (variable mixtures of KCl and NaCl). Potassium sulphate (K_2SO_4) is formed by treating KCl with sulphuric acid. It is more expensive than KCl but may be useful where sulphur is also required and/or chlorine is already in excess.

The world production of potash is between 35 and 40 million tonnes, but world resources are significant and it is unlikely that depletion of K^+ reserves will be a primary threat to future wheat production (USGS 2020). Environmental impacts of production relate to the physical aspects of mining and processing and include land disturbance, emissions of dust and exhaust gasses, water contamination, topsoil degradation, vegetation and wildlife disruption, noise and vibration, and waste production (UNEP-IFA 2001). Environmental concerns with regards to potash use are relatively small. Some leaching of K^+ may occur from the lighter soils, but the World Health Organization sets no guideline for potassium in drinking water because it invariably occurs at concentrations well below those of health concern (WHO 2017).

3.4 Land-Use Efficiency and Water

Crops require water to maintain nutrient uptake, growth, and metabolism, and to protect against heat stress. These functions need to be, and are, maintained whilst water is lost from the plants, particularly as stomata open to allow ingress of CO_2. Although the water requirement is substantial, wheat is relatively efficient at using water to produce grain dry matter compared with major crops from other plant families. This may be due to many factors, including the responsiveness of the grass-type stomata, efficient rooting, and the relatively high harvest index of modern wheat cultivars. The sustainability of wheat production in semi-arid areas, however, is under significant threat due to increasing frequencies of drought and irregularity of rainfall. Irrigated areas are particularly at risk due to falling water tables and depletion of groundwaters, irrespective of the additional challenge of salinisation covered in Chapter 2. If wheat production in water-limited areas is to continue, it is important to use available water as efficiently as possible. Significant improvements in efficiency are possible by using deficit irrigation strategies and technology to improve quantity, timing, and spatial distribution of irrigation water. By definition, deficit irrigation results in reduced land-use efficiency. However, the reductions can be minimised by providing sufficient moisture at particularly sensitive growth stages such as at meiosis and the early periods of grain filling. More generally, WUE is improved by ensuring that yield is not limited by other factors such as poor crop establishment, inadequate nutrition, lodging, and pests and diseases. Reducing loss of water through evaporation, ensuring adequate K^+ availability, and using soil management, agronomy, and cultivar choice to encourage adequate rooting at depth are particularly important. Reducing these limitations will increase land-use efficiency for a given level of water availability.

In more humid areas, many of the above factors are still relevant for reducing the impact of intermittent and/or early terminal droughts. In these areas, however, attention to excessive water is also required. Flooding and waterlogging need to be minimised through appropriate soil management and drainage. The risk of wet periods delaying crop management operations, particularly for sowing and harvest, needs to be mitigated by planning and appropriate machinery to make the best use of suitable weather windows and soil conditions.

References

Addy, J.W.G., Ellis, R.H., Macdonald, A.J. et al. (2020). Investigating the effects of inter-annual weather variation (1968–2016) on the functional response of cereal grain yield to applied nitrogen, using data from the Rothamsted Long-Term Experiments. *Agricultural and Forest Meteorology* 284: 107898. https://doi.org/10.1016/j.agrformet.2019.107898.

AHDB (2020). *Nutrient Management Guide (RB209) Section 1 Principle of Nutrient Management and Fertiliser Use*. Agriculture and Horticulture Development Board https://ahdb.org.uk/knowledge-library/rb209-section-1-principles-of-nutrient-management-and-fertiliser-use.

Alghory, A. and Yazar, A. (2019). Evaluation of crop water stress index and leaf water potential for deficit irrigation management of sprinkler-irrigated wheat. *Irrigation Science* 37: 61–77. https://doi.org/10.1007/s00271-018-0603-y.

Allan, R.G., Pereira, L.S., Raes, D. et al. (1998). *Crop Evapotranspiration – Guidelines for Computing Crop Water Requirements – FAO Irrigation and Drainage Paper 56*. Food and Agriculture Organization of the United Nations. http://www.fao.org/3/X0490E/x0490e00.htm

Angus, J.F. and van Herwaarden, A.F. (2001). Increasing water use and water use efficiency in dryland wheat. *Agronomy Journal* 93: 290–298. https://doi.org/10.2134/agronj2001.932290x.

Assuero, S.G. and Tognetti, J.A. (2010). Tillering regulation by endogenous and environmental factors and its agricultural management. *Americas Journal of Plant Science and Biotechnology* 4 (1): 35–48.

Atkinson, M.D., Kettlewell, P.S., Hollins, P.D. et al. (2005). Summer climate mediates UK wheat quality response to winter North Atlantic Oscillation. *Agricultural and Forest Meteorology* 130 (1–2): 27–37. https://doi.org/10.1016/j.agrformet.2005.02.002.

Balaghi, R., Tychon, B., Eerens, H. et al. (2008). Empirical regression models using NDVI, rainfall and temperature data for the early prediction of wheat grain yields in Morocco. *International Journal of Applied Earth Observation and Geoinformation* 10 (4): 438–452. https://doi.org/10.1016/j.jag.2006.12.001.

Barber, H., Carney, J., Alghabari, F. et al. (2015). Decimal growth stages for precision wheat production in changing environments? *Annals of Applied Biology* 166: 355–371. https://doi.org/10.1111/aab.12207.

Barnabás, B., Jäger, K., and Fehér, A. (2008). The effect of drought and heat stress on reproductive processes in cereals. *Plant, Cell & Environment* 31: 11–38. https://doi.org/10.1111/j.1365-3040.2007.01727.x.

Barraclough, P.B., Weir, A.H., and Kuhlmann, H. (1991). Factors affecting the growth and distribution of winter wheat roots under UK field conditions. *Developments in Agricultural and Managed Forest Ecology* 24: 410–417. https://doi.org/10.1016/B978-0-444-89104-4.50057-8.

Basso, B., Fiorentino, C., Cammarano, D. et al. (2012). Analysis of rainfall distribution on spatial and temporal patterns of wheat yield in Mediterranean environment. *European Journal of Agronomy* 41: 52–65. https://doi.org/10.1016/j.eja.2012.03.007.

Battaglin, W.A., Furlong, E.T., Burkhardt, M.R. et al. (2000). Occurrence of sulfonylurea, sulfonamide, imidazolinone, and other herbicides in rivers, reservoirs and ground water in the Midwestern United States, 1998. *Science of the Total Environment* 248 (2–3): 123–133. https://doi.org/10.1016/S0048-9697(99)00536-7.

Belford, R. (1981). Response of winter wheat to prolonged waterlogging under outdoor conditions. *Journal of Agricultural Science* 97 (3): 557–568. https://doi.org/10.1017/S0021859600036881.

Bennett, M.D., Chapman, V., and Riley, R. (1971). The duration of meiosis in pollen mother cells of wheat, rye and Triticale. *Proceedings of the Royal Society London B* 178: 259–275. http://doi.org/10.1098/rspb.1971.0065.

Bithell, S.L., Butler, R.C.C., McKay, A.C. et al. (2013). Influences of crop sequence, rainfall and irrigation, on relationships between Gaeumannomyces graminis var. tritici and take-all in New Zealand wheat fields. *Australasian Plant Pathology* 42: 205–217. https://doi.org/10.1007/s13313-012-0168-9.

Blum, A., Zhang, J., and Nguyen, H.T. (1999). Consistent differences among wheat cultivars in osmotic adjustment and their relationship to plant production. *Field Crops Research* 64 (3): 287–291. https://doi.org/10.1016/S0378-4290(99)00064-7.

Bracken, A.F. and Bailey, C.H. (1928). Effect of delayed harvesting on the quality of wheat. *Cereal Chemistry* 5: 128–145.

Brennan, J., Condon, A., Van Ginkel, M. et al. (2007). An economic assessment of the use of physiological selection for stomatal aperture-related traits in the CIMMYT wheat breeding programme. *Journal of Agricultural Science* 145 (3): 187–194. https://doi.org/10.1017/S0021859607007009.

Brisson, N., Rebière, B., Zimmer, D. et al. (2002). Response of the root system of a winter wheat crop to waterlogging. *Plant and Soil* 243: 43–55. https://doi.org/10.1023/A:1019947903041.

Brouwer, C., Prins, K., Kay, M. et al. (1988). *Irrigation Water Management Training Manual No. 5. Irrigation Methods.* FAO – Food and Agriculture Organization of the United Nations.

Camara, K.M., Payne, W.A., and Rasmussen, P.E. (2003). Long-term effects of tillage, nitrogen, and rainfall on winter wheat yields in the Pacific Northwest. *Agronomy Journal* 95: 828–835. https://doi.org/10.2134/agronj2003.8280.

Campbell, C.A. and Read, D.W.L. (1968). Influence of air temperature, light intensity and soil moisture on the growth, yield and some growth analysis characteristics of chinook wheat grown in the growth chamber. *Canadian Journal of Plant Science* 48: 299–311. https://doi.org/10.4141/cjps68-053.

Chandna, P., Khurana, M.L., Ladha, J.K. et al. (2011). Spatial and seasonal distribution of nitrate-N in groundwater beneath the rice–wheat cropping system of India: a geospatial analysis. *Environmental Monitoring and Assessment* 178: 545–562. https://doi.org/10.1007/s10661-010-1712-0.

Christy, B., Tausz-Posch, S., Tausz, M. et al. (2018). Benefits of increasing transpiration efficiency in wheat under elevated CO_2 for rainfed regions. *Global Change Biology* 24: 1965–1977. https://doi.org/10.1111/gcb.14052.

Clarke, M.P., Gooding, M.J., and Jones, S.A. (2004). The effects of irrigation, nitrogen fertilizer and grain size on Hagberg falling number, specific weight and blackpoint of winter wheat. *Journal of the Science of Food and Agriculture* 84: 227–236. https://doi.org/10.1002/jsfa.1657.

Clarke, C., Lukac, M., Gregory, P. et al. (2017a). Associating remotely sensed canopy traits with deep rooting in wheat. In: *Aspects of Applied Biology 135, Precision Systems in Agricultural and Horticultural Production* (eds. I. Grove and R. Kennedy), 9–18. Association of Applied Biologists.

Clarke, C.K., Gregory, P.J., Lukac, M. et al. (2017b). Quantifying rooting at depth in a wheat doubled haploid population with introgression from wild emmer. *Annals of Botany* 120 (3): 457–470. https://doi.org/10.1093/aob/mcx068.

Cook, R.J. and Veseth, R.J. (1991). *Wheat Health Management*. The American Phytopathological Society.

Cornish, E.A. (1950). The influence of rainfall on the yield of wheat in South Australia. *Australian Journal of Biological Sciences* 3: 178–218. https://doi.org/10.1071/BI9500178.

De Jong, R., Shields, J.A., and Sly, W.K. (1984). Estimated soil water reserves applicable to a wheat-fallow rotation for generalized soil areas mapped in southern Saskatchewan. *Canadian Journal of Soil Science* 64 (4): 667–680. https://doi.org/10.4141/cjss84-067.

Deery, D.M., Rebetzke, G.J., Jimenez-Berni, J.A. et al. (2016). Methodology for high-throughput field phenotyping of canopy temperature using airborne thermography. *Frontiers in Plant Science* 7: 1808. https://doi.org/10.3389/fpls.2016.01808.

Deines, J.M., Schipanski, M.E., Golden, B. et al. (2020). Transitions from irrigated to dryland agriculture in the Ogallala Aquifer: Land use suitability and regional economic impacts. *Agricultural Water Management* 233: 106061. https://doi.org/10.1016/j.agwat.2020.106061.

Derkx, A.P. and Mares, D.J. (2020). Late-maturity α-amylase expression in wheat is influenced by genotype, temperature and stage of grain development. *Planta* 251 (2): 1–14. https://doi.org/10.1007/s00425-020-03341-1.

Dickin, E. and Wright, D. (2008). The effects of winter waterlogging and summer drought on the growth and yield of winter wheat (*Triticum aestivum* L.). *European Journal of Agronomy* 28 (3): 234–244. https://doi.org/10.1016/j.eja.2007.07.010.

Dimmock, J.P.R.E. and Gooding, M.J. (2002). The effects of fungicides on Hagberg falling number and blackpoint in winter wheat. *Crop Protection* 21 (6): 475–487. https://doi.org/10.1016/S0261-2194(01)00135-1.

Dunn, J., Hunt, L., and Afsharinafar, M. (2019). Reduced stomatal density in bread wheat leads to increased water-use efficiency. *Journal of Experimental Botany* 70 (18): 4737–4748. https://doi.org/10.1093/jxb/erz248.

Easson, D.L., White, E.M., and Pickles, S.J. (1993). The effects of weather, seed rate and cultivar on lodging and yield in winter wheat. *Journal of Agricultural Science* 121 (2): 145–156. https://doi.org/10.1017/S0021859600077005.

EEA (2017). *Climate Change, Impacts and Vulnerability in Europe 2016: EEA Report No 1/2017.* European Environment Agency.

Evans, L.T., Wardlaw, I.F., and Fischer, R.A. (1975). Wheat. In: *Crop Physiology: Some Case Histories* (ed. L.T. Evans), 101–150. Cambridge University Press.

Evett, S.R., Colaizzi, P.D., Lamm, F.R. et al. (2020). Past, present and future of irrigation on the U.S. Great Plains. *Transactions of the American Society of Agricultural and Biological Engineers* 63 (3): 703–729. https://doi.org/10.13031/trans.13620.

Fang, Q., Yu, Q., Wang, E. et al. (2006). Soil nitrate accumulation, leaching and crop nitrogen use as influenced by fertilization and irrigation in an intensive wheat–maize double cropping system in the North China Plain. *Plant and Soil* 284: 335–350. https://doi.org/10.1007/s11104-006-0055-7.

Farquhar, G.D. and Sharkey, T.D. (1982). Stomatal conductance and photosynthesis. *Annual Revues of Plant Physiology* 33: 317–345.

Fernández-Gómez, J., Talle, B., Tidy, A.C. et al. (2020). Accurate staging of reproduction development in Cadenza wheat by non-destructive spike analysis. *Journal of Experimental Botany* 71 (12): 3475–3484. https://doi.org/10.1093/jxb/eraa156.

Fischer, R.A., Rees, D., Sayre, K.D. et al. (1998). Wheat yield progress associated with higher stomatal conductance and photosynthetic rate, and cooler canopies. *Crop Science* 38: 1467–1475. https://doi.org/10.2135/cropsci1998.0011183X003800060011x.

Fisher, R.A. and Russell, E.J. (1997). The influence of rainfall on the yield of wheat at Rothamsted. *Philosophical Transactions of the Royal Society London B* 213: 89–142. http://doi.org/10.1098/rstb.1925.0003.

Ford, K.E., Gregory, P.J., Gooding, M.J. et al. (2006). Genotype and fungicide effects on late-season root growth of winter wheat. *Plant and Soil* 284: 33–44. https://doi.org/10.1007/s11104-006-0028-x.

Foulkes, M.J., Sylvester-Bradley, R., Weightman, R. et al. (2007). Identifying physiological traits associated with improved drought resistance in winter wheat. *Field Crops Research* 103 (1): 11–24. https://doi.org/10.1016/j.fcr.2007.04.007.

French, R.J. and Schultz, J.E. (1984). Water use efficiency of wheat in a Mediterranean-type environment. II. Some limitations to efficiency. *Australian Journal of Agricultural Research* 35: 765–775. https://doi.org/10.1071/AR9840765.

Fujii, S. (2013). Chronology of the Jafr prehistory and protohistory: a key to the process of pastoral nomadization in the Southern Levant. *Syria* 90: 49–125. https://doi.org/10.4000/syria.1723.

Gajri, P.R.R. and Prihar, S.S. (1983). Effect of small irrigation amounts on the yield of wheat. *Agricultural Water Management* 6 (1): 31–41. https://doi.org/10.1016/0378-3774(83)90024-0.

Gallagher, J.N., Biscoe, P.V., and Wallace, J.S. (1979). Field studies of cereal leaf growth: IV. Winter wheat leaf extension in relation to temperature and leaf water status. *Journal of Experimental Botany* 30 (4): 657–668. https://doi.org/10.1093/jxb/30.4.657.

Gangopadhyaya, M. and Sarker, R.P. (1965). Influence of rainfall distribution on the yield of wheat crop. *Agricultural Meteorology* 2 (5): 331–350. https://doi.org/10.1016/0002-1571(65)90026-9.

Gao, W., Hodgkinson, L., Jin, K. et al. (2016). Deep roots and soil structure. *Plant, Cell and Environment* 39: 1662–1668. https://doi.org/10.1111/pce.12684.

Garrick, D.E., Hanemann, M., and Hepburn, C. (2020). Rethinking the economics of water: an assessment. *Oxford Review of Economic Policy* 36 (1): 1–23. https://doi.org/10.1093/oxrep/grz035.

Gooding, M.J., Ellis, R.H., Shewry, P.R. et al. (2003). Effects of restricted water availability and increased temperature on the grain filling, drying and quality of winter wheat. *Journal of Cereal Science* 37: 295–309. https://doi.org/10.1006/jcrs.2002.0501.

Goulding, K.W.T., Poulton, P.R., Webster, C.P. et al. (2000). Nitrate leaching from the Broadbalk Wheat Experiment, Rothamsted, UK, as influenced by fertilizer and manure inputs and the weather. *Soil Use and Management* 16 (4): 244–250. https://doi.org/10.1111/j.1475-2743.2000.tb00203.x.

Gregory, P., McGowan, M., Biscoe, P., and Hunter, B. (1978). Water relations of winter wheat: 1. Growth of the root system. *Journal of Agricultural Science* 91 (1): 91–102. https://doi.org/10.1017/S0021859600056653.

Gregory, P.J., Gooding, M.J., Ford, K.E. et al. (2005). *Managing Roots, Nitrogen and Fungicides to Improve Yield and Quality of Wheat. Project Report No. 359.* Home-Grown Cereals Authority https://pdfs.semanticscholar.org/9e87/551fe62c381bfeae9cf860dc5041d8152390.pdf.

Gupta, A.S., Berkowitz, G.A., and Pier, P.A. (1989). Maintenance of photosynthesis at low leaf water potential in wheat. *Plant Physiology* 89 (4): 1358–1365. https://doi.org/10.1104/pp.89.4.1358.

Hall, A.E., Richards, R.A., Condon, A.G. et al. (1994). Carbon isotope discrimination and plant breeding. In: *Plant Breeding Reviews*, vol. 12 (ed. J. Janik), 81–113. Wiley.

Hamada, A., Nitta, M., Nasuda, S. et al. (2012). Novel QTLs for growth angle of seminal roots in wheat (Triticum aestivum L.). *Plant and Soil* 354: 395–405. https://doi.org/10.1007/s11104-011-1075-5.

Hanks, R.J. and Thorp, F.C. (1956). Seedling emergence of wheat as related to soil moisture content, bulk density, oxygen diffusion rate, and crust strength. *Soil Science Society of America Journal* 20: 307–310. https://doi.org/10.2136/sssaj1956.03615995002000030003x.

Hay, R.K.M. and Porter, J.R. (2006). *The Physiology of Crop Yield*, 2e. Blackwell.

Hoad, S.P., Russell, G., Kettlewell, P.S. et al. (2004). *Root system management in winter wheat: practices to increase water and nitrogen use. Project Report 351.* Home-Grown Cereals Authority.

Hossain, M.A. and Uddin, S.N. (2011). Mechanisms of waterlogging tolerance in wheat: morphological and metabolic adaptations under hypoxia or anoxia. *Australian Journal of Crop Science* 5 (9): 1094–1101.

Hu, C., Delgado, J.A., Zhang, X. et al. (2005). Assessment of groundwater use by wheat (Triticum aestivum L.) in the Luancheng Xian Region and potential implications for water conservation in the Northwestern North China Plain. *Journal of Soil and Water Conservation* 60 (2): 80–88. https://www.jswconline.org/content/60/2/80/tab-article-info.

Huang, J., Ridoutt, B.G., Sun, Z. et al. (2020). Balancing food production within the planetary water boundary. *Journal of Cleaner Production* 253: 119900. https://doi.org/10.1016/j.jclepro.2019.119900.

Hudson, P.F. (2020). Water: how does agriculture impact freshwater resources? In: *Food and Sustainability* (eds. P. Behrens, T. Bosker and D. Ehrhardt), 71–101. Oxford University Press.

Hunt, J.R. and Kirkegaard, J.A. (2011). Re-evaluating the contribution of summer fallow rain to wheat yield in southern Australia. *Crop and Pasture Science* 62: 915–929. https://doi.org/10.1071/CP11268.

IPCC (2014). *Climate Change 2014: Synthesis Report. Contribution of Working Groups I, II and III to the Fifth Assessment Report of the Intergovernmental Panel on Climate Change.* Intergovernmental Panel on Climate Change. https://www.ipcc.ch/report/ar5/syr

Jalota, S.K., Jain, A.K., and Vashisht, B.B. (2018). Minimize water deficit in wheat crop to ameliorate groundwater decline in rice-wheat cropping system. *Agricultural Water Management* 208: 261–267. https://doi.org/10.1016/j.agwat.2018.06.020.

Jošt, M., Samobor-Galović, V., Vukobratović, Z. et al. (1997). Wheat of high bread-making quality and accompanying unfavorable effects in its production. In *Proceedings of the 1st Croatian Congress of Cereal Technologists* (pp. 28–36). Faculty of Food Technology, University of Josip Juraj Strossmayer.

Kahlown, M.A., Raoof, A., Zubair, M. et al. (2007). Water use efficiency and economic feasibility of growing rice and wheat with sprinkler irrigation in the Indus Basin of Pakistan. *Agricultural Water Management* 87 (3): 292–298. https://doi.org/10.1016/j.agwat.2006.07.011.

Kettlewell, P.S., Sothern, R.B., and Koukkari, W.L. (1999). UK wheat quality and economic value are dependent on the North Atlantic Oscillation. *Journal of Cereal Science* 29 (3): 205–209. https://doi.org/10.1006/jcrs.1999.0258.

Khah, E., Ellis, R., and Roberts, E. (1986). Effects of laboratory germination, soil temperature and moisture content on the emergence of spring wheat. *Journal of Agricultural Science* 107 (2): 431–438. https://doi.org/10.1017/S0021859600087232.

King, J., Gay, A., Sylvester-Bradley, R. et al. (2003). Modelling cereal root systems for water and nitrogen capture: towards an economic optimum. *Annals of Botany* 91 (3): 383–390. https://doi.org/10.1093/aob/mcg033.

Kumari, B., Madan, V.K., and Kathpal, T.S. (2008). Status of insecticide contamination of soil and water in Haryana, India. *Environmental Monitoring and Assessment* 136: 239–244. https://doi.org/10.1007/s10661-007-9679-1.

Latiri, K., Lhomme, J.P., Annabi, M. et al. (2010). Wheat production in Tunisia: Progress, inter-annual variability and relation to rainfall. *European Journal of Agronomy* 33 (1): 33–42. https://doi.org/10.1016/j.eja.2010.02.004.

Lawes, R.A., Oliver, Y.M., and Robertson, M.J. (2009). Integrating the effects of climate and plant available soil water holding capacity on wheat yield. *Field Crops Research* 113 (3): 297–305. https://doi.org/10.1016/j.fcr.2009.06.008.

Leuning, R., Condon, A.G., Dunin, F.X. et al. (1994). Rainfall interception and evaporation from soil below a wheat canopy. *Agricultural and Forest Meteorology* 67 (3–4): 221–238. https://doi.org/10.1016/0168-1923(94)90004-3.

Li, S., Wheeler, T., Challinor, A. et al. (2010). Simulating the impacts of global warming on wheat in China using a large area crop model. *Journal of Meteorological Research* 24 (1):

123–135. http://www.cmsjournal.net/qxxb_en/ch/reader/view_abstract.aspx?file_no=20100111&flag=1.

Li, X., Ingvordsen, C.H., Weiss, M. et al. (2019). Deeper roots associated with cooler canopies, higher normalized difference vegetation index, and greater yield in three wheat populations grown on stored soil water. *Journal of Experimental Botany* 70 (18): 4963–4974. https://doi.org/10.1093/jxb/erz232.

Limon-Ortega, A. and Sayre, K. (2012). Rainfall as a limiting factor for wheat grain yield in permanent raised beds. *Agronomy Journal* 104: 1171–1175. https://doi.org/10.2134/agronj2012.0053.

Liu, H., Yu, L., Luo, Y. et al. (2011). Responses of winter wheat (Triticum aestivum L.) evapotranspiration and yield to sprinkler irrigation regimes. *Agricultural Water Management* 98 (4): 483–492. https://doi.org/10.1016/j.agwat.2010.09.006.

Lopes, M.S. and Reynolds, M.P. (2010). Partitioning of assimilates to deeper roots is associated with cooler canopies and increased yield under drought in wheat. *Functional Plant Biology* 37: 147–156. https://doi.org/10.1071/FP09121.

MAFF (1986). *The Analysis of Agricultural Materials, MAFF Reference Book 427*, 3e. H. M. Stationary Office.

Mahmood, T., Ali, R., Malik, K. et al. (1998). Nitrous oxide emissions from an irrigated sandy-clay loam cropped to maize and wheat. *Biology and Fertility of Soils* 27: 189–196. https://doi.org/10.1007/s003740050419.

Malik, A.I., Colmer, T.D., Lambers, H. et al. (2002). Short-term waterlogging has long-term effects on the growth and physiology of wheat. *New Phytologist* 153: 225–236. https://doi.org/10.1046/j.0028-646X.2001.00318.x.

Mallya, G., Mishra, V., Niyogi, D. et al. (2016). Trends and variability of droughts over the Indian monsoon region. *Weather and Climate Extremes* 12: 43–68. https://doi.org/10.1016/j.wace.2016.01.002.

Manschadi, A.M., Christopher, J., deVoil, P. et al. (2006). The role of root architectural traits in adaptation of wheat to water-limited environments. *Functional Plant Biology* 33: 823–837. https://doi.org/10.1071/FP06055.

Mares, D.J. (1993). Pre-harvest sprouting in wheat. I. Influence of cultivar, rainfall and temperature during grain ripening. *Australian Journal of Agricultural Research* 44 (6): 1259–1272. https://doi.org/10.1071/AR9931259.

Mbava, N., Mutema, M., Zengeni, R. et al. (2020). Factors affecting crop water use efficiency: a worldwide meta-analysis. *Agricultural Water Management* 228: 105878. https://doi.org/10.1016/j.agwat.2019.105878.

Mega, R., Abe, F., Kim, J. et al. (2019). Tuning water-use efficiency and drought tolerance in wheat using abscisic acid receptors. *Nature Plants* 5: 153–159. https://doi.org/10.1038/s41477-019-0361-8.

Mithen, S. (2012). *Thirst: Water & Power in the Ancient World*. Orion Publishing Group.

Monteith, J.L. (1965). Evaporation and environment. *Symposium of the Society of Experimental Biology* 19: 205–234.

Moore, R., Clark, W.D., and Vodopich, D.S. (1998). *Botany*, 2e. McGraw-Hill.

Morgan, J. (1980). Possible role of abscisic acid in reducing seed set in water-stressed wheat plants. *Nature* 285: 655–657. https://doi.org/10.1038/285655a0.

Morgan, J.M. (1992). Osmotic components and properties associated with genotypic differences in osmoregulation in wheat. *Functional Plant Biology* 19: 67–76. https://doi.org/10.1071/PP9920067.

Morgan, J.M. and Condon, A.G. (1986). Water use, grain yield, and osmoregulation in wheat. *Functional Plant Biology* 13: 523–532. https://doi.org/10.1071/PP9860523.

Mpelasoka, F., Hennessy, K., Jones, R. et al. (2008). Comparison of suitable drought indices for climate change impacts assessment over Australia towards resource management. *International Journal Climatology* 28: 1283–1292. https://doi.org/10.1002/joc.1649.

Mphande, W., Kettlewell, P.S., Grove, I.G. et al. (2020). The potential of antitranspirants in drought management of arable crops: a review. *Agricultural Water Management* 236: 106143. https://doi.org/10.1016/j.agwat.2020.106143.

Mullan, D.J. and Reynolds, M.P. (2010). Quantifying genetic effects of ground cover on soil water evaporation using digital imaging. *Functional Plant Biology* 37: 703–712. https://doi.org/10.1071/FP09277.

Nunes, T.D.G., Zhang, D., and Raissig, M.T. (2020). Form, development and function of grass stomata. *Plant Journal* 101: 780–799. https://doi.org/10.1111/tpj.14552.

Olesen, J., Mortensen, J., Jørgensen, L. et al. (2000). Irrigation strategy, nitrogen application and fungicide control in winter wheat on a sandy soil. I. Yield, yield components and nitrogen uptake. *Journal of Agricultural Science* 134 (1): 1–11. https://doi.org/10.1017/S0021859699007352.

Oyanagi, A., Takahashi, H., and Suge, H. (1995). Interactions between hydrotropism and gravitropism in the primary seminal roots of Triticum aestivum L. *Annals of Botany* 75 (3): 229–235. https://doi.org/10.1006/anbo.1995.1016.

Parkinson, K.J. and Day, W. (1980). Temperature corrections to measurements made with continuous flow porometers. *Journal of Applied Ecology* 17 (2): 457–460.

Passioura, J.B. (1983). Roots and drought resistance. *Agricultural Water Management* 7 (1–3): 265–280. https://doi.org/10.1016/0378-3774(83)90089-6.

Peel, M.C., Finlayson, B.L., and Mcmahon, T.A. (2007). Updated world map of the Köppen-Geiger climate classification. *Hydrology and Earth System Sciences Discussions, European Geosciences Union* 4 (2): 439–473. https://hal.archives-ouvertes.fr/hal-00298818.

Pepler, S., Gooding, M.J., and Ellis, R.H. (2006). Modelling simultaneously water content and dry matter dynamics of wheat grains. *Field Crops Research* 95: 49–63. https://doi.org/10.1016/j.fcr.2005.02.001.

Percival, J. (1921). *The Wheat Plant: A Monograph*. Duckworth and Co.

Pereira, L.S., Oweis, T., and Zairi, A. (2002). Irrigation management under water scarcity. *Agricultural Water Management* 57 (3): 175–206. https://doi.org/10.1016/S0378-3774(02)00075-6.

Pettigrew, W.T. (2008). Potassium influences on yield and quality production for maize, wheat, soybean and cotton. *Physiologia Plantarum* 133: 670–681. https://doi.org/10.1111/j.1399-3054.2008.01073.x.

Portmann, F.T., Siebert, S., and Döll, P. (2010). MIRCA2000 – global monthly irrigated and rainfed crop areas around the year 2000: a new high-resolution data set for agricultural and hydrological modelling. *Global Biogeochemical Cycles* 24: GB1011. https://doi.org/10.1029/2008GB003435.

Rebetzke, G.J., Condon, A.G., Richards, R.A. et al. (2003). Gene action for leaf conductance in three wheat crosses. *Australian Journal of Agricultural Research* 54: 381–387. https://doi.org/10.1071/AR02151.

Regmi, A., Ladha, J., Pasuquin, E. et al. (2002). The role of potassium in sustaining yields in a long-term rice-wheat experiment in the Indo-Gangetic Plains of Nepal. *Biology and Fertility of Soils* 36: 240–247. https://doi.org/10.1007/s00374-002-0525-x.

Rodell, M., Velicogna, I., and Famiglietti, J. (2009). Satellite-based estimates of groundwater depletion in India. *Nature* 460: 999–1002. https://doi.org/10.1038/nature08238.

Ruane, A.C., Phillips, M.M., and Rosenzweig, C. (2018). Climate shifts within major agricultural seasons for +1.5 and +2.0 °c worlds: HAPPI projections and AGMIP modeling scenarios. *Agricultural and Forest Meteorology* 259: 329–344. https://doi.org/10.1016/j.agrformet.2018.05.013.

Saini, H.S. and Westgate, M.E. (1999). Reproductive development in grain crops during drought. *Advances in Agronomy* 68: 59–96. https://doi.org/10.1016/S0065-2113(08)60843-3.

Sayre, K.D., and Moreno Ramos, O.H. (1997). *Applications of Raised-Bed Planting Systems to Wheat. Wheat Special Report No. 31*. CIMMYT.

Schillinger, W.F., Donaldson, E., Allan, R.E. et al. (1998). Winter wheat seedling emergence from deep sowing depths. *Agronomy Journal* 90: 582–586. https://doi.org/10.2134/agronj1998.00021962009000050002x.

Semenov, M.A., Stratonovitch, P., Alghabari, F. et al. (2014). Adapting wheat in Europe for climate change. *Journal of Cereal Science* 59: 245–256. https://doi.org/10.1016/j.jcs.2014.01.006.

Senapati, N., Griffiths, S., Hawkesford, M. et al. (2020). Substantial increase in yield predicted by wheat ideotypes for Europe under future climate. *Climate Research* 80: 189–201. https://doi.org/10.3354/cr01602.

Setter, T. and Waters, I. (2003). Review of prospects for germplasm improvement for waterlogging tolerance in wheat, barley and oats. *Plant and Soil* 253: 1–34. https://doi.org/10.1023/A:1024573305997.

Shahinnia, F., Le Roy, J., Laborde, B. et al. (2016). Genetic association of stomatal traits and yield in wheat grown in low rainfall environments. *BMC Plant Biology* 16: 150. https://doi.org/10.1186/s12870-016-0838-9.

Siebert, S., Burke, J., Faures, J.M. et al. (2010). Groundwater use for irrigation – a global inventory. *Hydrology and Earth System Sciences* 14: 1863–1880. https://doi.org/10.5194/hess-14-1863-2010.

Sinclair, T.R. (2018). Effective water use required for improving crop growth rather than transpiration efficiency. *Frontiers in Plant Science* 9: 1442. https://doi.org/10.3389/fpls.2018.01442.

Skevas, T. (2020). Evaluating alternative policies to reduce pesticide groundwater pollution in Dutch arable farming. *Journal of Environmental Planning and Management* 63 (4): 733–750. https://doi.org/10.1080/09640568.2019.1606618.

Smith, S.P., and Davies, W.P. (1990). Response of wheat to high and low nitrogen and fungicide inputs on shallow limestone soil. *BCPC Monograph No. 45 Organic and Low Input Agriculture*, 227–230.

Smith, G.P. and Gooding, M.J. (1996). Relationships of wheat quality with climate and nitrogen application in regions of England (1974–1993). *Annals of Applied Biology* 129 (1): 97–108. https://doi.org/10.1111/j.1744-7348.1996.tb05735.x.

Smith, G.P. and Gooding, M.J. (1999). Models of wheat grain quality considering climate, cultivar and nitrogen effects. *Agricultural and Forest Meteorology* 94 (3–4): 159–170. https://doi.org/10.1016/S0168-1923(99)00020-9.

Stanhill, G. (1986). Irrigation in arid lands. *Philosophical Transactions of the Royal Society of London A* 316: 261–273. http://doi.org/10.1098/rsta.1986.0008.

Stark, J.C. and Longley, T.S. (1986). Changes in spring wheat tillering patterns in response to delayed irrigation. *Agronomy Journal* 78: 892–896. https://doi.org/10.2134/agronj1986.00021962007800050029x.

Sternberg, T. (2012). Chinese drought, bread and the Arab Spring. *Applied Geography* 34: 519–524. https://doi.org/10.1016/j.apgeog.2012.02.004.

Strzepek, K., Yohe, G., Neumann, J. et al. (2010). Characterizing changes in drought risk for the United States from climate change. *Environmental Research Letters* 5 (4): 044012. https://doi.org/10.1088/1748-9326/5/4/044012.

Tack, J., Barkley, A., and Hendricks, N. (2017). Irrigation offsets wheat yield reductions from warming temperatures. *Environmental Research Letters* 12: 114027. https://doi.org/10.1088/1748-9326/aa8d27.

Tripathi, S.C., Sayre, K.D., and Kaul, J.N. (2005). Planting systems on lodging behavior, yield components, and yield of irrigated spring bread wheat. *Crop Science* 45: 1448–1455. https://doi.org/10.2135/cropsci2003-714.

UNEP-IFA (2001). *Environmental Aspects of Phosphate and Potash Mining*. United Nations Environment Programme, International Fertilizer Industry Association.

USGS (2020). *Potash Statistics and Information*. Minerals Information Centre, U.S. Geological Survey https://www.usgs.gov/centers/nmic/potash-statistics-and-information.

Wang, X., Li, X., Xin, L. et al. (2016). Ecological compensation for winter wheat abandonment in groundwater over-exploited areas in the North China Plain. *Journal of Geographical Sciences* 26: 1463–1476. https://doi.org/10.1007/s11442-016-1338-4.

Warren, R., Yu, R., Osborn, T. et al. (2009). Future European drought regimes under mitigated and un-mitigated climate change. *IOP Conference Series: Earth and Environmental Science: Climate Change and Water Systems* 6: 292012. https://doi.org/10.1088/1755-1307/6/29/292012.

Wei, J., Li, C., Li, Y. et al. (2013). Effects of external potassium (k) supply on drought tolerances of two contrasting winter wheat cultivars. *PLoS One* 8 (7): e69737. https://doi.org/10.1371/journal.pone.0069737.

Westgate, M.E., Passioura, J.B., and Munns, R. (1996). Water status and aba content of floral organs in drought-stressed wheat. *Functional Plant Biology* 23: 763–772. https://doi.org/10.1071/PP9960763.

WHO (2017). *Guidelines for Drinking-Water Quality*, 4e. World Health Organization.

Wiik, L. and Ewaldz, T. (2009). Impact of temperature and precipitation on yield and plant diseases of winter wheat in southern Sweden 1983–2007. *Crop Protection* 28 (11): 952–962. https://doi.org/10.1016/j.cropro.2009.05.002.

Yadav, G. and Ellis, R.H. (2017). Effects of rain shelter or simulated rain during grain filling and maturation on subsequent wheat grain quality in the UK. *Journal of Agricultural Science* 155 (2): 300–316. https://doi.org/10.1017/S0021859616000411.

Yu, L., Zhao, X., Gao, X. et al. (2020). Improving/maintaining water-use efficiency and yield of wheat by deficit irrigation:sha a global meta-analysis. *Agricultural Water Management* 228: 105906. https://doi.org/10.1016/j.agwat.2019.105906.

Zadoks, J.C., Chang, T.T., and Konzak, C.F. (1974). A decimal code for the growth stages of cereals. *Weed Research* 14: 415–421.

Zampieri, M., Ceglar, A., Dentener, F. et al. (2017). Wheat yield loss attributable to heat waves, drought and water excess at the global, national and subnational scales. *Environmental Research Letters* 12 (6): 064008. https://doi.org/10.1088/1748-9326/aa723b.

Zaveri, E.B. and Lobell, D. (2019). The role of irrigation in changing wheat yields and heat sensitivity in India. *Nature Communications* 10: 4144. https://doi.org/10.1038/s41467-019-12183-9.

Zhang, H., Oweis, T.Y., Garabet, S. et al. (1998). Water-use efficiency and transpiration efficiency of wheat under rain-fed conditions and supplemental irrigation in a Mediterranean-type environment. *Plant and Soil* 201: 295–305. https://doi.org/10.1023/A:1004328004860.

Zhang, X., Pei, D., and Chen, S. (2004). Root growth and soil water utilization of winter wheat in the North China Plain. *Hydrological Processes* 18: 2275–2287. https://doi.org/10.1002/hyp.5533.

Zhang, X., Obringer, R., Wei, C. et al. (2017). Droughts in India from 1981 to 2013 and implications to wheat production. *Scientific Reports* 7: 44552. https://doi.org/10.1038/srep44552.

Zwart, S.J. and Bastiaanssen, W.G.M. (2004). Review of measured crop water productivity values for irrigated wheat, rice, cotton and maize. *Agricultural Water Management* 69 (2): 115–133. https://doi.org/10.1016/j.agwat.2004.04.007.

4

Mild Temperatures

Spring is sooner recognized by plants than by men

Chinese proverb

Temperature affects the rate of biochemical processes, the growth and functioning of plant organs, transpiration, and how quickly wheat passes through its growth stages (Slafer and Rawson 1994a). Wheat is often described as a temperate crop. *Temperate* literally means *mild temperature*. This description broadly distinguishes wheat from the more heat-tolerant, *tropical* crops (Fischer and Byerlee 1991). *Temperate* has more precise (but different) definitions when used by geographers and climatologists; these definitions are also relevant to understanding the distribution and production of wheat.

Wheat grows particularly well between 17 and 24 °C (Porter and Gawith 1999; Prasad et al. 2017). These are lower temperatures than often experienced in the summer days of, e.g. Mediterranean, continental, and subtropical areas, and over longer durations in the tropics. Wheat's need for mild temperatures is particularly evident for net photosynthesis, meiosis, anthesis, and grain filling, which are all variously more susceptible to higher temperatures than the more tropically adapted cereals, maize, sorghum, millet, and rice (Tashiro and Wardlaw 1989; Al-Khatib and Paulsen 1999; Prasad et al. 2017). In the warmer areas, therefore, where it is possible to grow more than one crop per year, wheat is often grown in the cooler seasons, i.e. it is known as a *cool-season* crop (Cook and Veseth 1991). In India and Pakistan wheat is the major *rabi* crop, (from the Arabic word for spring), because wheat grown on the Indo-Gangetic Plain is harvested before the high temperatures of summer. At the other extreme, wheat is more susceptible to low temperatures than its close relatives, rye and barley (e.g. Fowler et al. 1977).

Geographically, the broadest definition of the temperate zone is that which lies between 23.3° and 66.3° latitudes (Figure 1.2). This includes the subpolar (50° to 66.3°) and subtropical (23.3° to 35°) regions. Large areas of wheat are grown between 50° and 35° latitude, but production also extends into subpolar and subtropical areas. In the tropical zone, wheat production is mostly limited to cooler highland areas. The temperate zones are characterised by having distinct seasons based on daylength (or photoperiod), from short days in cool winters to long days in warm summers. Wheat detects and responds to changes in daylength and temperature in several ways to exploit the seasons with the most clement conditions. In particular, the timing of wheat growth and development in relation to the environmental changes of spring are important in most wheat growing areas.

As discussed in Chapter 1, the Köppen Climate Classification system (Peel et al. 2007) defines temperate areas as those where the mean temperature of the coldest month is above 0 but below 18 °C and the mean temperature of the hottest month is above 10 °C (Table 1.2). Although broad, this definition does not encompass all climates found between 23.3° and 66.3° latitudes (Figure 1.6), nor all the areas where wheat is grown (Figure 1.2). The wide distribution of wheat production depends principally on the adaptation of different wheat cultivars to complete the growth stages which are susceptible to heat and cold at clement times of the year. This includes the adaptation to different sowing and harvest dates, as well as the time taken to reach, for example, meiosis, anthesis, and grain filling.

The requirement for mild temperatures renders wheat vulnerable to global warming. There are many examples where higher than average temperatures have been associated with reduced grain yields (e.g. You et al. 2009; Li et al. 2010; Lobell et al. 2012; Dreccer et al. 2018). Heat stress events (unusually high temperatures during heat waves) have contributed to food shortages and increased food prices (Zampieri et al. 2017). We therefore need to understand the requirement for mild temperatures to identify strategies to mitigate against climate change.

4.1 The Temperature Requirement for Wheat

As previously stated, the optimum temperatures for wheat growth and development usually lie between 17 and 23 °C. Beyond this range there will be a minimum temperature below which there is no growth or development (the base temperature), and a maximum temperature above which progression will also be curtailed.

If temperatures remain between the lower and optimal limits, development may be more closely related to the passage of accumulated thermal time (degree days, °Cd), than to the passage of time (days, d) (McMaster 1997; Slafer et al. 2015). Thermal time can be calculated by summing the thermal time accumulated on each day, i.e.

$$°Cd = \left(\frac{Day\ max.temperature + Day\ min.temperature}{2} \right) - Base\ temperature \tag{4.1}$$

A closer correlation between thermal time and developmental progression is possible if the base temperature varies with growth stage. Thus, Porter and Gawith (1999) estimated base, optimum, and maximum temperatures (i.e. the *cardinal* temperatures) for a range of processes in wheat (Table 4.1). Overall, the base temperatures tend to increase for processes later in crop development.

Between the base and optimum temperatures, wheat passes through its developmental growth stages more quickly as the temperature rises (Midmore et al. 1982; Slafer et al. 2015). In this context, however, *optimal* temperatures for rates of developmental progression do not necessarily lead to the highest grain yields. This is because a shorter growth period leads to reduced time for resource capture, assimilate accumulation, and partitioning to the grain. There are many instances where an increase in temperature from base to optimum

Table 4.1 Base, optimum, and maximum temperatures (°C) for wheat growth and development.

	Base	Optimum	Maximum
Emergence from sowing	3.5	22	33
Leaf initiation	−1	22	24
Shoot growth	3	20	>21
Root growth	2	<16	>25
Vernalisation	−1	5	16
Terminal spikelet	1.5	11	>20
Meiosis	8	20	35
Anthesis	10	21	32
Grain filling	9	21	35

Source: From Porter and Gawith (1999), Subedi et al. (1998), and Semenov et al. (2014).

increases the *rate* but reduces the *duration* of growth and development. Above the optimum temperature it is common for both rates and durations of growth and developmental processes to be reduced.

If it is accepted that the temperatures for growth and development are universal, it follows that cultivars must vary in other ways to explain the wide distribution of wheat production. For example, in commercial production, the total thermal time taken for wheat to mature can range from around 1000 to 2500 °Cd (Batts et al. 1996; Saiyed et al. 2009; Barber et al. 2015; Hyles et al. 2020). This wide variation shows the extent to which individual cultivars can be adapted to a region, climate, and temperature profile. Cultivars not only need to avoid or acclimate to harsh conditions, but also to adapt their developmental profiles to exploit key periods for capturing light, water, and nutrients, and to increase the partitioning of biomass to the grain (Reynolds et al. 2009; Foulkes et al. 2011; Barber et al. 2015). Figure 4.1 demonstrates the importance of pre-anthesis duration, in thermal time, for increasing grain number in an oceanic climate. However, in this climate, it is possible to exploit high levels of light availability and moderate soil moisture contents with extended grain filling periods. Increased mean grain weights can, therefore be associated with earlier flowering (Figure 4.1; Austin 1999). The balance between achieving high numbers of grain and the availability of photosynthate to fill them contributes to relatively well-defined timings for anthesis in different environmental contexts. Finally, well-adapted cultivars will ripen to allow harvesting in drying conditions and in time for preparation for the following crop.

Variation in a range of traits contributes to the adaptation of cultivars; these include requirements for breaking seed dormancy, for passing through a cold period (vernalization) and/or for days to lengthen (long-day photoperiodism) before floral initiation and timely flowering, and for the ability to acclimate to cold. Dormancy, vernalization, and long-day photoperiodism are wild-type adaptations that ensure that wheat seeds do not germinate until conditions are cool, that reproductive development does not proceed until excessive cold periods have passed and light receipts are increasing, and that meiosis, anthesis, and grain filling are completed before temperatures are too high. Even when vernalisation and

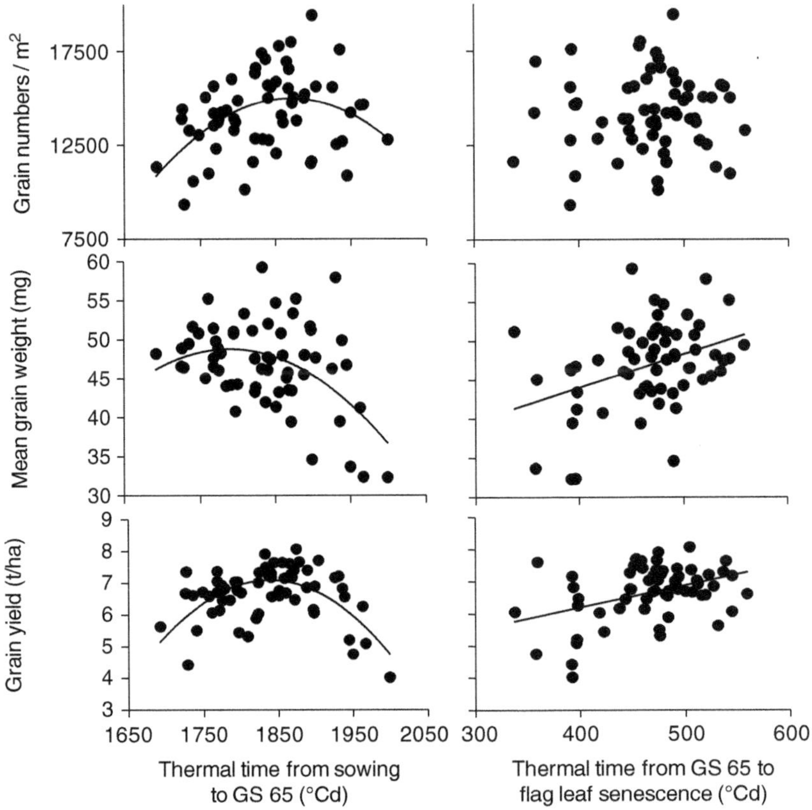

Figure 4.1 The effect of thermal time before and after mid-anthesis (GS 65) on grain yield components of field-grown winter wheat. Values are line means of 62 doubled haploid progeny of cv. Savannah × Renesansa (Germplasm Resource Unit, John Innes Centre) replicated twice in each of three years at Reading, UK. *Source:* Data from Barber (2017) and Carney (2016).

photoperiod requirements are fully met, developmental progression with thermal time still varies substantially between cultivars. These differences have been ascribed to allelic variation in *earliness* per se (*EPS*) genes (Hyles et al. 2020; Prieto et al. 2020).

4.2 'Waiting for Fine Times' (Snape et al. 2001)

4.2.1 Dormancy

Dormant grains will not germinate even though the conditions (e.g. with adequate oxygen, water, and temperature) are suitable for the germination of non-dormant grains (Lunn et al. 2002). Wheat seeds have 'high temperature dormancy', being less likely to germinate above 10 °C (Figure 4.2). Dormant seeds, therefore, will not germinate until the soil temperature is sufficiently low. For wild plants in the Middle East, this ensures that seeds do not germinate until the onset of cooler and wetter conditions, even though they may be shed in May and June (Cook and Veseth 1991).

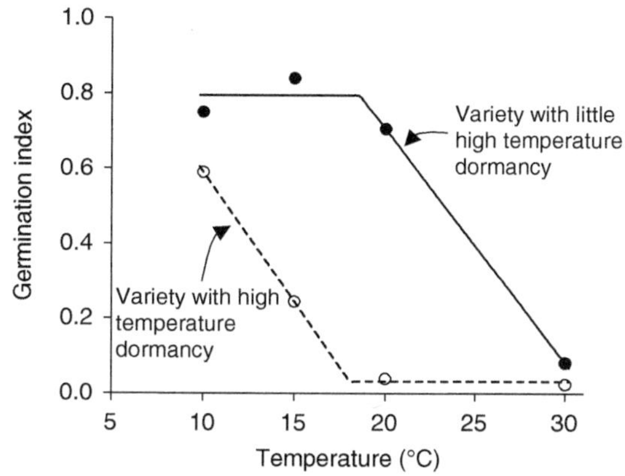

Figure 4.2 The effect of temperature and wheat cultivar on an index that accounts for both speed and level of germination. *Source:* Data from Nyachiro et al. (2002).

In many wheat-growing areas, wheat is not sown until the cooler months. In this case, even seeds with high temperature dormancy will germinate normally once they have imbibed sufficient moisture and temperatures are above a base minimum. However, there are situations where seed germination and establishment are required in warm seedbeds. Here, dormancy is undesirable. Cultivars of wheat can be bred with reduced levels of dormancy although the seeds of these cultivars are susceptible to pre-harvest sprouting in the ear if the conditions are wet.

The level of dormancy and risk of pre-harvest sprouting depend greatly on the wheat cultivar and the environment during grain filling (Lunn et al. 2002; Kindred et al. 2005; Biddulph et al. 2008; Gooding et al. 2012). If cultivars are to be grown that express high-temperature dormancy to prevent pre-harvest sprouting, but will also be sown into warm seedbeds, it is necessary to use grain where the dormancy has declined through several months of storage, or been broken with, for example, chilling treatments (e.g. 4 °C; Cook and Veseth 1991; Henry and Brennan 1988).

4.2.2 Cold Acclimation

Although vegetative growth stops at about 0 °C, wheat plants can acquire an ability to survive temperatures significantly below freezing while in the vegetative growth stages. This acclimation for *winter hardiness* and *frost tolerance* starts when temperatures fall to between 0 and 10 °C. Tolerance of low temperature has been associated with sucrose accumulation, synthesis of unsaturated phospholipids and protective proteins, small cell size, reduced water contents, and downregulation of growth-promoting hormones. Genes for low-temperature tolerance are highly expressed during the period before floral initiation. Low temperature tolerance can therefore be increased by increasing the vernalisation and photoperiod requirements, and by increasing the thermal time between the initiation of successive leaves (Limin and Fowler 2002). The combination of an appropriate cultivar with a suitable time for acclimation may allow wheat to survive short periods of air temperatures as low as −17 to −20 °C, particularly if there is a protective covering of snow (Percival 1921; Porter and Gawith 1999; Fowler and Limin 2007). Wheat plants have a greatly reduced ability to acclimate to cold after floral initiation, (Xu and Chong 2018) and susceptibility to cold increases significantly as the growing point is raised above ground level.

4.2.3 Vernalisation

Vernalisation, literally *'making ready for spring,'* (Snape et al. 2001) is an adaptation that allows wheat to persist over cold winters. The vernalisation requirement is for plants to pass through an extended period of cold before timely floral initiation. It prevents premature floral initiation and thus the detrimental effects on plant survival or ear sterility that would otherwise occur if the reproductive meristem experienced excessive cold (Limin and Fowler 2006; Fowler and Limin 2007). Effective vernalising temperatures commonly range from around 3 to 10 °C (Miralles and Slafer 1999) and the mean optimum of 5 °C in Table 4.1 hides wide variation between 0 and 12 °C (Porter and Gawith 1999). Plants can also become *devernalised* if the period of cold is interrupted by temperatures above 20–30 °C (Miralles and Slafer 1999).

A genetically determined vernalisation requirement is common in the wild relatives of wheat; it would have evolved as a survival mechanism in the colder areas of Turkey and beyond. The requirement for vernalisation is often used to distinguish winter wheats from spring wheats. Acevedo et al. (2002) found that their spring wheat cultivars required 7–18 °C for 5–15 days for floral initiation, while the winter wheats required 0–7 °C for 30–60 days. However, this distinction is not absolute and even spring wheats may still initiate reproductive development earlier after experiencing some cold (Miralles and Slafer 1999). Nonetheless, landraces with the winter wheat growth habit and a significant vernalisation requirement developed in areas where the mean temperature of the coolest month (i.e. January in the northern hemisphere) ranged from about −7 to 5 °C. Spring wheats are often drilled after winter to avoid frost injury in areas where winters are more severe. In areas with milder winters, spring cultivars may be sown in the autumn and cultivars that have a significant vernalisation requirement should be avoided (Iwaki et al. 2000, 2001).

Variation in the vernalisation requirement of wheat is largely associated with alleles at the *VRN1* loci on the long arms of the group 5 chromosomes, with recessive alleles at each locus being required for the full winter growth habit (*vrn-A1b*, *vrn-B1b* and *vrn-D1b*). *VRN1* is a flowering promoter but its transcription appears to be suppressed by epigenetic factors and protein binding until a period of cold (Hyles et al. 2020). Transcription of *VRN1* increases during cold and an accumulation of transcripts in the shoot apex is associated with a switch to reproductive development.

The effect of *VRN1* interacts with those of other genes. *VRN2* (also on chromosome 5A) is a repressor of flowering and prevents flowering in long days before winter. The *VRN1* protein downregulates *VRN2* by binding to its promoter region, thereby allowing plants to initiate flowering as days get longer after winter. Mutations in the promoter and first large intron of *VRN1* prevent its repression in the cold and thereby contribute to a spring wheat growth habit (Hyles et al. 2020). Further effects may result from increased copy numbers of the *VRN1* alleles (Díaz et al. 2012). In addition to the effects related to the onset of reproductive development, *VRN1* alleles are often linked to, and interact with, other genes conferring cold tolerance and acclimation (Reddy et al. 2006; Båga et al. 2007; Hyles et al. 2020) and therefore survival over winter.

Allard et al. (2012) emphasised the importance of differential rates of organ initiation at low temperatures. They reported that the time to floral initiation was *reduced* as temperatures *increased* from 1 °C to a maximum rate at 8 °C for the winter wheat phenotype (with recessive alleles at all *VRN1* loci). Flowering, however, was still delayed because more leaf primordia were produced at the higher temperatures and shorter durations of cold. Ultimately, flowering is delayed until the leaves initiated on the meristem have developed to produce emerged leaves, i.e. the flowering date is highly dependent on the final leaf number and how long each leaf takes to emerge.

The diversity in vernalisation loci, alleles, and copy numbers results in variation in the latest safe sowing dates amongst winter wheats. If winter wheat cultivars are sown later in autumn than recommended, they are at significant risk of low grain yields due to poor reproductive development. This is perhaps most relevant to oceanic climates where late sowing of winter wheat risks vernalisation requirements not being met, particularly if the winter, or early spring, is milder than usual. Conversely, cultivars with reduced vernalisation requirements can often be sown successfully in the autumn in the oceanic zone, while early autumn sowing of such *spring* cultivars risks poor acclimation against frost damage with premature floral initiation, particularly if the winter is colder than average.

4.2.4 Photoperiodism

In addition to vernalisation, premature floral development and flowering can be prevented by a requirement for days to lengthen. The need for long days also ensures that important growth stages for determining grain numbers and mean grain weights occur when light receipts are high. A requirement for long days can therefore be associated with improved grain yields due to increased resource capture before anthesis (Foulkes et al. 2004; Addisu et al. 2010), and increased numbers of ears, functional florets (Prieto et al. 2018), and grains, especially in higher latitudes and in areas with mild summers.

Photoperiod sensitivity in wheat is particularly associated with alleles at the *PPD1* loci on the short arms of the group 2 chromosomes. The wild-type alleles (*ppd-A1b*, *ppd-B1b*, *ppd-D1b*) are associated with accelerated flowering during long days. Dominant mutations are associated with progression to flowering irrespective of day length (Barber et al. 2015; Ochagavía et al. 2018; Pérez-Gianmarco et al. 2018). If the days are short, mutations at *PPD1* reduce the thermal time between the emergence of successive leaves and hasten the thermal time to stem extension and ear emergence (Table 4.2). In hexaploid wheats, *Ppd-D1a* has a greater accelerating effect compared with *Ppd-B1a* (Jones et al. 2017). Under field conditions in the UK, *Ppd-D1a* is associated with more rapid development from early growth stages (Figure 4.3), with plants flowering up to 14 days earlier than photoperiod-sensitive genotypes (Snape et al. 2001).

Ppd-D1a was introduced in foundational breeding programmes in the twentieth century. These programmes included the growing of southern European wheats from the Italian breeding programme of Nazareno Strampelli, as well as worldwide breeding programmes through the distribution of CIMMYT material originating with Norman Borlaug in Mexico (Salvi et al. 2013). Photoperiod insensitivity may be selected in a breeding programme through serendipity if successive generations are planted in different regions. This applies particularly when locations are chosen to facilitate the completion of more than one breeding cycle in each year. In many breeding programmes, however, earliness and therefore photoperiod insensitivity are desirable for the crop to escape stress when drought and heat

Table 4.2 The effect of mutations at the PPD1 loci on development of near isogenic lines of Paragon spring wheat under short days (mean day length = 9 h, mean temperature 13 °C). Lines provided by the Germplasm Resource Unit, John Innes Centre.

PPD1 alleles	Number of photoperiod insensitivity alleles	Leaf number after 1000 °Cd	Growth stage at 1500 °Cd		
			Tiller number	Node number	Median growth stage score (Zadoks et al. 1974)
A1b, B1b, D1b	0	4.6	2.1	1.4	First node detectable (DGS 31)
A1b, ***B1a***, *D1b*	1	4.9	2.7	3.3	Mid booting (DGS 45)
A1a, B1a, *D1b*	2	5.3	1.7	3.4	Mid ear emergence (DGS 55)
A1a, B1a, D1a	3	5.1	2.1	3.1	Full ear emergence (DGS 59)
S.E.D. (6 d.f.)		0.13	0.31	0.20	

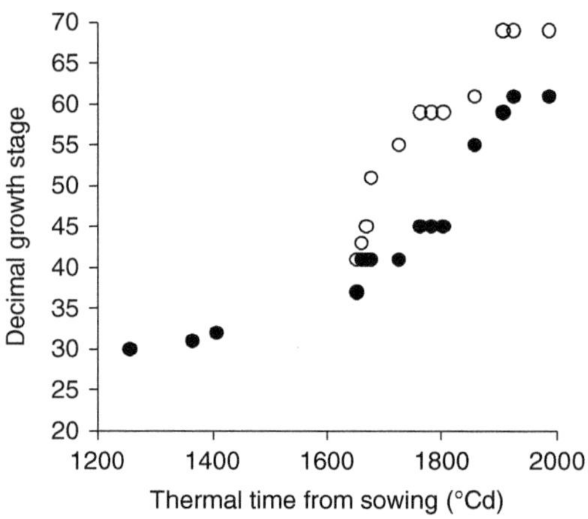

Figure 4.3 Modal decimal growth stage (DGS) scores for 64 doubled haploid progeny of Renesansa (*Ppd-D1a*, ○) and Savannah (*ppd-D1b*, ●) when field grown in the UK. *Source:* Data from Barber et al. (2015).

develop to excessive levels during summer (Bentley et al. 2013; Salvi et al. 2013; Sanchez-Garcia and Bentley 2019). This is particularly the case in Mediterranean climates. Avoidance of increasing temperatures is also relevant for wheat grown in humid subtropical climates. In these areas it is also necessary for the crop to mature before the early monsoon rains, which encourage pre-harvest sprouting and hinder harvesting.

The requirement for earliness and photoperiod insensitivity may become more widespread in climate change scenarios (Dowla et al. 2018) that predict warmer summer temperatures, and an increase in the frequency of heat waves and droughts during susceptible growth stages. However, the relevance of *PPD1* mutations to climate change adaptation requires further study. *PPD1* alleles influence the duration of flowering at the plant and crop levels (Jones et al. 2017); this in itself may affect resilience to heat stress depending on the length and frequency of high temperature events during susceptible growth stages (Lukac et al. 2012). Although *PPD1* mutations may help escape harsh summer conditions, their effect on tolerance of stress is less clear. For example, in one parental cross and environment *Ppd-D1a* has been associated with increased susceptibility to heat stress at anthesis (Barber et al. 2017).

In addition to adaptation related to temperature and water, *Ppd-D1a* has been associated with increased nitrogen capture early in the season in organic production systems, particularly if the wheat follows a legume-rich ley (Addisu et al. 2009; Gooding et al. 2012). Earliness may be beneficial in such systems, where nitrogen is likely to be plentiful initially but susceptible to loss, becoming deficient later in the season.

A second photoperiod locus, *PPD2*, is present on the long arms of the group 1 chromosomes (Zikhali et al. 2017). The causative gene is *FT-3*, which acts to promote heading during short days. After vernalisation is satisfied, this can be viewed as an early spring signal to promote the vegetative to floral transition before the days are long enough to activate *PPD1* (Zikhali et al. 2017).

4.2.5 Earliness per se

EPS alleles explain differences in developmental rate with thermal time when vernalisation and photoperiod requirements have been met (Zikhali and Griffiths 2015). They are considered to have smaller effects than variation at the *VRN1* and *PPD1* loci, but allow the fine-tuning of adaptation to regions and environments. Many *EPS* loci are postulated (Snape et al. 2001) but the effects of only a few have been characterised. Several *EPS* genes correspond to components of the circadian clock, enabling wheat to detect and respond to the diurnal cycle of night and day. Analysis of near-isogenic lines showed that *EPS* alleles appear to interact with both temperature and photoperiod (Zikhali and Griffiths 2015; Hyles et al. 2020; Prieto et al. 2020).

4.3 Vegetative Growth and Development

4.3.1 Germination and Emergence

In the absence of dormancy, germination can occur over a wide range of temperatures. Germination at temperatures of 4 °C and lower is possible but is slow and uneven. The fastest rate of germination happens at 20–25 °C (Figure 4.4; Evans et al. 1975).

Seedlings can progress from sowing to emergence over a wide range of temperatures (Table 4.1). For autumn sown wheats, however, it is important that sowing is early enough to facilitate germination and establishment before progressively colder weather delays emergence. Data provided by Percival (1921) and Hay and Porter (2006) from the UK show that once the mean monthly temperatures fall below 9 °C the speed of crop emergence falls substantially (Figure 4.5). This means that delaying the sowing of winter wheat beyond October risks emergence being slowed by several weeks as temperatures continue to decline into winter. In continental climates with colder winters, late sowing of autumn crops can result in wheat not emerging until the following spring.

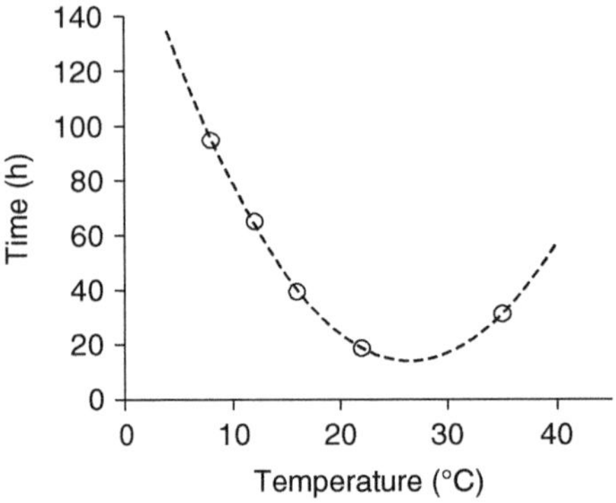

Figure 4.4 The effect of temperature on time to 50% germination. *Source:* From data in Percival (1921).

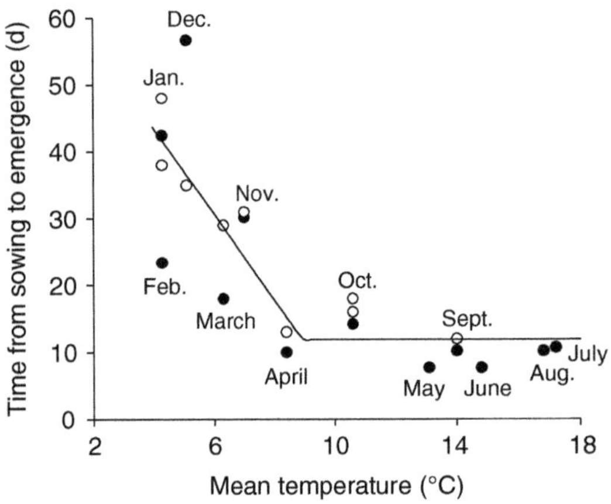

Figure 4.5 Relationship between mean temperatures of month of sowing in UK and time taken for seedlings to emerge. *Source:* Data from Percival (1921), ●; and Hay and Porter (2006), ○.

4.3.2 Leaves

By the time a wheat plant has emerged, six leaves have usually been initiated on the apical meristem. Four of these are differentiated in the embryo of the grain at sowing (Hay and Kirby 1991). Subsequent leaf initiation is highly dependent on the temperature of the meristem (Jamieson et al. 1995). The thermal time between the appearance of leaf primordia is known as the *plastochron* and remains near constant during the vegetative phase of a particular wheat crop. A typical quoted value is 50 °Cd (Miralles and Slafer 1999), although one detailed study reported variation between 46 and 72 °Cd (mean = 55 °Cd) among genotypes and experiments (Ochagavía et al. 2018). The further expansion and unfolding of leaves are also dependent on the temperature of the relevant tissues. The thermal time between the external emergence of leaves, (e.g. when the leaf tip or collar become visible) is known as the *phyllochron*. As with the plastochron, the phyllochron is close to constant within an experiment and genotype, i.e. successive leaves emerge after a similar thermal time when temperatures do not exceed an optimum value of about 22 °C (Porter and Gawith 1999; Hay and Porter 2006). Figure 4.6 shows the emergence of the later leaves of two winter wheat cultivars with a phyllochron of about 100 °Cd above a base temperature

Figure 4.6 The relationship between leaf emergence (collar visible) and thermal time for two cultivars of winter wheat. *Source:* Data from Barber (2017).

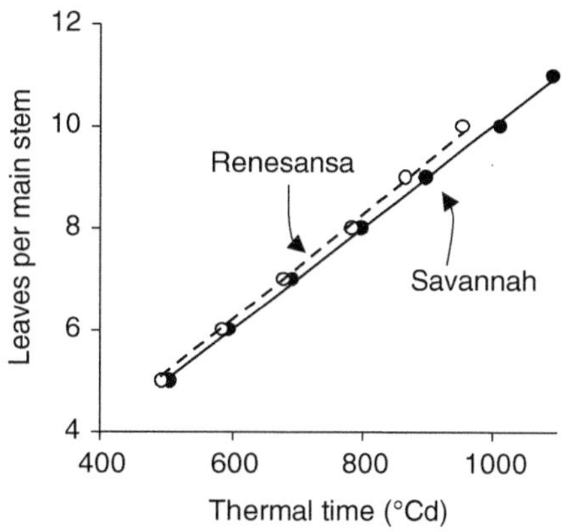

of 0 °C. This is within the range of 70–160 °Cd often quoted for wheat (Kirby and Perry 1987; Porter and Gawith 1999; Hay and Porter 2006). Despite the linearity apparent in Figure 4.6, some curvature may be evident with leaf number, i.e. the phyllochron can vary between earlier and later emerging leaves (McMaster 1997). Furthermore, the phyllochron is not consistent between cultivars (Cao and Moss 1989), nor over different environments (Slafer et al. 2015). The phyllochron is reported to vary with sowing date and latitude (Cao and Moss 1991; Hay and Porter 2006), although it is possible these apparent effects largely result from the disparity between the measured temperature (usually air temperature) and the temperatures of the vegetative meristem (mostly below ground level) and other plant tissues (Jamieson et al. 1995). Stresses such as aluminium toxicity, nutrient deficiency, and water stress (Cutforth et al. 1992; Longnecker and Robson 1994; Valle and Calderini 2010) all increase the phyllochron (i.e. reduce the rate of leaf emergence with thermal time).

The final leaf number will be the six already initiated by emergence, plus the production of any further subsequent initials. Leaf primordia will continue to increase at the rate of the plastochron while, e.g. the vernalisation and photoperiod requirements are being met (Brooking et al. 1995; Robertson et al. 1996). For vernalised plants grown in long days, the increase in the rate of production of leaf primordia as temperatures rise from 10 °C to around 25 °C is broadly matched by a reduction in the duration of production (i.e. floral initiation comes earlier). In these circumstances, therefore, temperature has little effect on final leaf number (Friend et al. 1963; Slafer and Rawson 1994a, b, 1995). When temperatures rise above 25 °C, however, the final leaf number is reduced because the production of leaf primordia is retarded (Friend et al. 1963).

The effect of temperature on final leaf size is complex. In controlled conditions, raising the temperature to around 20 °C can increase the area of leaf laminas through a lengthening associated with increased extension of leaf cells (Friend 1966). In the field, however, the effects of higher temperatures can be confounded by increased light intensity and water shortage, both of which are associated with reduced leaf areas. Additionally, although temperatures up to 20 °C can accelerate leaf extension, the duration of expansion can be reduced such that final leaf length may not be greatly affected by variation in mean temperature in field conditions (Gallagher et al. 1979). In addition to environmental effects, there are strong genetic and growth stage effects on leaf size. For example, Bos and Neuteboom (1998) found that one of the best predictors of leaf size was the size of the previous leaf.

In addition to its influence on the size and number of leaves, temperature also has significant effects on leaf function and longevity. In controlled conditions, temperatures above 25–30 °C are associated with reduced chlorophyll contents and lower rates of net photosynthesis (Friend 1966; Al-Khatib and Paulsen 1990, 1999; Zhao et al. 2007; Begcy et al. 2018); on the other hand, higher temperatures may increase photosynthesis if the CO_2 concentration is also increased (Alonso et al. 2008). More generically, higher temperatures are associated with a reduced duration of photosynthesis (Al-Khatib and Paulsen 1990). Senescence of the flag leaf occurs earlier as the temperature increases (Spiertz 1977; Al-Khatib and Paulsen 1984; Harding et al. 1990). In the field, and at regional levels, rapid crop senescence during grain filling has been associated with temperatures above 34 °C (Lobell et al. 2012).

4.3.3 Tillers

Tillers can be produced from axillary meristems in each leaf axil. The production rate of potential axillary meristems can therefore be related to the leaf plastochron (McMaster 1997). As with leaf initials, the production of axillary meristems ceases on floral initiation. The rate and duration of production of axillary meristems are therefore increased and reduced, respectively, as the temperatures increase. Production will also continue while the vernalisation and photoperiod requirements are being met. Whether the axillary buds develop to produce emerged tillers depends on the genotype and on environmental factors, including light intensity, light quality, and nutritional status.

The first tiller usually emerges with the appearance of the fourth leaf (Miralles and Slafer 1999); the subsequent tiller emergence can be correlated with the passage of leaf phyllochrons, provided the supply of resources is sufficient (Rodríguez et al. 1998; Miralles and Slafer 1999). Temperature, however, has opposing effects on the rate and duration of tiller production; there are varying requirements for vernalisation and photoperiod, and different levels of competition for resources between plants of different genotypes. There is not, therefore, a simple relationship between temperature and the total numbers of tillers produced. Hucl and Baker (1990), for example, report that raising mean daily temperatures from 8.3 to 16.7 °C led to increased, reduced, and the same number of tillers, depending on the cultivar observed. As with leaves, increasing temperatures may also hasten tiller death (Thorne and Wood 1987).

4.3.4 Roots

Root growth can proceed at lower temperatures than shoot growth (Evans et al. 1975; Table 4.1). Increasing temperatures can, therefore, reduce the root: shoot ratios of dry and fresh weights (Vincent and Gregory 1989; Equiza et al. 2001). Winter wheats often have larger root systems than spring wheats; this may partly result from a longer period of root growth for autumn-sown crops, even though much of this extra root growth is at relatively low temperature.

4.4 Reproductive Growth and Development

Floral initiation marks the point when the production of leaf primordia on the meristem ceases and the production of the ear structures begins. As already noted, when vernalisation and photoperiod requirements are satisfied, and as the temperature increases from 10 to 25 °C, the time to floral initiation is reduced such that final leaf number may be conserved.

4.4.1 Spikelet Formation and Stem Extension

Following floral initiation, the primordia for spikelets, as for leaves, are initiated according to a *plastochron* that is relatively constant with thermal time (Ochagavía et al. 2018). The plastochron for spikelet primordia is shorter than that for leaves (mean = 19 °Cd; range = 14–27 °Cd in Ochagavía et al. 2018). As for leaves, however, the duration of primordia production is reduced as the temperature increases up to 25 °C, so that the effects of lower temperatures on final spikelet number are often only small (Slafer and Rawson 1994b) or negative (Frank et al. 1987). The rates of subsequent stem extension and floral development to ear emergence are highly dependent on temperature; increasing to at least 30 °C (Friend et al. 1963). The duration of extension and development are again reduced by higher temperatures so that, for example, the plants ultimately become taller at lower temperatures (Friend 1965). The rapid development during this phase compromises grain yields due to reduction in grain numbers. This occurs when the accumulated light capture and carbon assimilation needed for spike growth and floret survival are reduced (Foulkes et al. 2011).

4.4.2 Meiosis and Anthesis

The period lasting from meiosis to anthesis is particularly vulnerable to both heat and cold stresses, with susceptibility lasting from the start of booting to the end of flowering (Figures 4.7 and 4.8). There is a short period of relative tolerance during late ear emergence (between meiosis and anthesis) (Figure 4.7). This tolerance becomes more diffuse as periods of stress are extended for longer than one day (e.g. Prasad et al. 2017), so its relevance to field-grown wheat, where heat waves may last for several days, may be minimal. In broad terms, significant reductions in grain numbers and yield can occur if day/night temperatures fall below 8/2 °C or exceed 32/28 °C during this period. Grain yield is reduced because any subsequent increases in mean grain weight are insufficient to compensate for poor grain set (Figures 4.7 and 4.8).

4.4.2.1 Heat Stress

Heat stress generally occurs above 32 °C, but there is significant variation associated with genotype, water availability, and growth stage. It results in reduced rates of photosynthesis and transpiration (Begcy et al. 2018), with consequences for pollen and stigma anatomy, sugar and lipid biochemistry, and oxidant levels. All these effects could contribute to poor pollen and ovule function (Figure 4.9), reduced grain set, and ultimately lower grain yield. Additionally, heat stress may reduce grain set by reducing the synchrony between pollen and ovule development (Saini et al. 1983; Lukac et al. 2012).

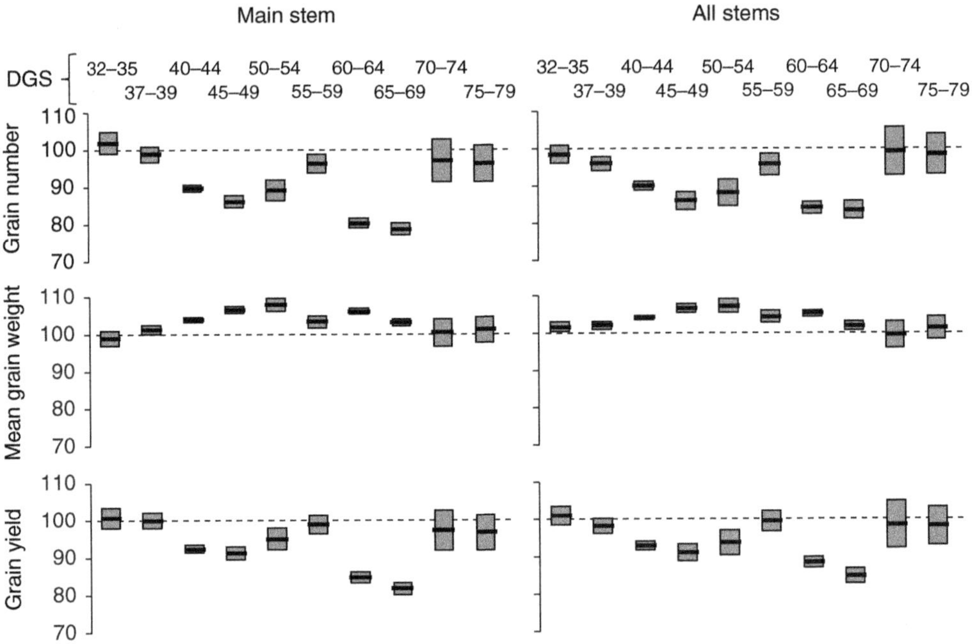

Figure 4.7 The effect of a one-day transfer to 35/30 °C day/night temperature (16 h day) during different growth stages of the main stem. Values are expressed as a percentage of the result of transferring to 20/15 °C. DGS are the decimal growth stage scores as per Table 1.4 (Zadoks et al. 1974) except that they are the mean of scores at the beginning and end of the day of transfer. Bars are + and − standard error of the difference to the 100% line. Data are from the three experiments described in Barber et al. (2017) including 65 wheat lines varying for photoperiod sensitivity (*Ppd*) and reduced height (*Rht*) alleles. All pots were at field capacity for water at the start of the day of transfer.

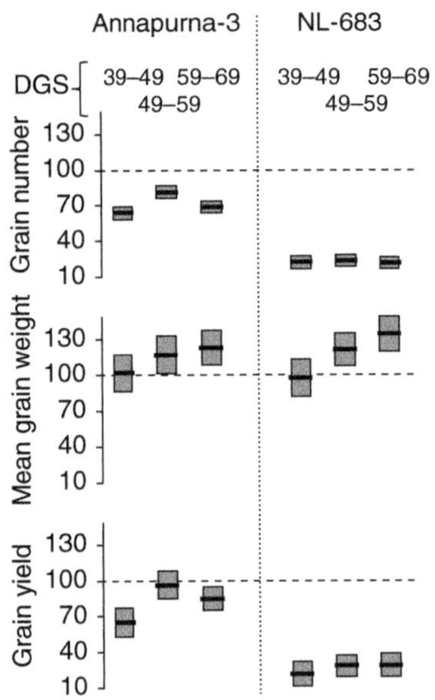

Figure 4.8 The effect of transfers to 8/2 °C day/night temperature (14 h day) during different growth stages of the main stem of two cultivars contrasting in susceptibility to cold temperatures. Values are expressed as a percentage of the results from plants maintained at ambient conditions. DGS are the decimal growth stage scores as per Table 1.4 (Zadoks et al. 1974). Bars are + and − standard error of the difference to the 100% line. *Source:* Data are from Subedi et al. (1998).

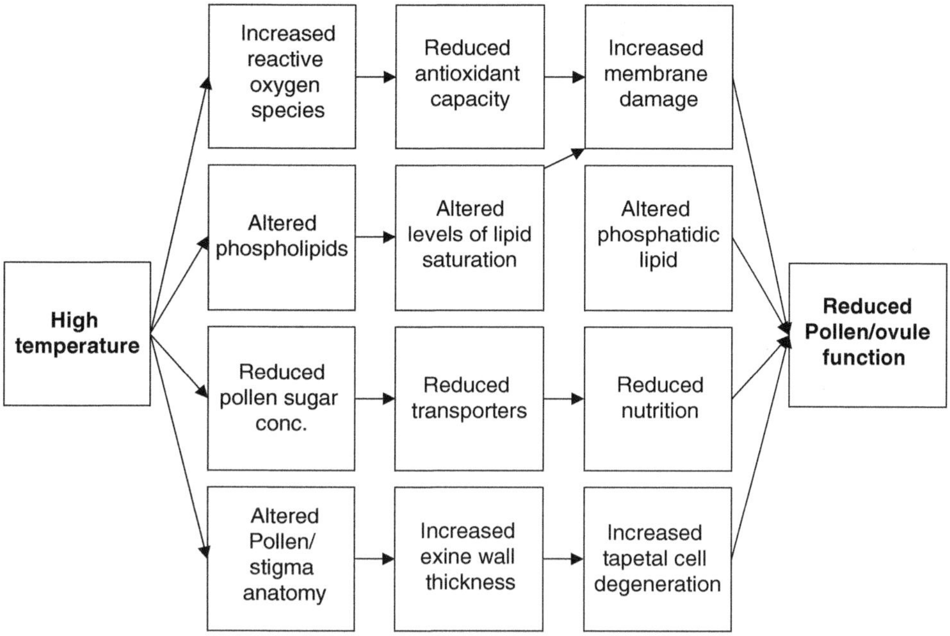

Figure 4.9 Reported responses to heat stress during meiosis and anthesis associated with reduced grain numbers being set. *Source:* Adapted from Prasad et al. (2017).

The effects of heat stress during booting depend greatly on the availability of water. In well-watered conditions, grain set may not be significantly affected until day temperatures exceed 35 °C (Figure 3.9; Alghabari et al. 2014; Semenov et al. 2014). However, under drought stress conditions, the grain set can be reduced by day temperatures above 25 °C. Even when water is not limited, high temperatures at booting may still lead to abnormal ovary development, disruption to meiosis in the pollen mother cells, disorientation of the developing pollen from the tapetum, and premature degradation of the tapetum (Saini et al. 1983, 1984; Draeger and Moore 2017).

The interactions between heat stress and drought are less at anthesis than at booting (Figure 3.9). The effects of excessive heat can, however, be particularly damaging at high levels of atmospheric humidity, which reduce evaporative cooling (Tashiro and Wardlaw 1990). The greatest susceptibility to heat stress occurs at flower opening (or glume loosening [Lukac et al. 2012]), which exposes the floral organs to the immediate microclimate (Prasad et al. 2017). This is broadly consistent with the peak susceptibility being during the two to five days prior to GS 65 (Tashiro and Wardlaw 1990; Wheeler et al. 2000; Barber et al. 2017).

There is genetic variation in heat tolerance during the period between pollen formation and anthesis (Barber et al. 2017; Begcy et al. 2018), but conflicting reports as to whether alleles conferring reduced sensitivity to gibberellic acid reduce heat stress tolerance. For example, the widely used semi-dwarfing alleles *Rht-B1b* and *Rht-D1b* have been reported to be associated with negative, zero, and positive effects on heat stress tolerance at anthesis (Alghabari et al. 2014; Barber et al. 2017). When compared under similar conditions,

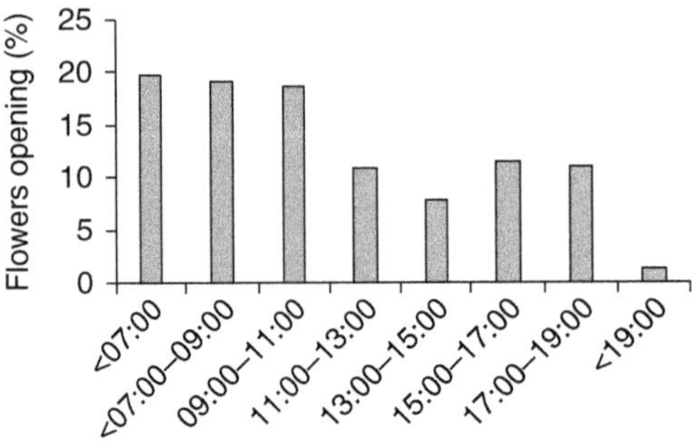

Figure 4.10 The proportion of flowers of a Hungarian wheat opening at different times of the day. *Source:* Data from Percival (1921).

cultivars adapted to regions with hotter summers do not necessarily have improved heat tolerance at anthesis. For example, wheats from south Europe can have similar cardinal temperatures to wheats from northwest Europe (Semenov et al. 2014; Barber et al. 2017). However, wheats from south Europe are less likely to be daylength sensitive and often carry the *Ppd-D1a* allele (i.e. they escape heat stress by flowering earlier in the year). A further strategy to *escape* heat stress is to flower during cooler parts of the day (Figure 4.10; Jagadish 2020).

4.4.2.2 Cold Stress

Day/night temperatures of 8 °C/2 °C and below at meiosis can result in reduced grain numbers (Subedi et al. 1998). The anthers from cold-stressed plants are small, pale, and shrunken, the process by which the mature anther splits open and releases pollen (anther dehiscence) is compromised, and the amount of viable pollen produced is significantly reduced (Subedi et al. 1998). However, cold stress at anthesis is rare in the major wheat-producing regions because both autumn-sown winter wheats and spring wheats enter meiosis when the temperatures have risen to relatively warm conditions. Nevertheless, field sterility can occur if the spring is unseasonably cool, and if the cultivars are either poorly adapted or are sown too early such that they enter meiosis before temperatures have risen. In other situations, such as at altitude, conditions may be routinely cold at meiosis (Subedi et al. 1998; Law 2000). In these cases, it is important to exploit the wide genetic variation in susceptibility to low temperatures (e.g. Figure 4.8) and only select cold-tolerant cultivars.

4.4.3 Grain Filling and Quality

Heat stress events (> 30 °C) occurring shortly after anthesis have a greater negative impact on final mean grain weight than on grain number (Tashiro and Wardlaw 1990), particularly if combined with drought stress (Barber et al. 2015). These effects may not be evident for short periods of stress (e.g. Figure 4.7), but an increase in mean temperatures during grain filling shortens the time to canopy senescence, increases the rate of grain drying (Gooding et al. 2003), and hastens the end of grain filling (Wheeler et al. 1996a). Controlled environment studies suggest that for mean temperatures up to around 25 °C, the decreases in the duration of grain filling can be partially (but never completely) compensated for by an increase in the rate of grain filling. Increases in temperatures above 25 °C reduce mean

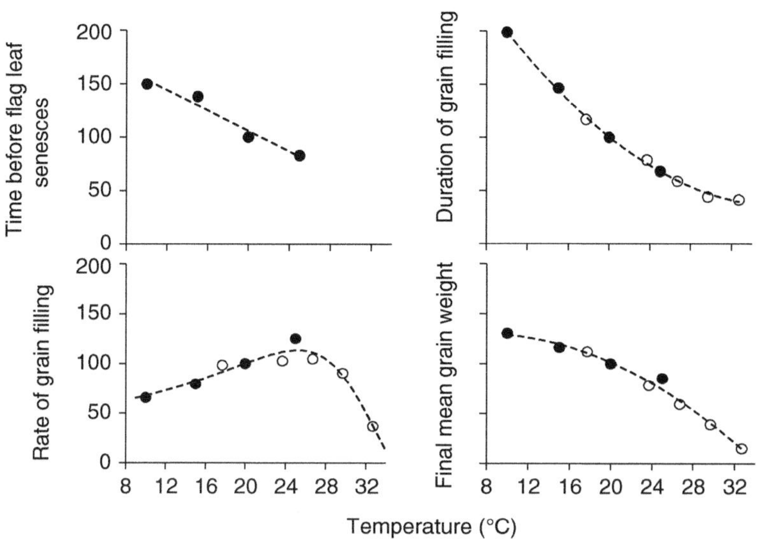

Figure 4.11 The effect of temperature on flag leaf senescence and grain filling. Derived from Spiertz (1977; ●), and Tashiro and Wardlaw (1989; ○). Values are relative to those of 20 °C = 100.

grain weights because both the rate and duration of grain filling are reduced (Figure 4.11). The effects of heat stress, however, become progressively less as the grain approaches maturity (Stone and Nicolas 1995a). They also vary significantly amongst genotypes (Stone and Nicolas 1995b), and may be less if associated with a period of gradually increasing heat to allow a degree of high-temperature acclimation (Stone and Nicolas 1995c). Wheeler et al. (1996a) showed that the negative effects of higher temperatures on the rate of grain filling in field-grown plots were eradicated at higher CO_2 concentrations. With current atmospheric CO_2 concentrations, however, it appears that the rapid senescence that occurs above 34 °C can severely limit grain filling (Lobell et al. 2012). This observation has been a major factor in the predictions of catastrophic yield losses in climate change scenarios for the warmer areas of wheat production (Asseng et al. 2011).

The effects of higher temperatures during grain filling on grain quality are often confounded with the effects of sunshine and water availability. For example, grain specific weight can be reduced if high temperatures exacerbate drought. However, specific weights can be improved if warmer conditions facilitate timely harvests before late rains cause grain weathering. Similarly, specific weights are increased with increased light capture during grain filling and sunny days are often associated with warmer conditions (Gooding 2017).

Increased temperatures during grain filling usually result in increases in grain nitrogen (protein) and sulphur concentrations. This is because high temperatures may be more damaging to dry matter accumulation than they are to nitrogen and sulphur accumulation. High protein concentrations are, therefore, often achieved in areas of the world where warmer (and drier) conditions result in shorter grain-filling periods. In cooler and wetter environments with longer grain filling periods, meeting grain protein thresholds for bread-making quality can be challenging, even when large quantities of nitrogen fertilizer are applied to increase grain N. Increased grain protein concentrations following warmer weather often result in improved breadmaking performance. However, excessive temperatures, above 32–35 °C, can lead to detrimental effects on protein and breadmaking quality (Chapter 8).

The effects of temperatures during grain filling on α-amylase activity and Hagberg falling number (HFN) depend greatly on cultivar. In cool (and wet) conditions, it is particularly important that cultivars develop sufficient dormancy to prevent pre-harvest sprouting. Similarly, cool conditions can lead to reduced HFN in cultivars predisposed to producing pre-maturity α-amylase (Gooding 2017; Derkx and Mares 2020).

4.5 Global Warming

The wide distribution of wheat production across the world would appear to belie its growth requirement for mild temperatures. This apparent contradiction, however, is largely resolved by the control and timing of critical growth stages to clement times of the year. The interactions between the numerous alleles influencing vernalisation requirements, photoperiodic responses, and *EPS*, and their varying effects on different growth stages and sowing dates, produce a vast range of maturation patterns and therefore adaptation to different temperate and continental climates.

Nonetheless, the mean temperatures, and the frequency of heat stress events, have increased in wheat-growing regions (Zampieri et al. 2017). Furthermore, these trends are predicted to continue (Easterling et al. 2000; Trnka et al. 2014; Ruane et al. 2018) with increasing temperatures expected to have negative effects on yields (Asseng et al. 2015). However, the effects of climate change on wheat yields are likely to be conditional on location; with predicted effects dependent on the interaction of higher temperatures with other changes in rainfall, atmospheric carbon dioxide concentration, cultivar development, and crop management (Semenov and Shewry 2011; Semenov et al. 2014; Liu et al. 2019; Harkness et al. 2020). For example, for wheat cultivars currently grown in the mildest climates, such as the oceanic areas, an increase in mean temperature alone will reduce yield due to faster development and reduced resource capture. However, this effect may be partially mitigated by beneficial effects of increased atmospheric CO_2 (Wheeler et al. 1996b; Senapati et al. 2020). In addition, increases in mean temperature advance the timing of susceptible growth stages, such as anthesis, so that they escape the higher predicted temperatures of late spring and summer (Rezaei et al. 2015). Other favourable effects are also predicted. For example, climate change and emission scenarios leading to hotter and drier summers in the UK could improve sowing and harvesting conditions and reduce lodging (Harkness et al. 2020). In Canada, with a more continental climate and colder winters, spring wheat yields are anticipated to increase with increasing temperatures because they will permit earlier sowing dates (Qian et al. 2019). However, even in the cooler areas, an increase in frequency of heat stress events (rather than just an increase in mean temperatures) will still threaten wheat yields (Trnka et al. 2014).

In warmer areas, further increases in mean temperature and the frequency of heat stress events are both predicted to cause significant reductions in wheat yields; these will challenge food security (Anwar et al. 2007; Ortiz et al. 2008; Li et al. 2010; Asseng et al. 2011;

Liu et al. 2019). Some mitigation through *escape* will be possible by the breeding of new cultivars and changing sowing dates; it may also be possible to improve the heat stress *tolerance* of new cultivars by exploiting genetic variation in heat tolerance of photosynthesis, meiosis, anthesis, and grain filling (Reynolds et al. 1998; Ortiz et al. 2008; Barber et al. 2017). Mitigating the effects of global warming on future wheat production will therefore be heavily dependent on the success of crop improvement programmes in optimizing wheat growth, development, and tolerance to the changing conditions (Semenov et al. 2014; Senapati et al. 2020).

References

Acevedo, E., Silva, P., and Silva, H. (2002). Wheat growth and physiology. In: *Bread Wheat Improvement and Production* (eds. B.C. Curtis, S. Rajaram and H. Gomez Macpherson), 39–70. Food and Agriculture Organization of the United Nations.

Addisu, M., Snape, J.W., Simmonds, J.R. et al. (2009). Reduced height (Rht) and photoperiod insensitivity (Ppd) allele associations with establishment and early growth of wheat in contrasting production systems. *Euphytica* 166: 249. https://doi.org/10.1007/s10681-008-9838-7.

Addisu, M., Snape, J.W., Simmonds, J.R. et al. (2010). Effects of reduced height (Rht) and photoperiod insensitivity (Ppd) alleles on yield of wheat in contrasting production systems. *Euphytica* 172: 169–181. https://doi.org/10.1007/s10681-009-0025-2.

Alghabari, F., Lukac, M., Jones, H.E. et al. (2014). Effect of Rht alleles on the tolerance of wheat grain set to high temperature and drought stress during booting and anthesis. *Journal of Agronomy and Crop Science* 200 (1): 36–45. https://doi.org/10.1111/jac.12038.

Al-Khatib, K. and Paulsen, G.M. (1984). Mode of high temperature injury to wheat during grain development. *Physiologia Plantarum* 61: 363–368. https://doi.org/10.1111/j.1399-3054.1984.tb06341.x.

Al-Khatib, K. and Paulsen, G.M. (1990). Photosynthesis and productivity during high-temperature stress of wheat genotypes from major world regions. *Crop Science* 30: 1127–1132 https://doi.org/10.2135/cropsci1990.0011183X003000050034x.

Al-Khatib, K. and Paulsen, G.M. (1999). High-temperature effects on photosynthetic processes in temperate and tropical cereals. *Crop Science* 39: 119–125. https://doi.org/10.2135/cropsci1999.0011183X003900010019x.

Allard, V., Veisz, O., Kõszegi, B. et al. (2012). The quantitative response of wheat vernalization to environmental variables indicates that vernalization is not a response to cold temperature. *Journal of Experimental Botany* 63 (2): 847–857. https://doi.org/10.1093/jxb/err316.

Alonso, A., Pérez, P., Morcuende, R. et al. (2008). Future CO_2 concentrations, though not warmer temperatures, enhance wheat photosynthesis temperature responses. *Physiologia Plantarum* 132: 102–112. https://doi.org/10.1111/j.1399-3054.2007.00997.x.

Anwar, M.R., O'Leary, G., McNeil, D. et al. (2007). Climate change impact on rainfed wheat in south-eastern Australia. *Field Crops Research* 104 (1–3): 139–147. https://doi.org/10.1016/j.fcr.2007.03.020.

Asseng, S., Foster, I., and Turner, N.C. (2011). The impact of temperature variability on wheat yields. *Global Change Biology* 17: 997–1012. https://doi.org/10.1111/j.1365-2486.2010.02262.x.

Asseng, S., Ewert, F., Martre, P. et al. (2015). Rising temperatures reduce global wheat production. *Nature Climate Change* 5: 143–147. https://doi.org/10.1038/nclimate2470.

Austin, R.B. (1999). Yield of wheat in the United Kingdom: recent advances and prospects. *Crop Science* 39 (6): 1604–1610. https://doi.org/10.2135/cropsci1999.3961604x.

Båga, M., Chodaparambil, S.V., Limin, A.E. et al. (2007). Identification of quantitative trait loci and associated candidate genes for low-temperature tolerance in cold-hardy winter wheat. *Functional and Integrative Genomics* 7: 53–68. https://doi.org/10.1007/s10142-006-0030-7.

Barber, H.M. (2017). Identifying beneficial traits for heat stress around reproductive phases of growth in wheat. PhD Thesis. The University of Reading

Barber, H., Carney, J., Alghabari, F. et al. (2015). Decimal growth stages for precision wheat production in changing environments? *Annals of Applied Biology* 166: 355–371. https://doi.org/10.1111/aab.12207.

Barber, H.M., Lukac, M., Simmonds, J. et al. (2017). Temporally and genetically discrete periods of wheat sensitivity to high temperature. *Frontiers in Plant Science* 8: 51. https://doi.org/10.3389/fpls.2017.00051.

Batts, G., Wheeler, T., Morison, J. et al. (1996). Developmental and tillering responses of winter wheat (Triticum aestivum) crops to CO_2 and temperature. *Journal of Agricultural Science* 127 (1): 23–35. https://doi.org/10.1017/S0021859600077340.

Begcy, K., Weigert, A., Egesa, A.O. et al. (2018). Compared to Australian cultivars, European summer wheat (Triticum aestivum) overreacts when moderate heat stress is applied at the pollen development stage. *Agronomy* 8: 99. https://doi.org/10.3390/agronomy8070099.

Bentley, A.R., Horsnell, R., Werner, C.P. et al. (2013). Short, natural, and extended photoperiod response in BC2F4 lines of bread wheat with different Photoperiod-1 (Ppd-1) alleles. *Journal of Experimental Botany* 64 (7): 1783–1793. https://doi.org/10.1093/jxb/ert038.

Biddulph, T.B., Plummer, J.A., Setter, T.L. et al. (2008). Seasonal conditions influence dormancy and preharvest sprouting tolerance of wheat (Triticum aestivum L.) in the field. *Field Crops Research* 107 (2): 116–128. https://doi.org/10.1016/j.fcr.2008.01.003.

Bos, H.J. and Neuteboom, J.H. (1998). Growth of individual leaves of spring wheat (Triticum aestivum l.) as influenced by temperature and light intensity. *Annals of Botany* 81 (1): 141–149. https://doi.org/10.1006/anbo.1997.0532.

Brooking, I.R., Jamieson, P.D., and Porter, J.R. (1995). The influence of daylength on final leaf number in spring wheat. *Field Crops Research* 41 (3): 155–165. https://doi.org/10.1016/0378-4290(95)00014-H.

Cao, W. and Moss, D.N. (1989). Temperature effect on leaf emergence and phyllochron in wheat and barley. *Crop Science* 29: 1018–1021. https://doi.org/10.2135/cropsci1989.0011183X002900040038x.

Cao, W. and Moss, D.N. (1991). Phyllochron change in winter wheat with planting date and environmental changes. *Agronomy Journal* 83 (2): 396–401. https://doi.org/10.2134/agronj1991.00021962008300020027x.

Carney, J. (2016). Combining early anthesis and extended canopy duration in UK winter wheat as a means to adapt to increased frequency of heat stress events caused by climate change. PhD Thesis. The University of Reading.

Cook, R.J. and Veseth, R.J. (1991). *Wheat Health Management*. The American Phytopathological Society.

Cutforth, H.W., Jame, Y.W., and Jefferson, P.G. (1992). Effect of temperature, vernalization and water stress on phyllochron and final main-stem leaf number of HY320 and Neepawa spring wheats. *Canadian Journal of Plant Science* 1992 (72): 1141–1151. https://doi.org/10.4141/cjps92-139.

Derkx, A.P. and Mares, D.J. (2020). Late-maturity α-amylase expression in wheat is influenced by genotype, temperature and stage of grain development. *Planta* 251 (2): 1–14. https://doi.org/10.1007/s00425-020-03341-1.

Díaz, A., Zikhali, M., Turner, A.S. et al. (2012). Copy number variation affecting the photoperiod-B1 and Vernalization-A1 genes is associated with altered flowering time in wheat (Triticum aestivum). *PLoS One* 7 (3): 33234. https://doi.org/10.1371/journal.pone.0033234.

Dowla, M.A.N., Edwards, I., O'Hara, G. et al. (2018). Developing wheat for improved yield and adaptation under a changing climate: optimization of a few key genes. *Engineering* 4 (4): 514–522. https://doi.org/10.1016/j.eng.2018.06.005.

Draeger, T. and Moore, G. (2017). Short periods of high temperature during meiosis prevent normal meiotic progression and reduce grain number in hexaploid wheat (Triticum aestivum L.). *Theoretical and Applied Genetics* 130: 1785–1800. https://doi.org/10.1007/s00122-017-2925-1.

Dreccer, M.F., Fainges, J., Whish, J. et al. (2018). Comparison of sensitive stages of wheat, barley, canola, chickpea and field pea to temperature and water stress across Australia. *Agricultural and Forest Meteorology* 248: 275–294. https://doi.org/10.1016/j.agrformet.2017.10.006.

Easterling, D.R., Meehl, G.A., and Parmesan, C. (2000). Climate extremes: observations, modeling, and impacts. *Science* 22: 2068–2074. https://doi.org/10.1126/science.289.5487.2068.

Equiza, M.A., Miravé, J.P., and Tognetti, J.A. (2001). Morphological, anatomical and physiological responses related to differential shoot vs. root growth inhibition at low temperature in spring and winter wheat. *Annals of Botany* 87 (1): 67–76. https://doi.org/10.1006/anbo.2000.1301.

Evans, L.T., Wardlaw, I.F., and Fischer, R.A. (1975). Wheat. In: *Crop Physiology: Some Case Histories* (ed. L.T. Evans), 101–150. Cambridge University Press.

Fischer, R.A. and Byerlee, D.R. (1991). Trends of wheat production in the warmer areas: Major issues and economic considerations. In: *Wheat for the Nontraditional, Warm Areas* (ed. Saunders), 3–27. CIMMYT.

Foulkes, M.J., Sylvester-Bradley, R., Worland, A.J. et al. (2004). Effects of a photoperiod-response gene Ppd-D1 on yield potential and drought resistance in UK winter wheat. *Euphytica* 135: 63–73. https://doi.org/10.1023/B:EUPH.0000009542.06773.13.

Foulkes, M.J., Slafer, G.A., Davies, W.J. et al. (2011). Raising yield potential of wheat. III. Optimizing partitioning to grain while maintaining lodging resistance. *Journal of Experimental Botany* 62 (2): 469–486. https://doi.org/10.1093/jxb/erq300.

Fowler, D.B. and Limin, A.E. (2007). Progress in breeding wheat with tolerance to low temperature in different phenological developmental stages. In: *Wheat Production in Stressed Environments* (eds. H.T. Buck, J.E. Nisi and N. Salomón), 301–314. Springer https://doi.org/10.1007/1-4020-5497-1_38.

Fowler, D., Dvorak, J., and Gusta, L. (1977). Comparative cold hardiness of several triticum species and Secale cereal. *Crop Science* 17: 941–943. https://doi.org/10.2135/cropsci1977.0011183X001700060031x.

Frank, A.B., Bauer, A., and Black, A.L. (1987). Effects of air temperature and water stress on apex development in spring wheat. *Crop Science* 27: 113–116. https://doi.org/10.2135/cropsci1987.0011183X002700010028x.

Friend, D.J.C. (1965). Ear length and spikelet number of wheat grown at different temperatures and light intensities. *Canadian Journal of Botany* 43: 345–353. https://doi.org/10.1139/b65-037.

Friend, D.J.C. (1966). The effects of light and temperature on the growth of cereals. In: *The Growth of Cereals and Grasses* (eds. F.L. Milthorpe and J.D. Ivins), 181–199. Butterworths.

Friend, D.J.C., Fisher, J.E., and Helson, V.A. (1963). The effect of light intensity and temperature on floral initiation and inflorescence development of marquis wheat. *Canadian Journal of Botany* 41: 1663–1674. https://doi.org/10.1139/b63-152.

Gallagher, J.N., Biscoe, P.V., and Wallace, J.S. (1979). Field studies of cereal leaf growth: IV. Winter wheat leaf extension in relation to temperature and leaf water status. *Journal of Experimental Botany* 30 (4): 657–668. https://doi.org/10.1093/jxb/30.4.657.

Gooding, M.J. (2017). The effects of growth environment and agronomy on grain quality. In: *Cereal Grains: Assessing and Managing Quality*, 2e (eds. C. Wrigley, I. Batey and D. Miskelly), 493–512. Woodhead Publishing.

Gooding, M.J., Ellis, R.H., Shewry, P.R. et al. (2003). Effects of restricted water availability and increased temperature on the grain filling, drying and quality of winter wheat. *Journal of Cereal Science* 37: 295–309. https://doi.org/10.1006/jcrs.2002.0501.

Gooding, M.J., Uppal, R.K., Addisu, M. et al. (2012). Reduced height alleles (Rht) and Hagberg falling number of wheat. *Journal of Cereal Science* 55: 305–311. https://doi.org/10.1016/j.jcs.2012.01.003.

Harding, S.A., Guikema, J.A., and Paulsen, G.M. (1990). Photosynthetic decline from high temperature stress during maturation of wheat. *Plant Physiology* 92 (3): 648–653. https://doi.org/10.1104/pp.92.3.648.

Harkness, C., Semenov, M.A., Areal, F. et al. (2020). Adverse weather conditions for UK wheat production under climate change. *Agricultural and Forest Meteorology* 282–283: 107862. https://doi.org/10.1016/j.agrformet.2019.107862.

Hay, R.K.M. and Kirby, E.J.M. (1991). Convergence and synchrony-a review of the coordination of development in wheat. *Australian Journal of Agricultural Research* 42: 661–700. https://doi.org/10.1071/AR9910661.

Hay, R.K.M. and Porter, J.R. (2006). *The Physiology of Crop Yield*, 2e. Blackwell.

Henry, R.J. and Brennan, P.S. (1988). Dormancy breaking procedures and the breeding of white-grained wheat with resistance to pre-harvest sprouting. *Euphytica* 39: 161–166. https://doi.org/10.1007/BF00039869.

Hucl, P. and Baker, R.J. (1990). Effect of seeding depth and temperature on tillering characteristics of four spring wheat cultivars. *Canadian Journal of Plant Science* 70: 409–417. https://doi.org/10.4141/cjps90-048.

Hyles, J., Bloomfield, M.T., Hunt, J.R. et al. (2020). Phenology and related traits for wheat adaptation. *Heredity* 125: 417–430. https://doi.org/10.1038/s41437-020-0320-1.

Iwaki, K., Nakagawa, K., Kuno, H. et al. (2000). Ecogeographical differentiation in east Asian wheat, revealed from the geographical variation of growth habit and Vrn genotype. *Euphytica* 111: 137–143. https://doi.org/10.1023/A:1003862401570.

Iwaki, K., Haruna, S., Niwa, T. et al. (2001). Adaptation and ecological differentiation in wheat with special reference to geographical variation of growth habit and Vrn genotype. *Plant Breeding* 120: 107–114. https://doi.org/10.1046/j.1439-0523.2001.00574.x.

Jagadish, S.V.K. (2020). Heat stress during flowering in cereals – effects and adaptation strategies. *New Phytologist* 226: 1567–1572. https://doi.org/10.1111/nph.16429.

Jamieson, P.D., Brooking, I.R., Porter, J.R. et al. (1995). Prediction of leaf appearance in wheat: a question of temperature. *Field Crops Research* 41 (1): 35–44. https://doi.org/10.1016/0378-4290(94)00102-I.

Jones, H.E., Lukac, M., Brak, M. et al. (2017). Photoperiod sensitivity affects flowering duration in wheat. *Journal of Agricultural Science* 155: 32–43. https://doi.org/10.1017/S0021859616000125.

Kindred, D.R., Gooding, M.J., and Ellis, R.H. (2005). Nitrogen fertilizer and seed rate effects on Hagberg falling number of hybrid wheats and their parents are associated with α–amylase activity, grain cavity size and dormancy. *Journal of the Science of Food and Agriculture* 85: 727–742. https://doi.org/10.1002/jsfa.2025.

Kirby, E.J.M. and Perry, M.W. (1987). Leaf emergence rates of wheat in a Mediterranean environment. *Australian Journal of Agricultural Research* 38: 455–464. https://doi.org/10.1071/AR9870455.

Law, C. (2000). *Sterility in Winter Wheat: Review of Occurrence in Different Varieties and Possible Causes*. Research Review No. 44. Home-grown Cereals Authority. https://pdfs.semanticscholar.org/86f9/790f0336360f570f6be5bec2637e2340358f.pdf

Li, S., Wheeler, T., Challinor, A. et al. (2010). Simulating the impacts of global warming on wheat in China using a large area crop model. *Journal of Meteorological Research* 24 (1): 123–135.

Limin, A.E. and Fowler, D.B. (2002). Developmental traits affecting low-temperature tolerance response in near-isogenic lines for the vernalization locus vrn-a1 in wheat (*Triticum aestivum* L. em Thell). *Annals of Botany* 89 (5): 579–585. https://doi.org/10.1093/aob/mcf102.

Limin, A.E. and Fowler, D.B. (2006). Low-temperature tolerance and genetic potential in wheat (Triticum aestivum L.): response to photoperiod, vernalization, and plant development. *Planta* 224: 360–366. https://doi.org/10.1007/s00425-006-0219-y.

Liu, B., Martre, P., Ewert, F. et al. (2019). Global wheat production with 1.5 and 2.0°C above pre-industrial warming. *Global Change Biology* 25: 1428–1444. https://doi.org/10.1111/gcb.14542.

Lobell, D., Sibley, A., and Ortiz-Monasterio, J. (2012). Extreme heat effects on wheat senescence in India. *Nature Climate Change* 2: 186–189. https://doi.org/10.1038/nclimate1356.

Longnecker, N. and Robson, A. (1994). Leaf emergence of spring wheat receiving varying nitrogen supply at different stages of development. *Annals of Botany* 74 (1): 1–7. https://doi.org/10.1006/anbo.1994.1087.

Lukac, M., Gooding, M.J., Griffiths, S. et al. (2012). Asynchronous flowering and within-plant flowering diversity in wheat and the implications for crop resilience to heat. *Annals of Botany* 109 (4): 843–850. https://doi.org/10.1093/aob/mcr308.

Lunn, G., Kettlewell, P., Major, B. et al. (2002). Variation in dormancy duration of the U.K. wheat cultivar Hornet due to environmental conditions during grain development. *Euphytica* 126: 89–97. https://doi.org/10.1023/A:1019651117813.

McMaster, G.S. (1997). Phenology, development, and growth of the wheat (Triticum aestivum L) shoot apex: a review. *Advances in Agronomy* 59: 63–118.

Midmore, D.J., Cartwright, P.M., and Fischer, R.A. (1982). Wheat in tropical environments. I. Phasic development and spike size. *Field Crops Research* 5: 185–200. https://doi.org/10.1016/0378-4290(82)90022-3.

Miralles, D.J. and Slafer, G.A. (1999). Wheat development. In: *Wheat: Ecology and Physiology of Yield Determination* (eds. E.H. Satorre and G.A. Slafer), 13–44. Food Products Press.

Nyachiro, J., Clarke, F., DePauw, R. et al. (2002). Temperature effects on seed germination and expression of seed dormancy in wheat. *Euphytica* 126: 123–127. https://doi.org/10.1023/A:1019694800066.

Ochagavía, H., Prieto, P., Savin, R. et al. (2018). Dynamics of leaf and spikelet primordia initiation in wheat as affected by Ppd-1a alleles under field conditions. *Journal of Experimental Botany* 69: 2621–2631. https://doi.org/10.1093/jxb/ery104.

Ortiz, R., Sayre, K.D., Govaerts, B. et al. (2008). Climate change: can wheat beat the heat? *Agriculture, Ecosystems & Environment* 126 (1–2): 46–58. https://doi.org/10.1016/j.agee.2008.01.019.

Peel, M.C., Finlayson, B.L., and Mcmahon, T.A. (2007). Updated world map of the Köppen-Geiger climate classification. *Hydrology and Earth System Sciences Discussions, European Geosciences Union* 4 (2): 439–473. https://hal.archives-ouvertes.fr/hal-00298818.

Percival, J. (1921). *The Wheat Plant: A Monograph*. Duckworth and Co.

Percival, J. (1936). *Agricultural Botany: Theoretical and Practical*, 8e. Duckworth.

Pérez-Gianmarco, T.I., Slafer, G.A., and González Fernanda, G. (2018). Wheat pre-anthesis development as affected by photoperiod sensitivity genes (Ppd-1) under contrasting photoperiods. *Functional Plant Biology* 45: 645–657. https://doi.org/10.1071/FP17195.

Porter, J.R. and Gawith, M. (1999). Temperatures and the growth and development of wheat: a review. *European Journal of Agronomy* 10 (1): 23–36. https://doi.org/10.1016/S1161-0301(98)00047-1.

Prasad, P.V.V., Bheemanahalli, R., and Jagadish, S.V.K. (2017). Field crops and the fear of heat stress – opportunities, challenges and future directions. *Field Crops Research* 200: 114–121. https://doi.org/10.1016/j.fcr.2016.09.024.

Prieto, P., Ochagavia, H., Savin, R. et al. (2018). Dynamics of floret initiation/death determining spike fertility in wheat as affected by Ppd genes under field conditions. *Journal of Experimental Botany* 69 (10): 2633–2645. https://doi.org/10.1093/jxb/ery105.

Prieto, P., Ochagavía, H., Griffiths, S. et al. (2020). Earliness per se × temperature interaction: consequences on leaf, spikelet, and floret development in wheat. *Journal of Experimental Botany* 71 (6): 1956–1968. https://doi.org/10.1093/jxb/erz568.

Qian, B., Zhang, X., Smith, W. et al. (2019). Climate change impacts on Canadian yields of spring wheat, canola and maize for global warming levels of 1.5°C, 2.0°C, 2.5°C and 3.0°C. *Environmental Research Letters* 14: 074005. https://doi.org/10.1088/1748-9326/ab17fb.

Reddy, L., Allan, R.E., and Campbell, K.A.G. (2006). Evaluation of cold hardiness in two sets of near-isogenic lines of wheat (Triticum aestivum) with polymorphic vernalization alleles. *Plant Breeding* 125: 448–456. https://doi.org/10.1111/j.1439-0523.2006.01255.x.

Reynolds, M.P., Singh, R.P., Ibrahim, A. et al. (1998). Evaluating the physiological traits to complement empirical selection for wheat in warm environments. *Euphytica* 100: 85–94 https://doi.org/10.1023/A:1018355906553.

Reynolds, M., Foulkes, M.J., Slafer, G.A. et al. (2009). Raising yield potential in wheat. *Journal of Experimental Botany* 60 (7): 1899–1918. https://doi.org/10.1093/jxb/erp016.

Rezaei, E.E., Siebert, S., and Ewert, F. (2015). Intensity of heat stress in winter wheat – phenology compensates for the adverse effect of global warming. *Environmental Research Letters* 10: 024012. https://doi.org/10.1088/1748-9326/10/2/024012.

Robertson, M.J., Brooking, I.R., and Ritchie, J.T. (1996). Temperature response of vernalization in wheat: modelling the effect on the final number of mainstem leaves. *Annals of Botany* 78 (3): 371–381. https://doi.org/10.1006/anbo.1996.0132.

Rodríguez, D., Pomar, M., and Goudriaan, J. (1998). Leaf primordia initiation, leaf emergence and tillering in wheat (Triticum aestivum L.) grown under low-phosphorus conditions. *Plant and Soil* 202: 149–157. https://doi.org/10.1023/A:1004352820444.

Ruane, A.C., Phillips, M.M., and Rosenzweig, C. (2018). Climate shifts within major agricultural seasons for +1.5 and +2.0 °c worlds: HAPPI projections and AGMIP modeling scenarios. *Agricultural and Forest Meteorology* 259: 329–344. https://doi.org/10.1016/j.agrformet.2018.05.013.

Saini, H.S., Sedgley, M., and Aspinall, D. (1983). Effect of heat stress during floral development on pollen tube growth and ovary anatomy in wheat (Triticum aestivum L.). *Functional Plant Biology* 10: 137–144. https://doi.org/10.1071/PP9830137.

Saini, H.S., Sedgley, M., and Aspinall, D. (1984). Development anatomy in wheat of male sterility induced by heat stress, water deficit or abscisic acid. *Functional Plant Biology* 11: 243–253. https://doi.org/10.1071/PP9840243.

Saiyed, I.M., Bullock, P.R., Sapirstein, H.D. et al. (2009). Thermal time models for estimating wheat phenological development and weather-based relationships to wheat quality. *Canadian Journal of Plant Science* 89 (3): 429–439. https://doi.org/10.4141/CJPS07114.

Salvi, S., Porfiri, O., and Ceccarelli, S. (2013). Nazareno Strampelli, the 'prophet' of the green revolution. *Journal of Agricultural Science* 151: 1–5. https://doi.org/10.1017/S0021859612000214.

Sanchez-Garcia, M. and Bentley, A.R. (2019). Global journeys of adaptive wheat genes. In: *Applications of Genetic and Genomic Research in Cereals* (eds. T. Miedaner and V. Korzun), 183–200. Woodhead Publishing https://doi.org/10.1016/B978-0-08-102163-7.00009-0.

Semenov, M. and Shewry, P. (2011). Modelling predicts that heat stress, not drought, will increase vulnerability of wheat in Europe. *Scientific Reports* 1: 66. https://doi.org/10.1038/srep00066.

Semenov, M.A., Stratonovitch, P., Alghabari, F. et al. (2014). Adapting wheat in Europe for climate change. *Journal of Cereal Science* 59: 245–256. https://doi.org/10.1016/j.jcs.2014.01.006.

Senapati, N., Griffiths, S., Hawkesford, M. et al. (2020). Substantial increase in yield predicted by wheat ideotypes for Europe under future climate. *Climate Research* 80: 189–201. https://doi.org/10.3354/cr01602.

Slafer, G.A. and Rawson, H.M. (1994a). Sensitivity of wheat phasic development to major environmental factors: a re-examination of some assumptions made by physiologists and modellers. *Functional Plant Biology* 21: 393–426. https://doi.org/10.1071/PP9940393.

Slafer, G.A. and Rawson, H.M. (1994b). Does temperature affect final numbers of primordia in wheat? *Field Crops Research* 39 (2–3): 111–117. https://doi.org/10.1016/0378-4290(94)90013-2.

Slafer, G.A. and Rawson, H.M. (1995). Photoperiod × temperature interactions in contrasting wheat genotypes: time to heading and final leaf number. *Field Crops Research* 44 (2–3): 73–83. https://doi.org/10.1016/0378-4290(95)00077-1.

Slafer, G.A., Kantolic, A.G., Appendino, M.L. et al. (2015). Genetic and environmental effects on crop development determining adaptation and yield. In: *Crop Physiology*, 2e (eds. V.O. Sadras and D.F. Calderini), 285–319. https://doi.org/10.1016/B978-0-12-417104-6.00012-1.

Snape, J., Butterworth, K., Whitechurch, E. et al. (2001). Waiting for fine times: genetics of flowering time in wheat. In: *Wheat in a Global Environment. Developments in Plant Breeding Volume 9* (eds. Z. Bedö and L. Láng), 67–74. Springer https://doi.org/10.1007/978-94-017-3674-9_7.

Spiertz, J.H.J. (1977). The influence of temperature and light intensity on grain growth in relation to the carbohydrate and nitrogen economy of the wheat plant. *Netherlands Journal of Agricultural Science* 25: 182–197. https://doi.org/10.18174/njas.v25i3.17131.

Stone, P.J. and Nicolas, M.E. (1995a). Effect of timing of heat stress during grain filling on two wheat varieties differing in heat tolerance. I. Grain growth. *Functional Plant Biology* 22: 927–934. https://doi.org/10.1071/PP9950927.

Stone, P.J. and Nicolas, M.E. (1995b). A survey of the effects of high temperature during grain filling on yield and quality of 75 wheat cultivars. *Australian Journal of Agricultural Research* 46: 475–492. https://doi.org/10.1071/AR9950475.

Stone, P.J. and Nicolas, M.E. (1995c). Comparison of sudden heat stress with gradual exposure to high temperature during grain filling in two wheat varieties differing in heat tolerance. I. Grain growth. *Functional Plant Biology* 22: 935–944. https://doi.org/10.1071/PP9950935.

Subedi, K.D., Gregory, P.J., Summerfield, R.J. et al. (1998). Cold temperatures and boron deficiency caused grain set failure in spring wheat (Triticum aestivum L.). *Field Crops Research* 57 (3): 277–288. https://doi.org/10.1016/S0378-4290(97)00148-2.

Tashiro, T. and Wardlaw, I.F. (1989). A comparison of the effect of high temperature on grain development in wheat and rice. *Annals of Botany* 64 (1): 59–65. https://doi.org/10.1093/oxfordjournals.aob.a087808.

Tashiro, T. and Wardlaw, I.F. (1990). The response to high temperature shock and humidity changes prior to and during the early stages of grain development in wheat. *Functional Plant Biology* 17: 551–561. https://doi.org/10.1071/PP9900551.

Thorne, G.N. and Wood, D.W. (1987). Effects of radiation and temperature on tiller survival, grain number and grain yield in winter wheat. *Annals of Botany* 59 (4): 413–426. https://doi.org/10.1093/oxfordjournals.aob.a087330.

Trnka, M., Rötter, R., Ruiz-Ramos, M. et al. (2014). Adverse weather conditions for European wheat production will become more frequent with climate change. *Nature Climate Change* 4: 637–643. https://doi.org/10.1038/nclimate2242.

Valle, S.R. and Calderini, D.F. (2010). Phyllochron and tillering of wheat in response to soil aluminum toxicity and phosphorus deficiency. *Crop and Pasture Science* 61: 863–872. https://doi.org/10.1071/CP09310.

Vincent, C.D. and Gregory, P.J. (1989). Effects of temperature on the development and growth of winter wheat roots. *Plant and Soil* 119: 87–97. https://doi.org/10.1007/BF02370272.

Wheeler, T., Batts, G., Ellis, R. et al. (1996a). Growth and yield of winter wheat (*Triticum aestivum*) crops in response to CO_2 and temperature. *Journal of Agricultural Science* 127 (1): 37–48. https://doi.org/10.1017/S0021859600077352.

Wheeler, T.R., Hong, T.D., Ellis, R.H. et al. (1996b). The duration and rate of grain growth, and harvest index, of wheat (Triticum aestivum L.) in response to temperature and CO_2. *Journal of Experimental Botany* 47 (5): 623–630. https://doi.org/10.1093/jxb/47.5.623.

Wheeler, T.R., Craufurd, P.Q., Ellis, R.H. et al. (2000). Temperature variability and the yield of annual crops. *Agriculture, Ecosystems & Environment* 82 (1–3): 159–167. https://doi.org/10.1016/S0167-8809(00)00224-3.

Xu, S. and Chong, K. (2018). Remembering winter through vernalisation. *Nature Plants* 4: 997–1009. https://doi.org/10.1038/s41477-018-0301-z.

You, L., Rosegrant, M.W., and Wood, S. (2009). Impact of growing season temperature on wheat productivity in China. *Agricultural and Forest Meteorology* 149 (6–7): 1009–1014 https://doi.org/10.1016/j.agrformet.2008.12.004.

Zadoks, J.C., Chang, T.T., and Konzak, C.F. (1974). A decimal code for the growth stages of cereals. *Weed Research* 14: 415–421.

Zampieri, M., Ceglar, A., Dentener, F. et al. (2017). Wheat yield loss attributable to heat waves, drought and water excess at the global, national and subnational scales. *Environmental Research Letters* 12 (6): 064008. https://doi.org/10.1088/1748-9326/aa723b.

Zhao, H., Dai, T., Jing, Q. et al. (2007). Leaf senescence and grain filling affected by post-anthesis high temperatures in two different wheat cultivars. *Plant Growth Regulation* 51: 149–158. https://doi.org/10.1007/s10725-006-9157-8.

Zikhali, M. and Griffiths, S. (2015). The effect of earliness per se (eps) genes on flowering time in bread wheat. In: *Advances in Wheat Genetics: From Genome to Field* (eds. Y. Ogihara, S. Takumi and H. Handa), 339–345. Springer https://doi.org/10.1007/978-4-431-55675-6_39.

Zikhali, M., Wingen, L.U., Leverington-Waite, M. et al. (2017). The identification of new candidate genes *Triticum aestivum* FLOWERING LOCUS T3-B1 (TaFT3-B1) and TARGET OF EAT1 (TaTOE1-B1) controlling the short-day photoperiod response in bread wheat. *Plant, Cell & Environment* 40 (11): 2678–2690. https://doi.org/10.1111/pce.13018.

Zohary, D., Harlan, J.R., and Vardi, A. (1969). The wild diploid progenitors of wheat and their breeding value. *Euphytica* 18: 58–65. https://link.springer.com/content/pdf/10.1007/BF00021982.pdf.

5

Sunshine

The sun provides not only the power of being seen for things seen, but, as I think you will agree, also their generation and growth and nurture

Plato

The importance of light in photosynthesis was introduced in Chapter 3. By the early 1800s, it was known that plant growth used both CO_2 and H_2O and that the process was associated with sunlight, green plant tissues, and the release of oxygen. In 1817, Pierre-Joseph Pelletier and Joseph-Bienaime isolated the green pigment in plants and called it chlorophyll, literally 'green leaf' from the classical Greek *chlorus phyllon*. By 1850, Julius Robert Mayer had proposed that plants convert sunlight energy into chemical energy. We now recognise that photosynthesis defines and permits nearly all life on earth. The few exceptions include microorganisms living in extreme environments that gain their energy from inorganic reactions (e.g. Shock 1997). For all other life, photosynthesis is the primary source of energy, fixed carbon, and oxygen for activity, growth, and respiration.

The twentieth century saw great progress in understanding how light energy was captured. This power is harnessed in the *light reactions*. Robert Hill described the direct relationship between chlorophyll, light, and the release of oxygen in his work on isolated chloroplasts (Hill 1939). Later, he and others elucidated how light energy drives the flow of electrons from water to compounds that can propel carbon fixation (e.g. Hill 1965). The light reactions produce chemical energy stored in the form of phosphate-to-phosphate bonds in adenosine triphosphate (ATP). They also produce a reducing agent (i.e. an electron donor), nicotinamide adenine dinucleotide phosphate (NADPH).

Light energy is captured in complexes of proteins and pigments, the latter mostly chlorophylls but with some carotenoids. These complexes called Photosystems I and 2 (Figure 5.1) are embedded in the thylakoid membranes inside chloroplasts. Photosystem 2 is the first in the chain but was the second to be discovered. It is responsible for most of the green colouration of plants and may account for a third of leaf protein (Thomas 2018). Photosystem 2 is associated with the oxygen evolving complex, generating oxygen together with protons (H^+) and electrons (e^-) as the products of splitting water (Figure 5.1).

The dark reactions, so called because they do not need light, use the energy (ATP) and reducing power (NADPH) from the light reactions to *fix* carbon from atmospheric CO_2 into organic compounds. Details of this process were revealed by Melvin Calvin and colleagues

Wheat: Environment, Food and Health, First Edition. Mike J. Gooding and Peter R. Shewry.

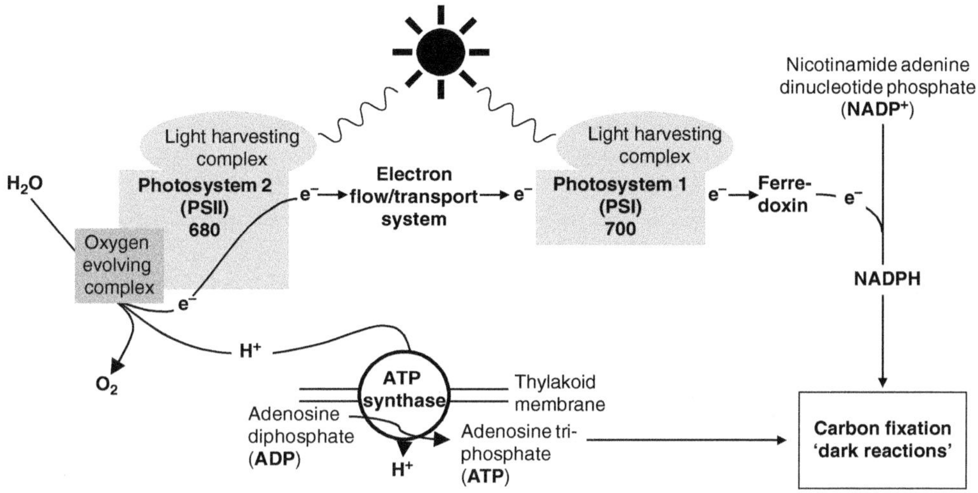

Figure 5.1 The light reactions of photosynthesis. *Source:* Adapted from Thomas (2018).

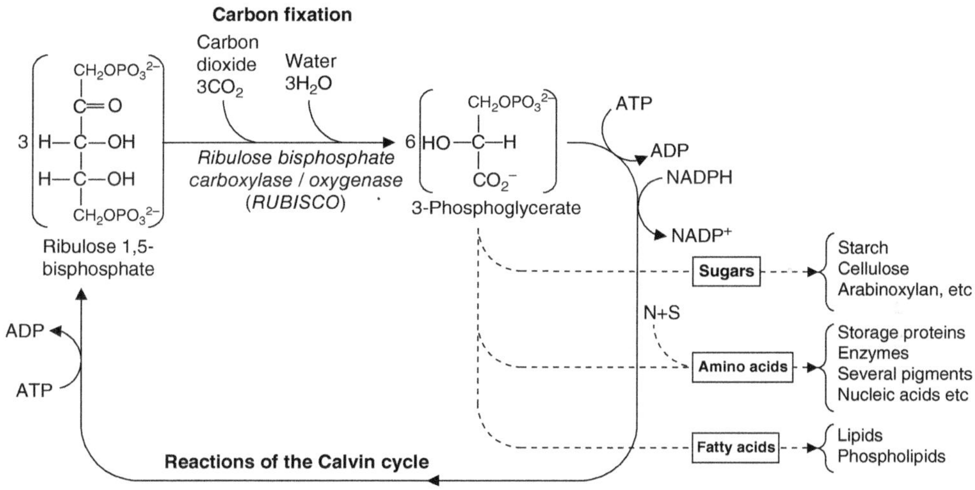

Figure 5.2 The dark reactions of photosynthesis.

(e.g. Bassham et al. 1950), who monitored the successive incorporation of C^{14} from labelled CO_2 into molecules of what became known as the Calvin or Calvin–Benson–Bassham cycle. In the Calvin cycle, the rate-limiting step is carbon fixation catalysed by the enzyme ribulose bisphosphate carboxylase/oxygenase, usually abbreviated to *RUBISCO* (Figure 5.2).

Carbon dioxide is combined with a five-carbon molecule, ribulose 1,5-bisphosphate, producing the three-carbon molecule 3-phosphoglycerate. If three molecules of ribulose 1,5-bisphosphate are combined with three molecules of carbon dioxide then six molecules of 3-phosphoglycerate are produced. Five of these are incorporated by the Calvin cycle to regenerate the three molecules of ribulose 1,5 bisphosphate. The single remaining molecule of 3-phosphoglycerate represents the net gain in fixed carbon. The fixed carbon is then used to produce sugars, fatty acids, and with the addition of nitrogen and sulphur, the amino acids. These are the building blocks from which wheat plants are formed and which ultimately determine yield and quality. Sugars are polymerised to form starch, the major component of the endosperm of wheat grain, and cell wall (fibre) polysaccharides, notably cellulose, arabinoxylan, and β-glucan. Polymerisation of amino acids forms proteins that include enzymes, transporters, structural proteins, defensive proteins, and the grain storage proteins. Amino acids are also precursors of several pigments including the chlorophylls, and the nucleic acids DNA and RNA. The fatty acids are components of the phospho- and glycolipids present in membranes and the storage oils (triacylglycerols) of the plant.

The first stable molecule produced following carbon fixation in wheat has three carbons; the process is, therefore, known as C3 photosynthesis. C3 photosynthesis is the form of carbon fixation found in most crops, and in all the temperate cereals such as barley, oats, rye, and triticale, as well as in tropical rice. Some tropical cereals and grasses, including maize, sorghum, millets, sugar cane, and miscanthus, are known as C4 crops. This is because they have a preliminary step whereby the first stable molecule after initial CO_2 fixation is oxaloacetate (a 4-carbon molecule). C3 photosynthesis can be more efficient than C4 in temperate conditions (Hay and Porter 2006).

The efficiency of CO_2 fixation by *RUBISCO* is limited; this is because the enzyme is also responsible for photorespiration, i.e. the absorption of oxygen and the release of CO_2 (Figure 5.3). Photorespiration can result in a significant loss of carbon and may reduce the efficiency of photosynthesis by 25% in ambient conditions (Timm and Hagemann 2020). Photorespiration rises as the concentration of O_2 relative to CO_2 increases close to the active site of *RUBISCO*; this occurs under drought and high light intensities. High temperature can also increase photorespiration, because it increases the solubility of O_2 relative to CO_2 and increases the affinity of *RUBISCO* for O_2.

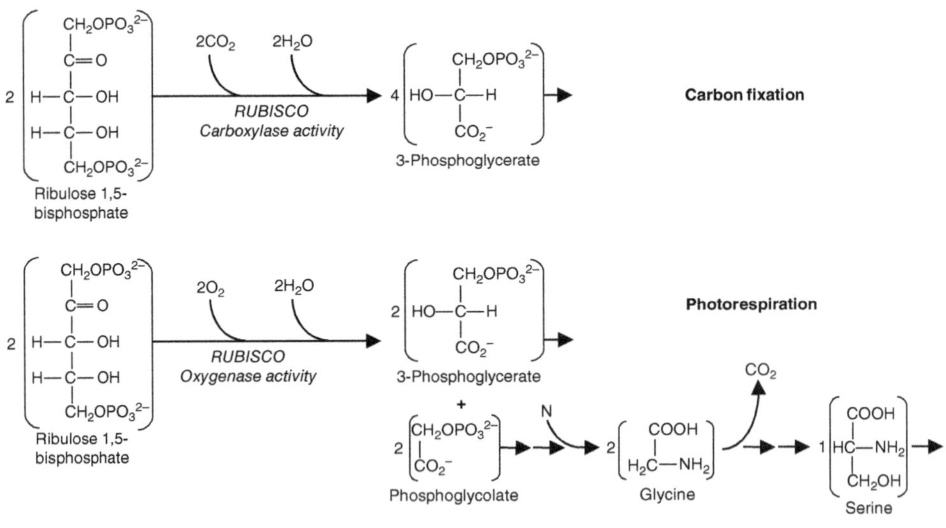

Figure 5.3 The two activities of ribulose bisphosphate carboxylase/oxygenase (*RUBISCO*) (only balanced for C).

Other inefficiencies are apparent in addition to photorespiration. In particular, only about 50% of the energy from sunlight can drive photosynthesis. It is evident, for example, that some of the energy in the green spectrum is reflected, rather than being absorbed by plants. Figure 5.4 shows the action spectrum for wheat, i.e. the rate at which carbon dioxide is taken up divided by the rate at which energy is received by the leaf for different wavelengths. Also shown is the quantum yield spectrum, i.e. the rate of photosynthesis per unit rate of absorption of quanta. The action spectrum of wheat is not significantly different from those of many other crop plants (McCree 1972) and lies between 400 and 700 nm. It is common to refer to light in this range as *photosynthetically active radiation* (PAR).

In addition to defining PAR, the absorptance spectrum of wheat can be used to estimate ground cover and vegetation indices. These measurements often exploit the high absorption by green canopies in mid red wavelengths (RED, e.g. 660 nm) and low absorption (high reflectance) of near infra-red (NIR, e.g. 770 nm) wavelengths (Figure 5.4). Normalized difference vegetation indices (NDVI) are thus calculated from reflection ratios (e.g. Eq. 5.1; Duan et al. 2017):

$$NDVI = \frac{NIR - RED}{NIR + RED} \tag{5.1}$$

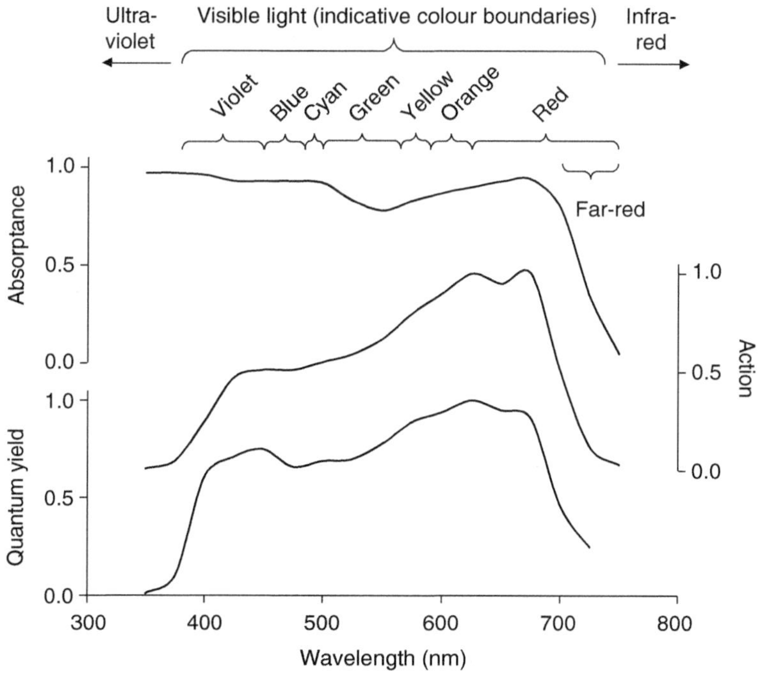

Figure 5.4 The absorbtance, action, and quantum yield spectra of wheat leaves against wavelengths of light. Values are relative to the maximum value across all wavelengths. *Source:* Data from McCree (1972).

5.1 The Light Requirement of Wheat

5.1.1 Light Quantity

We can routinely measure with sensors the amount of PAR intercepted by field-grown crops. This means we can relate biomass and grain yield empirically to the capture and radiation use efficiency (RUE) of PAR (Figure 5.5). In an analogous approach to that used for water, it is common for us to report linear relationships between components of light interception and grain yield in light-limited environments. As with the water relationships, however, it is also evident that there is significant scatter around lines of best fit (e.g. Figure 5.6). To identify possible areas of progress, we need to say a little more about the relationships between light interception and grain yield.

At the single leaf level, net photosynthesis (i.e. CO_2 uptake minus CO_2 lost in respiration and photorespiration) increases approximately linearly with light intensity until the leaf becomes *light saturated*. Figure 5.7a shows that wheat flag leaves reach a maximum rate of net photosynthesis of around 3 g $CO_2/m^2/h$ at about 110 W PAR/m^2 of leaf. This intensity of PAR may only represent 20–30% of the irradiance available in bright sunshine in many wheat growing areas (Evans et al. 1975; Hall and Rao 1981). For whole canopies, however, light saturation is rarer, and may not be approached until light receipts exceed 300 W PAR/m^2 of land (Figure 5.7b; Gallagher and Biscoe (1978)).

There are several reasons for the discrepancy between the responses of individual leaves and canopies to increasing PAR intensity. Photosynthesis in partially shaded, lower leaves may still be responsive to increases in light intensity while upper leaves become saturated. Upper leaves are also less liable to saturation (and less likely to shade lower leaves) if they are not perpendicular to the direction of light (Evans et al. 1975). Canopy and leaf photosynthesis can also be dependent on other factors, including CO_2 concentration, temperature at micro-climate scales, wind speed, acclimation, and demand for assimilate (Evans et al. 1975). Given the numerous interacting factors, curvature in the response of carbon fixation of wheat canopies to PAR absorption can disappear entirely if we consider cumulative effects over several days, during which PAR will be below 300 W/m^2 for a significant part of the time. The results of Gallagher and Biscoe (1978; Figure 5.7c) are particularly important because they reveal a mean RUE for wheat canopies of almost exactly 3 g/MJ PAR (or 1.5 g/MJ sunlight) when the regression is forced through the origin (precisely 3.01 g/MJ PAR, s.e. = 0.081 in Figure 5.7c). When using this value for comparative purposes, it should be remembered that Gallagher and Biscoe (1978) included estimates of root

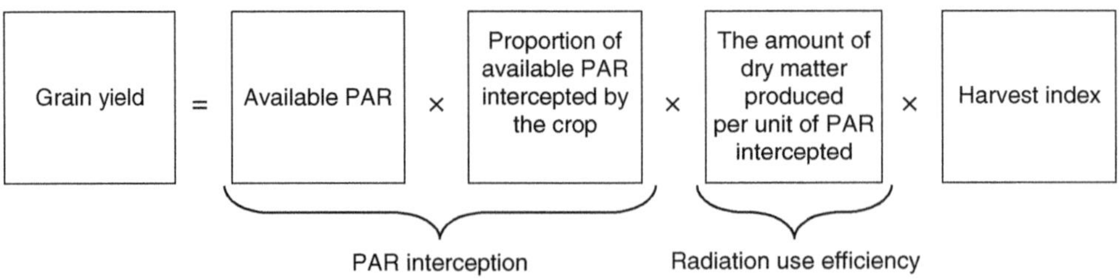

Figure 5.5 The components of PAR interception and use related to grain yield.

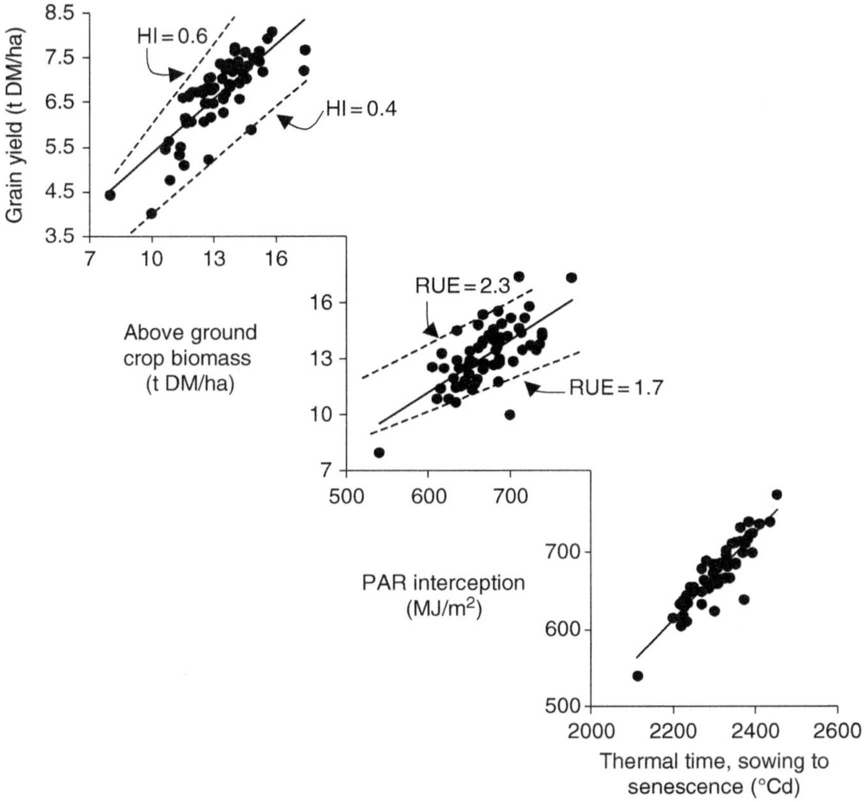

Figure 5.6 Relationship between canopy longevity, season-long PAR interception, above-ground crop biomass, and grain yield. The relationship between PAR interception and biomass depends on radiation use efficiency (RUE); the relationship between biomass and grain yield depends on harvest index (HI). *Source:* Data from Barber et al. (2015).

dry matter (reducing from 0.35 to 0.1 of crop biomass during development) in the calculation of crop growth rate. They also used PAR absorption rather than PAR interception.

PAR interception is the difference between incident PAR above the canopy and the PAR reaching soil level when weeds have been controlled and the canopy is green. PAR interception assessed in this way includes about 10% of PAR that will have been reflected away from the canopy. PAR absorption removes the reflected PAR from the denominator when calculating RUE. Many studies calculate RUE by relating the accumulation of above ground biomass to PAR intercepted (e.g. Calderini et al. 1997). On this basis, the equivalent RUE calculated from the Gallagher and Biscoe (1978) data would be 2.0–2.5 g above-ground crop DM/MJ PAR intercepted. Finally, they derived the value of 3 g/MJ PAR from assessments before ear emergence. RUE declines after ear emergence as the leaves and photosynthetic tissues of the ear become older, and the respiration per unit assimilation becomes greater (Gallagher and Biscoe 1978).

If we use the equation in Figure 5.5 and assume an RUE of 2.25 g of above-ground biomass/MJ PAR intercepted and a harvest index (HI) of 0.5, then each tonne of grain dry

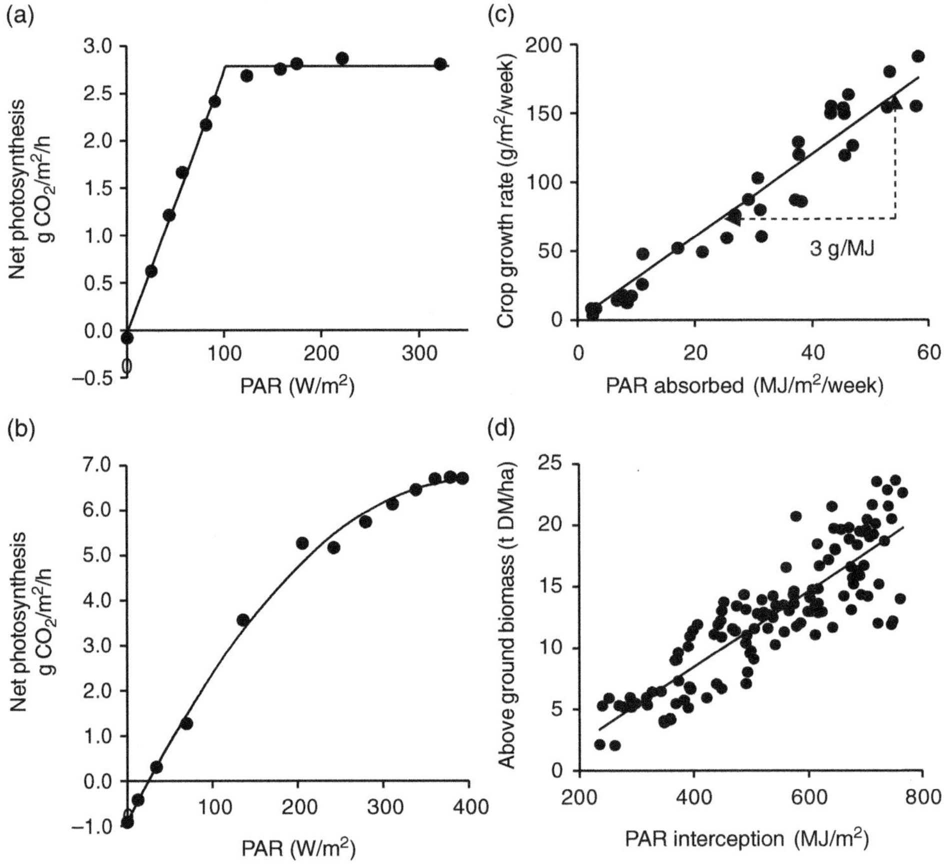

Figure 5.7 The relationship between dry matter accumulation and photosynthetically active radiation (PAR) at different spatial and temporal scales: (a) Flag leaf photosynthesis and PAR intensity (mean of two assessments on wheat flag leaves from Marshall and Biscoe (1980)); (b) Canopy photosynthesis and PAR intensity (mean of four cultivars during stem elongation from Gent and Kiyomoto (1992); (c) Crop growth rate and PAR absorption over two weekly periods before ear emergence over different sites and seasons (data from Gallagher and Biscoe 1978); and (d) Above-ground crop biomass at harvest and season-long interception of PAR (*Source:* Data amalgamated from Gooding et al. 2002; Kindred and Gooding 2005; Addisu et al. 2010; Uppal and Gooding 2013).

matter requires about 0.9 TJ PAR. As already noted, however, this value can vary considerably (as shown by the scatter around line of best fit in Figure 5.7d). Some of this variation is due to factors other than light interception limiting photosynthesis for some periods during crop growth. For example, RUE to produce biomass will be reduced if water shortage results in stomatal closure despite high levels of irradiance. Similarly, excessive temperatures can reduce net photosynthesis, particularly if photorespiration is increased.

The relationship between light interception and grain yield is further complicated by associated effects on HI. For example, tiller numbers increase with light intensity between seedling emergence and floral initiation (Friend 1966; Evans et al. 1975; Assuero and Tognetti 2010). However, in well-established crops with high seed rates and adequate nutrition, early season growth may be less important and be at the expense of HI (Barber et al. 2015). Extending the quantity and duration of light interception in the later stages of development can be more beneficial for grain yield; this is because of influences on specific yield components, dry matter partitioning, and RUE to produce grain (Slafer et al. 2009; Reynolds et al. 2012). For example, grain number per unit area limits grain yield in many areas of the world. Grain number is often positively related to light interception and RUE during a so-called critical period when the spikes are actively growing during the stem elongation phase (Reynolds et al. 2012; García et al. 2014). More specifically, Fischer (1985, 2011) identified the 25 days preceding anthesis as important in defining grain numbers per ear for the environments in his studies. Increasing light interception during this phase can increase the number of florets that become fertile and avoid floret death (Kirby 1988; Reynolds et al. 1999; Miralles et al. 2000; Gonzalez et al. 2003). Increased light interception during stem extension, and particularly around flag leaf emergence, can also increase the number of tillers that survive to produce ears (Thorne and Wood 1987).

After anthesis, photosynthetic rates and duration are important for grain filling and HI (Gooding et al. 2005; Sylvester-Bradley et al. 2005; Carmo-Silva et al. 2017). Various scientists have attempted to measure and model the photosynthetic contributions to grain yield of different organs of the wheat plant during this *yield forming* phase (e.g. Thorne 1965). One example is shown in Figure 5.8. Precise allocations will depend on the relative sizes of different leaves and ear structures, as well as their periodicity of activity relative to light receipts and other constraints. These allocations will therefore depend on both genetic and

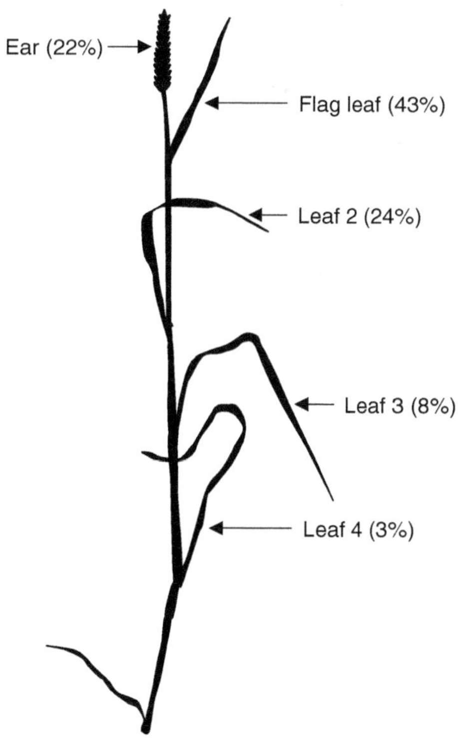

Figure 5.8 A typical allocation of the sources of photosynthate for grain yield in UK winter wheat. *Source:* Based on AHDB (2019).

environmental factors (e.g. Abbad et al. 2004). In a typical UK scenario, the flag leaf is particularly important (Figure 5.8). This is because it is long-lived (from mid-May to mid-July) and at the top of the canopy at the time of year with the highest light receipts and reasonable soil moisture levels. It is also the leaf in the closest vascular proximity to the developing grain.

We can discover the effects of the timing of light availability in field conditions with shading interventions. For example, in oceanic environments, cloudy conditions can last for days or even weeks during important periods for determining yield components (Beed et al. 2007). Shading during stem extension reduces ear numbers (Figure 5.9). Grain numbers are particularly susceptible to shading from flag leaf emergence to anthesis; the duration of this encompasses much of the *critical* period. Shading effects during these growth stages on yield can be partially mitigated by compensatory increases in mean grain weight. However, mean grain weight is significantly reduced if light availability is constrained during grain filling (Figure 5.9).

An impact of poor light availability on commercial wheat production was observed in 1987. An evidently poorly adapted wheat cultivar was grown over large areas of the UK and France. Catastrophic effects on grain set were at least partly attributed to low light levels around meiosis (Demotes-Mainard et al. 1995; Law 1999). In another practical example, the lowest wheat yields in the UK between 1990 and 2020 occurred in 2012 and were partly

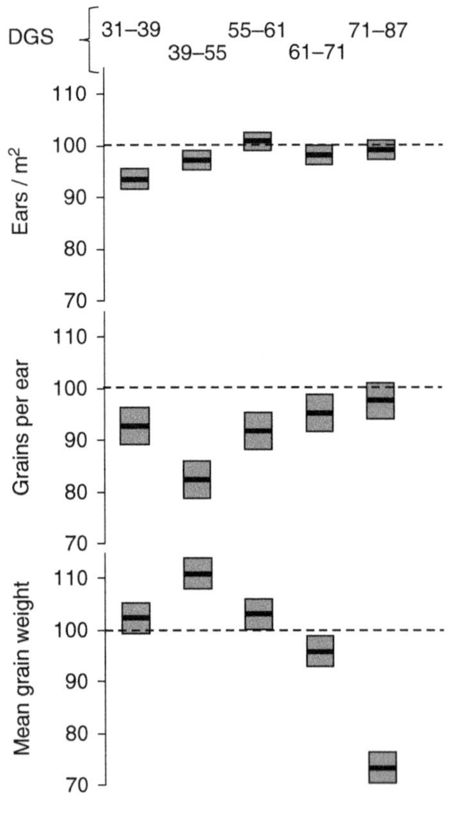

Figure 5.9 The effect of shading field grown wheat in England during different growth stages on grain yield components. Numbers in brackets are growth stage scores as per Zadoks et al. (1974). Bars are + and − standard error of the difference to the effect = zero line. Values are expressed as a percentage of the results from the unshaded crop. DGS are the decimal growth stage scores as per Table 1.4 (Zadoks et al. 1974). Bars are + and − standard error of the difference to the 100% line. Values are means of three years weighted for the inverse of the variance from Beed et al. (2007). Automated shading occluded 80% of incident light when solar radiation exceeded 250 W/m^2.

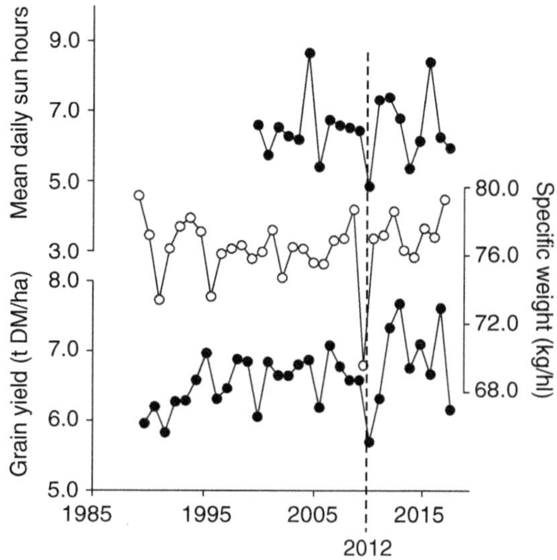

Figure 5.10 Annual UK wheat grain yields, grain specific weights, and mean daily sunshine hours for June and July. *Source:* Data from Department for Environment, Food and Rural Affairs (DEFRA), Agriculture and Horticulture Development Board (AHDB), and Office for National Statistics (ONS), respectively.

associated with exceptionally low light receipts during grain growth in June and July. The inadequate grain filling was indicated by record low specific weights (Figure 5.10).

5.1.2 Light Quality

As described in Figure 5.4, wheat receives, absorbs, and reflects a range of wavelengths of light. Plants contain several photo-reversible pigments that provide a means for detecting and responding to light quality. Green plants are adept at absorbing and using red light but strongly reflect far-red light. They also respond to changes in the ratio of red: far-red (R:FR) light due to the presence of phytochrome (Figure 5.11). This pigment occurs in two forms, termed P_R and P_{FR}. P_R absorbs red light, and in doing so is converted to P_{FR}, which in turn absorbs far-red light to become P_R. P_{FR} also reverts slowly to P_R in darkness. The changes in the molecular structure of the phytochrome protein associated with the P_{FR} chromophore trigger a cascade of biochemical reactions; these influence the development and phenology of the crop (Thomas 2018). The phytochrome system provides a means of detecting day-length (Chen et al. 2014). Additionally, as the R:FR ratio reduces as light passes through, or is reflected from green leaves, the phytochrome system enables the plant to *anticipate* and respond to competition from neighbouring plants. R:FR reduces more through canopies of closely spaced plants than through canopies where plants are more isolated. For example, an increase in far-red light and reduction in the R:FR ratio reduces tillering (Kasperbauer and Karlen 1986; Casal 1988). The R:FR ratio also reduces during canopy growth. Evers et al. (2006) found that tillering stopped when the proportion of PAR intercepted by the canopy dropped below 0.40–0.45, and the R:FR ratio at the base of the canopy was below 0.35–0.4. The start of tiller death also appears to be hastened by reduced R:FR, although this effect interacts with nitrogen availability (Sparkes et al. 2006). There are some other responses to reduced R:FR ratio because of inter-plant competition. Plants grow taller due to longer leaf sheaths and basal internodes; spike growth and development is delayed;

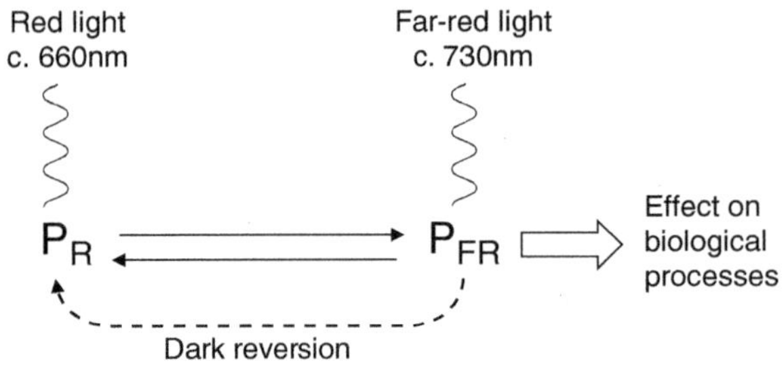

Figure 5.11 The photo-reversible forms of phytochrome (P).

fertile florets at anthesis are reduced; and grain yield per ear is impaired due to both fewer grain numbers and lower mean grain weight (Ugarte et al. 2010).

5.2 Light Interception

As already described, it is important that light is intercepted by green areas of the plant. The amount of light intercepted is a function of the area of green tissue (i.e. canopy size), its arrangement (canopy architecture), and its duration.

The size of a canopy is often reported as the Leaf Area Index (LAI) i.e. the area of one side of all the green leaves expressed as a ratio of the area of ground the canopy is growing on. For example, if the total area of the leaves growing on $1\,m^2$ of ground is $6\,m^2$ then LAI = 6. In theory, if all leaves were thick, horizontal, and interlocked perfectly to form a sheet over the soil all the light could be intercepted with LAI = 1. However, plant canopies are not like this! Wheat leaves are generally not parallel to the soil but are, on average, more erect. They are also thin and partially translucent, and do not interlock perfectly.

In young growth stages and with sparse canopies, a more horizontal growth habit may be advantageous in reducing the amount of light being *wasted* and directly reaching the soil. However, in dense and mature canopies receiving high light intensities, it can be advantageous for the upper leaves to be more erect, rather than horizontal. This reduces the propensity for the upper leaves to become saturated whilst shading leaves below (Ong and Monteith 1992; Parry et al. 2011). Furthermore, it is evident that in wheat, it is not just the leaf laminas that are green and photosynthesizing; significant carbon fixation can also derive from the leaf sheaths, stems, and ears whilst they are green. It is therefore more accurate to state the size of wheat canopies as a Green Area Index (GAI).

$$GAI = \frac{Green\ area\ of\ leaves,\ stems,\ and\ ears}{Ground\ area} \tag{5.2}$$

The proportion of available light that a canopy intercepts increases with canopy size approximately according to Beer's law. This relates the attenuation of light to the properties of the material (in this case, the wheat canopy) through which the light is travelling (Ong and Monteith 1992), where *k* is the extinction coefficient:

$$Proportion\ of\ light\ intercepted\ (\%) = 100.\left(1 - e^{-k.GAI}\right) \tag{5.3}$$

The extinction coefficient determines the GAI that is required to intercept a certain proportion of the available light (Figure 5.12). Crops with more horizontal leaves (planophile) or with leaves that are less translucent will tend to have a higher value of *k*. Values of *k* for wheat commonly range between 0.35 and 0.65 depending on environment, management, growth stage, and method of assessment (Hay and Porter 2006). The *k* value in wheat also tends to increase during maturation and with nitrogen availability (Sylvester-Bradley et al. 1997). Intercepting more than 95% of available light at anthesis, with *k* values around 0.5, requires a GAI of at least six (Figure 5.12).

It is clearly not possible to have a canopy of GAI = 6 in the field for the full duration of crop growth. Canopies take time to develop and light interception to increase. Additionally, as wheat is an annual plant, at some stage the canopy will die as resources are partitioned to the grain. It is also unnecessary or undesirable to have large canopies when growth is curtailed by extreme temperatures or by drought. In high latitudes, large canopies achieved in the autumn through earlier sowing of winter wheats are subject to greater pressures from weeds, pests, and diseases (e.g. Colbach et al. 1997). However, the benefits may be limited because radiation levels are low during late autumn and winter (Figure 5.13). Large canopies early in the season can also exacerbate drought if they extract excessive water so that there is insufficient water remaining to complete grain filling. In short, it is particularly important to intercept a high proportion of light when there is a large amount of light to be intercepted, when growth is not being curtailed by other environmental factors, and

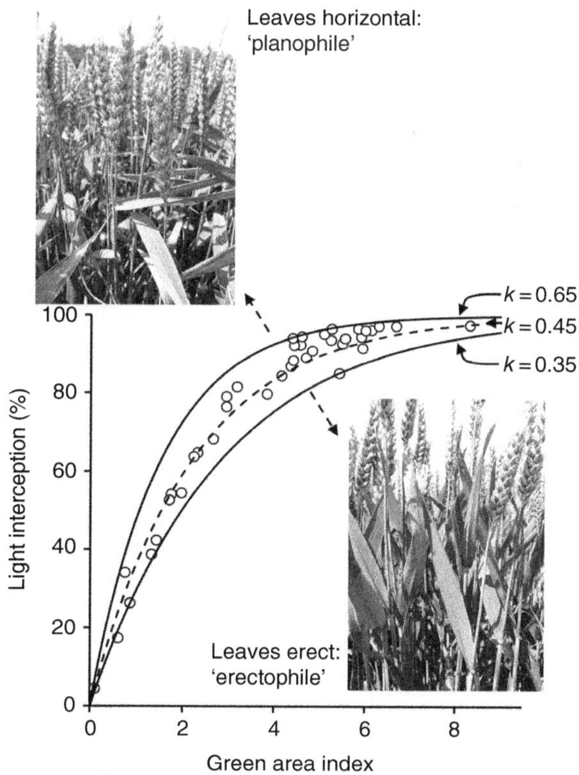

Figure 5.12 The relationship between green area index and light interception for different extinction coefficients according to Beer's law (Eq. 5.3). *Source:* Data are transformed from Thorne et al. (1988).

when light interception is particularly important for increasing grain numbers and the maintenance of grain filling.

Figure 5.13 shows a typical UK crop that yielded 9 t grain DM/ha from a season-long interception of 8 TJ PAR. The canopy is largest in May and June, when high light receipts are available, and temperatures are moderate. These months also include the critical period for grain number determination. The canopy reaches a maximum size around anthesis with a LAI of 5.5 and a light extinction coefficient $k = 0.45$. The attainment of maximal LAI during ear emergence and anthesis is a common observation (e.g. Gillett et al. 1999). During July, the green leaf area declines while the grains are filled with dry matter and the soil moisture deficits increase. The end of grain filling is often (Hanft and Wych 1982), but

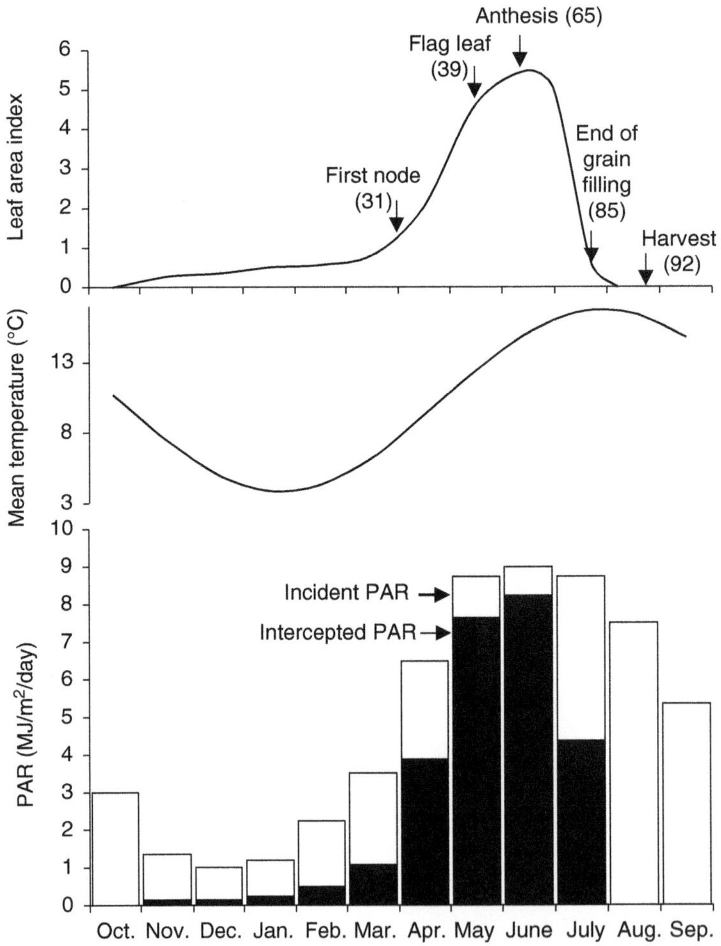

Figure 5.13 The relationship between growth stage, leaf area index, mean daily air temperature, incident photosynthetically active radiation (PAR), and the amount of PAR intercepted by a typical wheat crop in the UK sown in mid-October. *Source:* Data from the Crops Research Unit, University of Reading. Numbers in brackets are the decimal growth stage scores as per Table 1.4 (Zadoks et al. 1974).

not always (Dimmock and Gooding 2002; Pepler et al. 2005), coincident with the senescence of the flag leaf.

With reference to Figure 5.13, it is reasonable to suggest that a higher yield would have been achieved with further increases in canopy size during, for example, April, May, and June. However, the greatest opportunity for gains in PAR interception is for senescence to be delayed in July, and possibly even into August (AHDB 2018); this is because a large proportion of high incident PAR receipts are not currently intercepted by green tissue during this period. Carmo-Silva et al. (2017), therefore, identify the prolongation of the photosynthetic activity of the flag leaf to be a major opportunity for increasing yields.

There are, however, several barriers to delaying senescence and maintaining grain filling in this way. Firstly, cultivars would need to be adapted to such a scenario, e.g. the thermal time requirement for grain filling may need to be increased. Secondly, long-lived canopies often require intensive crop protection measures, particularly for late-season disease control (Pepler et al. 2005). Thirdly, crops with large canopies and high yield potential tend to be at greater risk of late-season droughts (AHDB 2018). Fourthly, once grain filling has been completed, the crop needs to dry during the higher temperatures of late July through to mid-August. Crops that are not harvested until the cooler conditions of September risk further delays and grain spoilage in the field. Even with current cultivars, UK farmers often resort to desiccating their wheat crops with herbicide to facilitate harvesting. Finally, harvest in August allows more timely preparation and drilling of the crops to follow.

5.3 Improving Radiation Use for Increased Land-Use Efficiency

The drive to produce and maintain wheat canopies able to maximise light interception and, therefore, land-use efficiency, has resulted in practices and the development of genotypes with major consequences for crop management and the environment. These are discussed in detail in Chapter 6. However, as producers have improved light interception (facilitated by advances in plant breeding, agronomy, and technology), further opportunities to significantly increase yields by intercepting yet more light have become increasingly limited. This is especially true when accounting for the constraints of the environment, environmental sustainability, and cropping system. Indeed, further improvements in yield through improving light interception may be restricted to small increments achieved with *fine-tuning* to local conditions (Sylvester-Bradley et al. 2005). There is, therefore, significant interest in increasing grain yields for a given amount of light intercepted. One strategy already exploited (Chapter 1) is to increase the proportion of biomass that is present in the grain at harvest, i.e. the HI (Figure 5.5). Shortening stems and increasing HI have resulted in dramatic increases in potential wheat yield (Austin et al. 1980; Reynolds et al. 1999). This achievement, and some of its implications, are further discussed in Chapter 6. For now, it is sufficient to recognise that many of our modern wheat crops already have HIs of over 50% and may well be approaching a theoretical maximum (Austin et al. 1980, 1989; Reynolds et al. 1999) (see also Chapter 1). Certainly, some genetic approaches to further shorten the wheat crop below heights commonly seen in modern wheat cultivars do not increase HI. Rather, crop establishment is compromised, light interception is reduced, canopy architecture is impaired, season-long RUE is lowered, disease spread is increased, and competition against weeds is significantly reduced (Addisu et al. 2009, 2010; Parry et al. 2011; Gooding et al. 2012; Uppal and Gooding 2013). Canopy radiation use-efficiency may be improved by more fine tuning of canopy architecture, for instance to better match leaf angles to developmental stages, production systems, and environments. However, if, as seems likely, both light interception and crop heights are close to optimal levels, then the best strategy to increase yield potential would be to increase the photosynthetic rate per unit leaf area (Parry et al. 2011). There are several ways in which this may be achieved.

Firstly, it is important that there is sufficient demand for assimilate by the grain late in the season (Fischer 2008). The RUE of leaves can decline as they age, and that of the canopy may drop significantly below 2 g/MJ PAR after anthesis. This is particularly the case if grain numbers are reduced (e.g. Acreche and Slafer 2009). This effect of grain numbers on RUE is often cited as evidence of *sink* limited yield formation during grain filling. Yield formation at this stage may be limited by the number and potential size of grain available to secure and contain yield (the *sink*), rather than by the light interception and potential RUE to provide assimilate to the developing grains (the *source*).

Increased grain numbers per ear and per unit area have been major achievements of wheat breeders and it is therefore likely that modern cultivars are less sink-limited than their predecessors (Acreche and Slafer 2009). Nevertheless, there may well be further scope for improving late-season RUE and yield by further increasing grain numbers (Foulkes et al. 2011). It is perhaps surprising that wheat plants do not set enough grains to fully exploit the potentially available photoassimilates during grain filling. It is likely that the balance reached between source and sink is conservative, i.e. the crop may always set fewer grains than could be potentially filled (Foulkes et al. 2011). Such a strategy may have had a selective advantage for the wild progenitors of wheat, with grain development occurring in harsh and variable conditions. Under these circumstances, the production of fewer grains with good viability and vigour and containing sufficient resources (carbohydrate and other nutrients, particularly nitrogen and phosphorus) to satisfy the requirements for germination and vigorous seedling growth (Marco and De Marco 1990) may have given better outcomes compared with spreading resources more thinly over more grains. This conservatism may result from hormonal signalling within the plants in response to heat and drought stresses (Foulkes et al. 2011).

It may also be possible to increase RUE by reducing photorespiration. This might be achieved, for example, by incorporating aspects of C4 metabolism into wheat (Parry et al. 2011; Hu et al. 2012). However, although photorespiration appears inefficient, the enzymes and metabolites involved in the pathway are important for nitrogen assimilation and metabolism. Their functions include amino acid and phospholipid synthesis, maintaining cellular balances of NADPH and ATP, and conferring tolerance to abiotic stresses (Busch 2020; Timm and Hagemann 2020). RUE could also be increased by changes to the photosynthetic machinery. Parry et al. (2011) suggest several targets that may be manipulated given further developments in the genetic modification of wheat. These include increasing the concentration, catalytic rate, affinity for CO_2, and reactivation after heat stress of *RUBISCO*; increasing the speed of regeneration of ribulose 1,5 bisphosphate in the Calvin cycle; and reducing damage to, and/or energy loss from, the photosynthetic machinery at high light levels.

References

Abbad, H., El Jaafari, S., Bort, J. et al. (2004). Comparison of flag leaf and ear photosynthesis with biomass and grain yield of durum wheat under various water conditions and genotypes. *Agronomie* 24 (1): 19–28. https://doi.org/10.1051/agro:2003056.

Acreche, M.M. and Slafer, G.A. (2009). Grain weight, radiation interception and use efficiency as affected by sink-strength in Mediterranean wheats released from 1940 to 2005. *Field Crops Research* 110 (2): 98–105. https://doi.org/10.1016/j.fcr.2008.07.006.

Addisu, M., Snape, J.W., Simmonds, J.R. et al. (2009). Reduced height (Rht) and photoperiod insensitivity (Ppd) allele associations with establishment and early growth of wheat in contrasting production systems. *Euphytica* 166 (2): 249. https://doi.org/10.1007/s10681-008-9838-7.

Addisu, M., Snape, J.W., Simmonds, J.R. et al. (2010). Effects of reduced height (Rht) and photoperiod insensitivity (Ppd) alleles on yield of wheat in contrasting production systems. *Euphytica* 172: 169–181. https://doi.org/10.1007/s10681-0090025-2.

AHDB (2018). *Wheat Growth Guide*. Agriculture and Horticulture Development Board.

AHDB (2019). *Wheat and Barley Disease Management Guide*. Agriculture and Horticulture Development Board.

Assuero, S.G. and Tognetti, J.A. (2010). Tillering regulation by endogenous and environmental factors and its agricultural management. *Americas journal of plant science and. Biotechnology* 4 (1): 35–48. http://www.globalsciencebooks.info/Online/GSBOnline/images/2010/AmJPSB_4(SI1)/AmJPSB_4(SI1)35-48o.pdf.

Austin, R.B., Bingham, J., Blackwell, R.D. et al. (1980). Genetic improvements in winter wheat yields since 1900 and associated physiological changes. *Journal of Agricultural Science* 94 (3): 675–689. https://doi.org/10.1017/S0021859600028665.

Austin, R.B., Ford, M.A., and Morgan, C.L. (1989). Genetic improvement in the yield of winter wheat: a further evaluation. *Journal of Agricultural Science* 112 (3): 295–301. https://doi.org/10.1017/S0021859600085749.

Barber, H., Carney, J., Alghabari, F. et al. (2015). Decimal growth stages for precision wheat production in changing environments? *Annals of Applied Biology* 166: 355–371. https://doi.org/10.1111/aab.12207.

Bassham, J.A., Benson, A.A., and Calvin, M. (1950). The path of carbon in photosynthesis. VIII. The role of malic acid. *Journal of Biological Chemistry* 185 (2): 781–787.

Beed, F.D., Paveley, N.D., and Sylvester-Bradley, R. (2007). Predictability of wheat growth and yield in light-limited conditions. *Journal of Agricultural Science* 145 (1): 63–79. https://doi.org/10.1017/S0021859606006678.

Busch, F.A. (2020). Photorespiration in the context of Rubisco biochemistry, CO_2 diffusion and metabolism. *The Plant Journal* 101 (4): 919–939. https://doi.org/10.1111/tpj.14674.

Calderini, D.F., Dreccer, M.F., and Slafer, G.A. (1997). Consequences of breeding on biomass, radiation interception and radiation-use efficiency in wheat. *Field Crops Research* 52 (3): 271–281. https://doi.org/10.1016/S0378-4290(96)03465-X.

Carmo-Silva, E., Andralojc, P.J., Scales, J.C. et al. (2017). Phenotyping of field-grown wheat in the UK highlights contribution of light response of photosynthesis and flag leaf longevity to grain yield. *Journal of Experimental Botany* 68 (13): 3473–3486. https://doi.org/10.1093/jxb/erx169.

Casal, J.J. (1988). Light quality effects on the appearance of tillers of different order in wheat (Triticum aestivum). *Annals of Applied Biology* 112: 167–173. https://doi.org/10.1111/j.1744-7348.1988.tb02052.x.

Chen, A., Li, C., Hu, W. et al. (2014). PHYTOCHROME C plays a major role in the acceleration of wheat flowering under long-day photoperiod. *Proceedings of the National Academy of Sciences* 111 (28): 10037–10044. https://doi.org/10.1073/pnas.1409795111.

Colbach, N., Lucas, P., and Meynard, J.M. (1997). Influence of crop management on take-all development and disease cycles on winter wheat. *Phytopathology* 87 (1): 26–32. https://doi.org/10.1094/PHYTO.1997.87.1.26.

Demotes-Mainard, S., Doussinault, G., and Meynard, J.M. (1995). Effects of low radiation and low temperature at meiosis on pollen viability and grain set in wheat. *Agronomie* 15 (6): 357–365. https://doi.org/10.1051/agro:19950603.

Dimmock, J.P.R.E. and Gooding, M.J. (2002). The effects of fungicides on rate and duration of grain filling in winter wheat in relation to maintenance of flag leaf green area. *Journal of Agricultural Science* 138 (1): 1–16. https://doi.org/10.1017/S0021859601001666.

Duan, T., Chapman, S.C., and Guo, Y. (2017). Dynamic monitoring of NDVI in wheat agronomy and breeding trials using an unmanned aerial vehicle. *Field Crops Research* 210: 71–80. https://doi.org/10.1016/j.fcr.2017.05.025.

Evans, L.T., Wardlaw, I.F., and Fischer, R.A. (1975). Wheat. In: *Crop Physiology: Some Case Histories* (ed. L.T. Evans), 101–150. Cambridge University Press.

Evers, J.B., Vos, J., Andrieu, B. et al. (2006). Cessation of tillering in spring wheat in relation to light interception and red : far-red ratio. *Annals of Botany* 97 (4): 649–658. https://doi.org/10.1093/aob/mcl020.

Fischer, R.A. (1985). Number of kernels in wheat crops and the influence of solar radiation and temperature. *Journal of Agricultural Science* 105 (2): 447–461. https://doi.org/10.1017/S0021859600056495.

Fischer, R.A. (2008). The importance of grain or kernel number in wheat: a reply to Sinclair and Jamieson. *Field Crops Research* 105 (1–2): 15–21. https://doi.org/10.1016/j.fcr.2007.04.002.

Fischer, R.A. (2011). Wheat physiology: a review of recent developments. *Crop and Pasture Science* 62 (2): 95–114. https://doi.org/10.1071/CP10344.

Foulkes, M.J., Slafer, G.A., Davies, W.J. et al. (2011). Raising yield potential of wheat. III. Optimizing partitioning to grain while maintaining lodging resistance. *Journal of Experimental Botany* 62 (2): 469–486. https://doi.org/10.1093/jxb/erq300.

Friend, D.J.C. (1966). The effects of light and temperature on the growth of cereals. In: *The Growth of Cereals and Grasses* (eds. F.L. Milthorpe and J.D. Ivins), 181–199. Butterworths.

Gallagher, J. and Biscoe, P. (1978). Radiation absorption, growth, and yield of cereals. *Journal of Agricultural Science* 91 (1): 47–60. https://doi.org/10.1017/S0021859600056616.

García, A., Serrago, R.A., González, F.G. et al. (2014). Wheat grain number: identification of favourable physiological traits in an elite doubled-haploid population. *Field Crops Research* 168: 126–135. https://doi.org/10.1016/j.fcr.2014.07.018.

Gent, M.P.N. and Kiyomoto, R.K. (1992). Canopy photosynthesis and respiration in winter wheat adapted and unadapted to Connecticut. *Crop Science* 32: 425–431. https://doi.org/10.2135/cropsci1992.0011183X003200020029x.

Gillett, A.G., Crout, N.M.J., Stokes, D.T. et al. (1999). Simple winter wheat green area index model under UK conditions. *Journal of Agricultural Science* 132: 263–271. https://doi.org/10.1017/S0021859699006346.

Gonzalez, F.G., Slafer, G.A., and Miralles, D.J. (2003). Grain and floret number in response to photoperiod during stem elongation in fully and slightly vernalized wheats. *Field Crop Research* 81: 17–27. https://doi.org/10.1016/S0378-4290(02)00195-8.

Gooding, M.J., Pinyosinwat, A., and Ellis, R.H. (2002). Responses of wheat grain yield and quality to seed rate. *Journal of Agricultural Science* 138: 317–331. https://doi.org/10.1017/S0021859602002137.

Gooding, M.J., Gregory, P.J., Ford, K.E. et al. (2005). Fungicide and cultivar affect post-anthesis patterns of nitrogen uptake, remobilization and utilization efficiency in wheat. *Journal of Agricultural Science* 143: 503–518. https://doi.org/10.1017/S002185960500568X.

Gooding, M.J., Addisu, M., Uppal, R.K. et al. (2012). Effect of wheat dwarfing genes on nitrogen-use efficiency. *Journal of Agricultural Science* 150 (1): 3–22. https://doi.org/10.1017/S0021859611000414.

Hall, D.O. and Rao, K.K. (1981). *Photosynthesis*. Arnold.

Hanft, J.M. and Wych, R.D. (1982). Visual indicators of physiological maturity of hard red spring wheat. *Crop Science* 22: 584–588. https://doi.org/10.2135/cropsci1982.0011183X002200030036x.

Hay, R.K.M. and Porter, J.R. (2006). *The Physiology of Crop Yield*, 2e. Blackwell.

Hill, R. (1939). Oxygen produced by isolated chloroplasts. *Proceedings of the Royal Society London B* 127: 192–210. http://doi.org/10.1098/rspb.1939.0017.

Hill, R. (1965). The biochemists' green mansions: the photosynthetic electron-transport chain in plants. *Essays in Biochemistry* 1: 121–151.

Hu, L., Li, Y., Xu, W. et al. (2012). Improvement of the photosynthetic characteristics of transgenic wheat plants by transformation with the maize C4 phosphoenolpyruvate carboxylase gene. *Plant Breeding* 131 (3): 385–391. https://doi.org/10.1111/j.1439-0523.2012.01960.x.

Kasperbauer, M.J. and Karlen, D.L. (1986). Light-mediated bioregulation of tillering and photosynthate partitioning in wheat. *Physiologia Plantarum* 66: 159–163. http://doi.org/10.1111/j.1399-3054.1986.tb01250.x.

Kindred, D.R. and Gooding, M.J. (2005). Heterosis for yield and its physiological determinants in wheat. *Euphytica* 142: 149–159. https://doi.org/10.1007/s10681-005-1250-y.

Kirby, E.J.M. (1988). Analysis of leaf, stem, and ear growth in wheat from terminal spikelet stage to anthesis. *Field Crops Research* 18: 127–140. https://doi.org/10.1016/0378-4290(88)90004-4.

Law, C.N. (1999). *Sterility in Winter Wheat: Review of Occurrence in Different Varieties and Possible Causes. HGCA Research Review No. 44*. Agriculture and Horticulture Development Board.

Marco, D. and De Marco, D.G. (1990). Effect of seed weight, and seed phosphorus and nitrogen concentrations on the early growth of wheat seedlings. *Australian Journal of Experimental Agriculture* 30 (4): 545–549. https://doi.org/10.1071/EA9900545.

Marshall, B. and Biscoe, P.V. (1980). A model for C3 leaves describing the dependence of net photosynthesis on irradiance: II. Application to the analysis of flag leaf photosynthesis. *Journal of Experimental Botany* 31 (1): 41–48. https://doi.org/10.1093/jxb/31.1.41.

McCree, K.J. (1972). The action spectrum, absorptance and quantum yield of photosynthesis in crop plants. *Agricultural Meteorology* 9: 191–216. https://doi.org/10.1016/0002-1571(71)90022-7.

Miralles, D.J., Richards, R.A., and Slafer, G.A. (2000). Duration of the stem elongation period influences the number of fertile florets in wheat and barley. *Australian Journal Plant Physiology* 27: 931–940. https://doi.org/10.1071/PP00021.

Ong, C.K. and Monteith, J.L. (1992). Canopy establishment: light capture and loss by crop canopies. In: *Crop Photosynthesis: Spatial and Temporal Determinants. Topics in Photosynthesis*, vol. 12 (eds. N.R. Baker and H. Thomas), 1–9. Elsevier.

Parry, M.A.J., Reynolds, M., Salvucci, M.E. et al. (2011). Raising yield potential of wheat. II. Increasing photosynthetic capacity and efficiency. *Journal of Experimental Botany* 62 (2): 453–467. https://doi.org/10.1093/jxb/erq304.

Pepler, S., Gooding, M.J., Ford, K.E. et al. (2005). A temporal limit to the association between flag leaf life extension by fungicides and wheat yields. *European Journal of Agronomy* 22 (4): 363–373. https://doi.org/10.1016/j.eja.2004.06.002.

Reynolds, M.P., Rajaram, S., and Sayre, K.D. (1999). Physiological and genetic changes of irrigated wheat in the post-green revolution period and approaches for meeting projected global demand. *Crop Science* 39: 1611–1621. https://doi.org/10.2135/cropsci1999.3961611x.

Reynolds, M., Foulkes, J., Furbank, R. et al. (2012). Achieving yield gains in wheat. *Plant Cell and Environment* 35: 1799–1823. https://doi.org/10.1111/j.1365-3040.2012.02588.x.

Shock, E.L. (1997). High-temperature life without photosynthesis as a model for Mars. *Journal of Geophysical Research: Planets* 102 (E10): 23687–23694. https://doi.org/10.1029/97JE01087.

Slafer, G.A., Kantolic, A.G., Appendino, M.L. et al. (2009). Crop development: genetic control, environmental modulation and relevance for genetic improvement of crop yield. In: *Crop Physiology: Applications for Genetic Improvement and Agronomy* (eds. V.O. Sadras and D.F. Calderini), 277–308. Academic Press.

Sparkes, D.L., Holme, S.J., and Gaju, O. (2006). Does light quality initiate tiller death in wheat? *European Journal of Agronomy* 24 (3): 212–217. https://doi.org/10.1016/j.eja.2005.08.003.

Sylvester-Bradley, R., Scott, R.K., Stokes, D.T. et al. (1997). The significance of crop canopies for N nutrition. In: *Aspects of Applied Biology 50 Optimising Cereal Inputs Part 1: Genetics and Nutrition* (eds. M.J. Gooding and P.R. Shewry), 103–116. Association of Applied Biologists.

Sylvester-Bradley, R., Foulkes, J., and Reynolds, M.P. (2005). Future wheat yields: evidence, theory and conjecture. In: *Yields of Farmed Species: Constraints and Opportunities in the 21st Century* (eds. R. Sylvester-Bradley and J. Wiseman), 233–260. Nottingham University Press.

Thomas, H. (2018). *The Tale of Three Little Pigments*. Howard Thomas.

Thorne, G.N. (1965). Photosynthesis of ears and flag leaves of wheat and barley. *Annals of Botany* 29 (3): 317–329. https://doi.org/10.1093/oxfordjournals.aob.a083954.

Thorne, G.N. and Wood, D.W. (1987). Effects of radiation and temperature on tiller survival, grain number and grain yield in winter wheat. *Annals of Botany* 59 (4): 413–426. https://doi.org/10.1093/oxfordjournals.aob.a087330.

Thorne, G.N., Pearman, I., Day, W. et al. (1988). Estimation of radiation interception by winter wheat from measurements of leaf area. *Journal of Agricultural Science* 110 (1): 101–108. https://doi.org/10.1017/S0021859600079740.

Timm, S. and Hagemann, M. (2020). Photorespiration – how is it regulated and how does it regulate overall plant metabolism? *Journal of Experimental Botany* 71 (14): 3955–3965. https://doi.org/10.1093/jxb/eraa183.

Ugarte, C.C., Trupkin, S.A., Ghiglione, H. et al. (2010). Low red/far-red ratios delay spike and stem growth in wheat. *Journal of Experimental Botany* 61 (11): 3151–3162. https://doi.org/10.1093/jxb/erq140.

Uppal, R.K. and Gooding, M.J. (2013). Gibberellin-responsive and -insensitive dwarfing alleles on wheat performance in contrasting tillage systems. *Field Crops Research* 141: 55–62. https://doi.org/10.1016/j.fcr.2012.11.001.

Zadoks, J.C., Chang, T.T., and Konzak, C.F. (1974). A decimal code for the growth stages of cereals. *Weed Research* 14: 415–421.

6

Canopy Management

Wisdom, power and goodness meet in the bounteous field of wheat

Hannah Flagg Gould

In Chapter 5, we presented the requirements for light interception over important growth stages. We also introduced some aspects of canopy size and architecture (e.g. green area index, leaf angle, and crop height) that are related to grain yield improvement. In this chapter we focus on how crop management has developed alongside changes in cultivars to influence light interception and utilization. We focus on aspects of crop establishment, crop nutrition, and the control of fungal diseases. In other words, we address some of the most important genotype × management interactions that have facilitated the intensification and increased land-use efficiency of wheat production. The impact of the parallel changes in cultivars and agronomy on the environment is significant. We therefore discuss the major environmental risks and hazards that accompany efforts to increase light interception by wheat crops.

6.1 Crop Establishment

6.1.1 Sowing Date

Clearly, earlier crop establishment increases the duration of light interception if the date of senescence remains constant (Stapper and Fischer 1990a). The drilling of autumn-sown wheats becomes earlier as latitudes increase (Gooding and Davies 1997) and, because there is frequently only one crop per year (or less in wheat: fallow systems), it is here that there is greater potential for bringing forward drilling dates. In the oceanic areas at least, clement conditions may persist for an extended duration after the harvest of preceding crops and before the onset of winter. Early drilling results in earlier canopy formation in the autumn, and more rapid expansion in the following spring (Figure 6.1; Dennett et al. 1999). Similarly, early sown autumn crops can have more extensive root systems for nutrient and water capture in the following spring and summer, while increased tillering in the autumn may help tolerate pest attacks on stem growing points. Within the limits imposed by the climate, rotation, and cultivar adaptation (Stapper and Fischer 1990a, 1990b; Dennett et al. 1999)

Wheat: Environment, Food and Health, First Edition. Mike J. Gooding and Peter R. Shewry.

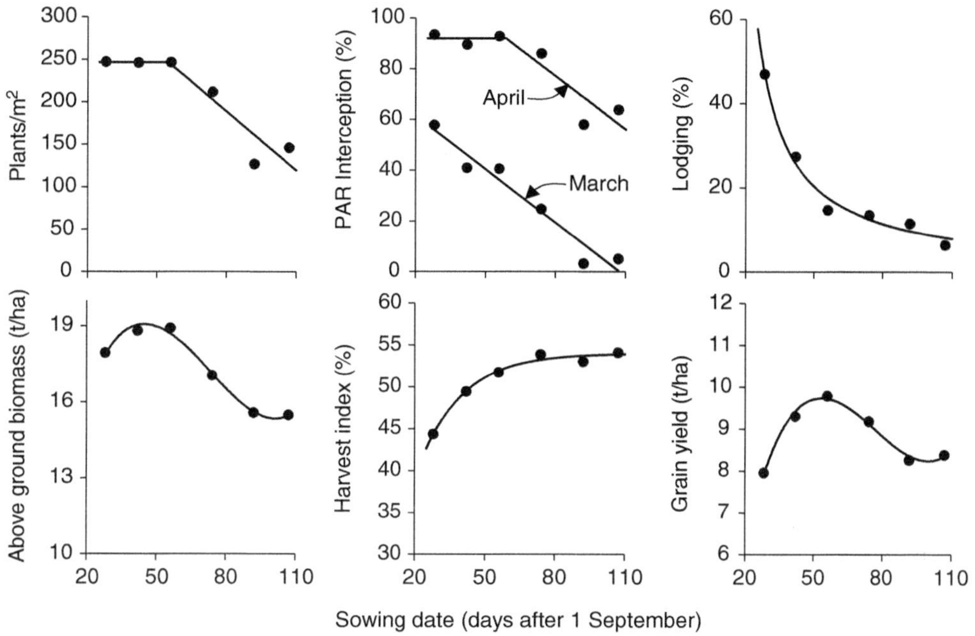

Figure 6.1 The effect of sowing date on plants established and surviving to March, interception of photosynthetically active radiation in spring, lodging in August, and above-ground biomass, harvest index, and grain yield at harvest of winter wheat in the UK. Points are means of six cultivars in each of two years. *Source:* Data from Spink et al. (2000a).

there is, therefore, often a perception that earlier drilling of appropriately adapted cultivars can increase yield (Spink et al. 2000a). Indeed, Wibberley (1989) recounts the views of progressive farmers all the way back to Jethro Tull in 1731 advocating the benefits of early sowing. More latterly, within the UK there was a move from spring- to autumn-sown cereals in the 1970s followed by a trend towards earlier drilling (in September rather than October [Dennett et al. 1999]) from the 1980s. Irrespective of effects on potential yield, early drilling also helps to ensure that large areas can be sown at a time when weather and soil conditions may be variable and, in any case, deteriorating into winter. Delaying drilling until soil conditions are more difficult may result in lower plant establishment, light interception, biomass production, and grain yields (Figure 6.1). The need for speed and timeliness under these pressures has therefore led to the greater popularity of more minimal cultivations and rapid seedbed preparation techniques, with associated challenges for control of weeds, pests, and stubble-borne diseases (Chapter 2).

The greater light interception from early sown crops would be expected to increase biomass production. This relationship is, however, modified by other effects associated with early sowing. Early sowing of wheat can lead to taller, weaker straw, such that the plants become more susceptible to falling over (i.e. becoming *laid* or *lodged*) before harvest (Figure 6.1; Stapper and Fischer 1990b). Lodged crops have reduced net photosynthesis (Berry and Spink 2012), and therefore reduced biomass production, in addition to grain spoilage in the field and greatly reduced harvesting efficiency. For earlier sown crops there is increased risk of disease and pest carryover from one crop to the next, due to an extended period in the autumn suitable for the start of infestations (e.g. Dewar and Carter 1984; Gladders et al. 2005). Weed control is also more difficult as significantly greater numbers of weeds germinate and grow after drilling compared to later sown crops (Cosser et al. 1997a; Lutman et al. 2013). The effects of early sowing on grain yield may also be disappointing because early sowing and light interception can be at the expense of harvest index (Figure 6.1; Stapper and Fischer 1990b; Barber et al. 2015). Overall, therefore, to benefit from early sowing at high latitudes it is important that cultivars are short and stiff-strawed with good lodging resistance; that they are late maturing with, for example, significant vernalisation, and photoperiod requirements; and that they have good levels of disease resistance. Despite cultivar improvement, early sowing of wheat at high latitudes has often increased the reliance on agrochemicals for crop protection and growth regulation. The change from spring to early autumn sowing changes the arable weed flora from spring-germinating to autumn-germinating species. Having a standing crop on the land during autumn and winter, rather than it being left as stubble, has also been a significant factor in the decline of farmland bird populations in Europe (e.g. Donald and Vickery 2000; Donald et al. 2001).

In lower latitudes and warmer climates, the cropping period is often restricted to the cooler winter months and may in any case be limited by the temporal positioning of other crops grown in the same year. It may be particularly important to ensure that meiosis and anthesis coincide with the coolest time of the year. Additionally, early sowing could expose the seedlings to excessive temperatures leading to reduced tiller numbers, vigour, and ultimately grain yield (Gooding and Davies 1997). Conversely, in the humid sub-tropics, late-planted crops can experience excessive heat and rainfall from anthesis to maturity, resulting in poor grain set, impaired grain filling, and increased foliar diseases.

6.1.2 Plant Populations and Sowing Rate

It is intuitive that light interception per unit land area should increase as the number of plants increases. This relationship cannot be linear because as more plants are added there is increasing likelihood that they will shade and crowd each other. The relationship between light interception and the number of plants established per unit area (ρ) can, therefore, usually be approximated by simple asymptotic relationships (Figure 6.2). Those presented here are an application of the population density equations used by e.g. Khah et al. (1989) and Ellis et al. (1999). Thus, the light intercepted by individual plants (ω) declines according to:

$$\omega = \frac{1}{a + b\rho} \tag{6.1}$$

i.e. the inverse of light intercepted per plant is a straight line (Figure 6.2b):

$$\frac{1}{\omega} = a + b\rho \tag{6.2}$$

and light intercepted per unit area (Figure 6.2c) is:

$$\omega\rho = \rho\left[\frac{1}{a + b\rho}\right] \tag{6.3}$$

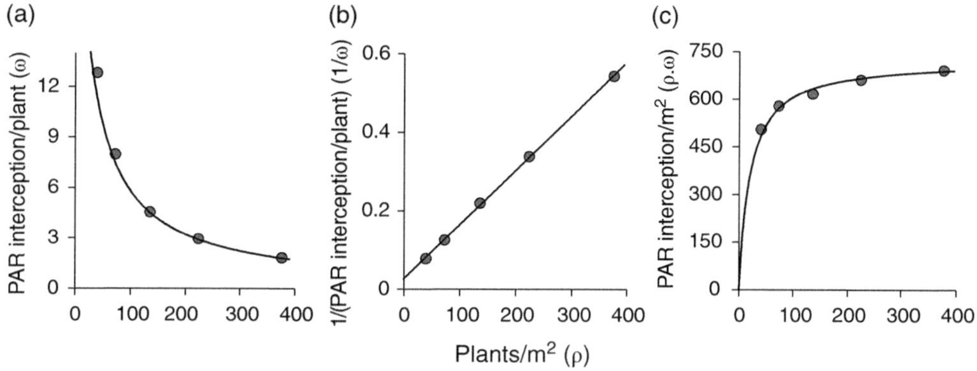

Figure 6.2 Relationships between interception of photosynthetically active radiation (PAR, MJ) and plant population density. (a) Interception per plant ($\omega = 1/(a + b\,\rho)$); (b) Inverse of interception per plant ($1/\omega = a + b\rho$); and (c) Interception per m^2 ($\omega\rho = \rho/(a + b\rho)$). In all three graphs $a = 0.027$ and $b = 0.00138$. *Source:* Data from Gooding et al. (2002).

This approach can be extended to approximate the relationship between population density and biomass and grain yield. In these cases, however, closer fits can be obtained by the addition of a further parameter (Eq. 6.4; Bleasdale 1984) to account for confounding effects on radiation use efficiency and harvest index as plant population increases (as in Figure 6.3).

$$grain\ yield\ per\ unit\ area = \rho\left[\frac{1}{a+b\rho}\right]^{\frac{1}{q}} \tag{6.4}$$

Low population densities are associated with delayed increases in PAR interception (Figure 6.3; Gooding et al. 2002), and lower maximal leaf area indices (Figure 6.3). Wheat does, however, have several compensatory mechanisms to respond to sparse establishment, which account for much of the curvature in response to plant population density seen in Figures 6.2 and 6.3. Some of these mechanisms are a result of the effects of light intensity and light quality, as described in Chapter 5. Additionally, at low plant populations each plant has greater access to nutrients and water with less competition from its neighbours.

Plants established at low densities therefore compensate through increased duration of tiller production, increased green area per shoot, and longer duration of tiller survival. They also often (Whaley et al. 2000; Gooding et al. 2002), but not always (e.g. Figure 6.3), have increased radiation use efficiencies. In terms of yield components, low-density crops have higher numbers of ears per plant and grains per ear (Figure 6.4). Excessive reliance on compensatory growth can, however, lead to several problems. The importance of the survival of individual plants is increased and pests that kill the growing points of relatively few plants at the seedling stage can leave large gaps, resulting in poor utilization of light and nutrients. Late tillering of thin crops may also be the cause of variable and delayed maturation (Thompson et al. 1993; Gooding et al. 2002), making the crop uneven and difficult to manage and harvest.

In addition to their simplicity, Eqs. (6.1)–(6.3) provide further utility in the interpretation of a and b. The inverse of a is an assessment of the amount of light that would be

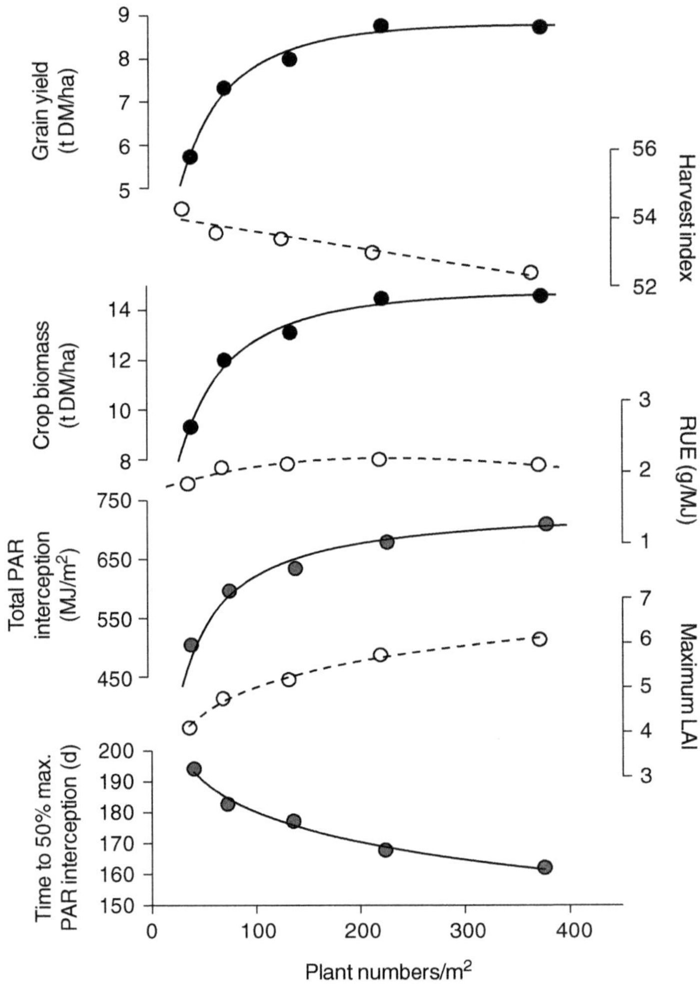

Figure 6.3 The relationships between plant population density and interception of photosynthetically active radiation (PAR), crop biomass, and grain yield, as modified by effects on radiation-use efficiency and harvest index. *Source:* Data from Gooding et al. (2002).

intercepted by a single plant, free of any competition from neighbours. For the experiment in Figure 6.2, $1/a = 37$ MJ PAR. The inverse of b is a measure of light interception when not limited by plant population density, i.e. it is defined by environmental and temporal factors such as irradiance, canopy duration, and availability and use of nutrients and water by the crop. Here, $1/b = 725$ MJ PAR/m^2.

This is an appropriate point to emphasise that characteristics of wheat cultivars that maximise the light interception and yield of individual, spaced plants (high $1/a$), should also compensate for low plant population densities. However, these same characteristics are often undesirable when the aim is to maximise the yield per unit area of a community of densely spaced plants (high $1/b$). Plants that are competitive individually (high $1/a$) have a *spreading* growth habit, i.e. they tiller profusely, have lax leaves, and grow tall

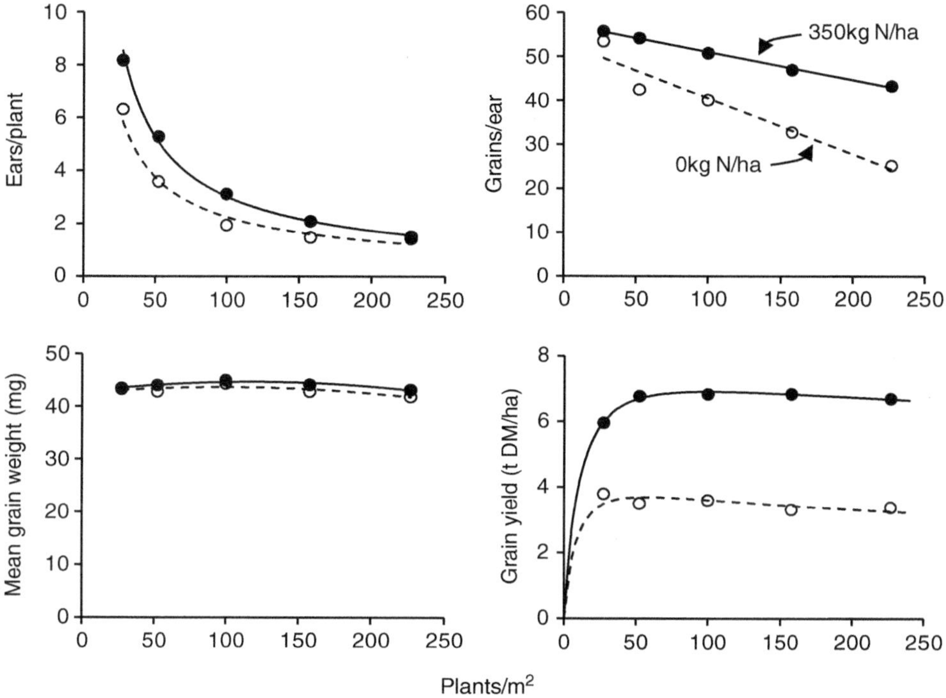

Figure 6.4 Effects of plant population density and nitrogen fertiliser on yield components of winter wheat. *Source:* Data from Gooding et al. (2002).

(Donald 1968). This investment in vegetative growth means that ears are small and variable, and the plants have a low harvest index (Figure 6.5B). Conversely, *communal* plants compete less with their neighbours, have fewer tillers, have leaves that are more erect, and stems that are shorter. Less investment in vegetative growth results in larger, more evenly sized ears and a higher harvest index (Figure 6.5A). The reduced investment in vegetative growth on a per plant basis can be mitigated by establishing more plants per unit area to exploit a higher value of $1/b$, particularly for grain yield. The change from competitive to communal wheat types has been a major target of wheat breeding from the 1960s to the present day (Figure 6.5), confirming the early analysis of Donald (1968). The success and maintenance of the communal ideotype requires that these types are grown as pure cultivars, or at least in blends where all components have near equal competitivity. If wheats of varying genotypes and competitive abilities are grown together, the competitive (more *wild*) types tend to prevail over the communal types (Knapp et al. 2020).

For researchers studying individual plants, e.g. in pots in glasshouses or growth cabinets, it should always be remembered that it is unlikely that increases in yield per plant will translate to improved crop yields from a dense population of plants in a field. Indeed, the reverse may well be true (Donald 1968; Knapp et al. 2020).

The adoption of the more communal ideotype not only reduces competition against neighbouring crop plants, but also competition against weeds (Cosser et al. 1997a, 1997b; Addisu et al. 2010; Gooding et al. 2012a). This can be mitigated by increasing plant

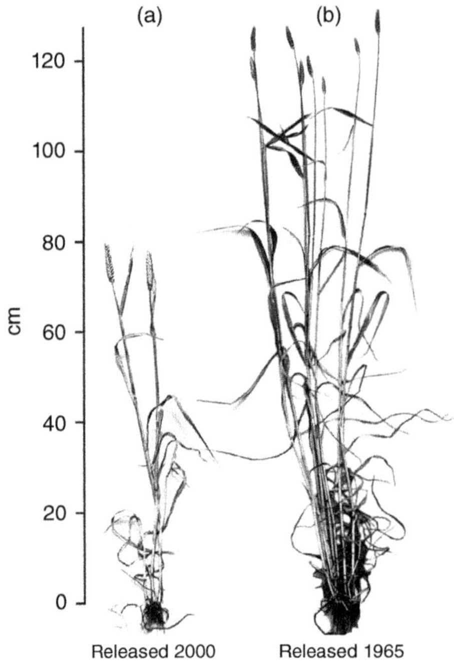

Figure 6.5 The effect of plant breeding on the shift to more communal (a), rather than competitive (b) wheat phenotypes.

population density (Korres and Froud-Williams 2002; Lutman et al. 2013) but growing dense stands has other implications for crop protection. For example, high population densities can be associated with increased prevalences of some foliar (Tompkins et al. 1992, 1993) and root and stem-base diseases (Glynne 1951). Increasing the plant population density also increases lodging risks (Glynne 1951; Stapper and Fischer 1990b). Except for weed control, therefore, increasing light interception by increasing plant population density generally increases reliance on crop protection measures such as using resistant cultivars and/or agrochemicals. For a given plant population density, selecting communal types also increases reliance on herbicides.

Farmers principally manipulate plant population density by varying their seed rates (Figure 6.6a). The seed rate required to achieve a target plant population will depend on the establishment rate (Figure 6.6a). This in turn depends on the quality of the seed with regards to viability and seedling vigour (Ellis 1992); the seed bed conditions in terms of tilth fineness, structure, and moisture content; the accuracy of seed placement at target depths and spatial separation with the available sowing equipment; possible losses due to pests and diseases; and, if autumn sowing, losses due to severe conditions over winter. The establishment rate also tends to decline as sowing rate increases (Figure 6.6a). Given the uncertainties of establishment rates, and the greater penalties for having too few plants compared with too many (e.g. Figure 6.6b), farmers often sow more seeds than may prove to be economically optimal in a single season (Wibberley 1989).

There can be strong interactions between sowing date and sowing rate on yield. This is because crops sown early into clement conditions have longer to compensate for low population densities through increased tillering and are more liable to lodge at higher sowing

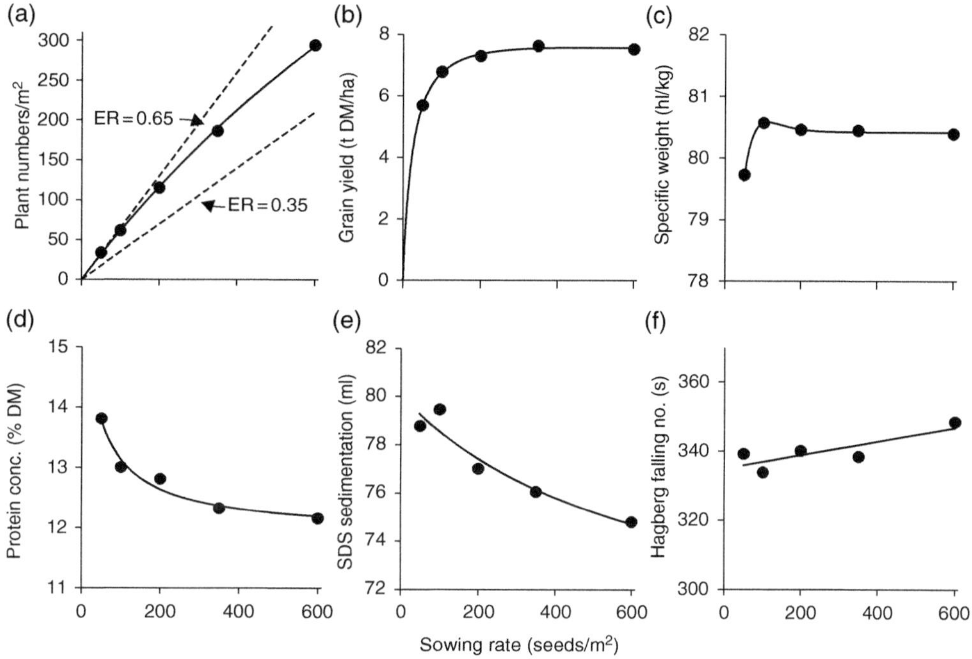

Figure 6.6 The effect of sowing density of winter wheat on the plant establishment rate (ER) by the following spring, grain yield, and grain qualities. Points are means of three experiments described in Gooding et al. (2002).

rates. In these circumstances, sowing rates can be significantly reduced (Spink et al. 2000b). Similarly, optimal sowing rates are often higher for spring-sown crops compared with autumn sowings.

Variation in seed rates within ranges commonly used in commercial practice have little impact on point-of-sale measures of grain quality. At low population densities, however, grain protein concentrations often rise, with associated effects on sedimentation volume (Figure 6.6d, e; Kindred and Gooding 2004). This is because wheat plants at low population densities are more efficient at capturing the available N than they are at intercepting and utilizing the available light to accumulate starch (Gooding et al. 2001). At exceptionally low seed and population densities, specific weight may decline if very late tillers survive to produce severely shrivelled grain. In the absence of lodging, there are often small reductions in α-amylase activity and increases in Hagberg falling number as plant populations increase (Figure 6.6f; Kindred et al. 2005), which possibly result from more even ripening and reduced contribution from late, green tillers (Lunn et al. 2001).

6.2 Crop Nutrition

Producing canopies with a green area of six (Chapter 5) requires large quantities of nutrients. Here we focus on four macro-nutrients, nitrogen (N), phosphorus (P), sulphur (S), and magnesium (Mg), due to their major and direct roles in the photosynthetic machinery.

Potassium and calcium are also relevant, but have been discussed in other chapters because of their respective roles in the water relations of the crop (Chapter 3) and pH of the soil (Chapter 2).

6.2.1 Nitrogen

Nitrogen is often identified as the single most important crop nutrient defining the yield and quality of wheat crops. As well as being of direct importance to wheat production, N is a critical nutrient determining the environmental impact of farming. This is because there are environmental costs in producing N in forms available to crop plants, and increased availability of N may lead to additional problems of diffuse pollution and greenhouse gas emissions. Increasing N availability affects the growth of other organisms in wheat fields and has a major impact on the need for crop protection measures. Before discussing these issues further, however, it is important to emphasise the importance of N for photosynthesis, and hence for all food production.

6.2.1.1 The Requirement for Nitrogen

Nitrogen is an essential component of photosynthetic capacity (Figure 6.7). Kumar et al. (2002) estimate that 60–80% of the leaf N of C3 plants is invested in the photosynthetic apparatus. The amino acids polymerised in proteins account for the bulk of the N. Nitrogen is present in large quantities in the proteins and enzymes of the photosystems and Calvin cycle, with *RUBISCO* alone accounting for 20–25% of all leaf N in wheat (Evans 1989). Nitrogen is also present in the purine and pyrimidine building blocks of the nucleic acids DNA and RNA. Nitrogen is, therefore, essential for the coding, decoding, regulation, and expression of genes to produce proteins. Nitrogen is also an important constituent of some pigments including chlorophyll and phytochrome and is thus essential for the capture of light energy and responses to light quality. Nitrogen is also a component of ATP (adenosine triphosphate) and NADPH (nicotinamide adenine dinucleotide phosphate), and so is vital for powering the Calvin cycle.

Given the above, it is unsurprising that important components of the photosynthetic machinery in wheat leaves increase as N concentration increases. For example, chlorophyll and *RUBISCO* concentrations increase linearly with increasing N concentrations (Evans 1989; Lawlor et al. 1989), with the levels of *RUBISCO* being particularly responsive to N availability (Kumar et al. 2002). When N concentrations are high, the higher chlorophyll content makes the crop greener. Conversely, growers will often ascribe yellowing of a crop to N deficiency. This is particularly true of the older leaves because the plant prioritizes and relocates N to the younger and emerging leaves at the top of the canopy at the expense of older leaves in more shady positions (Moreau et al. 2012). It is the older leaves, therefore, which senesce prematurely when N is remobilized early in times of shortage. Similarly, at the end of the season N depletion can hasten senescence because the plant prioritises N for grain filling (Distelfeld et al. 2014). By harvest 65–80% of above-ground N will be in the grain (Gooding et al. 2005, 2007) and under fertile conditions 60–75% of this will have been remobilized from the canopy, with the remainder deriving from post-anthesis uptake from the roots and soil (Gooding et al. 2005, 2012a). When N is in short supply, over 80% of N in the grain at harvest may have been derived from the N in the canopy at anthesis (Gooding et al. 2012a).

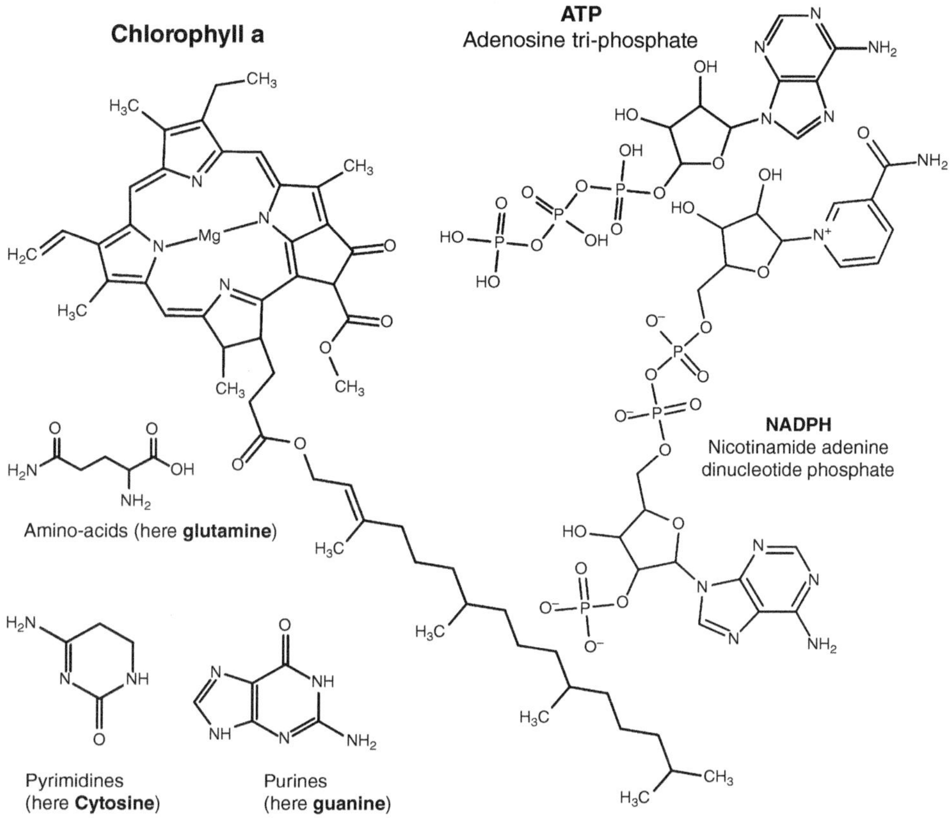

Figure 6.7 Some of nitrogen-containing compounds important in photosynthesis.

The rate of light-saturated photosynthesis in individual leaves increases with leaf lamina N content. Figure 6.8a demonstrates an increase in maximal photosynthesis until there is about 2 g N/m^2 of leaf. This effect can be seen when the N availability to N-deficient crops is increased (Figure 6.8b). At lower levels of light intensity, however, the effect of N availability on leaf photosynthesis is much reduced.

The largest contribution of N to biomass accumulation is by increasing light interception. This is achieved through increasing the expansion rate and ultimate size of the canopy with both leaf size and tiller number increasing in response to rising N availability (Figure 6.9a, b). Applying N to N-deficient crops thus increases the amount of PAR that is intercepted early in the season and the maximum proportion of PAR that is intercepted in mature canopies (Figure 6.9c). Nitrogen applications also delay crop senescence (Figure 6.9d), as the grain demand for N can be partially met by remobilisation and post-anthesis uptake of N, rather than senescence of the canopy.

The effects of N on canopy components contribute to near linear relationships between the green area index (GAI) of a canopy and the amount of N in the canopy. The regression coefficient in Figure 6.10 indicates, that on average, each unit increase in GAI is associated with an additional 30 kg N/ha in the canopy. There is clearly variation around this

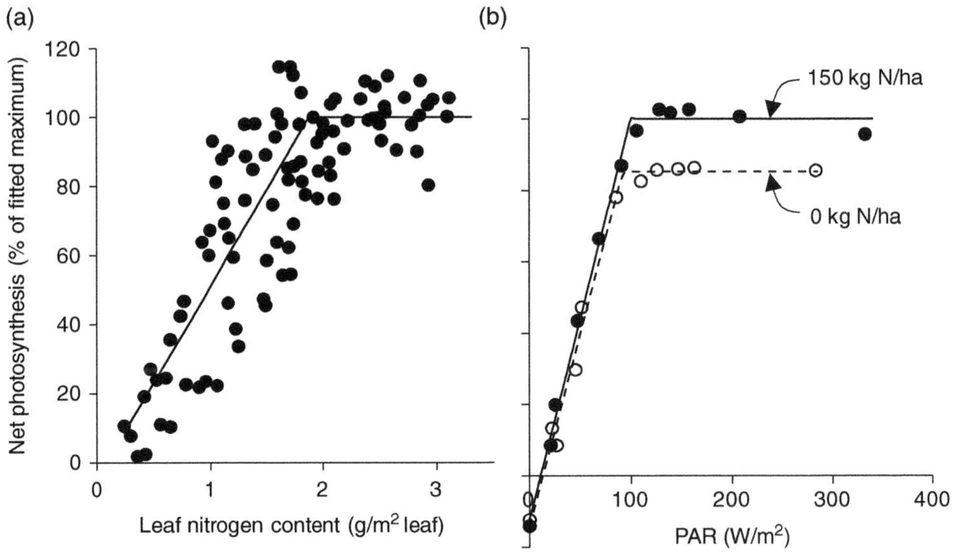

Figure 6.8 The effect of nitrogen on leaf photosynthesis. (a) The relationship between nitrogen content and photosynthetic rate of wheat leaves under high irradiance. *Source:* Data from Evans (1989) and Makino et al. (1988). (b) The effect of applying nitrogen fertiliser to N-deficient wheat, and of increasing the intensity of photosynthetically active radiation on photosynthesis of flag leaves. *Source:* Data Gregory et al. (1981).

relationship, but a similar mean coefficient has been derived from other data sets (Sylvester-Bradley et al. 1997). The 30 kg N/ha/per unit GAI equates to 3 g N/m^2 of leaf, which is clearly greater than that which would be required to maximise photosynthesis if it was restricted to the leaves (as per Figure 6.8a). At the canopy level, however, N is also required for other structures such as stems, leaf sheaths, and developing ears.

It follows that an average of around 180 kg N will be present in canopies of GAI = 6, i.e. the canopy size required to intercept 95% of the available PAR with an extinction coefficient of $k \approx 0.5$. Historically, enabling wheat plants to incorporate 180 kg N/ha into the canopy was problematic. For example, improving light interception through increasing N availability, principally by increasing ammonium and nitrate in soil solution (Figure 6.11), has several deleterious consequences. High N availability increases lodging risk (e.g. Figure 6.12; Berry et al. 2000) because it increases the height of the centre of gravity of the crop, increases the size of the ears, reduces mean stem diameter, and reduces stem wall thickness (Berry et al. 2000). High N availability also significantly increases the severity of numerous fungal pathogens (e.g. Olesen et al. 2003a; Neumann et al. 2004; Robert et al. 2005; Walters and Bingham 2007), the growth of weeds (Kristensen et al. 2008), and the growth and multiplication of wheat aphids (Aqueel and Leather 2011). Therefore, as for sowing date and seed rate, increasing light interception by increasing N availability places greater reliance on crop protection measures and increases the need for cultivars that are resistant to lodging, pests, and diseases.

Regarding cultivar development, farmers have long recognised the benefit of growing shorter wheats to reduce lodging in high N conditions and to increase the harvest index for

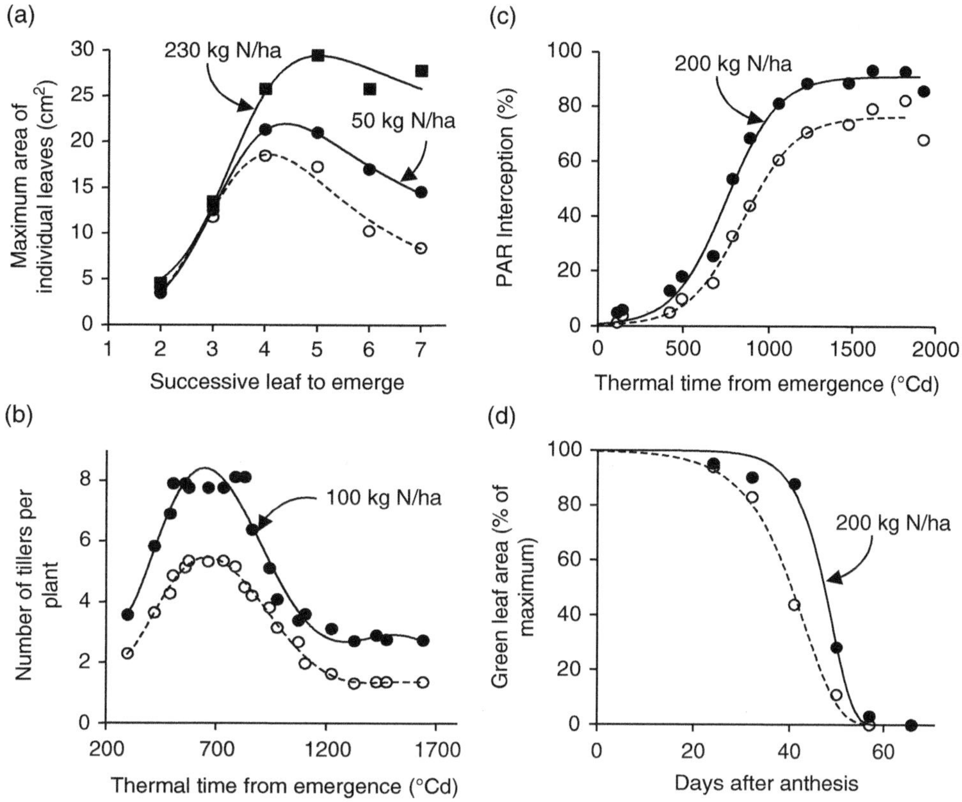

Figure 6.9 The effect of applying nitrogen fertiliser on canopy components of N-deficient field grown wheat. (a) The size of individual leaves (Bunting and Drennan 1966). (b) The number of tillers per plant (Alzueta et al. 2012). (c) Interception of photosynthetically active radiation (Kindred and Gooding 2005). (d) Senescence of flag leaves (Kindred and Gooding 2005). ○ = no nitrogen fertilizer.

Figure 6.10 The relationship between nitrogen in the canopy and canopy size. Data from field experiments in the UK (●, Scott et al. (1994); +, White et al. (1998); ▲, Critchley (2001)), France (■, Lemaire et al. (2007)), and Australia (◆, Lemaire et al. (2007)). Depending on source, points are treatment means varying for growth stage, sowing rate, N-fertiliser application rate, and genotype.

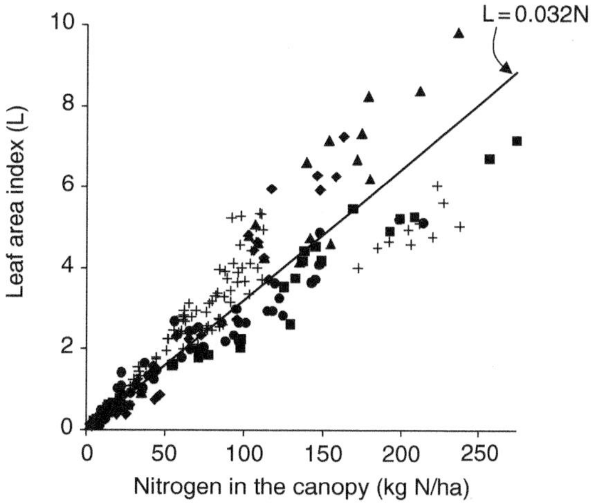

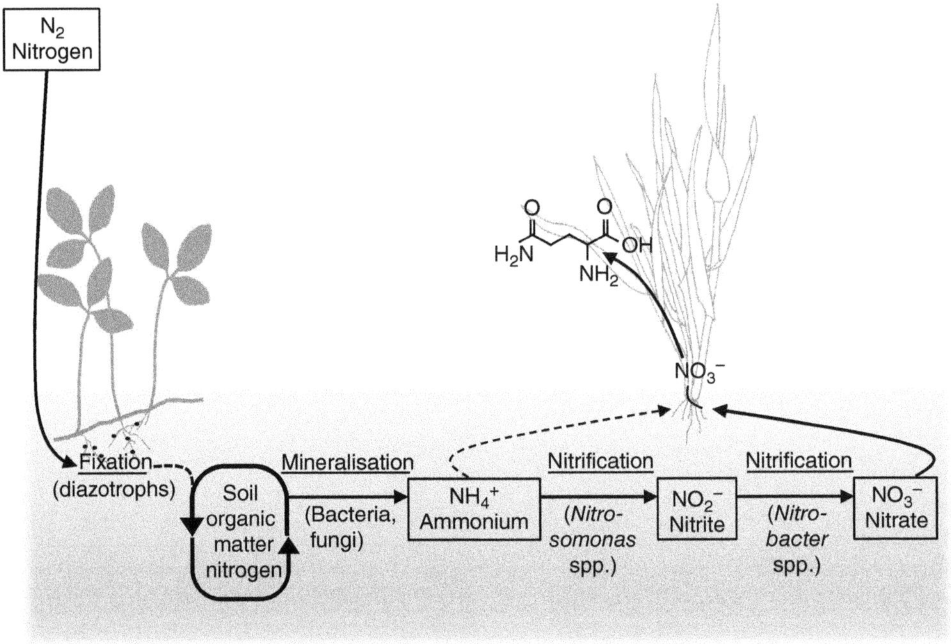

Figure 6.11 A pathway of biologically fixed nitrogen from atmospheric N_2 to wheat plants. In this case via *Rhizobia* in root nodules of a forage legume, with the proximal separation from wheat more likely to be temporal rather than spatial.

grain production (Roberts 1847; Garnett 1883). To meet this need plant breeders have incorporated major alleles for reducing height (*Rht*) into wheat breeding programmes. These alleles make a major contribution to the communal plant phenotype presented in Section 6.1.2 and were introduced into western breeding programmes principally from short wheats grown in Japan (Borojevic and Borojevic 2005). In 1913, the Italian breeder Nazareno Strampelli introduced the reduced height allele *Rht8* on chromosome 2D from the Japanese cultivar Akakomugi. *Rht8* shortens height by about 10%, reduces lodging risk (Kowalski et al. 2016), and is still found in many Central and Southern European wheats.

Of greater impact was the introduction of the *RHT1* semi-dwarfing alleles, *Rht-B1b* and *Rht-D1b* (Figures 6.13 and 6.14), from the Japanese cultivar Daruma. These are located on the long arms of chromosomes 4B and 4D. Daruma was crossed with wheats from the USA to produce Norin 10, which Orville Vogel used to produce semi-dwarf wheats for the USA in the 1950s. Dr. Vogel supplied seeds of a Norin-10 × Brevor cross to Norman Borlaug at CIMMYT in Mexico, who in turn developed widely adapted, photoperiod-insensitive, rust-resistant semi-dwarf lines that could be used in breeding programmes in developing countries (Reynolds and Borlaug 2006; Swaminathan 2006).

RHT1 codes for a DELLA protein. DELLA proteins are repressors of growth (Li et al. 2018) and this repression is alleviated by the growth hormone gibberellic acid (GA; Hirano et al. 2008). Mutations in the genes encoding for DELLA proteins, such as in *Rht-B1b* and *Rht-D1b*, can therefore cause an accumulation of DELLA proteins and reduce GA-responsive growth (Pearce et al. 2011; Van de Velde et al. 2017). Over 70% of the cultivars

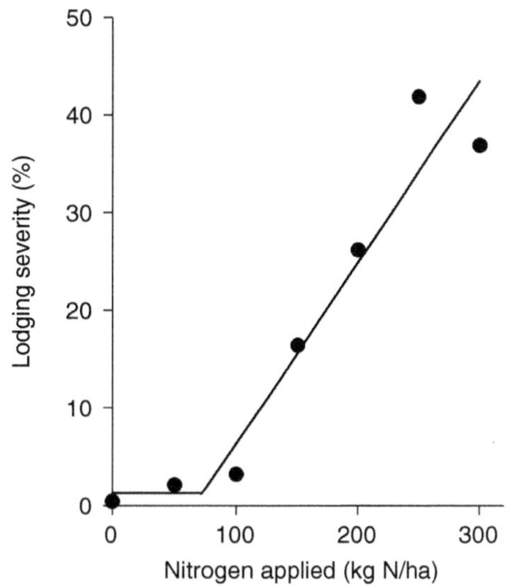

Figure 6.12 The effect of nitrogen fertiliser application rate on the lodging severity. Points are means of three winter wheat cultivars. *Source:* Data from Gooding et al. (1986).

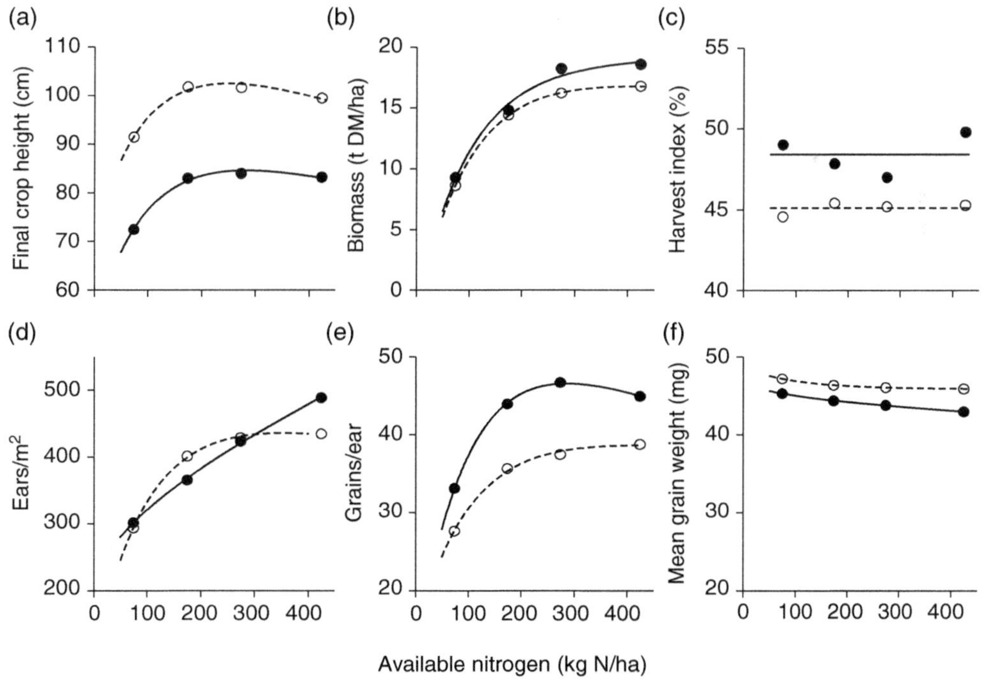

Figure 6.13 The effect of nitrogen and a semi-dwarfing reduced height allele on the crop height and yield components of winter wheat. Points are means of near-isogenic lines in three parental backgrounds. Solid symbols are with *Rht-B1b* (semi-dwarf) and open symbols are *Rht-B1a* (tall). *Source:* Data from Loddo and Gooding (2012).

grown in the world today carry mutated genes for DELLA proteins (Vergauwen and De Smet 2017), while others may carry *Rht8* or a combination of minor alleles to reduce height (e.g. Mercia in Figure 6.14). *Rht-B1b* and *Rht-D1b* can each reduce crop height by about 15–20% (Figure 6.13a), reduce lodging (Figure 6.14), and can increase harvest index (Figure 6.13c). The increase in harvest index due to semi-dwarfing alleles of *RHT1* results from a higher grain yield, which is largely at the expense of dry matter in the shorter stems. The increase in grain yield (Figure 6.15a) is due to more grain being set in each ear (Figure 6.13e) with more of the distal florets in the spikelet surviving to anthesis in *Rht-B1b* and *Rht-D1b* lines. Grains set in these distal sites have lower potential grain weights compared to the basal grain sites, so the mean grain weight is often reduced by semi-dwarfing alleles (Figure 6.13f; Miralles et al. 1998; Casebow et al. 2016).

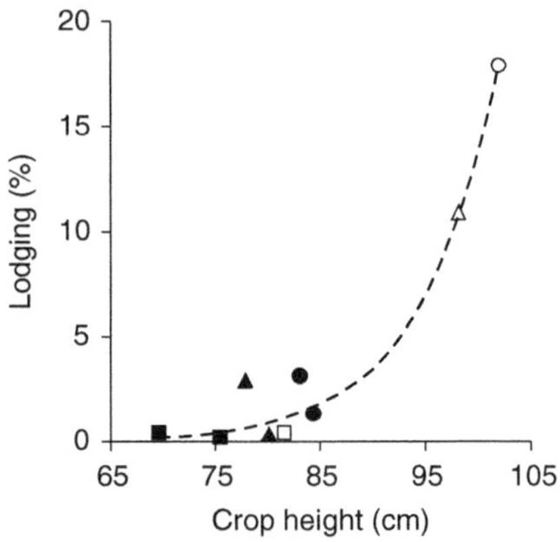

Figure 6.14 The relationship between lodging and crop height as affected by parental background (circles = Maris Widgeon, triangles = Maris Huntsman, squares = Mercia), and *RHT1* allele (open = *rht-B1a*, tall; closed = *Rht-B1b* or *Rht-D1b*, semi-dwarf). All plots received 200 kg N/ha as ammonium nitrate. *Source:* Data from Gooding et al. (2012b).

The reduced risk of lodging has enabled farmers to increase N availability to produce canopies large enough to intercept near maximal amounts of PAR. This has been a major driver to increasing crop yields and land-use efficiency in temperate, humid, and irrigated systems. The release of semi-dwarf types of wheat and rice underpinned the *Green Revolution*. This term was originally coined by William Gaud in 1968 to describe the rapid increases in yield per unit area in South Asia during the 1960s but is more widely applied to describe the increased yields achieved across the globe where the adoption of shorter cereals has been accompanied by improved availability and technology for crop nutrition, irrigation, and protection.

In the absence of lodging, when diseases are well controlled, and for well-adapted cultivars, the positive effect of N applications on grain yield (Figure 6.15a) are principally through effects on ears/m^2 (Figure 6.13d) and/or grains per ear (Figure 6.13e). Both effects are accounted for by increased light interception and photosynthesis during critical periods governing the production and survival of both tillers and florets (Fischer 2008). Often there are only small effects of N on mean grain weight (Figure 6.13f). Indeed, the relative constancy of mean grain weight for crops that have widely varying canopy sizes and grain populations in response to N availability is striking and is further evidence of a conservative balance between the size of the sink (number of grains) and the potential to fill it with photosynthate (source) after anthesis. Even allowing for this conservatism, there are occasions when mean grain weights can be significantly reduced by increased N availability. Large reductions in mean grain weight can occur if high N availability and large canopy sizes early in the season exacerbate droughts during grain filling (Gooding and Davies 1997). Similarly, if high N availability has encouraged late-season foliar disease, reductions in mean grain weight would be expected.

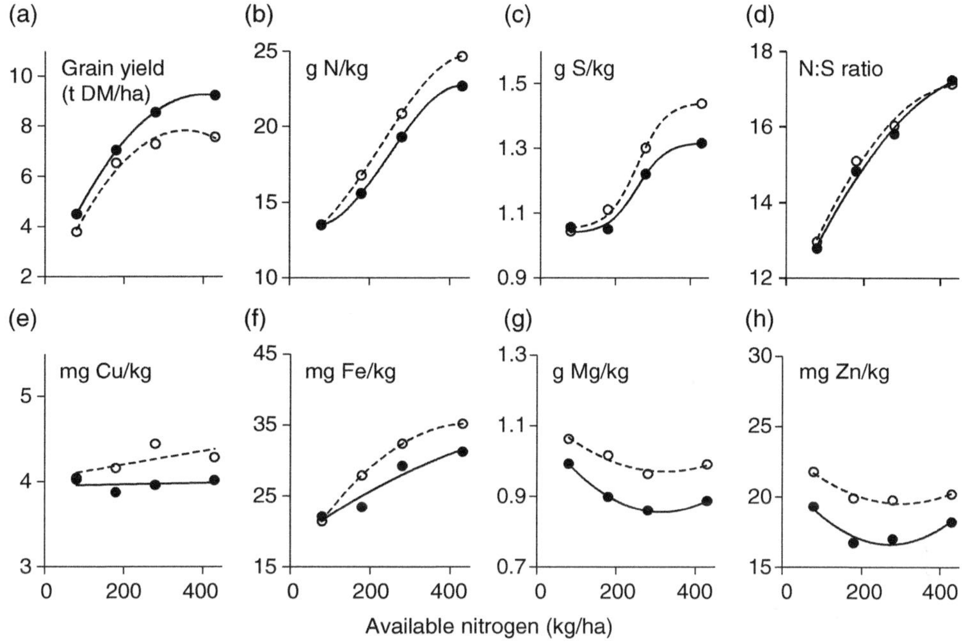

Figure 6.15 The effect of nitrogen and a semi-dwarfing reduced height allele on the yield and mineral content of winter wheat. Points are means of near-isogenic lines in three parental backgrounds. Solid symbols are with *Rht-B1b* (semi-dwarf) and open symbols are *Rht-B1a* (tall). *Source:* Data from Gooding et al. (2012c).

The increasing yield potential and N responsiveness of shorter, more communal wheats has had implications for grain quality. For example, the higher grain yields associated with the adoption of *RHT1* semi-dwarfing alleles lead to a dilution of N (and therefore grain protein content), sulphur, copper, iron, magnesium, and zinc (Figure 6.15). Conversely, the dilution of protein and iron can be rectified by the increasing N availability (Figure 6.15; Casebow et al. 2016; Gooding et al. 2012c). The lower mean grain weights associated with *RHT1* semi-dwarfing alleles occur alongside reductions in grain-specific weight (Figure 6.16a). More positively, *RHT1* semi-dwarfing alleles, and increased N fertilizer applications when lodging does not occur, both reduce grain α-amylase activity and increase Hagberg falling numbers (Figure 6.16b; Pushman and Bingham 1976; Gooding et al. 1986, 2012b; Kettlewell 1999).

6.2.1.2 Nitrogen Fixation

Nitrogen is an abundant element: dinitrogen gas (N_2) makes up 78.1% of the volume of the atmosphere. This is not immediately available to wheat plants. The triple N≡N bond is difficult to break so most organisms cannot use atmospheric N directly. *Fixed* N, however, can be made available to wheat via several routes: through biological fixation, natural atmospheric reactions, and through artificial synthesis. Additionally, significant quantities of fixed N can derive from anthropogenic pollution.

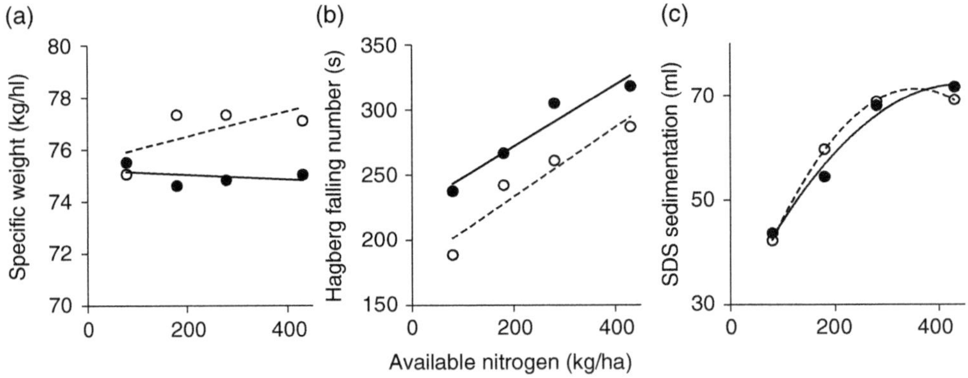

Figure 6.16 The effect of nitrogen and a semi-dwarfing reduced height allele on grain qualities of winter wheat. Points are means of near-isogenic lines in three parental backgrounds. Solid symbols are with *Rht-B1b* (semi-dwarf) and open symbols are *Rht-B1a* (tall). *Source:* Data from Casebow et al. (2016).

Biological nitrogen fixation: Atmospheric N is fixed by diazotrophic (literally 'dinitrogen utilizing') bacteria and Archaea, whereby ammonia is produced from N_2 via catalysis with nitrogenase enzymes. The diazotrophs can either be free living in soil and oceans or in symbiotic relationships with plants. Free living diazotrophs fix useful amounts of N in wheat fields (Ladha et al. 2016), but of greater agricultural importance are the symbiotic relationships between the Fabales (legumes such as clovers, medics, vetches, lucerne, and pulses) and rhizobia bacteria (Figure 6.11). It has been estimated that biological N fixation adds c. 50–70 million t of N to agricultural systems worldwide each year (Herridge et al. 2008). Biologically fixed N contributes to a reservoir of N in the soil. This reservoir is predominantly in the organic fraction such that agricultural soils often contain several tonnes of fixed N per hectare. The ratio of soil organic carbon: soil N commonly ranges between 9 and 14 and an arable soil with 1% carbon will therefore often have 3–5 t N/ha in just the topsoil. This is many times greater than the 0.18 t N/ha that is required in canopies for a green area index of six. This soil N, however, only becomes available to plants at a slow rate. In warm and moist conditions soil organic N is mineralised by bacteria and fungi to produce ammonium (NH_4^+) (Figure 6.11). Plants can take up ammonium ions via the roots but the movement of NH_4^+ in most soils is slow due to its positive charge. In aerobic conditions, ammonium is readily oxygenated by nitrifying bacteria to yield nitrate which is highly soluble and mobile. In wheat, the nitrate is transported via the roots and xylem to the leaves. Here it is assimilated as glutamine and glutamate before further amino-acid biosynthesis.

Atmospheric reactions: Lightning can generate nitric oxide (NO) and nitrogen dioxide (NO_2) (Tie et al. 2002; Shepon et al. 2007). NO_2 is oxidised to produce nitric acid (HNO_3), which is deposited on the ground or dissolved in clouds. In soils, nitric acid is further converted to nitrate. Global deposition from fixation in lightning is probably less than 5% of the biologically fixed N (Fowler et al. 2013) and deposition is concentrated in the tropical to mid latitudes (30°N to 30°S), due to the greater intensity of lightning in these areas (Shepon et al. 2007)

Anthropogenic pollution: Industry, home heating, and vehicle emissions are responsible for several N-containing pollutants. Goulding et al. (1998) estimated that c. 50 kg N/ha could be deposited each year on an arable site within 50 km of the centre of London, UK. Most of this was as dry-deposited oxidized forms, i.e. nitrogen dioxide and nitric acid.

Artificial nitrogen fixation: The amount of N made available to wheat through the above processes is usually insufficient for crops to produce canopies big enough to intercept over 95% of the light anthesis. In the UK, on sandy soils after cereals winter wheat may only recover 40–50 kg N/ha in the above-ground biomass by harvest if no further N supplementation is given (Webb et al. 1997, 1998). This could be associated with a GAI = 1.5 at anthesis, with only 50% of light being intercepted at that time. More N is recovered from medium to heavier soils, around 60–70 kg N/ha (Vaidyanathan et al. 1987), but this is still significantly deficient. Increased global food production has, therefore, been associated with a marked increase in the amount of N that is fixed artificially and applied to agricultural land as fertilizers (Figure 6.17; Smil 2004; Galloway et al. 2008; Jensen et al. 2011). In the UK, a 60% increase in wheat yields from 1970 to 1990 was associated with a doubling of the amount of N applied as fertilizer (Figure 6.17a). For global agriculture, the amount of artificially fixed N is now about twice (i.e. c. 120 million tonnes) that supplied by biological fixation (Fowler et al. 2013). Most of the artificial fertilizer applied to wheat is as ammonium nitrate ($NH_4NO_3^-$) or urea ($CO(NH_2)_2$) (Figure 6.18). Straight ammonium nitrate fertilizer contains 34.5% N, but its use has been restricted in several countries because of its explosive properties. The risk of fire or explosion is partially reduced by prilling, i.e. coating granules, which also reduces water absorption in store. Greater beneficial effects can be achieved by mixing ammonium nitrate with sources of calcium such as limestone or

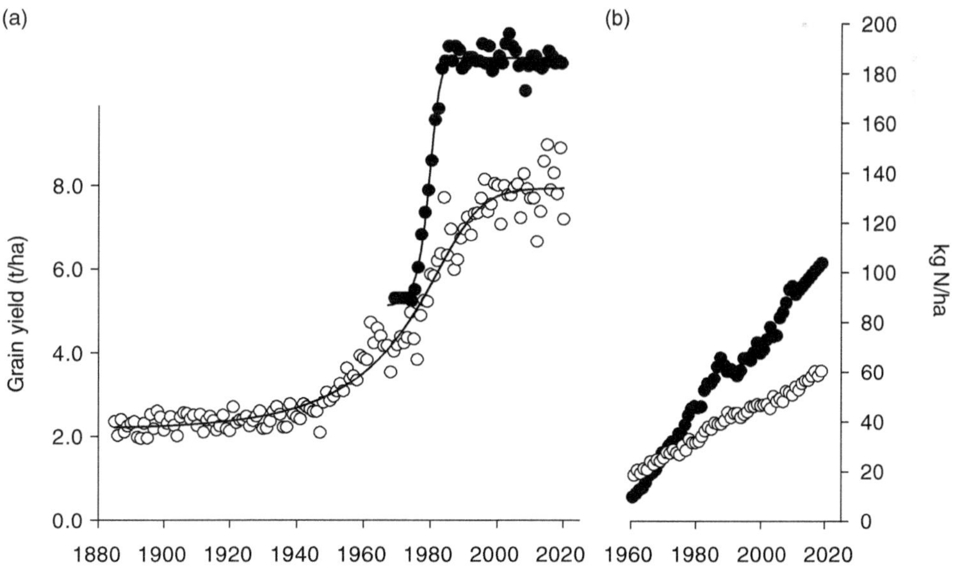

Figure 6.17 Estimates of grain yields (open symbols) and nitrogen fertilizer applications (closed symbols) for (a) UK and (b) global wheat production. *Source:* Based on data from DEFRA, the British Survey of Fertilizer Practice, and FAOSTAT.

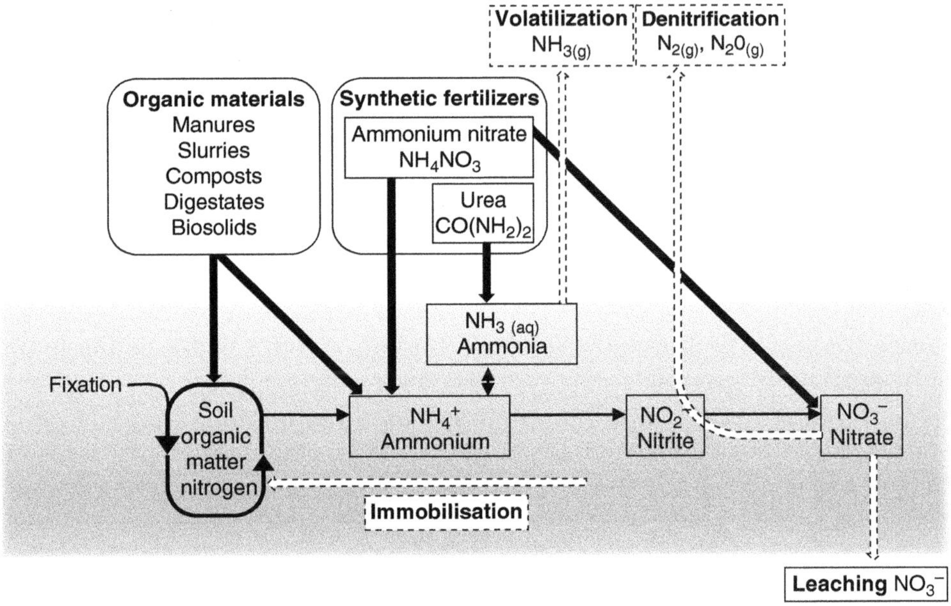

Figure 6.18 Sources of nitrogen to increase availability to crops (heavy arrows), and principal pathways of nitrogen loss (dashed arrows and boxes).

calcium nitrate to produce calcium ammonium nitrate (CAN). These mixtures typically contain 21–27% N. With CAN there are greater handling and application costs for supplying a given amount of N but, as well as reducing water adsorption and fire risk, they are less acidifying than ammonium nitrate.

Urea $CO(NH_2)_2$ contains 46.6% N and is the most concentrated form of nitrogenous fertiliser available as a solid. On application to the soil, urea is rapidly converted to ammonia and carbon dioxide by urease. Urease is an extracellular enzyme that is produced by bacteria and widely distributed in soils (Cordero et al. 2019). In solution the ammonia can rapidly become the ammonium ion (Figure 6.18).

The production of both ammonium nitrate and urea require synthetically produced ammonia. Over 99% of the artificially fixed N for fertilizers is produced by the Haber-Bosch process (Smil 2004). This uses high pressure (15–30 MPa), high temperature (350–500 °C), and catalysts to combine hydrogen and nitrogen gasses (Eq. 6.5; Royal Society 2020).

$$N_2 + 3H_2 \rightarrow 2NH_3 \tag{6.5}$$

The hydrogen is produced by splitting hydrogen from carbon in fossil fuels. This hydrogen production also requires high temperature, high pressure, and catalysis. The current use of fossil fuels to provide energy and hydrogen for the Haber-Bosch process is divided between natural gas (50%), oil (31%), or coal (19%) (Smith et al. 2020). The price of N fertilizer is, therefore, intricately linked to the cost and availability of fossil fuels. Depending on the product and process optimisation, each kilogram of N fixed requires from 25 to 42 MJ energy (Mudahar and Hignett 1985; Kuesters and Lammel 1999; Smil 2004). In terms of greenhouse gas emissions, ammonia production generates about 1·5–2·5 kg CO_2/kg of ammonia N. Additionally, ammonium nitrate production requires the conversion of ammonia to nitric acid, which can release nitrous oxide (N_2O) at an average rate of about 1 g of N_2O-N per kg of HNO_3-N (Sylvester-Bradley et al. 2014). Nitrous oxide is a powerful greenhouse gas and depletes ozone (see Section 6.2.1.3, *Denitrification*). Ladha et al. (2016) estimate that 18% of the global synthesis of fixed N through the Haber-Bosch process is used for wheat production, compared with 16% each for maize and rice.

In wheat systems relying on 180+ kg N/ha as artificial fertilizer (e.g. Figure 6.17a), over 50% of the total support energy (i.e. in addition to sunlight) for wheat production is accounted for by N fertilizer. The support energy for this amount of N fertilizer is at least equivalent to the sum of all the support energy to produce seeds and plant protection substances, and the energy for machinery, transport from factory to field, and all on-farm activities (Kuesters and Lammel 1999).

Globally, ammonia production for fertilisers is estimated to account for about 1.5% of all anthropogenic CO_2 emissions. This is significant but may yet seem to be a small price to pay for the impact of the Haber-Bosch process on improving food security (Smil 2004). Nonetheless, reliance on fossil fuels increases price volatility and is ultimately unsustainable. There is, therefore, significant interest in reducing the reliance of ammonia production on fossil fuels, and on increasing the energy efficiency of its production. For example, electricity from renewables can provide energy and the production of hydrogen through electrolysis of water (Smith et al. 2020). There are also potential developments in the use of wind energy, biogas, and biomass for localised production of ammonia (e.g. Tunå et al. 2014; Tallaksen et al. 2015; Reese et al. 2016).

6.2.1.3 Nitrogen Efficiencies and Losses

Traditionally, wheat has been regarded as an *exhaustive* crop. This means that, compared to many other crops, particularly non-cereals, wheat leaves less fertility in the soil after it has been harvested. Indeed, when N levels are low, wheat is highly efficient at recovering what little N is available. This seems particularly the case for older, taller cultivars (Foulkes et al. 1998). Compared to several other arable crops wheat has a rapidly developing and extensive root system (Barraclough et al. 1991), which, in well-structured soils, often quickly exceeds that required to extract mobile solutes such as nitrate. In fact, wheat is so efficient at extracting N at low N availability that agronomists often take crop N at zero fertilizer N (when weeds, pests, and diseases have been controlled) as being a good approximation of soil-available N (e.g. Sylvester-Bradley and Kindred 2009). Wheat can be grown with grain legumes in intercrops to significantly enhance the capture of soil N (compared to monocrop grain legumes) and increase the reliance of the grain legume on biologically fixed N (Jensen et al. 2020).

However, the performance of wheat without artificial N fertilizers is less relevant to modern commercial production. In the UK, average N fertilizer application rates to winter wheat are around 185 kg N/ha. It is notable that this rate varied little between 1985 and 2020 (Figure 6.17a). The fact that N applications have not increased for 35 years broadly indicates that increasing N availability would not provide a sufficient gain in yield (or increased light interception) to cover the direct and associated costs of the extra fertilization. However, this is a simplified view because risk management (Hoogmoed et al. 2018), legislation (DEFRA 2018), and effects of N on quality (Baker et al. 2004) also affect farmer decision making. Nonetheless, in contrast to the situation in the UK, and more widely in northwestern Europe, there are still areas where wheat yields appear to be significantly constrained by insufficient N input (Schils et al. 2018). Globally, N fertilizer rates to wheat continue to rise (Figure 6.17b) but remain significantly below those used in high yielding humid (or irrigated) environments. This difference is to be expected as much wheat is grown in areas where water is a greater constraint to yield than is light interception. Nonetheless, the observation that global N application rates to wheat do not appear to be approaching a maximum suggests that the nutrient is still thought to limit yield in many areas.

It is a convenient, but glib observation that the amount of fertilizer-N applied to UK wheat crops is about the same as the amount required to form canopies to intercept more than 95% of the light. The superficiality of this observation is exposed because it disregards all the other sources of N that become available to wheat crops. When combined these additional sources contribute, on average, 80 kg N/ha to the UK wheat crop (Sylvester-Bradley and Kindred 2009). As N availability increases, the proportion of that N that is taken up declines (Figure 6.19; Sylvester-Bradley and Kindred 2009). This measure is known as the N uptake efficiency (*NUpE*, Eq. 6.6):

$$NUpE = \frac{N\ in\ crop\ biomass}{Soil + fertilizer\ N} \tag{6.6}$$

Concurrently the efficiency with which the plant *uses* the N to produce grain (g) dry matter, i.e. the N utilization efficiency (*NUtEg*) also declines where:

$$NUtEg = \frac{DM\ yield\ of\ grain}{N\ in\ crop} \tag{6.7}$$

Effects on uptake and utilization efficiencies are combined (Moll et al. 1982) to assess the N use efficiency to produce grain (*NUEg*):

$$NUEg = \frac{DM\ yield\ of\ grain}{Soil + fertilizer\ N} = NUpE \times NUtEg \tag{6.8}$$

Maximising the margin between the value of grain/ha and the cost of N fertilizer is often used to justify N-fertilizer use on a field scale (Figure 6.19). This margin is maximised when the two lines in Figure 6.19a have the same slope. That is, for this definition of optimum N (N_{opt}), the incremental cost of the fertilizer should at least be equal to the incremental increase in value of grain/ha. N_{opt} therefore depends on the price ratio of N to grain, each on a per unit weight basis. Sylvester-Bradley and Kindred (2009) report that this ratio varied from two to nine over several years in the UK. At time of writing the cost of N as ammonium nitrate in the UK is c. £600/t and the value of wheat grain is c. £200/t @ 86% DM, giving an N-cost:grain value ratio of three. For the experiment in Figure 6.19, N_{opt} with a N:grain price ratio of 5 was achieved, with 330 kg N/ha provided by 250 kg N/ha as ammonium nitrate fertilizer and 80 kg N/ha from other sources. At this level of N availability

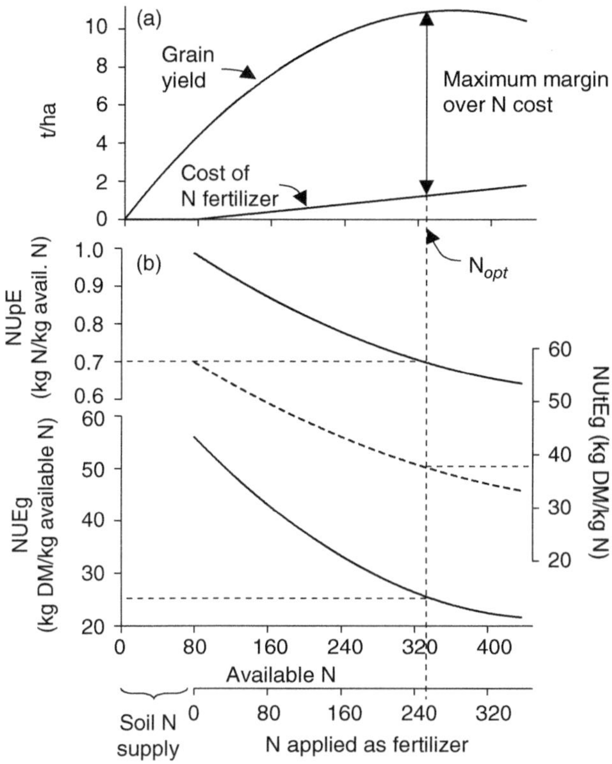

Figure 6.19 The responses of (a) grain yield, and (b) nitrogen use efficiencies to available nitrogen. Nitrogen uptake efficiency (NUpE), utilisation efficiency to produce grain (NUtEg), and use efficiency to produce grain (NUEg) calculated with Eqs. (6.6)–(6.8). The maximum margin over N cost in (a) assumes N costs five times as much as grain is worth on a kg/kg basis. Curves fitted to data in Loddo and Gooding (2012).

NUpE, *NUtEg*, and *NUEg* were 0.7, 38, and 26 kg/kg, respectively. For the UK crop Sylvester-Bradley and Kindred (2009) calculate the same ratios to be 0.65, 39, and 27 for wheat produced for animal feed, and 0.65, 33, and 22 for breadmaking wheat. The lower *NUtEg* and *NUEg* for breadmaking wheats derive from the lower average yields achieved, despite more N being made available to increase grain protein concentrations. Using the data of Ladha et al. (2016), and similar assumptions to those used by Sylvester-Bradley and Kindred (2009), global *NUpE*, *NUtEg*, and *NUEg* for current wheat production can be estimated to be around 0.5, 33, and 17.

N_{opt} is highly dependent on lodging resistance, harvest index, crop protection, and water availability. In terms of cultivar development, N_{opt} for a N:grain price ratio of 5 is 90 kg N/ha higher for the semi-dwarf line than for the tall line in Figure 6.15a (Loddo and Gooding 2012). The trend for higher N-rates to be justified economically continues as the yield potential of modern cultivars increases (Sylvester-Bradley and Kindred 2009). At a single quantity of available N, semi-dwarf and modern cultivars clearly have higher *NUEg*, particularly on account of their higher harvest indices and lodging resistance

(Ortiz-Monasterio et al. 1997). However, as these lines also have a high N_{opt}, and because *NUEg* declines as N availability increases, the increase in land-use efficiency has not necessarily been associated with an increase in *NUEg* (Sylvester-Bradley and Kindred 2009; Loddo and Gooding 2012).

NUEg is reduced if fertilizers are subsidised as part of national programmes to increase food security because N_{opt} is increased (Swarbreck et al. 2019). Conversely, *NUEg* is increased if N applications are constrained by legislation (Zhang et al. 2015; Swarbreck et al. 2019). In the latter case, reductions in yield can be mitigated if reductions in N applications are accompanied by more focussed agronomy to improve the efficiency of N use. i.e. more attention is applied to precision with regards to rates, timings, formulations, and spatial distribution of N fertilizer (see Section 6.2.1.5). In this respect, there should be more scrutiny of the above definition of N_{opt}. An obvious criticism is that it does not include all the costs of extra N applications. For example, requirements for crop protection often increase with fertilizer N-rate. It has been argued that profitability can be maintained by simultaneously reducing N and agrochemical inputs and adopting more systems-orientated approaches (e.g. Jordan et al. 1997). Conversely, if farmers are paid a premium for high grain N concentrations, this effectively reduces the N:grain cost ratio as N applications increase, and thereby increases N_{opt}.

As the proportion of available N taken up (NU_PE) is only 0.5–0.65 at commercial application rates of fertilizer-N, significant quantities of available N are at high risk of being lost from the soil (Zhang et al. 2015; Ladha et al. 2016). For the experiment in Figure 6.15, 110 kg mineral N/ha was present in the soil after harvest when 350 kg N had been applied and it is estimated that globally only about 50% of the fertilizer N applied to wheat is recovered in the crop (Ladha et al. 2016). The main processes of loss are denitrification, ammonia volatilisation, leaching of nitrate, and immobilisation (Figure 6.18).

Denitrification occurs when nitrate or nitrite is reduced, ultimately back to nitrogen gas. In certain conditions, the loss of $N_{2(g)}$ via denitrification can account for a significant proportion of fertilizer N applied to wheat fields (e.g. Goulding et al. 1993; Matson et al. 1998). This loss, therefore, increases the carbon footprint of wheat production assuming that fossil fuels are used to produce synthetic N-fertilizers. In addition, however, the intermediate steps in denitrification involve the production of nitric oxide (NO) and nitrous oxide (N_2O). Of particular importance is the leakage of $N_2O_{(g)}$ to the atmosphere (Figure 6.18). Average N losses as N_2O are only about 2 kg N/ha/yr from the European arable area (Lugato et al. 2017), and for wheat fields a simple average conversion of 1 kg N loss as N_2O per 100 kg N applied may be appropriate. However, conditions favouring denitrification can result in significantly higher values and the environmental impact of N_2O is disproportionate to the amount of N that is lost.

N_2O is a significant greenhouse gas, being the third most important of the *well mixed greenhouse gasses* (WMGHG) after CO_2 and methane (CH_4), and is long-lived (c. 120 years). The nitrous oxide in the atmosphere is responsible for about 6% of the radiative forcing (*rf*) of the WMGHG (Myhre et al. 2013); where *rf* is the difference between solar radiation absorbed by the earth and the energy radiated out to space such that positive values cause global warming. Nitrous oxide also significantly depletes ozone from the atmosphere (Myhre et al. 2013). The amount of N_2O in the atmosphere has increased by c. 25% since pre-industrial levels (from c. 270 ppb in 1750 to c. 335 ppb in 2020). Crop lands are the largest anthropogenic source of atmospheric N_2O, and over 80% of the increase in terrestrial sources of N_2O is associated with increased use of N fertilizers (Tian et al. 2020; Wang et al. 2020). On a per weight basis, N_2O has about 260 times the global warming potential of CO_2. The global warming potential of wheat production can, therefore, be dominated by effects associated with N_2O emissions, rather than by the support energy required to grow the crop (Williams et al. 2010).

Denitrification from wheat fields is particularly severe when aerobic and high N conditions favouring nitrate formation are followed by a decrease in oxygen availability. High rainfall events, wetting–drying cycles, freeze–thawing cycles, irrigation, poor drainage, high temperatures, and compaction can therefore all result in reduced oxygen levels and increased denitrification (Colbourn and Harper 1987; Goulding et al. 1993; Powlson 1993; Ball et al. 2007; Wagner-Riddle et al. 2017; Millar et al. 2018; Hansen et al. 2019; Thilakarathna et al. 2020). High C : N ratios encourage denitrification by *Pseudomonas* spp., *Paracoccus* spp., and *Thiobacillus denitrificans* and the establishment of wheat with zero tillage has been associated with increased populations of these denitrifying organisms, possibly due to the denser surface soil and higher moisture retention (Aulakh and Rennie 1986).

Strategies to mitigate against denitrification include ensuring that the timing and amount of N fertilizer applied correspond to crop requirements, and that applications are delayed until after any pre-planting irrigation (Matson et al. 1998; Ortiz-Monasterio and Raun 2007). Ensuring the soil is well-structured with good drainage to maintain aerobic conditions is therefore important. Formulations of fertilizer that ultimately slow the production of NO_3^-, such as the use of urease and nitrification inhibitors and polymer coatings have reduced N_2O loss (Thilakarathna et al. 2020). Nitrification inhibitors, however, lead to greater retention of NH_4^+ in the soil, which can lead to greater N losses as volatilised ammonia (Pan et al. 2016).

Volatilisation: Volatilization of N as ammonia gas has been estimated to account for an average loss of 10–14% of N from artificial fertilizers. Volatilisation also indirectly leads to increased N_2O emissions (Bøckman and Olfs 1998), while its deposition on waters and non-agricultural land can lead to significant perturbations in water quality, ecosystem function, and biodiversity (e.g. Simkin et al. 2016). Ammonia reacts with other atmospheric components, contributing to haze, cloud seeding, and fine particulate matter that can be harmful to health (Tao and Murphy 2018). Volatilisation is more likely to occur from urea-based fertilizers, than from ammonium nitrate. A meta-analysis of European studies suggests that losses of N from urea fertilizers applied to arable soils average about 15% (Hutchings et al. 2019). Significant losses are particularly likely if applications are made to dry or drying soils (Ni et al. 2014), conditions are warm, soils are alkaline (Hutchings et al. 2019), or there has been recent lime application. Where possible, volatilisation losses from urea can be reduced by incorporating the fertilizer into the soil, by irrigating or timing application to coincide with wet conditions (Sylvester-Bradley et al. 2014), with the use of urease inhibitors, and using other fertilizer formulations that slow the release of N (Black et al. 1989; Sylvester-Bradley et al. 2014; Pan et al. 2016). Some ammonia can be lost from wheat leaves, particularly as they senesce and/or are diseased (Harper et al. 1987; Parton et al. 1988).

Leaching of N as nitrate can lead to significant losses of N and pollution of the environment. There are also possible but contested implications for human health. The European

Union has set a limit of 50 mg NO_3^-/l (equivalent to 11.3 mg nitrate-N/l) in water intended for human consumption (EU 2015) and the WHO has used the same level in its guidelines (WHO 2017). This limit has been set to protect bottle-fed infants up to three to six months of age from methaemoglobinaemia, i.e. blue-baby syndrome (WHO 2017). Nitrate is readily reduced to nitrite on consumption and excessive levels of nitrite entering the bloodstream can impair the ability of haemoglobin to carry oxygen. However, blue-baby syndrome is mostly associated with microbial contamination of water and associated gastrointestinal infections. Microbial contamination can be combined with high nitrate content, but a direct role of nitrate in methaemoglobinaemia has been refuted (Addiscott and Benjamin 2004). Indeed, in adult humans the greatest exposure to nitrate and nitrite is from consuming vegetables and meats where nitrite or nitrate has been used as a preservative e.g. in cured meats (Van Grinsven et al. 2006). It has also been suggested that adults may benefit from nitrate consumption due to protection against gastrointestinal pathogens (Addiscott and Benjamin 2004). There are, however, additional concerns over the role of nitrate/nitrite in water in the development of specific cancers (Van Grinsven et al. 2006).

The environmental hazards associated with nitrate leaching include the *eutrophication* of surface waters such as lakes and estuaries. When present with phosphate, nitrate availability accelerates the growth of algae and higher plants. Harmful algal blooms can proliferate, and oxygen deficiency develops when the excessive growth decays. These effects can significantly impair water quality and biodiversity (Boeykens et al. 2017).

Nitrate is highly mobile in soils due to its negative charge and high solubility. Leaching is increased when the nitrate quantities in the soil exceed the crop demand and there is sufficient rainfall and drainage to *wash* the nitrate beyond the rooting zone and into groundwaters and water catchment systems of ditches, streams, and rivers. For example, in autumn-sown winter wheats in oceanic environments, tillage to produce a seed bed can aerate and mix the soil to encourage mineralisation and nitrification of soil organic N while temperatures are still mild. However, autumn-sown crops remain small with low nitrogen-demands through winter while there is mass flow of water down through the soil profile. Consequently, the leaching risk in the autumn-winter period is increased, particularly if significant quantities of N have been left in the soil from the previous crop. In addition, high soluble-N contents in autumn soils, and therefore high leaching risks, can follow the destruction and incorporation of leys rich in forage legumes (Watson et al. 1993). Leaching can still occur after winter, particularly if heavy rainfall or irrigation follow mineralization or fertilizer application (Goulding et al. 2000; WHO 2017). Leaching from the root zone can be particularly rapid from freely draining sandy soils and/or shallow soils on permeable rock such as limestone or chalk.

Attempts to reduce nitrates in surface waters and ground waters have included legislation. Nitrate Vulnerable Zones (NVZ) are identified in the European Commission Nitrates Directive (EU 2008) as those where surface waters or ground waters are at risk of exceeding 50 mg NO_3^-/l. For example, at the time of writing 55% of the area of England is in an NVZ. In these areas, farmers must demonstrate planning of their N applications with due regard to crop demand, soil available N, and risks of leaching and surface runoff. More specifically, for example, an autumn-sown winter wheat destined for the feed wheat market with an expected grain yield of 8 t/ha is limited to receiving a maximum of 220 kg N/ha applied as available N from synthetic plus organic fertilizers. No synthetic fertilizer can be applied from the 1 September to 15 January and, depending on soil type, no organic manure or slurry with high available N can be applied between 1 October and 31 January. Further strategies to reduce nitrate leaching may include earlier sowing of autumn drilled wheat to capture greater quantities of N in the crop that would otherwise be leached during autumn and winter (Cosser et al. 1994; Goulding et al. 2000). The possibility of earlier sowing, however, is subject to suitable adaptation of the cultivars used, and to the strictures of the cropping rotation. As already discussed, early autumn sowing is also likely to increase the need for crop protection measures against higher pressures from weeds, pests, and diseases. Alternatively, for late-sown winter wheat or spring-sown wheat it might be possible to grow a cover crop over autumn and early winter rather than leave the soil bare or as fallow (Plaza-Bonilla et al. 2015).

Immobilisation: Just as microbial activity of the soil is essential for the mineralisation of N from soil organic matter, the microbial populations also accumulate mineral N and incorporate, or *immobilise*, it into their own organic chemistry. The microbial population is, therefore, both immobilising and mineralising N at the same time. Whether the combined effects of these processes lead to net immobilisation or mineralisation depends on several factors (Bengtsson et al. 2003). Broadly, the soil C : N ratio should be important. If microbial cells have a C : N ratio of about 10 and respire about 50% of their C uptake, they may be N limited above a soil C : N ratio of around 20. These circumstances could be expected to lead to net immobilisation of N. At lower soil C : N ratios, soil microbes may be expected to become increasingly C limited, and therefore produce a surplus of mineralised ammonium (Tate 1995).

High C : N ratios can occur, at least temporarily and in localised areas of the soil, when producers incorporate organic materials with a high C : N ratio. Incorporation of wheat straw, for example, can lead to net immobilisation of N (Shindo and Nishio 2005). Incorporation of straw before autumn sowings of winter wheat can, therefore, reduce N availability to establishing seedlings, but might also reduce nitrate leaching during the autumn-winter period (Powlson et al. 1985). As the straw decomposes its impact on net immobilisation can decline, to the point of being negligible by the following spring–summer (Bhogal et al. 2007).

6.2.1.4 Recovering and Recycling Fixed Nitrogen

Nitrogen recovery: It has long been recognised that wheat crops grown following non-cereal arable crops have higher yields than with wheat grown following a cereal. In temperate areas these *break* crops (as in a *break* from cereals) include combinable crops such as oilseed rape, field peas, and field beans (Vaidyanathan et al. 1987) and root crops such as potatoes and sugar beet (Webb et al. 1997). Some of the yield benefit to wheat is undoubtedly due to the break crop disrupting weed, pest, and disease cycles and some break crops may also improve soil structure. In addition, however, more N is available to wheat after a break crop than after a cereal (Vaidyanathan et al. 1987; Webb et al. 1997). This may be due to the breakdown of N-rich crop residues, such as from the leguminous crops, and/or to their less efficient utilisation of soil and fertilizer N resulting in more remaining in the soil for the following wheat (Sylvester-Bradley et al. 2001).

Recommendations for UK farmers imply that medium soils following peas and beans (which would not have received fertilizer N), or oilseed rape, sugar beet, and potatoes (which would have received fertilizer N) would supply around 20 kg N/ha more to following winter wheat crops, than would be the case if wheat had followed wheat. This difference is reduced in high rainfall areas and depends on soil type (AHDB 2020). The amount of N left in the soil for wheat is higher after well-fertilized grass leys, or after forage legumes such as white clover, red clover, and lucerne. A 3-to-5 year grazed ley receiving high levels of N may leave 40–100 kg N/ha more for a following wheat than a preceding wheat crop. This higher amount would be relevant for areas with heavier soils and with lower rainfalls. Beneficial but progressively decreasing effects are still seen for several years after the ley (AHDB 2020).

Nitrogen recycling: There are many opportunities to return fixed N and other nutrients that have been consumed by livestock and humans back to agricultural land. Exploiting these cycles is a fundamental component of sustainability. Traditionally, cycling of nutrients within farms has been facilitated by running livestock and arable enterprises alongside each other in a mixed system. In such systems, manures and slurries from housed livestock can be applied to arable land. In grazed systems, arable crops will benefit from following temporary grass leys. As farms have become more specialized and more intensive, manures and slurries on livestock farms have increasingly accumulated to excess, presenting significant storage and disposal challenges, and contributing to the considerable impact that livestock systems can have on the environment. Nonetheless, there are still scenarios where mixed systems have improved sustainability and N use efficiency (Schiere et al. 2002; Fageria and Baligar 2005). Livestock can be used to *move* some nutrients they consume from grazed pastures, which may also be unsuitable for arable cropping, to arable fields. They can also help recycle and return nutrients from crop residues. When the plant (i.e. organic) N is passed through an animal some of the N (and other nutrients) becomes more immediately available to crop plants as urea (and uric acid from poultry) and ammonic forms. In addition to the content of readily available nutrients, the application of livestock wastes can also increase soil organic matter content with the associated benefits described in Chapter 2 (Bhattacharyya et al. 2007; Johnston et al. 2009; Ghimire et al. 2018).

Increasingly, there are possibilities of cycling nutrients from other enterprises. These include the use of sewage sludge and biosolids; digestate from anaerobic digesters; and municipal and other composts (Table 6.1). *Biosolids* describes treated sewage sludge. This can provide a range of nutrients, particularly P and N. Its use can, however, be limited, as sewage often contains heavy metals that can accumulate to levels toxic to humans when crops are grown on heavily contaminated soils. Current use typically requires that the topsoil and the biosolids are assessed for heavy metal contamination such that permitted levels are not exceeded.

Digestate is a product of anaerobic digestion, i.e. the controlled biodegradation of an organic feedstock in the absence of oxygen. As well as digestate, the anaerobic digestion produces biogas (principally methane and carbon dioxide) for use as a fuel. The nutrient value of the digestate will depend on the feedstock which could include food waste, crops such as maize, and animal manures (Holm-Nielsen et al. 2009). Food-based systems generally produce digestates with higher N contents than crop + manure-based systems.

When utilizing organic materials it is important to remember that livestock and humans do not *make* N but consume it from crops (in livestock pastures and concentrate, and human food) for which it has already been biologically or synthetically fixed. However, in

Table 6.1 Typical nutrient contents of different organic sources applied to medium soils. Nutrient availabilities are for winter wheat following applications in the autumn.

		kg/t DM					
		N		P_2O_5		SO_3	MgO
	Dry matter (%)	**Total**	**Avail.**	**Total**	**Avail.**	**Total**	**Total**
Manures							
Cattle	25	24	2	13	8	10	7
Pig	25	28	3	24	14	14	7
Sheep	25	28	3	13	8	16	11
Poultry	40	48	12	30	18	14	11
Slurries							
Cattle	6	43	11	20	10	12	10
Pig	4	90	32	38	19	18	18
Thermally dried biosolids							
	95	42	6	58	29	24	6
Green compost							
	60	13	1	5	3	6	6
Whole digestate from anaerobic digestion							
Food	4	117	18	27	16	5	17
Farm	6	65	10	31	28	11	15

Source: Adapted from AHDB (2020).

attempting to cycle plant N consumed by animals back to crops such as wheat, there are many potential sources of N loss that need to be addressed to improve sustainability. Loss of N can be diffuse in terms of gaseous emissions or as leachate, or more acute in water-polluting events due to surface run-off or to failures in storage and distribution. In addition to N contamination, potential water pollution from the use of animal manures, slurry, and sludges includes losses of phosphate and the spread of infectious protozoa (*Cryptosporidium*, *Gardia*) and bacteria (*Salmonella*) (Hooda et al. 2000).

It can be difficult to predict the nutrient values and characteristics of N release from organic applications, particularly when significant additions are repeated year after year (e.g. Petersen 2003). Farmyard manure (FYM), for example, consists of dung, urine, and bedding. Its nutrient value depends on the type of animal, the nutrient content of the feed, the bedding (usually straw) content, the manner of storage and transport, and the timing and method of application in the field (Table 6.1). The nutrient value of slurry (liquid waste) will depend strongly on its dilution with rainwater or water from cleaning hoses. Some N will be in an available form for immediate release, but other N components will become available over longer timescales and not necessarily coincident with crop demand.

Any temporal mismatch can contribute to significant nutrient losses (Goulding et al. 2000; Thomsen 2005). In particular, available N in organic sources is susceptible to volatilisation. Before application, ammonia can be volatilised while materials are moved, stored, and or composted (Gibbs et al. 2002). After application, volatilisation is particularly problematic when material is left on the soil surface, notably on dry soils and/or in warm conditions. Losses can be significantly reduced by incorporating manures within a few hours of application, or by injecting slurries and liquids shallowly below the soil surface. Leaching of nitrate can be a problem if manure is exposed to rain. After application, leaching can be significant if materials are applied before autumn-sown winter wheat, particularly on lighter soils. Leaching losses can be greatly reduced by applications ahead of rapid wheat growth in the spring, particularly if the net flow of water through the soil profile has ceased (AHDB 2020). The greatest amounts of N from FYM and slurry are recovered by crops when storage periods are short, dry matter contents are high, soil types and application timings are less conducive to leaching, and when slurries and manures are not left on the soil surface for any significant time.

The levels of different nutrients within organic wastes (e.g. between N, P, and K) do not reflect crop demand. It is common for the N : P_2O_5 ratio to be much lower in organic materials compared to the ratio required by crops. This is particularly true of the biosolids (Table 6.1). Similarly, livestock manures and slurries have a comparatively low N : K_2O ratio. Often, therefore, integration with artificial N application is necessary to achieve an appropriate balance (e.g. Bhattacharyya et al. 2007). Improving the accurate application of the correct amount and balance of available nutrients can be facilitated by analysis on-farm (e.g. for ammonium N content in slurries (Hooda et al. 2000; Mateo-Marín et al. 2020), and in the laboratory (e.g. for total N, nitrate-N, and uric acid N; phosphate [P_2O_5], potash [K_20], sulphite [SO_3], and magnesium oxide [MgO]; AHDB 2020).

6.2.1.5 Optimising Nitrogen Application

The impact of N on wheat production and the environment is clear and many consider current practices to be unsustainable. There are strong reasons to reduce losses of fixed reactive N to the environment and to ensure that applications that are made are effectively used by the crop (Swarbreck et al. 2019). Bruulsema et al. (2009) described the components of fertiliser nutrient optimisation in terms of four *rights*. Nutrients should be applied from the *right* source, at the *right* rate, at the *right* time, and in the *right* place. In the previous sections we have discussed the sources of fixed N available to wheat crops and briefly noted the use of nitrification and urease inhibitors and other slow-release formulations. In addition, however, there is considerable research on the optimisation of the rate, time, and placement of N fertilizers.

Rate optimisation: We have so far addressed the rate of N availability in terms of the amount that is needed to be taken up by the crop and thus contribute to light interception and photosynthesis. The maximum yield that can be achieved by increasing N is partly defined by the balance between the asymptotic response of N in the canopy and light interception and the negative effects of N-rate on e.g. N uptake, lodging, and infection pressures from pests and diseases. As already discussed, optimum rates depend on the cost of N fertiliser, N losses, the value of grain, and the extent to which yield is limited by other factors such as other nutrients, soil conditions, water availability, temperature, associated costs of crop protection, and the N-responsiveness of the wheat cultivar being grown (Swarbreck et al. 2019). At the very least the amount of artificial N fertiliser applied should be adjusted to account for estimates of N available from previous cropping and management as affected by soil type and rainfall and the application of manures (Swarbreck et al. 2019; AHDB 2020). Alternatively, soil mineral N (NH_4-N plus NO_3-N) from laboratory analysis of soil samples can be used to estimate soil N supply (Bundy and Andraski 2004; Sylvester-Bradley et al. 2008). For autumn-sown wheats, soil N analysis is most informative if sampling is after the autumn–winter leaching period, but before the onset of stem extension. For spring sown, rainfed wheats, where subsequent leaching is likely to be less, soil analysis before planting can be appropriate.

Even after making these assessments, the yield response to fertiliser N is still variable (e.g. Kindred et al. 2015) and the uncertainty can lead to farmers increasing the N-rates because the yield and financial penalties of applying too little N are often greater than applying too much. This tendency is particularly encouraged if fertiliser-N is subsidized due to national programmes to address food security. Attempts to reduce supra-optimal applications with in-season canopy measurements include assessments of green area index following application of only a proportion of the anticipated full rate of N fertiliser, facilitating a more precise later application (Sylvester-Bradley and Stokes 2001). Rates can also be based on measurements including visual assessment of leaf colour (Thind et al. 2012), optical sensing of leaves to assess chlorophyll contents (Fox et al. 1994), optical sensing of the canopy with calibration against N-rich strips or N-response plots (Ortiz-Monasterio and Raun 2007; Cao et al. 2017; Aula et al. 2020), and assessments of N concentration in the wheat foliage (Yue et al. 2012). In terms of simplicity and cost, the utility of the comparison of leaf colour with a colour chart has clearly been demonstrated in reducing N rates where there has been a history of supra-optimal applications (Thind et al. 2012; Swarbreck et al. 2019). After harvest, the assessment of grain N concentration can be used to update the proposed rates required for subsequent crops grown in similar spatial and rotational locations (Sylvester-Bradley and Clarke 2009).

Spatial optimisation: There has long been interest in being able to vary the amount of fertiliser applied within variable fields (Fairchild 1988; Fiez et al. 1994). Relationships between light interception, yield, soil N supply, and N requirement are as relevant to spatial variation within fields as they are between fields. For example, Figure 6.20 shows the results from mini-nitrogen response experiments located at different locations within the same field. The highest rates of additional N required to maximise yield were in those parts of the field that had lower levels of PAR interception before application. Observational and sensor-based technologies used to vary N fertiliser application rates between fields can therefore be used to vary N application rates within fields (Kindred et al. 2017; Aula et al. 2020). However, the same caveats apply as those which inform whole field rates, e.g. that canopy size and yield are limited by N availability rather than other biotic or abiotic factors. The precision of application in location and rate within a field can be improved, therefore, by understanding soil variation as well as assessing variation in the size and colour of the canopy (e.g. Walley et al. 2002). In the past, the sampling and laboratory analyses required to provide maps of soil physical and chemical properties across a field have been time consuming and expensive. Increasingly, however, soil sensor technologies are being developed to measure within field variation rapidly and cheaply (Diacono et al. 2013). Soil

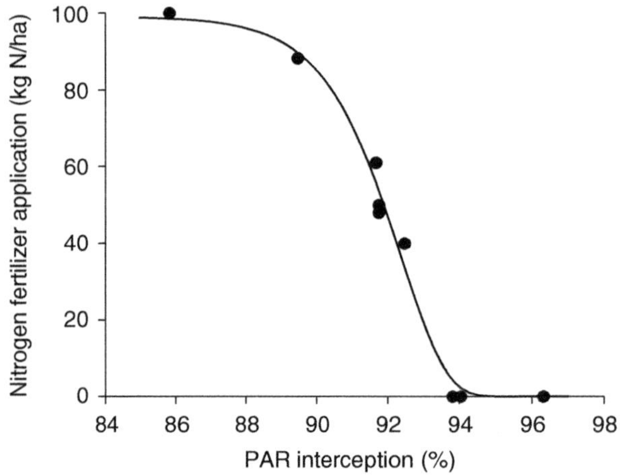

Figure 6.20 The relationship between the additional amount of nitrogen required at the end of booting in different locations within a 6 ha field to maximise wheat grain yield, and the interception of photosynthetically active radiation (PAR) prior to application. Points represent each location, all of which had previously received 200 kg N/ha at stem extension. *Source:* Data from Gooding et al. (1999).

information combined with canopy-based sensing, and maps of previous yields and grain N concentrations, can inform the spatial distribution of N fertiliser to be applied within a field. Such approaches have often increased *NUEg* compared with uniform rates of N applied across whole fields as in standard farm practice (Diacono et al. 2013).

As well as precision of spatial application over a field, there is also the possibility of applying fertiliser more precisely within the soil profile. As already discussed, losses of N to the atmosphere can be reduced by incorporating or placing fertiliser into the soil. This is difficult when applications are to be made to an established crop, such as in the spring to autumn-sown wheat. There are, however, potential technological solutions. For example, if irrigation is also required, N can be delivered in drip irrigation water below ground, i.e. subsurface fertigation. Such systems permit frequent and accurate placement of N directly into the root zone, allowing the N application to be closely matched to crop demands over time. Initial capital costs are high, but if combined with zero-tillage and conservation agriculture approaches, subsurface irrigation can be used to reduce N and water application rates for a succession of crops (Sidhu et al. 2019).

Time optimisation: The risk of N losses means that the correct timing of N application is crucial. Large quantities of N need to be available for periods of rapid canopy expansion and the development and survival of yield components. In oceanic environments, the greatest yield responses are often when the main N-fertiliser dressings are applied to winter wheat at the start of stem extension (GS 30-31; Figure 6.21). Yield increases following such applications are mostly associated with increased numbers of florets and tillers surviving to produce grains. The application of some N earlier (c. 50 kg N/ha) to stimulate tillering may help to compensate for poor or patchy establishment, but excessive tillering will increase the risk of lodging. Delaying some of the N application until after the first node detectable growth stage allows some tactical adjustment of N-rates in response to, for example, canopy assessments. The longer the delay in applying some of the N (e.g. until flag leaf emergence), the greater the benefits to canopy survival during grain filling. Delaying some of the N applied also reduces risks of lodging and severity of some diseases (Sylvester-Bradley et al. 1997; Neumann et al. 2004). However, delaying the start of N top-dressing until flag leaf emergence significantly reduces grain numbers and grain yields (Figure 6.18). The negative effect

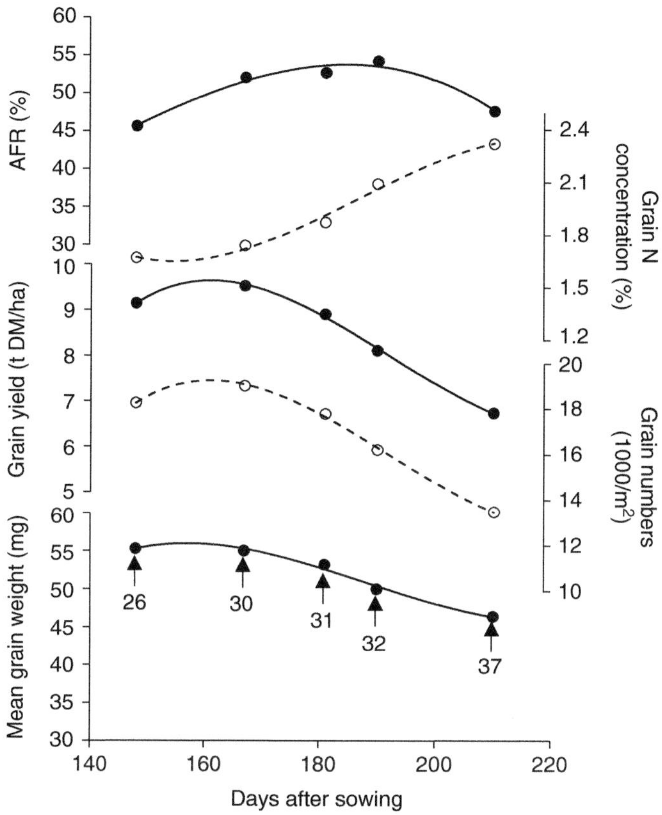

Figure 6.21 The effect of time of nitrogen fertiliser application to winter wheat in Ireland. Numerals with arrows are decimal growth stages as per Table 1.4 (Zadoks et al. 1974) when first nitrogen application was made. *Source:* Data from Efretuei et al. (2016).

of delaying N applications until after the start of stem extension is partly mitigated by an increase in N concentration in the grain. This is helpful to achieve quality thresholds, e.g. for producing breadmaking wheat. In the UK, where a high yielding environment is not conducive to high grain N concentrations, many farmers targeting the breadmaking market will apply c 50 kg N/ha above the N_{opt} for a feed wheat crop, solely to increase grain N concentrations. These applications are often applied as late as anthesis or during the early stages of grain filling when the grain N response is particularly high (Gooding and Davies 1992). This is because N accumulated after most of the canopy has been formed is less likely to be immobilised in structurally related compounds and more likely to be translocated to the grain compared with N taken up earlier (Powlson et al. 1987, 1989). Coincidentally, late N uptake is less associated with yield improvements and hence less likely to become diluted with starch in the grain. Unfortunately, however, the apparent fertiliser recovery of N applications (*AFR*, Eq. 6.9) made at and after anthesis, made in addition to previous yield-based applications, can be lower than 20% (Gooding et al. 2007).

$$AFR(\%) = 100 \times \frac{(N\ in\ fertilized\ crop - N\ in\ unfertilized\ crop)}{N\ fertilizer\ applied} \quad (6.9)$$

6.2.2 Phosphorus

Phosphorus plays a vital role in the reactions of photosynthesis (e.g. Figures 5.2, 6.7). It is also essential for a range of other processes, including nucleic acid synthesis, energy release and transfer, respiration, membrane structure and function, enzyme regulation, and cell signalling (Herold and Walker 1979; Vance et al. 2003). Phosphorus deficiency in wheat leads to poor seedling establishment, reduces the maximum rates of photosynthesis in light-saturated leaves, delays leaf and tiller emergence through increasing the plastochron and phyllochron, reduces the expansion rates and final sizes of leaves, and reduces the production and survival of tillers (Marco and de Marco 1990; Rodríguez et al. 1998, 1999; Valle and Calderini 2010). In extreme cases, P-deficient leaves develop a dull, bluish-green colour with a purple or bronze hue. The leaves may also die back from the tips. As for N deficiency, reduced biomass due to P insufficiency is associated with lower leaf area indices and PAR interception (Rodríguez et al. 2000). Similarly, reduced grain numbers and yields in P-deficient plants can be related to poorer light interception during the spike growth period of stem extension (Lázaro et al. 2010). Below ground, root growth tends to proliferate in areas of available P (Flavel et al. 2014), while P deficiency is associated with reduced root biomass and root length densities (Teng et al. 2013). In contrast to N, increasing the availability of P to deficient plants can help to suppress some fungal pathogens (e.g. Sweeney et al. 2000; Walters and Bingham 2007).

Lynch (2011) estimates that plant growth on over half of the global agricultural land area is limited by P availability or would be without fertilisation. Even with recent fertiliser practices, MacDonald et al. (2011) estimate that 30% of the global crop lands are still P-deficient. Although both P and N are fundamental to photosynthesis and light interception, there has often been a tendency for farmers to focus on N. This is partly due to the solubility and mobility of NO_3^- in the soil compared to that of phosphate. Visible responses of deficient crops to applied N are rapid and obvious, whereas additions of P may not alleviate deficiency in the short term. This has contributed to a previously unchallenged view in many wheat-growing areas that 'high N equals high yield' with insufficient emphasis placed on the requirement for other macronutrients. Consequently, large imbalances in the amounts of N and P available to wheat plants can develop leading to poor N responses and NUE (Duncan et al. 2018; Chen et al. 2020a).

The amount of soil P that can become available to wheat is low compared to the total soil P content. Most available P is originally derived from the weathering and degradation of rock material. This releases dihydrogen phosphate ($H_2PO_4^-$), which is poorly soluble and readily immobilised by being bound to soil minerals and charged sites or fixed into organic forms (Ryan et al. 2001). Phosphate, therefore, moves slowly in soil and is easily depleted in the root zone. Wheat roots can exude phosphatase (Ciereszko et al. 2011) and organic acids (Pearse et al. 2006; Nobile et al. 2019) in response to P deficiency to help mobilise soil P. Exudates also increase microbial activity that may increase P availability. Wheat plants growing in soils deficient in available P often form associations with mycorrhizal and other fungi that can increase P supply to the crop (Wakelin et al. 2004; Li et al. 2006).

Low availability of P is more likely in acidic ($pH < 5.5$) or alkaline/calcareous (pH 7.5–8.2) soils (Figure 2.5). Deficiency is also more likely in low organic matter soils and/or in situations where root growth is constrained, for example by compaction and waterlogging (Archer 1985; Wibberley 1989). If these constraints have been eliminated, farmers can increase P availability to crops over several years by adding ground rock phosphate. The nutrient content of ground rock phosphate is usually expressed in terms of P_2O_5, which contains 43% P. It can be applied directly to soils where the P will become slowly available to crops. However, if wheat land is deficient, it is better to apply products containing rock phosphate that has been treated with acid to increase the amount of phosphate that is, at least initially, water soluble. Sulphuric acid is used to produce superphosphate (18–21% water-soluble P_2O_5), and phosphoric acid to produce triple superphosphate (c 47% water-soluble P_2O_5). Even when water-soluble phosphate is applied, much of it can become immobilised in the soil. Due to the difficulty in correcting deficiencies quickly, commercial practice is often to maintain levels of soil available P at levels of sufficiency.

Several soil tests can be used to determine the amount of available P (Speirs et al. 2013). A common method used for agricultural soils is that of Olsen et al. (1954), whereby P is extracted with sodium bicarbonate. Figure 6.22 relates wheat yield to available phosphate levels and the soil indices of AHDB (2020). Sufficiency is broadly achieved at a soil index of 2 (Olsen-P = 16–25 mg/l). The approach for soils in this range is to apply sufficient P to balance that which will be removed by the crop, meaning that deficiency does not develop for future crops. In favourable conditions, 80–90% of P in the above-ground biomass of wheat crops is present in the grain by harvest (Batten 1992; Takahashi and Anwar 2007). In situations of sufficiency, each tonne of grain at 86% DM removes about 6.5 kg P_2O_5 equivalents.

Due to the low mobility of phosphate in the soil, the timing of application is not critical at an index of 2. At lower indices, however, there may be benefits in applying phosphate close to the seed at drilling. This can be achieved with combine drills where fertiliser and seed are placed in the same row at the same time (Peterson et al. 1981; Matar and Brown 1989). Seedlings are at particular risk of deficiency due to their small root systems and this risk is exacerbated if the amount of P in each seed is low because the seed crop was harvested from deficient land (Marco and de Marco 1990).

There are several reasons why a soil index of 2 should not be exceeded. As applications of fertiliser P have increased, there has been a significant increase in the diffuse pollution of water courses and bodies with P (Farmer 2018). Phosphorus can enter water courses via the surface runoff of soluble phosphate or of phosphate attached to sediment (particulate P). Smaller amounts of soluble P may also be leached to below the root zone (Mekonnen and Hoekstra 2018). Increased P in freshwater can lead to eutrophication, particularly when algal growth is not otherwise constrained by N availability (See Section 6.2.1.3 *Leaching*). The contamination of sediments in lakes and water courses with P from agricultural sources can result in release of phosphate long after the initial polluting events (Redshaw et al. 1990). Losses of phosphate from wheat fields are particularly problematic when the availability is above that required for crop uptake. More specifically, do Carmo Horta and Torrent (2007) identify increased risks when Olsen-P values exceed 20. Other measures to protect waters from P pollution include soil erosion control (Iho and Laukkanen 2012), growing perennial buffer strips of grass between production fields and water courses (Dorioz et al. 2006), using cover crops where appropriate, and adopting conservation tillage systems (Herzog et al. 2008).

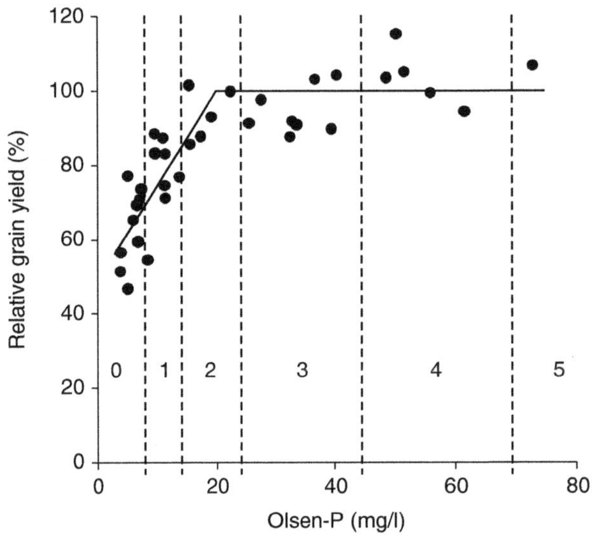

Figure 6.22 The relationship between available phosphorus and wheat grain yield. Data from Teng et al. (2013). Vertical lines and numerals denote soil indices (AHDB 2020) as related to the Olsen-P method (Olsen et al. 1954).

In addition to the threats to water quality, the sustainability of fertilisation is challenged by the finite supply of high-grade rock phosphate (Ulrich and Frossard 2014). Ground rock phosphate is extracted from sedimentary (87%) or igneous (13%) deposits, mostly by open cast mining. Over 80% of rock phosphate production is for agricultural fertilizers (Geissler et al. 2018). Sedimentary reserves are extracted from ancient marine continental shelves, mostly in Morocco, the US, and China, while most mining of igneous deposits is in Russia, China, Brazil, and South Africa (Steiner et al. 2015). As for potassium (see Chapter 3), the environmental impacts of phosphate mining include land disturbance, emissions of dust and exhaust gases, water contamination, topsoil degradation, vegetation and wildlife disruption, noise and vibrations, and waste production (UNEP-IFA 2001).

Estimates of high-grade rock phosphate reserves vary widely. Ulrich and Frossard (2014) review 28 estimates from the past 100 years that give a mean of 35 500 million tonnes but with a standard deviation of 30 000 million tonnes. The authors point out that as well as the estimates being highly variable, the mean is likely to be a significant underestimate of future supply given the likelihood of new discoveries and improved extraction technologies. More recent estimates (USGS 2020) suggest world production at 220 million tonnes against known reserves of 63 000 million tonnes (i.e. c. 300 years of supply). However, even when taking an optimistic view of future reserves, it is still inescapable that rock phosphate is a finite resource.

Rock phosphate is also used inefficiently, with significant losses (up to 95%) between reserves in the ground and the amount consumed in food by humans (Scholz and Wellmer 2015). This results partly from the immobilisation of P in agricultural soils. There is therefore significant scope for reducing system losses by recycling P recovered in animal wastes, digestates, and biosolids (Table 6.1). As for N, however, it is necessary to achieve the correct quantities and balance of nutrients, and to prevent the pollution of water courses.

An additional concern about the use of ground rock phosphate is its contamination with heavy metals. The nature and level of contamination depend greatly on the source (Kpomblekou and Tabatabai 1994; McLaughlin et al. 1996; Carnelo et al. 1997; Mehmood et al. 2009; Chen et al. 2020b), but continuous use may increase the accumulation of heavy metals in soils and transfer harmful metals into the human food chain.

At indices above 2, therefore, no application of P should be made until the available phosphate levels decline to an index of 2. This threshold could also be used to vary the application rates of phosphate fertilizers across spatially variable fields. As for N, there is increasing development and use of measurement and sensor technology to assess the P-status of soils and crops to inform more precise approaches to P nutrition (Mallarino and Schepers 2005; Mulla 2013).

6.2.3 Sulphur

In common with N and P, S has many essential functions in the plant. Most of the S (about 80%) in wheat is present as a component of proteins, in two sulphur-containing amino-acids: cysteine and methionine (Figure 6.23). The S in cysteine is present as a thiol (-SH) group, which plays a role in the active sites of some enzymes. The thiol groups on two cysteine molecules can also be oxidised to form a stable cross-linking disulphide bond, forming a dimeric form: cystine. Disulphide bonds are formed between cysteine residues present in proteins, where they play crucial roles in stabilising the folded 3-D structures. They also form stable cross-links between protein subunits present in oligomers and polymers (such as glutenin, see Chapter 7). Sulphur is also present in the protein thioredoxin and the peptide glutathione (glutamate-cysteine-glycine) that regulate reduction/oxidation reactions within the cell, in ferredoxin, which is part of the electron flow system of the light reactions of photosynthesis (Figure 5.1), in several co-enzymes (molecules that aid or initiate enzyme functions) including biotin, co-enzyme A, thiamine pyrophosphate, and lipoic acid (Zhao et al. 1999), and in sulpholipids that may help regulate ion transport across chloroplast thylakoid membranes (Hawkesford et al. 2012).

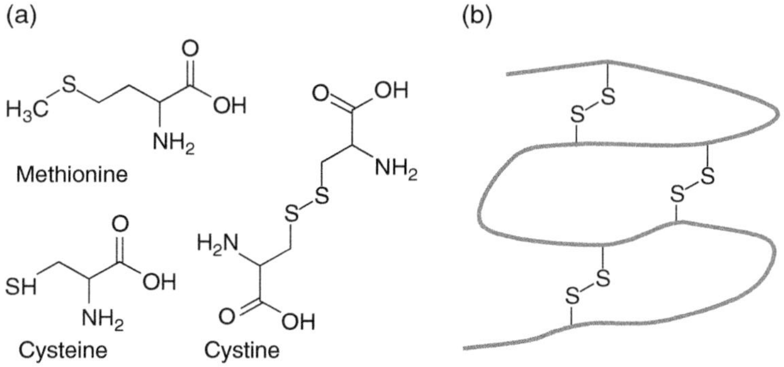

Figure 6.23 (a) Sulphur-containing amino acids. (b) A representation of protein folding around disulphide bridges.

Given the multiple roles of S in proteins, it is unsurprising that some of the symptoms of S deficiency are like those for N deficiency. Sulphur deficiency also reduces responses to, and recovery of, N fertiliser (Salvagiotti et al. 2009). Sulphur deficiency therefore leads to stunted plants, with smaller, paler leaves with lower chlorophyll contents. Increasing S availability to deficient plants increases biomass production with the most S responsive yield component usually being grain numbers per ear (Haneklaus et al. 1995, 2016; Zhao et al. 1996, 1999; Hoel 2011).

Wheat requires an N:S ratio of around 15:1 to satisfy the requirements for protein synthesis (Zhao et al. 1999). Therefore, a GAI of 6 achieved with 180 kg N/ha in the canopy should require 12 kg S/ha for c. 95% light interception by anthesis. There are, however, additional considerations. In contrast to N deficiency, it is the young and emerging leaves that are particularly pale and chlorotic when wheat canopies have insufficient S. That is, wheat appears to be less able to remobilise S to prioritise the younger tissues. Furthermore, by harvest, a smaller proportion of the crop S is partitioned to the grain compared to that of N. This is illustrated in Figure 6.24, which shows harvest indices for N and S of 0.65 and 0.55, respectively. This results in a significantly higher N:S ratio in the grain compared to that in the above-ground biomass.

The implication is that a wheat crop can have sufficient S for *normal* protein synthesis in the canopy, but insufficient S for protein synthesis in the grain. Sulphur deficiency can, therefore, result in changes in the amino acid and protein composition of the grain without obvious deficiency symptoms in the foliage (Byers and Bolton 1979; Naeem and MacRitchie 2003; Shewry et al. 2009). These are discussed in more detail in relation to impacts on grain quality in Chapter 8. Indicators that S deficiency may have occurred are if the grain N:S ratio exceeds 17:1 and/or the grain S concentration is less than 1.2 g/kg (Naeem and MacRitchie 2003). With the latter threshold and an S harvest index of 50%, it follows that each tonne of wheat grain requires at least 2.4 kg S. For a UK crop of 8 t/ha, the minimum amount of S required to be taken up by the crop will be around 20 kg S/ha, significantly more than that required for light interception.

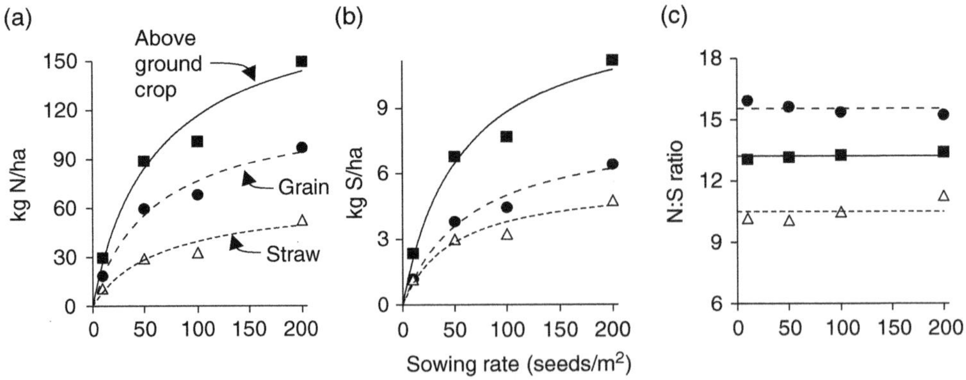

Figure 6.24 The effect of sowing rate on the partitioning of nitrogen and sulphur at harvest of field grown winter wheat. *Source:* Data from Daraja (2009).

Sulphur is taken up by wheat roots as sulphate (SO_4^{2-}). This can be derived from the microbial mineralisation of organic matter but, groundwater can also be a significant source of S for those crops that can access it directly or via irrigation (Haneklaus et al. 2016). Sulphate is also deposited on the soil from atmospheric sulphur dioxide (SO_2) reacting with water to produce H_2SO_4. Atmospheric SO_2 occurs naturally from volcanic activity. Much, however, is anthropogenic from industry; particularly from using S-rich coal.

Wheat has a lower requirement of S compared with combinable crops of the Cruciferеae (e.g. the Brassicas such as oilseed rape) and the Fabaceae (the pulses) (Aarabi et al. 2020), which reflects the requirements of these crops for synthesising glucosinolates and S-rich seed proteins, respectively. Nonetheless, S deficiency has become more widespread in wheat over the past 40 years. There are several reasons for this. The use of ammonium sulphate as a N fertiliser has declined significantly, as has the amount of S applied as a contaminant in P fertilisers. With greater concerns for the environment, S emissions from power stations and industry are being reduced in many areas. In temperate regions, an increased prevalence of oilseed rape in rotation with wheat has increased the total S demand of the arable system and demand has further intensified as wheat yields have increased. Over longer timescales, reduced soil organic matter contents have reduced the amount of S that would otherwise be slowly released from this material (Zhao et al. 1995; Győri 2005). The sulphate ion is both soluble and negatively charged. It is therefore highly mobile in soils and susceptible to leaching, and deficiency is more likely on lighter soils, on soils with low organic matter, and in high rainfall areas where industrial emissions have been reduced (Hoel 2011; AHDB 2020).

As already discussed, wheat crops can be deficient in S before symptoms on the foliage are easily visible. It is also the case that delaying remedial applications of S until symptoms are obvious is too late to prevent reductions in yield and/or quality (Haneklaus et al. 1995). Applications of S-containing fertilisers are, therefore, often made in response to assessing risk in terms of soil type, winter rainfall, and likely deposition. Grain analysis after harvest may also inform requirements for future crops in a similar rotational context. Alternatively, several leaf tests are available such as for N : S ratio, sulphate content, and glutathione concentration. Of some promise for early detection (during tillering) is the malate : sulphate ratio as described by Blake-Kalff et al. (2000, 2001).

The S contents of fertiliser products are usually expressed as either S or SO_3 (40% S) equivalents. Where deficiency is anticipated or indicated, the most effective remediation appears to be sulphate-containing fertilisers (such as ammonium-, potassium-, or magnesium sulphate) applied to the soil during stem extension. Elemental S, either mined as brimstone or as a by-product of oil and gas extraction, can increase S availability over the longer term. Sulphur can also be applied in organic materials (Table 6.1), although less than half of the content will be immediately available, and much sulphate is lost as leachate if applied in the autumn for winter wheats (AHDB 2020). The environmental impact of S fertilisation is considered relatively benign, although ammonium sulphate is highly acidifying, and the sulphate levels present in drinking water are not considered to represent a health concern (WHO 2017). High levels, however, will affect taste resulting in rejection by consumers, so most countries have limits for sulphate in drinking water of between 250 and 500 mg/l (WHO 2004).

Elemental S has long been recognised as a fungicide. There are additional potential implications of S nutrition on the disease status of crops, although the evidence is mostly from work on the S-demanding brassica crops (Walters and Bingham 2007). Nonetheless, the decline of atmospheric deposition in the UK does appear to be associated with a change in the relative abundance of different foliar fungal pathogens of wheat (Bearchell et al. 2005).

6.2.4 Magnesium

Magnesium is an important component of chlorophyll (Figure 6.7) and is also involved in the activation of many enzymes, including *RUBISCO*, and for the synthesis and function of nucleic acids and ATP. Furthermore, it is involved in the loading of the phloem with sucrose, and therefore the transport of photosynthate from mature leaves to emerging tissue, roots, and grain (Verbruggen and Hermans 2013; Tränkner et al. 2018). When Mg in the soil is sufficient wheat removes about 1.0–1.5 kg Mg/ t DM in the grain with a harvest index of around 0.6, i.e. a grain yield of 8 t DM/ha removes about 10 kg Mg/ha in the grain and 7 kg Mg/ha in the straw (Oury et al. 2006; Gooding et al. 2012c).

When crops are deficient in Mg the leaves become striped with interveinal chlorosis, starting on the older leaves. Increasing the availability of Mg to deficient wheat plants increases chlorophyll contents, leaf area indices, PAR interception (Singh and Gill 1987), root : shoot ratios, and tolerance to heat stress (Mengutay et al. 2013). Grain yields are improved through increases in numbers of ears, grains per ear, and mean grain weight (Singh and Gill 1987; Ceylan et al. 2016).

Magnesium is taken up as the Mg^{2+} ion. This becomes available via the weathering of soil mineral material, and via mineralisation of organic matter. Despite the positive charge, Mg^{2+} is not held as tightly to the cation exchange sites in soils compared with other ions such as H^+, Al^{3+}, Ca^{2+}, K^+, and Na^+ (Ceylan et al. 2016). In agricultural contexts, heavy applications of potassium and calcium not only displace Mg^{2+} leading to potential leaching, but also inhibit root uptake of Mg^{2+}. Soils will vary in their Mg content according to the Mg content of parental material (Lark et al. 2019) and, deficiency is most likely on acidic soils, light soils, and on soils with low organic matter and poor structure.

Available Mg can be determined using soil tests, for example, following extraction with 1 M ammonium nitrate (MAFF 1986). Using this method, Mg^{2+} is deemed to be sufficient at 50–100 mg/l in UK conditions (AHDB 2020). Below this level, additional Mg can be provided from several sources. Magnesium contents of fertilisers are often expressed in terms of magnesium oxide (MgO) equivalents (60% Mg). The choice of fertiliser may well depend on what other nutrients are required. For example, mined sources include the potassium-rich kainit (11% K_2O, 5% MgO, 26% Na_2O, 10% SO_3) and magnesium sulphate in kieserite (25% MgO, 50% SO_3). Where available, magnesian limestone (15% MgO) or dolomite can be used as a liming material. Organic materials can also supply Mg (Table 6.1), although the high availability of potassium in some of these materials may hinder Mg recovery in the short term.

6.3 Diseases and their Control

As a canopy is established and constructed, it becomes necessary to maintain its activity. Canopy size and function are directly challenged by numerous diseases. Wiese (1987) describes over 40 pathogens of wheat including fungi, viruses, bacteria, mycoplasmas, and nematodes. Even with the existing crop protection measures, diseases are estimated to reduce global wheat production by over 12% (Oerke 2006). The wheat pathogens most often cited as threatening food security are fungi.

Fungi exist as branched chains of cells forming filamentous hyphae. The hyphal mass constitutes the mycelium. Reproduction and dispersal are chiefly via spores. Rapid rates of reproduction during epidemics usually involve the production and spread of asexual spores, although sexual stages and spores often develop as fungal colonies mature. Wheat diseases have been favoured by rotations and in regions with a high prevalence of wheat. In particular, monocropping wheat within a field and across larger areas encourages the spread of disease. Depending on the pathogen, disease spread may also be increased by higher nitrogen fertilizer applications, by the earlier drilling of autumn-sown crops, and by wider use of minimal cultivation methods. Splash dispersal of some diseases may also have been favoured by the reduced crop heights of the N-responsive cultivars (Saville et al. 2012; Kollers et al. 2013; Miedaner et al. 2013). The semi-dwarfing *RHT1* alleles have been associated with increased disease susceptibility either through effects on plant architecture or pleiotropic effects of altered DELLA activity (Saville et al. 2012) (Chapter 5).

Fungal diseases that attack wheat canopies are the most important pathogens in areas with high levels of attainable yield (Oerke 2006). The canopy can be damaged by infection as and when it is being formed, but it is premature senescence that usually accounts for most of the grain yield depression (e.g. Cook et al. 1999). The most vulnerable yield component is, therefore, mean grain weight as grain filling is curtailed with the early death of the canopy (Dimmock and Gooding 2002a). Foliar pathogens not only reduce light interception and radiation use efficiency, but also the efficiency with which assimilates are partitioned to the grain (Bancal et al. 2007). N use efficiency is also reduced as diseased crops are less able to take up N and therefore leave more mineral N in the soil after harvest, which is susceptible to loss (Ruske et al. 2003a). They are also less efficient at remobilising N from the foliage to the grain (Ruske et al. 2003a; Gooding et al. 2005) and can lose more N from the canopy as volatilised ammonia (Sadler and Scott 1974).

The major foliar pathogens of wheat travel long distances, either via wind or in droplets of mist. This renders foliar pathogens less amenable to control via crop rotation within a field compared with the pathogens of roots and stem bases that are soil or stubble borne (Oerke 2006). Control, therefore, has relied principally on a combination of continual breeding to introduce wheat cultivars with resistance to the pathogens, alongside the development, and greater use of synthetic fungicides. At the same time, the pathogen populations have responded to the selection pressures to overcome cultivar resistance and fungicidal activity. It is a constant battle to prevent catastrophic losses due to disease in the context of trying to maximise light interception by the crop. It is far from clear *who will win the war*.

Figueroa et al. (2018) consider the most important foliar pathogens currently threatening global wheat production to be the rusts (*Puccinia* spp.) and the principal blotch diseases, namely Septoria tritici blotch (*Zymoseptoria tritici*), Septoria nodorum blotch (*Parastagonospora nodorum*), and tan spot (*Pyrenophora tritici-repentis*).

6.3.1 The Rusts

The rusts have been associated with wheat production for at least 3300 years (Roelfs 1992). They are members of the Basidiomycota division of fungi and have complex life cycles involving five different spore types and alternate hosts (Schwessinger 2017; Figueroa et al. 2018). Epidemics on wheat are principally due to multiple generations of asexual spores (urediniospores) spread from pustules erupting through leaf surfaces (uredinia). The rusts are obligate biotrophs (literally 'life-feeding'), i.e. they gain their nourishment from the living tissue of the host, cannot feed from dead tissue, and are difficult to grow on artificial media. Spores of the rust fungi can move great distances and land on wheat leaves via wind or rain. Free water (e.g. from extended periods of dew) on the leaf surface is essential for the germination of urediniospores and the penetration of the leaf surface (Roelfs 1992). After entering the leaf, the hyphae ramify between and through the leaf cells and specialised structures (haustoria) penetrate leaf cells to extract nutrients and manipulate host metabolism. After a short time (typically 7–14 days at optimum temperatures) uredinia are produced, which release further spores (Parry 1990; Chen et al. 2014). The three rust diseases of wheat are yellow (or stripe) rust (*P. striiformis* f. sp. *tritici*), leaf (or brown) rust (*P. triticina*), and stem (or black) rust (*P. graminis* f. sp. *tritici*) (Roelfs 1992).

6.3.1.1 Yellow Rust

Yellow rust may well be the single most damaging disease of wheat (Beddow et al. 2015). Infection pressures increase with N availability and earlier sowing (Neumann et al. 2004; Chen 2020). Areas of infection in a field are usually first seen as discreet patches of plants exhibiting yellow/orange uredinia, about 0.5–1.0 mm in diameter on the leaves. The uredinia are often aligned on mature tissue, causing characteristic yellow stripes, while on younger tissue the pustules may appear to be more randomly scattered.

Photosynthesis declines in infected wheat plants because of the reduction of green leaf area due to the coverage of pustules. Radiation use efficiency is also lowered as respiration increases and carbon is utilized in fungal structures. Additionally, the loss of integrity of the leaf cuticle reduces resistance to water loss and exacerbates drought effects. The rate of expansion of new leaves is reduced and, although the strategy of the pathogen is to parasitise living tissue, the crop ultimately senesces prematurely. Reductions in yield can be largely accounted for by reduced green area index, canopy duration, and therefore poorer light interception (Bryson et al. 1997). The early senesce of the canopy is particularly associated with a premature end to grain filling and therefore a reduction in mean grain weight (Dimmock and Gooding 2002a).

The pathogen is highly mobile and readily develops new races that can overcome resistance in wheat cultivars (Schwessinger 2017). Epidemics are favoured by widespread monocropping of wheat cultivars with the same source of disease resistance genes. In such circumstances there is high selection pressure for the development of new virulent races of the pathogen. Due to the rapid progress of yellow rust in fields in favourable conditions, farmers are less confident of being able to control yellow rust with fungicide applications. They will, therefore, readily change cultivars once a previously resistant cultivar has become susceptible to new races of the fungus. There are, therefore, continuing breeding efforts and genetic approaches to develop cultivars with more durable resistance (Schwessinger 2017).

Optimum temperatures for the germination, penetration, growth, and sporulation of yellow rust on wheat have been between 8 and 15 °C (Roelfs 1992). Traditionally, therefore, yellow rust has been concentrated in the cooler areas of wheat production such as northwest Europe and the northwestern US (Chen 2020). Within these areas, epidemics are most common in coastal areas typified by cool summer temperatures and regular mists and extended periods of dew on the leaf surface. In North America, particularly favourable cool and humid conditions develop at high elevations between mountain ranges. New strains, however, appear to be contributing to the movement of yellow rust into warmer areas of wheat production (Hovmøller et al. 2008). This expansion led Beddow et al. (2015) to calculate that 88% of the world's wheat is now grown in locations susceptible to yellow rust infection. This rise in importance and extent of the pathogen has required increased use of fungicides in many regions (e.g. Carmona et al. 2020).

6.3.1.2 Leaf Rust

Wheat leaf rust produces uredinia that are larger (1×1–2 mm; Parry 1990) than those of yellow rust and are orange-brown. Rather than being arranged in stripes, the pustules appear to erupt randomly over mature leaf surfaces. The disease is found nearly everywhere the crop is grown. Epidemics are often less dramatic than those of yellow rust, but the wide range of the disease means that annual global losses are substantial (Huerta-Espino et al. 2011). Temperatures favouring development on wheat (15–25 °C; Roelfs (1992)) are higher than those for yellow rust. Therefore, in areas where wheat matures as temperatures rise, wheat leaf rust develops later in the season than yellow rust and will tend to be more severe on later maturing crops, such as on spring, rather than autumn-sown crops. The adaptation to warmer climates appears to have contributed to increased severity of leaf rust in oceanic areas that have experienced decadal rises in temperature, and this trend of greater infection pressures is predicted to continue in climate change scenarios (Caubel et al. 2017). Consistent with the effects of yellow rust, reductions in yield mostly result from decreases in chlorophyll content, green leaf area, and canopy duration (Robert et al. 2005; Serrago et al. 2009; Carretero et al. 2011). Grain yields are reduced, mainly because of lower mean grain weights resulting from shorter grain filling durations (Gooding et al. 2005; Huerta-Espino et al. 2011; Serrago and Miralles 2014).

As for yellow rust, the disease is favoured by N fertiliser application (Tiedemann 1996; Simón et al. 2020). Similarly, control of leaf rust depends on the continuing development of resistant cultivars in the face of the selection and migration of virulent races of the pathogen. Despite breeding efforts, fungicides principally targeted at controlling leaf rust are used in many parts of the world, either routinely or as an emergency measure (Huerta-Espino et al. 2011).

6.3.1.3 Stem Rust

Stem rust has, historically, been found in all major wheat-growing areas. The dark uredinia are comparatively large, typically 2×5 mm, and occur mainly on the leaf sheaths and stems. It is, however, the most heat-loving of the wheat rusts, with optimum temperatures for different stages of development ranging from 15 to 30 °C (Roelfs 1992). Greatest risks are, therefore, identified for wheat in the warmer growing areas, and/or where wheat matures at high temperatures. In favourable conditions and on susceptible cultivars, disease spread is rapid and difficult to contain. For effective control with fungicides, applications need to be made as soon as the pathogen is seen (Loughman et al. 2005). This may not always be possible due to logistics or the weather being unsuitable for spraying.

As with other rusts, yields can be reduced following premature loss of chlorophyll, shorter duration of leaf photosynthesis, and reduced mean grain weights (Moerschbacher et al. 1994; Loughman et al. 2005). In addition, the disease causes the stems to weaken and break. At the field level crop loss can be total. Severe epidemics have caused food shortages and significant price rises (e.g. Roelfs 1982). Irradicating an alternative host of stem rust (common barberry, *Berberis vulgaris*) can be partially successful in reducing the disease (Roelfs 1982). Compared to the other wheat rusts, however, resistance genes have been more durable.

The semi-dwarfed wheats released by Norman Borlaug carried resistance to stem rust, and by the mid-1990s breeding efforts had rendered stem rust insignificant for most wheat production (Singh et al. 2008). In the past two decades, however, new races have appeared in East Africa. In particular, the race Ug99 caused serious losses of wheat lines planted in Uganda in 1998 and has since spread further in Africa and into the Middle East (Singh et al. 2011). Experts still identify stem rust Ug99 as a major threat to world wheat production, citing the lack of resistance to this race and its variants in the wheat cultivars grown in major wheat-growing areas (Lidwell-Durnin and Lapthorn 2020). Recent outbreaks of stem rust in Europe and Russia have been ascribed to the comparative lack of focus on maintaining stem rust resistance in areas where the disease has been minimal for decades, the reduced efforts on control of barberry (the alternative host required to complete the life cycle), the emergence of new races, and the warming climate (Figueroa et al. 2018; Saunders et al. 2019).

6.3.2 The Blotch Diseases

The three principal blotch diseases are members of the Ascomycota division of fungi. In contrast to the rusts, they gain their nutrition from dead and decaying tissue and hence are known as necrotrophs ('death feeding'). They are facultative pathogens because they can be grown relatively easily on media outside the host. In killing the green tissue, light interception is reduced directly by symptom expression.

6.3.2.1 *Septoria tritici* Blotch

Septoria tritici blotch (STB) can enter the crop via windborne sexual spores (ascospores), or from splash-dispersed asexual spores (pycnidiospores). Primary infection of winter wheats often occurs in the autumn with spores from crop debris and volunteer plants (Eyal 1999). After entering the leaf, principally through the stomata, the pathogen has an extended latent period during which there are no visible symptoms. This phase lasts about 250 °Cd above a base temperature of −2.5 °C (Lovell et al. 2004). During this period, the hyphae grow slowly between leaf cells, whilst subverting the host defences (Figueroa et al. 2018). As the base temperature is lower than that for crop growth, lower temperatures favour the development of the pathogen relative to the ability of the crop to *grow away* from infection on mature leaves (Lovell et al. 2004). Some commentators describe this latent phase as a form of biotrophic development and therefore class *Z. tritici* as a hemibiotroph. Once the latent period has ended, however, development is very much necrotrophic. The leaf cells rupture and their contents become available to the pathogen. Small chlorotic/yellow flecks appear on the older leaves, the flecks expand, and their centres become red-brown and dead. These necrotic areas tend to be parallel sided and surrounded by expanding regions of chlorosis. Yield is reduced due to effects on both light interception and radiation use efficiency (Cornish et al. 1990; Olesen et al. 2003b). Relationships with green leaf area can vary with wheat cultivar. In some breeding programmes, cultivars with high yield potential have tended to be less tolerant to losing green leaf area due to STB infection compared with lower yielding lines (Parker et al. 2004).

Distinctive black dots develop in the necrotic areas, in the sites of stomata (Eyal 1999). These are the pycnidia, i.e. flask-like structures embedded within the lesion which produce the pycnidiospores. Rapid, secondary, spread of these asexual spores is then via water droplets, spreading up the canopy in rebounding energetic droplets from heavy rain events (splash dispersal) and/or through spores being washed down a leaf laminar on to emerging leaves (Lovell et al. 1997). Spread within and beyond fields may be through pycnidiospores in mist droplets, but also via further generations of ascospores carried on the wind. Sexual reproduction is slower than that via pycnidiospores, but can occur throughout crop growth, and on post-harvest stubble (Hunter et al. 1999).

Overall, disease spread appears optimal between 15 and 20 °C. The disease is thus prevalent in cool, damp, temperate areas of production. It has been the most frequent serious foliar disease on wheat in northwestern Europe since the early 1980s (Bearchell et al. 2005; Fones and Gurr 2015; Figueroa et al. 2018), where its spread and development have been facilitated by wheat crop residues being left on the soil surface, earlier drilling and reduced time between susceptible crops, the adoption of shorter cultivars, and in some studies, increased N fertilizer applications.

The development of the disease in the latent, asymptomatic phase means that best levels of disease control are achieved through fungicide applications made before the pathogen is visible. The need for applications can be informed by assessment of disease on lower leaves, cultivar susceptibility, and the intensity and duration of rainfall (Lovell et al. 1997; Fones and Gurr 2015). The latent period and nature of disease spread, however, encourage a more prophylactic, routine approach to fungicide use in vulnerable regions of production.

6.3.2.2 *Septoria nodorum* Blotch

Septoria nodorum blotch (SNB) has similarities with STB. Infection can be via windborne ascospores and/or splash-dispersed asexual pycnidiospores. Significant differences are, however, also evident. Ascospore dispersal is probably less important than for STB, while dispersal in seed is more important for SNB (Shaw et al. 2008). Foliar infection can be directly into the leaf cuticle and then between or into leaf cells, as well as through the stomata (Solomon et al. 2006). There is no initial *biotrophic* latent phase and the pathogen quickly releases toxins (or effector proteins), which result in the death of the host cells and the disruption of host defences (Figueroa et al. 2018). The lesions are lens-shaped rather than parallel-sided and the pycnidia are less obvious, being small, pinky/brown, and translucent. The time from infection to release of another generation of pycnidiospores can be as little as seven days (Solomon et al. 2006). Rapid spread requires temperatures of 20–27 °C with 100% relative humidity (Polley and Clarkson 1978). Optimum temperatures are, therefore, higher than for STB, but *S. nodorum* blotch can occur in most areas where wheat is grown (Stukenbrock et al. 2006). In Australia, with commercial control practices in place, Murray and Brennan (2009) calculate SNB to be one of the top three most damaging diseases of wheat (with yellow rust and tan spot). In areas of western Europe, however, SNB has been largely replaced by STB as the major blotch disease of wheat. High N availability often reduces the severity of SNB (Tiedemann 1996) and tends to favour STB over SNB (Eyal 1999). Additionally, over a 160-year time course, Bearchell et al. (2005) found SNB to be more prevalent than STB during periods of high atmospheric emissions of sulphur dioxide.

6.3.2.3 Tan Spot

Tan spot (TS) occurs throughout the major wheat-producing areas (Moffat and Santana 2019), although optimum temperatures are relatively high (20–28 °C). The pathogen can be spread with the seed or may persist on stubble and other crop debris from which ascospores can be released and dispersed by wind and splashing. Following the infection of lower leaves, leaf lesions appear in 7–14 days, initially as small brown flecks. These expand to form diamond-shaped spots with a yellow, chlorotic margin. The spots then extend and coalesce to form irregular shapes of dead tissue, often with a small dark brown *eyespot* near the centre of each original lesion. Conidia are released from these spots and splash dispersed to cause secondary infections. Infection, lesion development, and conidia production are all significantly increased by prolonged periods of rain (18 hours or more) and/or dews (Wiese 1987). Tan spot has become more widespread with the increased use of minimum and zero-tillage techniques (Faris et al. 2013). As with SNB, infection is associated with effector proteins that lead to the death of the host cells. It appears likely that one of these effector proteins was acquired from SNB; an acquisition that may well also have contributed to a recent rise in the importance of TS.

6.3.3 Diseases Contributing to Mycotoxins in the Grain

In addition to pathogens that influence light capture, radiation use efficiency, and therefore grain yield, there are a number that have specific implications for animal and human health because they produce poisons, i.e. mycotoxins. The effects of mycotoxins on health and processing are discussed in Chapter 9, but here we describe two diseases that frequently lead to the rejection or down-grading of wheat.

6.3.3.1 Ergot

Ergot is caused by the fungus *Claviceps purpurea*, which can infect many grasses. Wheat, rye, triticale, and blackgrass are particularly susceptible. An ergot is a purple-black fungal body, 2–25 mm long (a sclerotium), that replaces a grain site on the ear. Sclerotia are dense, hard masses of hyphae, that contain a number of mycotoxins including the alkaloid ergotamine. The sclerotia fall to the soil surface before or at harvest or are harvested with the grain and spread with contaminated seed lots. The ergots germinate in the following spring to produce spherical stromata, about 2 mm in diameter, which release ascospores that are then spread by wind onto the flowers of cereals and weed grasses, where they infect the ovaries. The infected sites exude sweet *honey dew*, which contain asexual conidia that can be further spread by rain splash or insects to other open florets.

Infection on wheat relies on the coincidence of the release of ascospores, or subsequent spread of conidia, with the opening of the flowers. The probability of this happening is increased when the flowering period is extended, which may occur in cooler conditions, if there are many late tillers due to low plant populations, or if the genotype is predisposed to flowering over a longer duration. Secondary infections by conidia released from infected weed grasses greatly increase the likelihood of temporal coincidence, and the weeds themselves will extend the infection period as they act as susceptible hosts in the field for initial infection by ascospores.

Methods to supress ergot include good grass weed control, burial of ergots after harvest with inversion tillage such as ploughing, having a break from cereals in the rotation, and using clean seed.

6.3.3.2 *Fusarium* Head Blight (FHB)

Head blight or scab can be caused by several species of *Fusarium*; most commonly *Fusarium graminearum* (sexual stage *Gibberrella zeae*) in warmer areas and *Fusarium culmorum* in cooler areas (Wegulo et al. 2015). The fungi persist on stubble and crop debris on or in the soil. This applies to any cereal stubble, although that left by maize can harbour particularly large quantities of inoculum. *Fusarium* spp. can also be spread on seeds, whereby the seedlings suffer from seedling blight and can then also act as a source of inoculum. Seedborne inoculum also infects the roots, stem base, and crown. Ascospores are released from the crop residue in warm and moist weather, and are wind dispersed to infect florets at flowering. The fungus enters the plant through openings at the base of the lemma and palea, or through degenerating anthers (Trail 2009). Secondary infections can arise from splash dispersal of asexual conidia. Initially individual spikelets become bleached, but bleaching can spread to a proportion, or all of, the ear. Pinkish/orange spore-bearing structures can be seen at the base of the glumes and in warm and moist conditions pinkish mycelium is evident. Grain infected at flowering become severely shrivelled and chalky-coloured. Grain infected later have less severe visible symptoms but may still contain mycotoxins. Although yields can be significantly reduced through increasing floret sterility and effects on mean grain weights, it is contamination of the grain with mycotoxins such as deoxynivalenol (DON) and zearalenone that causes most economic damage and harm to health. DON is a virulence factor and causes death of wheat cells and facilitates the spread of the pathogen within the ear (Trail 2009).

FHB is increasing in many wheat-growing areas (Wegulo et al. 2015). This can be partly attributed to climate change, adoption of shorter cultivars, and increased use of reduced soil tillage methods, which leave more cereal crop residue in the surface layers (Wegulo et al. 2015). Control can thus include inversion tillage and crop rotation, whereby cereals are avoided for at least two years. In irrigated areas, irrigation should be avoided just prior to and during flowering.

Some control is also possible through cultivar selection (e.g. Bai et al. 2000). Quantitative contributions to resistance include resistance to initial infection, resistance to spread in infected tissue, resistance to kernel infection, tolerance, and resistance to accumulation of toxins (Wegulo et al. 2015). An application of a DMI fungicide (Table 6.2) targeted at the spikes between ear emergence and the end of flowering can help suppress the disease but the application of QoI fungicides (Table 6.2) should be avoided as they can increase mycotoxin production by *Fusarium*.

6.3.4 Fungicides and Fungicide Use

Fungicides can be divided into those that target several biochemical pathways in the pathogen (multi-site fungicides) and those that target one biochemical pathway (single-site fungicides). As a broad generalisation, multi-site fungicides tend only to be protectant and do not move within the plant. They cannot control the pathogen once the infection process has started, and they cannot be transported into young, emergent tissue. A great advantage of the multi-site fungicides is, however, that pathogens rarely develop resistance to them. By contrast, the single-site fungicides are more likely to be able to move within the leaves and to emergent tissue, and to have some curative, as well as protectant activity. The major groups of single-site fungicides for controlling foliar diseases of wheat are those that inhibit demethylation in the sterol biosynthesis pathway (e.g. the DMIs); those that prevent respiration as quinone outside inhibitors (the QoIs); and those that prevent respiration through succinate dehydrogenase inhibition (the SDHIs) (Table 6.2).

Fungicide use to control foliar pathogens on wheat has been increasing throughout the world (Oliver and Hewitt 2014), encouraged by conditions that favour disease spread. As we have seen, many of these conditions are provided by farmers when they maximise light interception (irrigation, high N availability, early sowing, high seed rates), and when they reduce cultivation intensity. These practices have been adopted alongside an increase in the yield potential of the cultivars grown, particularly through increases in harvest index. Fungicide use has, therefore, not only been encouraged due to conducive conditions, but also by the increasing returns in terms of grain yield resulting from protecting cultivars with high yield potential. In the UK for example, the use of fungicides on wheat crops was negligible until the mid-1970s (Figure 6.25). Since then, there have been significant increases in the numbers of fungicide spray passes and of active ingredient sprays (sum of number of active ingredients in all the passes) applied to the crop (Figure 6.25). In the UK wheat is currently sprayed either three or four times a season, with each spray pass applying an average of three active ingredients. Typical timings include (AHDB 2019):

1) The second node detectable growth stage (DGS32) to control disease on the second penultimate leaf and prevent further spread up the plant.
2) Flag leaf emergence (DGS37/39) to protect the flag leaf and the penultimate leaf.
3) Ear emergence (DGS59) to further control disease on the flag leaf, and also to protect the ear.

The current level of fungicide use reflects increases that have continued while N applications and mean grain yields have remained comparatively static since 2000 (Figure 6.17). Indeed, despite the increase in agrochemical applications, national yields appear to have become less stable over the past two decades (Figure 6.17). Foliar fungal diseases, therefore, pose a significant threat to the sustainability of current methods of production and cultivar development.

Table 6.2 Fungicide groups commonly used against foliar fungal pathogens of wheat

Major mode of action	Example of specific mode of action	Example of important chemical group	Example of a specific active ingredient	Structure of specific active ingredient. Redrawn from FRAC (2020).
Sterol Biosynthesis Inhibitors (SBI)	Demethylation inhibitors (DMI)	Triazoles	Mefentriflu-conazole	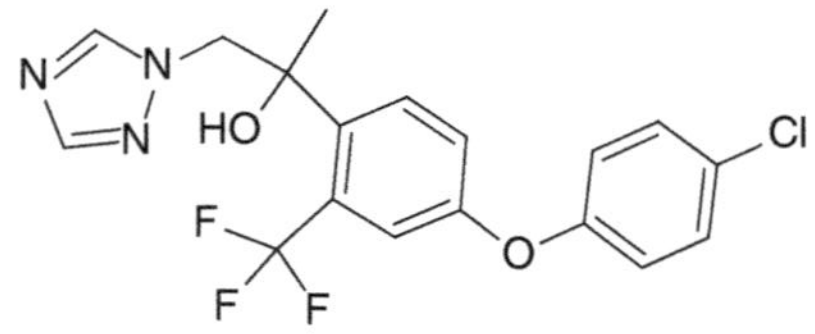
Respiration inhibitors	Succinate dehydrogenase inhibitors (SDHI)	Pyrazole - 4 -carboxamides	Fluxapyroxad	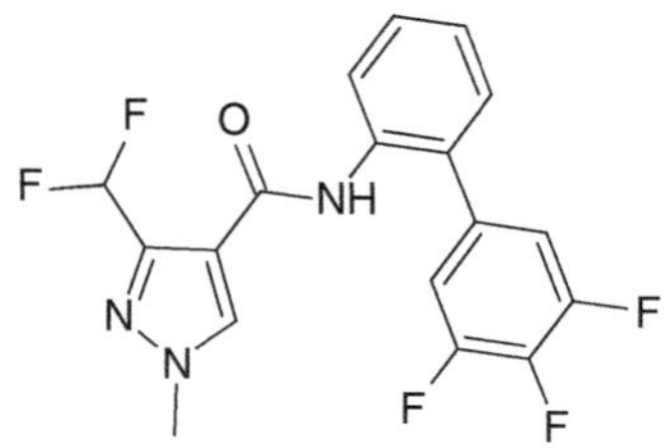
Respiration inhibitors	Quinone outside inhibitors (QoI)	Strobilurins	Pyraclostrobin	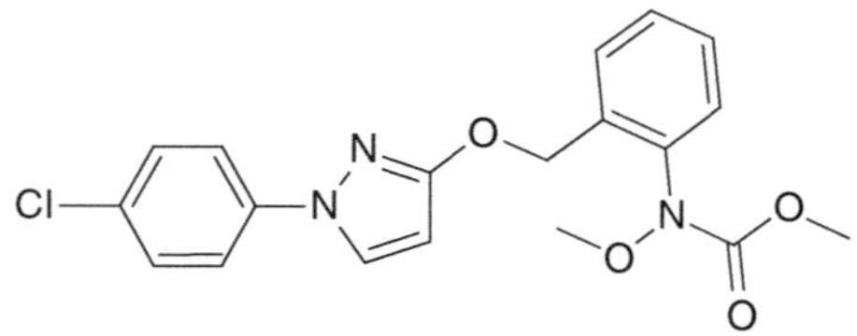
Multi-site	Several modes of action	Phthalimides	Folpet	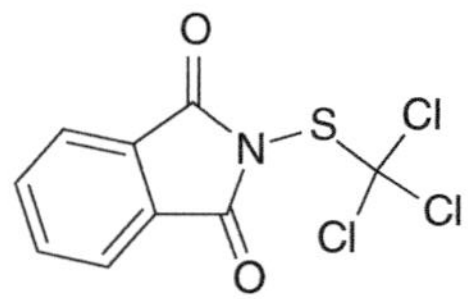

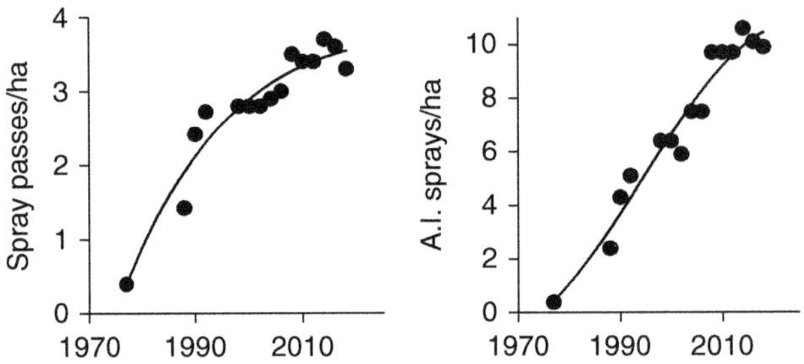

Figure 6.25 Applications of fungicides to UK wheat in terms of the number of spray passes per crop, and the number active ingredient (A.I.) sprays (sum of A.I. in each pass). *Source:* Based on data from Garthwaite et al. (2019) and previous editions.

Europe has become particularly dependant on fungicide use: the continent has accounted for over 60% of worldwide fungicide sales for temperate cereals. Germany, France, and the UK are the most important countries in terms of volume because use is often justified economically by a combination of high potential yields and high infection pressures: both deriving from temperate conditions with adequate moisture, large applications of nitrogenous fertilizers, and rotations dominated by cereals.

The rise in fungicide use has occurred alongside breeding efforts for disease resistance. The relationship between the reliance on breeding efforts and that on fungicide use for disease control is complex and the dependence of wheat farmers on fungicides, even when there appear to be resistant cultivars available, has been ascribed to many factors (Vanloqueren and Baret 2008). These include a preference for cultivars that are more disease susceptible when they have high yields, lodging resistance, and good end use quality when comprehensive fungicide spray programmes are employed. This is particularly true when producers are convinced that they have the armoury of fungicides, spray technology, and logistics to remain on top of disease progress. This has been relevant to some of the splash dispersed foliar pathogens such as STB where cultivars with high yield potential have frequently had high levels of disease susceptibility. Farmers have persisted with such cultivars with confidence that they can adequately suppress the disease with continuing use of available chemistry. Reliance on fungicides, rather than on the adoption of resistant cultivars, is also practiced because genetic resistance is rarely total and that the occurrence of epidemics as cultivar resistance breaks down is unpredictable. Furthermore, it is rare that a recommended cultivar will be resistant to all common diseases.

When fungicides do reduce the loss of yield caused by fungal diseases, this is usually associated with effects on light interception, canopy size, and canopy duration (e.g. Paveley et al. 2001; Audsley et al. 2005). In particular, the duration of the life of the flag leaf is often a good predictor of the effects of fungicide on grain yield (Figures 6.26 and 6.27). This association not only reflects the importance of the flag leaf itself for providing photosynthate to the grain (Figure 5.8), but presumably because its health status, being the last leaf to senesce, is associated with the previous health of other areas of the canopy affecting light capture, translocation, and remobilization of reserves (Bancal et al. 2007). The decline in the green area of the flag leaf is adequately captured by a modified Gompertz curve

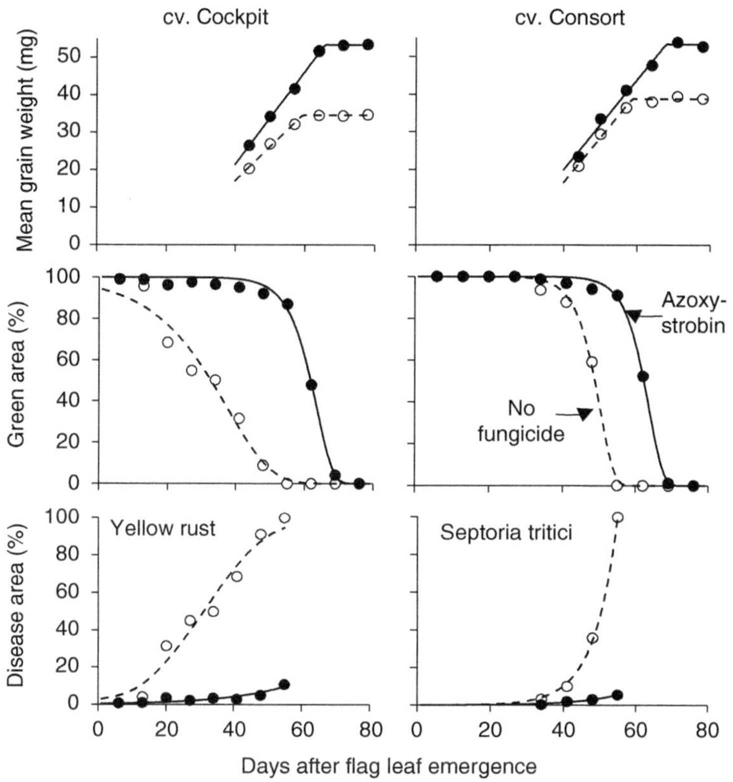

Figure 6.26 The effect of applying 0.125g azoxystrobin, a QoI fungicide (Table 6.2) at flag leaf emergence and again at ear emergence on the diseased and green area of the flag leaf, and on grain filling of two cultivars of winter wheat contrasting in disease susceptibility. *Source:* Data from Dimmock and Gooding (2002a).

(Gooding et al. 2000, as fitted in Figures 6.9d, 6.26) where the percentage of the flag leaf that is green ($GLA\%$) on day (d) after flag leaf emergence is:

$$GLA\% = 100.e^{-e^{-k(d-m)}} \tag{6.10}$$

In this form it is only necessary to fit the Gompertz rate (k) and time (m) scalars. The time scalar, m, i.e. the day when flag leaf green area has declined to 37% of the maximum, is an arbitrary but adequate and convenient measure of the duration of leaf life.

As the life of the flag leaf is extended by controlling diseases, grain filling is also extended (Figure 6.26). The improvement in grain yield associated with increased disease control therefore mainly results from increased mean grain weight (Figure 6.27). However, grain number may also be increased (Figure 6.27) resulting from a direct effect, as severe early epidemics may reduce floret and grain survival. However, diseases may also have indirect effects on the number of harvested grain due to loss of shrivelled grain during combining. The reduced shrivelling following fungicidal disease control is shown by improved grain-specific weight (Figure 6.28).

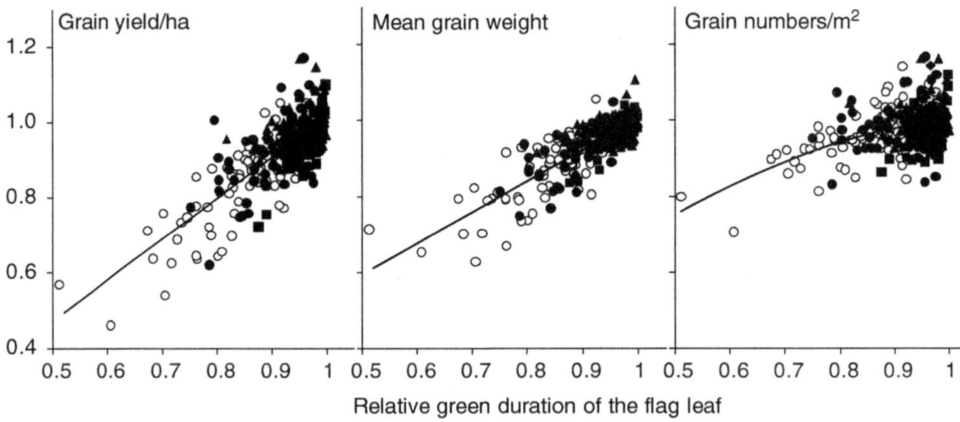

Figure 6.27 The relationship between maintaining flag leaf life with fungicides and grain yield of winter wheat. Points are from replicated (usually 4) plots within 120 cultivar × field experiment combinations in the UK. *Source:* Data mostly from Gooding et al. (2000), Dimmock and Gooding (2002a), Ruske et al. (2003b), Pepler et al. (2005a). All values are relative to the fungicide treatment giving the greatest duration of flag leaf life in each experiment × cultivar combination (not plotted). Duration of flag leaf life is defined as the time to reach 37% green leaf area (the Gompertz time scalar, *m*, as in Eq. 6.10). Fungicide treatments applied at and after flag leaf emergence are: ○, no fungicide; ●, DMI; ■, QoI; ▲, DMI + QoI; ◆, DMI + multisite.

The effects of fungicide use on grain protein concentration are small and inconsistent, depending on whether the disease has had larger or smaller effects on N recovery compared with effects on dry matter yield. Controlling severe disease routinely increases the recovery of N and S in the grain (Figure 6.29; Gooding et al. 2005; Pepler et al. 2005b). In general (but with exceptions) controlling rusts results in greater benefits to N yield than to dry matter yield so that grain protein concentration and grain yield are increased simultaneously. By contrast, there is often a small dilution in protein concentration when controlling the blotch diseases (Dimmock and Gooding 2002b; Simón et al. 2020). This is because the blotch pathogens are more damaging to dry matter accumulation than they are to N accumulation. The small negative relationship between fungicide use and grain protein concentration (and SDS-sedimentation volume) in Figure 6.28 occurs because the main pathogen affecting the untreated plots was more frequently STB than yellow rust or leaf rust. In several studies, controlling severe disease resulted in increased α-amylase activity (Salmon and Cook 1987) and/or reduced Hagberg falling number (Figure 6.28; Ruske et al. 2003b). The reason for these effects remains unclear, although low Hagberg falling number following fungicide application has been correlated with delayed grain drying before harvest and increased mean grain weights (Kettlewell 1997; Dimmock and Gooding 2002c).

Dependency on fungicides may be short-lived as reports of fungicide resistance in wheat fungal populations are widespread (Oliver and Hewitt 2014). For example, the mainstay for systemic fungicidal control of STB in northwestern Europe since the 1980s has been the DMIs (Table 6.2). However, sensitivity of the pathogen to a range of DMI fungicides has been reducing, sometimes gradually, sometimes more rapidly (Cools et al. 2005; Cools and Fraaije 2008). Of significant promise, therefore, was the introduction of the QoI fungicides

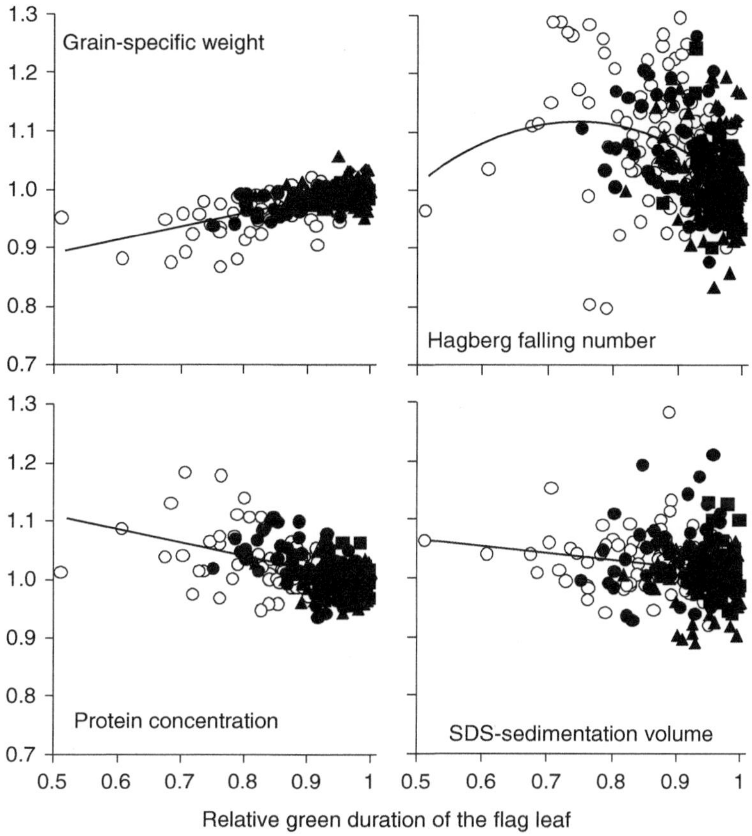

Figure 6.28 The relationship between maintaining flag leaf life with fungicides and grain qualities of wheat crops. Points are means from replicated (usually 4) plots within 120 cultivar × field experiment combinations in the UK. All values are relative to the fungicide treatment giving the greatest duration of flag leaf life in each experiment × cultivar combination. *Source:* Data sources and key are as for Figure 6.27.

in the late 1990s, which initially gave good control of STB (e.g. Figure 6.27). Widespread use of QoI fungicides to control STB in northwestern Europe, however, led to resistance in the pathogen. Resistant isolates were first found in the UK in 2001 but resistance then spread rapidly and is now ubiquitous in several northern European countries, resulting in the failure of QoI fungicides to adequately control STB in the field (Cheval et al. 2017). More recently, the SDHI (Table 6.2) fungicides have been used to improve STB control in DMI-based fungicide programmes. There are, however, increasing concerns over the development of resistance to the new SDHIs (Yamashita and Fraaije 2018). Strategies to delay the build-up of resistance include:

1) Reducing reliance on fungicides by using resistant cultivars (where available) and other cultural methods of control.
2) Diversifying the fungicide groups used over the course of a cropping season.
3) Using mixtures of fungicide groups in each spray pass.
4) Adding a multi-site fungicide to mixtures otherwise based on single-site systemic chemistry.

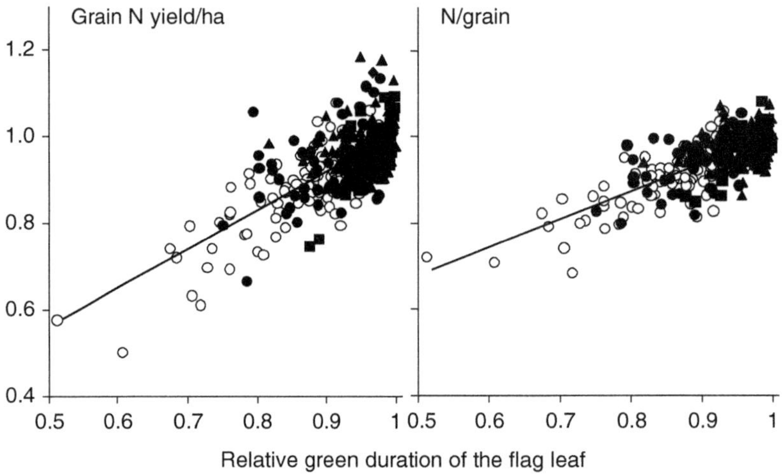

Figure 6.29 The relationship between maintaining flag leaf life with fungicides and nitrogen recovery by wheat crops. Points are from replicated (usually 4) plots within 120 cultivar × field experiment combinations in the UK. All values are relative to the fungicide treatment giving the greatest duration of flag leaf life in each experiment × cultivar combination (not plotted). *Source:* Data sources and key are as for Figure 6.27.

The continued use of fungicides on wheat is not only threatened by the development of resistance in the pathogen populations, but also by regulatory and approval processes.

For example, the EU is considered to have the most stringent processes for the initial and recurrent approval of pesticides. Regulation (EC) No. 1107/2009 (EU 2009) requires industry to demonstrate that substances 'shall not have any harmful effects on human health, including that of vulnerable groups, or animal health, taking into account known cumulative and synergistic effects, or on groundwater . . . they shall not have any unacceptable effect on the environment'. A number of fungicides have had approval withdrawn by the EU, the most notable being chlorothalonil, previously one of the most widely used fungicides on wheat. This resulted from several concerns about the possible effects of the chemical and/or its breakdown products on health and the environment (EFSA 2018). Additionally, a number of the DMI fungicides (in common with a large number of other pesticides) can potentially interfere with the hormone systems of animals including humans, i.e. they are possible *endocrine disruptors* (Mnif et al. 2011). These concerns have contributed to EU approval being withdrawn from certain DMI fungicides that were once frequently applied to wheat (EFSA 2017).

6.4 Land-use Efficiency and Canopy Management

This chapter has identified some of the key factors responsible for the significant increases in grain yield per unit area that have been achieved since the 1960s. Plant breeders have been successful in producing wheat with a communal phenotype, with lodging resistance, and with high harvest index. This has meant that crops, rather than individual plants, can be grown to maximise the capture of resources. In the case of light, wheat in appropriate environments can be sown earlier, at adequate seed rates, and respond to sufficient nutrients to maximise radiation interception during critical growth stages. The high harvest index further ensures that photo-assimilates are directed to the maximum grain yield.

This chapter has also identified some of the threats to the sustainability of a light-capture approach to increasing land-use efficiency. In the case of N-use, there are two sets of direct issues. The first is in the sustainability of its synthetic production, and particularly the need to increase the use of renewable energy sources and reduce reliance on fossil fuels. The second is in the sustainability of N use. Agriculture is the main driver to increasing reactive nitrogen in the hydro-, bio- and atmospheres, whether that nitrogen comes from synthetic or biological fixation. Diffuse pollution from agricultural N-use perturbs natural and agro-ecosystems and is a significant contributor to global warming and ozone depletion. We can significantly increase the efficiency of N-use for wheat production (kg grain/kg available N) by reducing the amount that farmers apply (by price or legislation), such that levels are sub-optimal for yield. This is analogous to the deficit-irrigation approaches discussed in Section 3.3.3. By definition, at least in the short term, this reduces land-use efficiency, but as in the case for water, it would further encourage the adoption of knowledge and development of technology to maximize N-use efficiency, and fertiliser N recovery. Nonetheless, such an approach would make it particularly challenging for wheat farmers in high yielding areas to achieve the minimum concentration of grain protein for breadmaking.

For all nutrients, but particularly for N and P, there is significant potential to increase the efficiency with which nutrients are recycled, both within farms and after human consumption. Greater use of sewage sludge (or biosolids), in particular, would reduce the rate of depletion of mined sources of P.

As we have seen, growing and maintaining large canopies for light interception poses significant challenges for crop protection. The greater infection pressures from several diseases threatens the sustainability of producing large canopies. The increasing restrictions on fungicide use in some parts of the world, and the tendency to reduce tillage intensity, will place even greater reliance on plant breeders and the strategies that they have available (Chapter 10), to maintain adequate levels of resistance as pathogens evolve and as the climate changes.

Finally, in many areas of the world it can be argued that methods for increasing yield through improved canopy management and increased harvest indices have been exhausted. Figure 6.17a, for example, shows average UK yields to have plateaued, and to become less predictable. Many farmers achieve greater than average yields, and the record national wheat yield was set relatively recent (in 2014), and it could therefore be argued that there is still scope to increase average wheat yields in the UK by improving canopy management on more farms. It is, however, salutary to reflect that national wheat yields remain the same despite breeding efforts to improve yield potential, continuing high rates of fertilizer inputs, and increasing number of herbicide (Figure 2.15) and fungicide sprays (Figure 6.25). This leads to one of three conclusions: that the yield potential is no longer being improved by plant breeders; that yield potential is not limiting yield on farmer's fields; or that constant efforts to increase yield potential are necessary *just to stand still* whilst land is degraded and biotic and abiotic stresses increase. All of these conclusions raise questions about the sustainability of wheat production and the ability to meet future food demands based on these intensive systems. Globally, the picture is somewhat different and although many of the issues of sustainability still apply, yields, N fertiliser rates, and associated canopy sizes continue to increase (Figure 6.17b) and have not yet reached levels that would maximize light interception at critical growth stages. In terms of the global wheat crop, however, this may never happen, as many wheat crops will be limited more by water availability than by light interception.

References

Aarabi, F., Naake, T., Fernie, A.R. et al. (2020). Coordinating sulfur pools under sulfate deprivation. *Trends in Plant Science* 25 (12): 1227–1239. https://doi.org/10.1016/j.tplants.2020.07.00.

Addiscott, T.M. and Benjamin, N. (2004). Nitrate and human health. *Soil Use and Management* 20 (2): 98–104. https://doi.org/10.1111/j.1475-2743.2004.tb00344.x.

Addisu, M., Snape, J.W., Simmonds, J.R. et al. (2010). Effects of reduced height (Rht) and photoperiod insensitivity (Ppd) alleles on yield of wheat in contrasting production systems. *Euphytica* 172: 169–181. https://doi.org/10.1007/s10681-0090025-2.

AHDB (2019). *Wheat and Barley Disease Management Guide*. Agriculture and Horticulture Development Board.

AHDB (2020). *Nutrient Management Guide (RB209) Section 2 Organic Materials*. Agriculture and Horticulture Development Board. https://ahdb.org.uk/knowledge-library/rb209-section-2-organic-materials

Alzueta, I., Abeledo, L.G., Mignone, C.M. et al. (2012). Differences between wheat and barley in leaf and tillering coordination under contrasting nitrogen and sulfur conditions. *European Journal of Agronomy* 41: 92–102. https://doi.org/10.1016/j.eja.2012.04.002.

Aqueel, M.A. and Leather, S.R. (2011). Effect of nitrogen fertilizer on the growth and survival of Rhopalosiphum padi (L.) and Sitobion avenae (F.) (Homoptera: Aphididae) on different wheat cultivars. *Crop Protection* 30 (2): 216–221. https://doi.org/10.1016/j.cropro.2010.09.013.

Archer, J. (1985). *Crop Nutrition and Fertiliser Use*. Farming Press.

Audsley, E., Milne, A., and Paveley, N. (2005). A foliar disease model for use in wheat disease management decision support systems. *Annals of Applied Biology* 147 (2): 161–172. https://doi.org/10.1111/j.1744-7348.2005.00023.x.

Aula, L., Omara, P., Nambi, E. et al. (2020). Review of active optical sensors for improving winter wheat nitrogen use efficiency. *Agronomy* 10 (8): 1157. https://doi.org/10.3390/agronomy10081157.

Aulakh, M.S. and Rennie, D.A. (1986). Nitrogen transformations with special reference to gaseous N losses from zero-tilled soils of Saskatchewan, Canada. *Soil and Tillage Research* 7 (1–2): 157–171. https://doi.org/10.1016/0167-1987(86)90015-2.

Bai, G.H., Shaner, G., and Ohm, H. (2000). Inheritance of resistance to *Fusarium graminearum* in wheat. *Theoretical and Applied Genetics* 100 (1): 1–8. https://doi.org/10.1007/PL00002902.

Baker, D.A., Young, D.L., Huggins, D.R. et al. (2004). Economically optimal nitrogen fertilization for yield and protein in hard red spring wheat. *Agronomy Journal* 96 (1): 116–123. https://doi.org/10.2134/agronj2004.1160.

Ball, B.C., Watson, C.A., and Crichton, I. (2007). Nitrous oxide emissions, cereal growth, N recovery and soil nitrogen status after ploughing organically managed grass/clover swards. *Soil Use and Management* 23 (2): 145–155. https://doi.org/10.1111/j.1475-2743.2006.00072.x.

Bancal, M.O., Robert, C., and Ney, B. (2007). Modelling wheat growth and yield losses from late epidemics of foliar diseases using loss of green leaf area per layer and pre-anthesis reserves. *Annals of Botany* 100 (4): 777–789. https://doi.org/10.1093/aob/mcm163.

Barber, H.M., Carney, J., Alghabari, F. et al. (2015). Decimal growth stages for precision wheat production in changing environments? *Annals of Applied Biology* 166 (3): 355–371. https://doi.org/10.1111/aab.12207.

Barraclough, P.B., Weir, A.H., and Kuhlmann, H. (1991). Factors affecting the growth and distribution of winter wheat roots under UK field conditions. *Developments in Agricultural and Managed Forest Ecology* 24: 410–417. https://doi.org/10.1016/B978-0-444-89104-4.50057-8.

Batten, G.D. (1992). A review of phosphorus efficiency in wheat. *Plant and Soil* 146 (1–2): 163–168. https://doi.org/10.1007/BF00012009.

Bearchell, S.J., Fraaije, B.A., Shaw, M.W. et al. (2005). Wheat archive links long-term fungal pathogen population dynamics to air pollution. *Proceedings of the National Academy of Sciences* 102 (15): 5438–5442. https://doi.org/10.1073/pnas.0501596102.

Beddow, J.M., Pardey, P.G., Chai, Y. et al. (2015). Research investment implications of shifts in the global geography of wheat stripe rust. *Nature Plants* 1 (10): 15132. https://doi.org/10.1038/nplants.2015.132.

Bengtsson, G., Bengtson, P., and Månsson, K.F. (2003). Gross nitrogen mineralization-, immobilization-, and nitrification rates as a function of soil C/N ratio and microbial activity. *Soil Biology and Biochemistry* 35 (1): 143–154. https://doi.org/10.1016/S0038-0717(02)00248-1.

Berry, P.M. and Spink, J. (2012). Predicting yield losses caused by lodging in wheat. *Field Crops Research* 137: 19–26. https://doi.org/10.1016/0378-4290(87)90038-4.

Berry, P.M., Griffin, J.M., Sylvester-Bradley, R. et al. (2000). Controlling plant form through husbandry to minimise lodging in wheat. *Field Crops Research* 67 (1): 59–81. https://doi.org/10.1016/S0378-4290(00)00084-8.

Bhattacharyya, R., Chandra, S., Singh, R.D. et al. (2007). Long-term farmyard manure application effects on properties of a silty clay loam soil under irrigated wheat–soybean rotation. *Soil and Tillage Research* 94 (2): 386–396. https://doi.org/10.1016/j.still.2006.08.014.

Bhogal, A., Young, S.D., and Sylvester-Bradley, R. (2007). Straw incorporation and immobilization of spring-applied nitrogen. *Soil Use and Management* 13 (3): 111–116. https://doi.org/10.1111/j.1475-2743.1997.tb00568.x.

Black, A.S., Sherlock, R.R. et al. (1989). Ammonia volatilisation from urea broadcast in spring on to autumn-sown wheat. *New Zealand Journal of Crop and Horticultural Science* 17 (2): 175–182. https://doi.org/10.1080/01140671.1989.10428028.

Blake-Kalff, M.M.A., Hawkesford, M.J., Zhao, F.J. et al. (2000). Diagnosing sulfur deficiency in field-grown oilseed rape (Brassica napus L.) and wheat (Triticum aestivum L.). *Plant and Soil* 225 (1–2): 95–107. https://doi.org/10.1023/A:1026503812267.

Blake-Kalff, M.M., Zhao, F.J., Hawkesford, M.J. et al. (2001). Using plant analysis to predict yield losses caused by sulphur deficiency. *Annals of Applied Biology* 138 (1): 123–127. https://doi.org/10.1111/j.1744-7348.2001.tb00093.x.

Bleasdale, J.K.A. (1984). *Plant Physiology in Relation to Horticulture*, 2e. MacMillan.

Bøckman, O.C. and Olfs, H.W. (1998). Fertilizers, agronomy and N_2O. *Nutrient Cycling in Agroecosystems* 52 (2–3): 165–170. https://doi.org/10.1023/A:1009736327495.

Boeykens, S.P., Piol, M.N., Legal, L.S. et al. (2017). Eutrophication decrease: phosphate adsorption processes in presence of nitrates. *Journal of Environmental Management* 203: 888–895. https://doi.org/10.1016/j.jenvman.2017.05.026.

Borojevic, K. and Borojevic, K. (2005). The transfer and history of "reduced height genes" (Rht) in wheat from Japan to Europe. *Journal of Heredity* 96 (4): 455–459. https://doi.org/10.1093/jhered/esi060.

Bruulsema, T., Lemunyon, J., and Herz, B. (2009). Know your fertilizer rights. *Crops and Soils* 42 (2): 13–18.

Bryson, R.J., Paveley, N.D., Clark, W.S. et al. (1997). Use of in-field measurements of green leaf area and incident radiation to estimate the effects of yellow rust epidemics on the yield of winter wheat. In: *Perspectives for Agronomy: Adopting Ecological Principles and Managing Resource Use* (eds. M.K. van Ittersum and S.C. van de Geij), 77–86. Elsevier https://doi.org/10.1016/S0378-519X(97)80010-4.

Bundy, L.G. and Andraski, T.W. (2004). Diagnostic tests for site-specific nitrogen recommendations for winter wheat. *Agronomy Journal* 96 (3): 608–614. https://doi.org/10.2134/agronj2004.0608.

Bunting, A.H. and Drennan, D.S.H. (1966). Some aspects of the morphology and physiology of cereals in the vegetative phase. In: *The Growth of Cereals and Grasses* (eds. F.L. Milthorpe and J.D. Ivins), 20–38. Butterworths.

Byers, M. and Bolton, J. (1979). Effects of nitrogen and sulphur fertilisers on the yield, N and S content, and amino acid composition of the grain of spring wheat. *Journal of the Science of Food and Agriculture* 30 (3): 251–263. https://doi.org/10.1002/jsfa.2740300308.

Cao, Q., Miao, Y., Li, F. et al. (2017). Developing a new crop circle active canopy sensor-based precision nitrogen management strategy for winter wheat in North China plain. *Precision Agriculture* 18 (1): 2–18. https://doi.org/10.1007/s11119-016-9456-7.

Carmona, M., Sautua, F., Pérez-Hérnandez, O. et al. (2020). Role of fungicide applications on the integrated management of wheat stripe rust. *Frontiers in Plant Science 11*: 733. https://doi.org/10.3389/fpls.2020.00733.

Carnelo, L.G.L., de Miguez, S.R., and Marbán, L. (1997). Heavy metals input with phosphate fertilizers used in Argentina. *Science of the Total Environment* 204 (3): 245–250. https://doi.org/10.1016/S0048-9697(97)00187-3.

Carretero, R., Bancal, M.O., and Miralles, D.J. (2011). Effect of leaf rust (*Puccinia triticina*) on photosynthesis and related processes of leaves in wheat crops grown at two contrasting sites and with different nitrogen levels. *European Journal of Agronomy* 35 (4): 237–246. https://doi.org/10.1016/j.eja.2011.06.007.

Casebow, R., Hadley, C., Uppal, R. et al. (2016). Reduced height (Rht) alleles affect wheat grain quality. *PLoS One* 11 (5): e0156056. https://doi.org/10.1371/journal.pone.0156056.

Caubel, J., Launay, M., Ripoche, D. et al. (2017). Climate change effects on leaf rust of wheat: implementing a coupled crop-disease model in a French regional application. *European Journal of Agronomy* 90: 53–66. https://doi.org/10.1016/j.eja.2017.07.004.

Ceylan, Y., Kutman, U.B., Mengutay, M. et al. (2016). Magnesium applications to growth medium and foliage affect the starch distribution, increase the grain size and improve the seed germination in wheat. *Plant and Soil* 406 (1–2): 145–156. https://doi.org/10.1007/s11104-016-2871-8.

Chen, X. (2020). Pathogens which threaten food security: *Puccinia striiformis*, the wheat stripe rust pathogen. *Food Security* 12: 239–251. https://doi.org/10.1007/s12571-020-01016-z.

Chen, W., Wellings, C., Chen, X. et al. (2014). Wheat stripe (yellow) rust caused by *Puccinia striiformis* f. sp. *tritici*. *Molecular Plant Pathology* 15 (5): 433–446. https://doi.org/10.1111/mpp.12116.

Chen, Y., Zhang, P., Wang, L. et al. (2020a). Interaction of nitrogen and phosphorus on wheat yield, n use efficiency and soil nitrate nitrogen distribution in the North China plain. *International Journal of Plant Production* 14: 415–426. https://doi.org/10.1007/s42106-020-00093-6.

Chen, X.X., Liu, Y.M., Zhao, Q.Y. et al. (2020b). Health risk assessment associated with heavy metal accumulation in wheat after long-term phosphorus fertilizer application. *Environmental Pollution* 262: 114348. https://doi.org/10.1016/j.envpol.2020.114348.

Cheval, P., Siah, A., Bomble, M. et al. (2017). Evolution of QoI resistance of the wheat pathogen Zymoseptoria tritici in Northern France. *Crop Protection* 92: 131–133. https://doi.org/10.1016/j.cropro.2016.10.017.

Ciereszko, I., Szczygła, A., and Żebrowska, E. (2011). Phosphate deficiency affects acid phosphatase activity and growth of two wheat varieties. *Journal of Plant Nutrition* 34 (6): 815–829. https://doi.org/10.1080/01904167.2011.544351.

Colbourn, P. and Harper, I.W. (1987). Denitrification in drained and undrained arable clay soil. *Journal of Soil Science* 38 (3): 531–539. https://doi.org/10.1111/j.1365-2389.1987.tb02288.x.

Cook, R.J., Hims, M.J., and Vaughan, T.B. (1999). Effects of fungicide spray timing on winter wheat disease control. *Plant Pathology* 48 (1): 33–50. https://doi.org/10.1046/j.1365-3059.1999.00319.x.

Cools, H.J. and Fraaije, B.A. (2008). Are azole fungicides losing ground against Septoria wheat disease? Resistance mechanisms in Mycosphaerella graminicola. *Pest Management Science* 64 (7): 681–684. https://doi.org/10.1002/ps.1568.

Cools, H.J., Fraaije, B.A., and Lucas, J.A. (2005). Molecular examination of Septoria tritici isolates with reduced sensitivities to triazoles. In: *Modern Fungicides and Antifungal Compounds IV: 14th International Reinhardsbrunn Symposium* (eds. H.W. Dehne, U. Gisi, K.H. Kuck, et al.), 103–114. British Crop Protection Council.

Cordero, I., Snell, H., and Bardgett, R.D. (2019). High throughput method for measuring urease activity in soil. *Soil Biology and Biochemistry* 134: 72–77.

Cornish, P.S., Baker, G.R., and Murray, G.M. (1990). Physiological responses of wheat (Triticum aestivum) to infection with Mycosphaerella graminicola causing Septoria tritici blotch. *Australian Journal of Agricultural Research* 41 (2): 317–327. https://doi.org/10.1071/AR9900317.

Cosser, N.D., Thompson, A.J., Gooding, M.J. et al. (1994). Influences of variety and establishment date on nitrogen accumulation during winter in the shoot system of winter wheat. *Aspects of Applied Biology* 39: 195–195.

Cosser, N.D., Gooding, M.J., Thompson, A.J. et al. (1997a). Competitive ability and tolerance of organically grown wheat cultivars to natural weed infestations. *Annals of Applied Biology* 130 (3): 523–535. https://doi.org/10.1111/j.1744-7348.1997.tb07679.x.

Cosser, N.D., Gooding, M.J., Davies, W.P. et al. (1997b). Cultivar and Rht gene influences on the competitive ability, yield and the breadmaking quality of organically grown winter wheat. In: *Optimising Cereal Inputs Part 1: Genetics and Nutrition*, Aspects of Applied Biology, vol. 50 (eds. M.J. Gooding and P.R. Shewry), 39–51. Association of Applied Biologists.

Critchley, C.S. (2001). *A physiological explanation for the canopy nitrogen requirement of winter wheat* (Doctoral dissertation, University of Nottingham). https://core.ac.uk/download/pdf/33565781.pdf

Daraja, Y.B. (2009). *The effects of intercropping, nitrogen and sulphur nutrition on the performance of wheat (Triticum aestivum) and faba bean (Vicia faba) in different production systems*. PhD thesis. The University of Reading.

DEFRA (2018). *Nitrate vulnerable zones*. Department for Environment, Food and Rural Affairs. https://www.gov.uk/government/collections/nitrate-vulnerable-zones

Dennett, M.D., Satorre, E.H., and Slafer, G.A. (1999). Effects of sowing date and the determination of optimum sowing date. In: *Wheat, Ecology and Physiology of Yield Determination* (eds. E.H. Sattore and G.S. Slafer), 123–140. CRC Press.

Dewar, A.M. and Carter, N. (1984). Decision trees to assess the risk of cereal aphid (Hemiptera: Aphididae) outbreaks in summer in England. *Bulletin of Entomological Research* 74 (3): 387–398. https://doi.org/10.1017/S0007485300015650.

Diacono, M., Rubino, P., and Montemurro, F. (2013). Precision nitrogen management of wheat. A review. *Agronomy for Sustainable Development* 33 (1): 219–241. https://doi.org/10.1007/s13593-012-0111-z.

Dimmock, J.P.R.E. and Gooding, M.J. (2002a). The effects of fungicides on rate and duration of grain filling in winter wheat in relation to maintenance of flag leaf green area. *Journal of Agricultural Science* 138 (1): 1–16. https://doi.org/10.1017/S0021859601001666.

Dimmock, J.P.R.E. and Gooding, M.J. (2002b). The influence of foliar diseases, and their control by fungicides, on the protein concentration in wheat grain: a review. *Journal of Agricultural Science* 138 (4): 349–366. https://doi.org/10.1017/S0021859602002058.

Dimmock, J.P.R.E. and Gooding, M.J. (2002c). The effects of fungicides on Hagberg falling number and blackpoint in winter wheat. *Crop Protection* 21 (6): 475–487. https://doi.org/10.1016/S0261-2194(01)00135-1.

Distelfeld, A., Avni, R., and Fischer, A.M. (2014). Senescence, nutrient remobilization, and yield in wheat and barley. *Journal of Experimental Botany* 65 (14): 3783–3798. https://doi.org/10.1093/jxb/ert477.

do Carmo Horta, M. and Torrent, J. (2007). The Olsen P method as an agronomic and environmental test for predicting phosphate release from acid soils. *Nutrient Cycling in Agroecosystems* 77 (3): 283–292. https://doi.org/10.1007/s10705-006-9066-2.

Donald, C.T. (1968). The breeding of crop ideotypes. *Euphytica* 17 (3): 385–403. https://doi.org/10.1007/BF00056241.

Donald, P.F. and Vickery, J.A. (2000). The importance of cereal fields to breeding and wintering Skylarks *Alauda arvensis* in the UK. In: *Ecology and Conservation of Lowland Farmland Birds* (eds. N.J. Aebischer, A.D. Evans, P.V. Grice and J.A. Vickery), 140–150. British Ornithologists Union.

Donald, P.F., Green, R.E., and Heath, M.F. (2001). Agricultural intensification and the collapse of Europe's farmland bird populations. *Proceedings of the Royal Society of London. Series B: Biological Sciences* 268 (1462): 25–29. https://doi.org/10.1098/rspb.2000.1325.

Dorioz, J.M., Wang, D., Poulenard, J. et al. (2006). The effect of grass buffer strips on phosphorus dynamics—a critical review and synthesis as a basis for application in agricultural landscapes in France. *Agriculture, Ecosystems & Environment* 117 (1): 4–21. https://doi.org/10.1016/j.agee.2006.03.029.

Duncan, E.G., O'Sullivan, C.A., Roper, M.M. et al. (2018). Influence of co-application of nitrogen with phosphorus, potassium, and sulphur on the apparent efficiency of nitrogen fertiliser use, grain yield and protein content of wheat. *Field Crops Research* 226: 56–65. https://doi.org/10.1016/j.fcr.2018.07.010.

Efretuei, A., Gooding, M., White, E. et al. (2016). Effect of nitrogen fertilizer application timing on nitrogen use efficiency and grain yield of winter wheat in Ireland. *Irish Journal of Agricultural and Food Research* 55 (1): 63–73. http://hdl.handle.net/11019/1143.

EFSA (2017). Peer review of the pesticide risk assessment of the active substance propiconazole. *EFSA Journal* 15 (7): 04887. https://doi.org/10.2903/j.efsa.2017.4887.

EFSA (2018). Peer review of the pesticide risk assessment of the active substance chlorothalonil. *EFSA Journal* 16 (1): e05126. https://doi.org/10.2903/j.efsa.2018.5126.

Ellis, R.H. (1992). Seed and seedling vigour in relation to crop growth and yield. *Plant Growth Regulation* 11 (3): 249–255. https://doi.org/10.1007/BF00024563.

Ellis, R.H., Salahi, M., and Jones, S.A. (1999). Yield-density equations can be extended to quantify the effect of applied nitrogen and cultivar on wheat grain yield. *Annals of Applied Biology* 134 (3): 347–352. https://doi.org/10.1111/j.1744-7348.1999.tb05275.x.

EU (2008). *Consolidated text: Council Directive of 12 December 1991 concerning the protection of waters against pollution caused by nitrates from agricultural sources (91/676/EEC)*. European Union. http://data.europa.eu/eli/dir/1991/676/2008-12-11

EU (2009). *Regulation (EC) No 1107/2009 of the European Parliament and of the Council of 21 October 2009 concerning the placing of plant protection products on the market and repealing Council Directives 79/117/EEC and 91/414/EEC. Official Journal of the European Union*, L 309/1. https://eur-lex.europa.eu/legal-content/EN/TXT/?uri=CELEX:32009R1107

EU (2015). *Consolidated text: Council Directive 98/83/EC of 3 November 1998 on the quality of water intended for human consumption*. European Union. http://data.europa.eu/eli/dir/1998/83/2015-10-27

Evans, J.R. (1989). Photosynthesis and nitrogen relationships in leaves of C 3 plants. *Oecologia* 78 (1): 9–19. https://www.jstor.org/stable/4218825.

Eyal, Z. (1999). The Septoria tritici and Stagonospora nodorum blotch diseases of wheat. *European Journal of Plant Pathology* 105 (7): 629–641. https://doi.org/10.1023/A:1008716812259.

Fageria, N.K. and Baligar, V.C. (2005). Enhancing nitrogen use efficiency in crop plants. *Advances in Agronomy* 88: 97–185. https://doi.org/10.1016/S0065-2113(05)88004-6.

Fairchild, D.S. (1988). Soil information system for farming by kind of soil. In: *Proceedings, International interactive workshop on soil resources: their inventory, analysis and interpretations for use in the 1990s*, 159–164. University of Minnesota.

Faris, J.D., Liu, Z., and Xu, S.S. (2013). Genetics of tan spot resistance in wheat. *Theoretical and Applied Genetics* 126 (9): 2197–2217. https://doi.org/10.1007/s00122-013-2157-y.

Farmer, A.M. (2018). Phosphate pollution: a global overview of the problem. In: *Phosphorus: Polluter and Resource of the Future* (ed. C. Schaum), 35–55. IWA Publishing.

Fiez, T.E., Miller, B.C., and Pan, W.L. (1994). Assessment of spatially variable nitrogen fertilizer management in winter wheat. *Journal of Production Agriculture* 7 (1): 86–93. https://doi.org/10.2134/jpa1994.0086.

Figueroa, M., Hammond-Kosack, K.E., and Solomon, P.S. (2018). A review of wheat diseases—a field perspective. *Molecular Plant Pathology* 19 (6): 1523–1536. https://doi.org/10.1111/mpp.12618.

Fischer, R.A. (2008). The importance of grain or kernel number in wheat: a reply to Sinclair and Jamieson. *Field Crops Research* 105 (1–2): 15–21. https://doi.org/10.1016/j.fcr.2007.04.002.

Flavel, R.J., Guppy, C.N., Tighe, M.K. et al. (2014). Quantifying the response of wheat (Triticum aestivum L) root system architecture to phosphorus in an Oxisol. *Plant and Soil* 385 (1–2): 303–310. https://doi.org/10.1007/s11104-014-2191-9.

Fones, H. and Gurr, S. (2015). The impact of Septoria tritici Blotch disease on wheat: an EU perspective. *Fungal Genetics and Biology* 79: 3–7. https://doi.org/10.1016/j.fgb.2015.04.004.

Foulkes, M.J., Sylvester-Bradley, R., and Scott, R.K. (1998). Evidence for differences between winter wheat cultivars in acquisition of soil mineral nitrogen and uptake and utilization of applied fertilizer nitrogen. *Journal of Agricultural Science* 130 (1): 29–44. https://doi.org/10.1017/S0021859697005029.

Fowler, D., Coyle, M., Skiba, U. et al. (2013). The global nitrogen cycle in the twenty-first century. *Philosophical Transactions of the Royal Society B: Biological Sciences 368* (1621): 20130164. https://doi.org/10.1098/rstb.2013.0164.

Fox, R.H., Piekielek, W.P., and Macneal, K.M. (1994). Using a chlorophyll meter to predict nitrogen fertilizer needs of winter wheat. *Communications in Soil Science and Plant Analysis* 25 (3–4): 171–181. https://doi.org/10.1080/00103629409369027.

FRAC (2020). *FRAC Classification of Fungicides*. Fungicide Resistance Action Committee. https://www.frac.info/docs/default-source/publications/frac-mode-of-action-poster/frac-moa-poster-2020v2.pdf?sfvrsn=a48499a_2

Galloway, J.N., Townsend, A.R., Erisman, J.W. et al. (2008). Transformation of the nitrogen cycle: recent trends, questions, and potential solutions. *Science* 320 (5878): 889–892. https://doi.org/10.1126/science.1136674.

Garnett, T. (1883). *Essays in Natural History and Agriculture*. Chiswick Press.

Garthwaite, D., Ridley, L., Mace, A. et al. (2019). *Pesticide Usage Survey Report 284 Arable Crops in the United Kingdom 2018*. Fera. https://secure.fera.defra.gov.uk/pusstats/surveys/documents/arable2018.pdf

Geissler, B., Hermann, L., Mew, M.C. et al. (2018). Striving toward a circular economy for phosphorus: the role of phosphate rock mining. *Minerals* 8 (9): 395. https://doi.org/10.3390/min8090395.

Ghimire, R., Machado, S., and Bista, P. (2018). Decline in soil organic carbon and nitrogen limits yield in wheat-fallow systems. *Plant and Soil* 422: 423–435. https://doi.org/10.1007/s11104-017-3470-z.

Gibbs, P.A., Parkinson, R.J., Misselbrook, T.H. et al. (2002). Environmental impacts of cattle manure composting. In: *Microbiology of Composting* (eds. H. Insam, N. Riddech and S. Klammer), 445–456. Springer https://doi.org/10.1007/978-3-662-08724-4_37.

Gladders, P., Paveley, N.D., Barrie, I.A. et al. (2005). Agronomic and meteorological factors affecting the severity of leaf blotch caused by *Mycosphaerella graminicola* in commercial wheat crops in England. *Annals of Applied Biology* 138 (3): 301–311. https://doi.org/10.1111/j.1744-7348.2001.tb00115.x.

Glynne, M.D. (1951). Effects of cultural treatments on wheat and on the incidence of eyespot, lodging, take-all and weeds. Field experiments 1945–8. *Annals of Applied Biology* 38 (3): 665–688. https://doi.org/10.1111/j.1744-7348.1951.tb07837.x.

Gooding, M.J. and Davies, W.P. (1992). Foliar urea fertilization of cereals: a review. *Fertilizer Research* 32 (2): 209–222. https://doi.org/10.1007/BF01048783.

Gooding, M.J. and Davies, W.P. (1997). *Wheat Production and Utilization: Systems, Quality and the Environment*. CAB international.

Gooding, M.J., Kettlewell, P.S., Davies, W.P. et al. (1986). Effects of spring nitrogen fertilizer on the Hagberg falling number of grain from breadmaking varieties of winter wheat. *Journal of Agricultural Science* 107 (2): 475–477. https://doi.org/10.1017/S002185960008730X.

Gooding, M.J., Salahi, M., and Ellis, R.H. (1999). Spatial variation in the response of wheat grain yield and quality to late nitrogen. In: *Precision Agriculture '99* (ed. J.V. Stafford), 531–538. Sheffield Academic Press.

Gooding, M.J., Dimmock, J.P.R.E., France, J. et al. (2000). Green leaf area decline of wheat flag leaves: the influence of fungicides and relationships with mean grain weight and grain yield. *Annals of Applied Biology* 136 (1): 77–84. https://doi.org/10.1111/j.1744-7348.2000.tb00011.x.

Gooding, M.J., Pinyosinwat, A., and Ellis, R.H. (2001). The response of wheat grain nitrogen concentration to plant population density. In: *Aspects of Applied Biology, 64, Wheat Quality* (eds. M.J. Gooding, S.A. Barton and G.P. Smith), 157–162. Association of Applied Biologists.

Gooding, M.J., Pinyosinwat, A., and Ellis, R.H. (2002). Responses of wheat grain yield and quality to seed rate. *Journal of Agricultural Science* 138: 317–331. https://doi.org/10.1017/S0021859602002137.

Gooding, M.J., Gregory, P.J., Ford, K.E. et al. (2005). Fungicide and cultivar affect post-anthesis patterns of nitrogen uptake, remobilization and utilization efficiency in wheat. *Journal of Agricultural Science* 143: 503–518. https://doi.org/10.1017/S002185960500568X.

Gooding, M.J., Gregory, P.J., Ford, K.E. et al. (2007). Recovery of nitrogen from different sources following applications to winter wheat at and after anthesis. *Field Crops Research* 100 (2–3): 143–154. https://doi.org/10.1016/j.fcr.2006.06.002.

Gooding, M.J., Addisu, M., Uppal, R.K. et al. (2012a). Effect of wheat dwarfing genes on nitrogen-use efficiency. *Journal of Agricultural Science* 150 (1): 3–22. https://doi.org/10.1017/S0021859611000414.

Gooding, M.J., Uppal, R.K., Addisu, M. et al. (2012b). Reduced height alleles (Rht) and Hagberg falling number of wheat. *Journal of Cereal Science* 55 (3): 305–311. https://doi.org/10.1016/j.jcs.2012.01.003.

Gooding, M.J., Fan, M., McGrath, S.P. et al. (2012c). Contrasting effects of dwarfing alleles and nitrogen availability on mineral concentrations in wheat grain. *Plant and Soil* 360 (1–2): 93–107. https://doi.org/10.1007/s11104-012-1203-x.

Goulding, K.W.T., Webster, C.P., Powlson, D.S. et al. (1993). Denitrification losses of nitrogen fertilizer applied to winter wheat following ley and arable rotations as estimated by acetylene inhibition and 15N balance. *Journal of Soil Science* 44 (1): 63–72. https://doi.org/10.1111/j.1365-2389.1993.tb00434.x.

Goulding, K.W.T., Bailey, N.J., and Bradbury, N.J. (1998). A modelling study of nitrogen deposited to arable land from the atmosphere and its contribution to nitrate leaching. *Soil Use and Management* 14 (2): 70–77. https://doi.org/10.1111/j.1475-2743.1998.tb00617.x.

Goulding, K.W.T., Poulton, P.R., Webster, C.P. et al. (2000). Nitrate leaching from the Broadbalk Wheat Experiment, Rothamsted, UK, as influenced by fertilizer and manure inputs and the weather. *Soil Use and Management* 16 (4): 244–250. https://doi.org/10.1111/j.1475-2743.2000.tb00203.x.

Gregory, P.J., Marshall, B., and Biscoe, P.V. (1981). Nutrient relations of winter wheat: 3. Nitrogen uptake, photosynthesis of flag leaves and translocation of nitrogen to grain. *Journal of Agricultural Science* 96 (3): 539–547. https://doi.org/10.1017/S0021859600034493.

Győri, Z. (2005). Sulphur content of winter wheat grain in long term field experiments. *Communications in Soil Science and Plant Analysis* 36 (1–3): 373–382. https://doi.org/10.1081/CSS-200043098.

Haneklaus, S., Murphy, D.P.L., Nowak, G. et al. (1995). Effects of the timing of sulphur application on grain yield and yield components of wheat. *Zeitschrift für Pflanzenernährung und Bodenkunde* 158 (1): 83–85. https://doi.org/10.1002/jpln.19951580117.

Haneklaus, S., Bloem, E., Schnug, E. et al. (2016). Sulfur. In: *Handbook of Plant Nutrition* (eds. A.V. Barker and D.J. Pilbeam), 183–238. CRC Press.

Hansen, S., Berland Frøseth, R., Stenberg, M. et al. (2019). Review of causes and sources of N_2O emissions and NO_3 leaching from organic arable crop rotations. *Biogeosciences* 16 (14): 2795–2819. https://doi.org/10.5194/bg-16-2795-2019.

Harper, L.A., Sharpe, R.R., Langdale, G.W. et al. (1987). Nitrogen cycling in a wheat crop: soil, plant, and aerial nitrogen transport. *Agronomy Journal* 79 (6): 965–973. https://doi.org/10.2134/agronj1987.00021962007900060004x.

Hawkesford, M., Horst, W., Kichey, T. et al. (2012). Functions of macronutrients. In: *Marschner's Mineral Nutrition of Higher Plants*, 3e (ed. P. Marschner), 135–189. Academic Press https://doi.org/10.1016/B978-0-12-384905-2.00006-6.

Herold, A. and Walker, D.A. (1979). Transport across chloroplast envelopes. The role of phosphate. In: *Membrane Transport in Biology II. Transport across Single Biological Membranes* (ed. D.C. Tosteson), 411–439. Springer-Verlag.

Herridge, D.F., Peoples, M.B., and Boddey, R.M. (2008). Global inputs of biological nitrogen fixation in agricultural systems. *Plant and Soil* 311: 1–18. https://doi.org/10.1007/s11104-008-9668-3.

Herzog, F., Prasuhn, V., Spiess, E. et al. (2008). Environmental cross-compliance mitigates nitrogen and phosphorus pollution from Swiss agriculture. *Environmental Science & Policy* 11 (7): 655–668. https://doi.org/10.1016/j.envsci.2008.06.003.

Hirano, K., Ueguchi-Tanaka, M., and Matsuoka, M. (2008). GID1-mediated gibberellin signaling in plants. *Trends in Plant Science* 13: 192–199. https://doi.org/10.1016/j.tplants.2008.02.005.

Hoel, B.O. (2011). Effects of sulphur application on grain yield and quality, and assessment of sulphur status in winter wheat (Triticum aestivum L.). *Acta Agriculturae Scandinavica, Section B-Soil & Plant Science* 61 (6): 499–507. https://doi.org/10.1080/09064710.2010.510120.

Holm-Nielsen, J.B., Al Seadi, T., and Oleskowicz-Popiel, P. (2009). The future of anaerobic digestion and biogas utilization. *Bioresource Technology* 100 (22): 5478–5484. https://doi.org/10.1016/j.biortech.2008.12.046.

Hooda, P.S., Edwards, A.C., Anderson, H.A. et al. (2000). A review of water quality concerns in livestock farming areas. *Science of the Total Environment* 250 (1–3): 143–167. https://doi.org/10.1016/S0048-9697(00)00373-9.

Hoogmoed, M., Neuhaus, A., Noack, S. et al. (2018). Benchmarking wheat yield against crop nitrogen status. *Field Crops Research* 222: 153–163. https://doi.org/10.1016/j.fcr.2018.03.013.

Hovmøller, M.S., Yahyaoui, A.H., Milus, E.A. et al. (2008). Rapid global spread of two aggressive strains of a wheat rust fungus. *Molecular Ecology* 17 (17): 3818–3826. https://doi.org/10.1111/j.1365-294X.2008.03886.x.

Huerta-Espino, J., Singh, R.P., German, S. et al. (2011). Global status of wheat leaf rust caused by *Puccinia triticina*. *Euphytica* 179 (1): 143–160. https://doi.org/10.1007/s10681-011-0361-x.

Hunter, T., Coker, R.R., and Royle, D.J. (1999). The teleomorph stage, *Mycosphaerella graminicola*, in epidemics of septoria tritici blotch on winter wheat in the UK. *Plant Pathology* 48 (1): 51–57. https://doi.org/10.1046/j.1365-3059.1999.00310.x.

Hutchings, N., Webb, J., and Amon, B. (2019). *Crop production and agricultural soils*. In: *EMEP/EEA Emission Inventory Guidebook*. European Environment Agency https://www.eea.europa.eu/publications/emep-eea-guidebook-2019/part-b-sectoral-guidance-chapters/4-agriculture/3-d-crop-production-and/view.

Iho, A. and Laukkanen, M. (2012). Precision phosphorus management and agricultural phosphorus loading. *Ecological Economics* 77: 91–102. https://doi.org/10.1016/j.ecolecon.2012.02.010.

Jensen, L.S., Schjørring, J.K., and van Der Hoek, K.W. (2011). Benefits of nitrogen for food, fibre and industrial production. In: *European Nitrogen Assessment: Sources, Effects and Policy Perspectives* (eds. M.A. Sutton, C.M. Howard, J.W. Erisman, et al.), 32–61. Cambridge University Press.

Jensen, E.S., Carlsson, G., and Hauggaard-Nielsen, H. (2020). Intercropping of grain legumes and cereals improves the use of soil N resources and reduces the requirement for synthetic fertilizer N: a global-scale analysis. *Agronomy for Sustainable Development* 40 (1): 5. https://doi.org/10.1007/s13593-020-0607-x.

Johnston, A.E., Poulton, P.R., and Coleman, K. (2009). Soil organic matter: its importance in sustainable agriculture and carbon dioxide fluxes. *Advances in Agronomy* 101: 1–57. https://doi.org/10.1016/S0065-2113(08)00801-8.

Jordan, V.W.L., Hutcheon, J.A., Donaldson, G.V. et al. (1997). Research into and development of integrated farming systems for less-intensive arable crop production: experimental progress (1989–1994) and commercial implementation. *Agriculture, Ecosystems & Environment* 64 (2): 141–148. https://doi.org/10.1016/S0167-8809(97)00032-7.

Kettlewell, P.S. (1997). Seasonal variation in the response of Hagberg falling and liquefaction numbers to propiconazole fungicide in wheat. *Annals of Applied Biology* 130 (3): 569–580. https://doi.org/10.1111/j.1744-7348.1997.tb07683.x.

Kettlewell, P.S. (1999). The response of alpha-amylase activity during wheat grain development to nitrogen fertiliser. *Annals of Applied Biology* 134 (2): 241–249. https://doi.org/10.1111/j.1744-7348.1999.tb05260.x.

Khah, E.M., Roberts, E.H., and Ellis, R.H. (1989). Effects of seed ageing on growth and yield of spring wheat at different plant-population densities. *Field Crops Research* 20 (3): 175–190. https://doi.org/10.1016/0378-4290(89)90078-6.

Kindred, D.R. and Gooding, M.J. (2004). Heterotic and seed rate effects on nitrogen efficiencies in wheat. *Journal of Agricultural Science* 142: 639–657. https://doi.org/10.1017/S0021859605004843.

Kindred, D.R. and Gooding, M.J. (2005). Heterosis for yield and its physiological determinants in wheat. *Euphytica* 142: 149–159. https://doi.org/10.1007/s10681-005-1250-y.

Kindred, D.R., Gooding, M.J., and Ellis, R.H. (2005). Nitrogen fertilizer and seed rate effects on Hagberg falling number of hybrid wheats and their parents are associated with α-amylase activity, grain cavity size and dormancy. *Journal of the Science of Food and Agriculture* 85 (5): 727–742. https://doi.org/10.1002/jsfa.2025.

Kindred, D.R., Milne, A.E., Webster, R. et al. (2015). Exploring the spatial variation in the fertilizer-nitrogen requirement of wheat within fields. *Journal of Agricultural Science* 153 (1): 25–41. https://doi.org/10.1017/S0021859613000919.

Kindred, D.R., Sylvester-Bradley, R., Milne, A.E. et al. (2017). Spatial variation in nitrogen requirements of cereals, and their interpretation. *Advances in Animal Biosciences* 8 (2): 303–307. https://doi.org/10.1017/S2040470017001327.

Knapp, S., Döring, T.F., Jones, H.E. et al. (2020). Natural selection towards wild-type in composite cross populations of winter wheat. *Frontiers in Plant Science 10*: 1757. https://doi.org/10.3389/fpls.2019.01757.

Kollers, S., Rodemann, B., Ling, J. et al. (2013). Genetic architecture of resistance to Septoria tritici blotch (Mycosphaerella graminicola) in European winter wheat. *Molecular Breeding* 32 (2): 411–423. https://doi.org/10.1007/s11032-013-9880-6.

Korres, N.E. and Froud-Williams, R.J. (2002). Effects of winter wheat cultivars and seed rate on the biological characteristics of naturally occurring weed flora. *Weed Research* 42 (6): 417–428. https://doi.org/10.1046/j.1365-3180.2002.00302.x.

Kowalski, A.M., Gooding, M., Ferrante, A. et al. (2016). Agronomic assessment of the wheat semi-dwarfing gene Rht8 in contrasting nitrogen treatments and water regimes. *Field Crops Research* 191: 150–160. https://doi.org/10.1016/j.fcr.2016.02.026.

Kpomblekou, A.K. and Tabatabai, M.A. (1994). Metal contents of phosphate rocks. *Communications in Soil Science and Plant Analysis* 25 (17–18): 2871–2882. https://doi.org/10.1080/00103629409369231.

Kristensen, L., Olsen, J., and Weiner, J. (2008). Crop density, sowing pattern, and nitrogen fertilization effects on weed suppression and yield in spring wheat. *Weed Science* 56 (1): 97–102. https://doi.org/10.1614/WS-07-065.1.

Kuesters, J. and Lammel, J. (1999). Investigations of the energy efficiency of the production of winter wheat and sugar beet in Europe. *European Journal of Agronomy* 11 (1): 35–43. https://doi.org/10.1016/S1161-0301(99)00015-5.

Kumar, P.A., Parry, M.A., Mitchell, R.A. et al. (2002). Photosynthesis and nitrogen-use efficiency. In: *Photosynthetic Nitrogen Assimilation and Associated Carbon and Respiratory Metabolism* (eds. C.H. Foyer and G. Noctor), 23–34. Springer https://link.springer.com/chapter/10.1007/0-306-48138-3_2.

Ladha, J.K., Tirol-Padre, A., Reddy, C.K. et al. (2016). Global nitrogen budgets in cereals: a 50-year assessment for maize, rice and wheat production systems. *Scientific Reports* 6 (1): 1–9. https://doi.org/10.1038/srep19355.

Lark, R.M., Ander, E.L., and Broadley, M.R. (2019). Combining two national-scale datasets to map soil properties, the case of available magnesium in England and Wales. *European Journal of Soil Science* 70 (2): 361–377. https://doi.org/10.1111/ejss.12743.

Lawlor, D.W., Kontturi, M., and Young, A.T. (1989). Photosynthesis by flag leaves of wheat in relation to protein, ribulose bis phosphate carboxylase activity and nitrogen supply. *Journal of Experimental Botany* 40 (1): 43–52. https://doi.org/10.1093/jxb/40.1.43.

Lázaro, L., Abbate, P.E., Cogliatti, D.H. et al. (2010). Relationship between yield, growth and spike weight in wheat under phosphorus deficiency and shading. *Journal of Agricultural Science* 148 (1): 83–93. https://doi.org/10.1017/S0021859609990402.

Lemaire, G., van Oosterom, E., Sheehy, J. et al. (2007). Is crop N demand more closely related to dry matter accumulation or leaf area expansion during vegetative growth? *Field Crops Research* 100 (1): 91–106. https://doi.org/10.1016/j.fcr.2006.05.009.

Li, H., Smith, S.E., Holloway, R.E. et al. (2006). Arbuscular mycorrhizal fungi contribute to phosphorus uptake by wheat grown in a phosphorus-fixing soil even in the absence of positive growth responses. *New Phytologist* 172 (3): 536–543. https://doi.org/10.1111/j.1469-8137.2006.01846.x.

Li, S., Tian, Y., Wu, K. et al. (2018). Modulating plant growth–metabolism coordination for sustainable agriculture. *Nature* 560 (7720): 595–600. https://doi.org/10.1038/s41586-018-0415-5.

Lidwell-Durnin, J. and Lapthorn, A. (2020). The threat to global food security from wheat rust: ethical and historical issues in fighting crop diseases and preserving genetic diversity. *Global Food Security* 26: 100446. https://doi.org/10.1016/j.gfs.2020.100446.

Loddo, S. and Gooding, M. (2012). Semi-dwarfing (Rht-B1b) improves nitrogen-use efficiency in wheat, but not at economically optimal levels of nitrogen availability. *Cereal Research Communications* 40 (1): 116–121. https://doi.org/10.1556/crc.40.2012.1.13.

Loughman, R., Jayasena, K., and Majewski, J. (2005). Yield loss and fungicide control of stem rust of wheat. *Australian Journal of Agricultural Research* 56 (1): 91–96. https://doi.org/10.1071/AR04126.

Lovell, D.J., Parker, S.R., Hunter, T. et al. (1997). Influence of crop growth and structure on the risk of epidemics by Mycosphaerella graminicola (Septoria tritici) in winter wheat. *Plant Pathology* 46 (1): 126–138.

Lovell, D.J., Hunter, T., Powers, S.J. et al. (2004). Effect of temperature on latent period of septoria leaf blotch on winter wheat under outdoor conditions. *Plant Pathology* 53 (2): 170–181. https://doi.org/10.1111/j.0032-0862.2004.00983.x.

Lugato, E., Paniagua, L., Jones, A. et al. (2017). Complementing the topsoil information of the Land Use/Land Cover Area Frame Survey (LUCAS) with modelled N2O emissions. *PloS One* 12 (4): 0176111. https://doi.org/10.1371/journal.pone.0176111.

Lunn, G.D., Kettlewell, P.S., Major, B.J. et al. (2001). Effects of pericarp alpha-amylase activity on wheat (*Triticum aestivum*) Hagberg falling number. *Annals of Applied Biology* 138 (2): 207–214. https://doi.org/10.1111/j.1744-7348.2001.tb00104.x.

Lutman, P.J.W., Moss, S.R., Cook, S. et al. (2013). A review of the effects of crop agronomy on the management of *Alopecurus myosuroides*. *Weed Research* 53 (5): 299–313. https://doi.org/10.1111/wre.12024.

Lynch, J.P. (2011). Root phenes for enhanced soil exploration and phosphorus acquisition: tools for future crops. *Plant Physiology* 156 (3): 1041–1049. https://doi.org/10.1104/pp.111.175414.

MacDonald, G.K., Bennett, E.M., Potter, P.A. et al. (2011). Agronomic phosphorus imbalances across the world's croplands. *Proceedings of the National Academy of Sciences* 108 (7): 3086–3091. https://doi.org/10.1073/pnas.1010808108.

MAFF (1986). *The Analysis of Agricultural Materials, MAFF Reference Book 427*, 3e. H. M. Stationary Office.

Makino, A., Mae, T., and Ohira, K. (1988). Differences between wheat and rice in the enzymic properties of ribulose-1, 5-bisphosphate carboxylase/oxygenase and the relationship to photosynthetic gas exchange. *Planta* 174 (1): 30–38. https://doi.org/10.1007/BF00394870.

Mallarino, A.P. and Schepers, J.S. (2005). Role of precision agriculture in phosphorus management practices. In: *Phosphorus: Agriculture and the Environment* (eds. J.T. Sims, A.N. Sharpley and T.C. Daniel), 881–908. Wiley https://doi.org/10.2134/agronmonogr46.c27.

Marco, D. and De Marco, D.G. (1990). Effect of seed weight, and seed phosphorus and nitrogen concentrations on the early growth of wheat seedlings. *Australian Journal of Experimental Agriculture* 30 (4): 545–549. https://doi.org/10.1071/EA9900545.

Matar, A.E. and Brown, S.C. (1989). Effect of rate and method of phosphate placement on productivity of durum wheat in Mediterranean environments. *Fertilizer Research* 20 (2): 75–82. https://doi.org/10.1007/BF01055431.

Mateo-Marín, N., Isla, R., Guillén, M. et al. (2020). Agronomic and environmental implications of substituting pig slurry for synthetic nitrogen in Mediterranean wheat systems. *Agronomy* 10 (10): 1498. https://doi.org/10.3390/agronomy10101498.

Matson, P.A., Naylor, R., and Ortiz-Monasterio, I. (1998). Integration of environmental, agronomic, and economic aspects of fertilizer management. *Science* 280 (5360): 112–115. https://doi.org/10.1126/science.280.5360.112.

McLaughlin, M.J., Tiller, K.G., Naidu, R. et al. (1996). The behaviour and environmental impact of contaminants in fertilizers. *Soil Research* 34 (1): 1–54. https://doi.org/10.1071/SR9960001.

Mehmood, T., Chaudhry, M.M., Tufail, M. et al. (2009). Heavy metal pollution from phosphate rock used for the production of fertilizer in Pakistan. *Microchemical Journal* 91 (1): 94–99. https://doi.org/10.1016/j.microc.2008.08.009.

Mekonnen, M.M. and Hoekstra, A.Y. (2018). Global anthropogenic phosphorus loads to freshwater and associated grey water footprints and water pollution levels: a high-resolution global study. *Water Resources Research* 54 (1): 345–358. https://doi.org/10.1002/2017WR020448.

Mengutay, M., Ceylan, Y., Kutman, U.B. et al. (2013). Adequate magnesium nutrition mitigates adverse effects of heat stress on maize and wheat. *Plant and Soil* 368 (1–2): 57–72. https://doi.org/10.1007/s11104-013-1761-6.

Miedaner, T., Zhao, Y., Gowda, M. et al. (2013). Genetic architecture of resistance to Septoria tritici blotch in European wheat. *BMC Genomics* 14 (1): 858. https://doi.org/10.1186/1471-2164-14-858.

Millar, N., Urrea, A., Kahmark, K. et al. (2018). Nitrous oxide (N_2O) flux responds exponentially to nitrogen fertilizer in irrigated wheat in the Yaqui Valley, Mexico. *Agriculture, Ecosystems & Environment* 261: 125–132. https://doi.org/10.1016/j.agee.2018.04.003.

Miralles, D.J., Katz, S.D., Colloca, A. et al. (1998). Floret development in near isogenic wheat lines differing in plant height. *Field Crops Research* 59 (1): 21–30. https://doi.org/10.1016/S0378-4290(98)00103-8.

Mnif, W., Hassine, A.I.H., Bouaziz, A. et al. (2011). Effect of endocrine disruptor pesticides: a review. *International Journal of Environmental Research and Public Health* 8 (6): 2265–2303. https://doi.org/10.3390/ijerph8062265.

Moerschbacher, B.M., Vander, P., Springer, C. et al. (1994). Photosynthesis in stem rust-infected, resistant and susceptible near-isogenic wheat leaves. *Canadian Journal of Botany* 72 (7): 990–997. https://doi.org/10.1139/b94-124.

Moffat, C.S. and Santana, F.M. (2019). Diseases affecting wheat: tan spot. In: *Integrated Disease Management of Wheat and Barley* (ed. R. Oliver), 92–106. Burleigh Dodds https://doi.org/10.1201/9780429201219.

Moll, R.H., Kamprath, E.J., and Jackson, W.A. (1982). Analysis and interpretation of factors which contribute to efficiency of nitrogen utilization. *Agronomy Journal* 74 (3): 562–564. https://doi.org/10.2134/agronj1982.00021962007400030037x.

Moreau, D., Allard, V., Gaju, O. et al. (2012). Acclimation of leaf nitrogen to vertical light gradient at anthesis in wheat is a whole-plant process that scales with the size of the canopy. *Plant Physiology* 160 (3): 1479–1490. https://doi.org/10.1104/pp.112.199935.

Mudahar, M.S. and Hignett, T.P. (1985). Energy efficiency in nitrogen fertilizer production. *Energy in Agriculture* 4: 159–177. https://doi.org/10.1016/0167-5826(85)90014-2.

Mulla, D.J. (2013). Twenty-five years of remote sensing in precision agriculture: key advances and remaining knowledge gaps. *Biosystems Engineering* 114 (4): 358–371. https://doi.org/10.1016/j.biosystemseng.2012.08.009.

Murray, G.M. and Brennan, J.P. (2009). Estimating disease losses to the Australian wheat industry. *Australasian Plant Pathology* 38 (6): 558–570. https://doi.org/10.1071/AP09053.

Myhre, G., Shindell, D., Bréon, F.-M. et al. (2013). Anthropogenic and natural radiative forcing. In: *Climate Change 2013: The Physical Science Basis. Contribution of Working Group I to the Fifth Assessment Report of the Intergovernmental Panel on Climate Change* (eds. T.F. Stocker, D. Qin, G.-K. Plattner, et al.), 659–740. Cambridge University Press.

Naeem, H.A. and MacRitchie, F. (2003). Effect of sulphur nutrition on agronomic and quality attributes of wheat. In: *Sulphur in Plants* (eds. Y.P. Abrol and A. Ahmad), 305–322. Springer https://doi.org/10.1007/978-94-017-0289-8_17.

Neumann, S., Paveley, N.D., Beed, F.D. et al. (2004). Nitrogen per unit leaf area affects the upper asymptote of Puccinia striiformis f. sp. tritici epidemics in winter wheat. *Plant Pathology* 53 (6): 725–732. https://doi.org/10.1111/j.1365-3059.2004.01107.x.

Ni, K., Pacholski, A., and Kage, H. (2014). Ammonia volatilization after application of urea to winter wheat over 3 years affected by novel urease and nitrification inhibitors. *Agriculture, Ecosystems & Environment* 197: 184–194. https://doi.org/10.1016/j.agee.2014.08.007.

Nobile, C., Houben, D., Michel, E. et al. (2019). Phosphorus-acquisition strategies of canola, wheat and barley in soil amended with sewage sludges. *Scientific Reports* 9 (1): 14878. https://doi.org/10.1038/s41598-019-51204-x.

Oerke, E.C. (2006). Crop losses to pests. *Journal of Agricultural Science* 144 (1): 31–43. https://doi.org/10.1017/S0021859605005708.

Olesen, J.E., Jørgensen, L.N., Petersen, J. et al. (2003a). Effects of rate and timing of nitrogen fertilizer on disease control by fungicides in winter wheat. 1. Grain yield and foliar disease control. *Journal of Agricultural Science* 140 (1): 1–13. https://doi.org/10.1017/S0021859602002885.

Olesen, J.E., Jørgensen, L.N., Petersen, J. et al. (2003b). Effects of rates and timing of nitrogen fertilizer on disease control by fungicides in winter wheat. 2. Crop growth and disease development. *Journal of Agricultural Science* 140 (1): 15–29. https://doi.org/10.1017/S0021859602002897.

Oliver, R.P. and Hewitt, H.G. (2014). *Fungicides in Crop Protection*. CABI.

Olsen, S., Cole, C., Watanabe, F. et al. (1954). *Estimation of available phosphorus in soils by extraction with sodium bicarbonate. USDA Circular Nr 939*, US Government Print Office.

Ortiz-Monasterio, J.I. and Raun, W. (2007). Reduced nitrogen and improved farm income for irrigated spring wheat in the Yaqui Valley, Mexico, using sensor based nitrogen management. *Journal of Agricultural Science* 145 (3): 215–222. https://doi.org/10.1017/S0021859607006995.

Ortiz-Monasterio, J.I., Sayre, K.D., Rajaram, S. et al. (1997). Genetic progress in wheat yield and nitrogen use efficiency under four nitrogen rates. *Crop Science* 37 (3): 898–904. https://doi.org/10.2135/cropsci1997.0011183X003700030033x.

Oury, F.X., Leenhardt, F., Remesy, C. et al. (2006). Genetic variability and stability of grain magnesium, zinc and iron concentrations in bread wheat. *European Journal of Agronomy* 25 (2): 177–185. https://doi.org/10.1016/j.eja.2006.04.011.

Pan, B., Lam, S.K., Mosier, A. et al. (2016). Ammonia volatilization from synthetic fertilizers and its mitigation strategies: a global synthesis. *Agriculture, Ecosystems & Environment* 232: 283–289. https://doi.org/10.1016/j.agee.2016.08.019.

Parker, S.R., Welham, S., Paveley, N.D. et al. (2004). Tolerance of septoria leaf blotch in winter wheat. *Plant Pathology* 53 (1): 1–10. https://doi.org/10.1111/j.1365-3059.2004.00951.x.

Parry, D.W. (1990). *Plant Pathology in Agriculture*. Cambridge University Press.

Parton, W.J., Morgan, J.A., Altenhofen, J.M. et al. (1988). Ammonia volatilization from spring wheat plants. *Agronomy Journal* 80 (3): 419–425. https://doi.org/10.2134/agronj1988.00021962008000030008x.

Paveley, N.D., Sylvester-Bradley, R., Scott, R.K. et al. (2001). Steps in predicting the relationship of yield on fungicide dose. *Phytopathology* 91: 708–716. https://doi.org/10.1094/PHYTO.2001.91.7.708.

Pearce, S., Saville, R., Vaughan, S.P. et al. (2011). Molecular characterization of Rht-1 dwarfing genes in hexaploid wheat. *Plant Physiology* 157: 1820–1831. https://doi.org/10.1104/pp.111.183657.

Pearse, S.J., Veneklaas, E.J., Cawthray, G.R. et al. (2006). Carboxylate release of wheat, canola and 11 grain legume species as affected by phosphorus status. *Plant and Soil* 288 (1): 127–139. https://doi.org/10.1007/s11104-006-9099-y.

Pepler, S., Gooding, M.J., Ford, K.E. et al. (2005a). A temporal limit to the association between flag leaf life extension by fungicides and wheat yields. *European Journal of Agronomy* 22 (4): 363–373. https://doi.org/10.1016/j.eja.2004.06.002.

Pepler, S., Gooding, M.J., Ford, K.E. et al. (2005b). Delaying senescence of wheat with fungicides has interacting effects with cultivar on grain sulphur concentration but not with sulphur yield or nitrogen: sulphur ratios. *European Journal of Agronomy* 22 (4): 405–416. https://doi.org/10.1016/j.eja.2004.06.006.

Petersen, J. (2003). Nitrogen fertilizer replacement value of sewage sludge, composted household waste and farmyard manure. *Journal of Agricultural Science* 140 (2): 169–182. https://doi.org/10.1017/S0021859603003010.

Peterson, G.A., Sander, D.H., Grabouski, P.H. et al. (1981). A new look at row and broadcast phosphate recommendations for winter wheat. *Agronomy Journal* 73 (1): 13–17. https://doi.org/10.2134/agronj1981.00021962007300010004x.

Plaza-Bonilla, D., Nolot, J.M., Raffaillac, D. et al. (2015). Cover crops mitigate nitrate leaching in cropping systems including grain legumes: field evidence and model simulations. *Agriculture, Ecosystems & Environment* 212: 1–12. https://doi.org/10.1016/j.agee.2015.06.014.

Polley, R.W. and Clarkson, J.D.S. (1978). Forecasting cereal disease epidemics. In: *Plant Disease Epidemiology* (eds. P.R. Scott and A. Bainbridge), 141–151. Blackwell Scientific Publications.

Powlson, D.S. (1993). Understanding the soil nitrogen cycle. *Soil Use and Management* 9 (3): 86–93. https://doi.org/10.1111/j.1475-2743.1993.tb00935.x.

Powlson, D.S., Jenkinson, D.S., Pruden, G. et al. (1985). The effect of straw incorporation on the uptake of nitrogen by winter wheat. *Journal of the Science of Food and Agriculture* 36 (1): 26–30. https://doi.org/10.1002/jsfa.2740360105.

Powlson, D.S., Poulton, P.R., Penny, A. et al. (1987). Recovery of 15N-labelled urea applied to the foliage of winter wheat. *Journal of the Science of Food and Agriculture* 41 (3): 195–203. https://doi.org/10.1002/jsfa.2740410302.

Powlson, D.S., Poulton, P.R., Møller, N.E. et al. (1989). Uptake of foliar-applied urea by winter wheat (Triticum aestivum): the influence of application time and the use of a new 15N technique. *Journal of the Science of Food and Agriculture* 48 (4): 429–440. https://doi.org/10.1002/jsfa.2740480405.

Pushman, F.M. and Bingham, J. (1976). The effects of a granular nitrogen fertilizer and a foliar spray of urea on the yield and bread-making quality of ten winter wheats. *Journal of Agricultural Science* 87 (2): 281–292. https://doi.org/10.1017/S0021859600027556.

Redshaw, C.J., Mason, C.F., Hayes, C.R. et al. (1990). Factors influencing phosphate exchange across the sediment-water interface of eutrophic reservoirs. *Hydrobiologia* 192 (2–3): 233–245. https://doi.org/10.1007/BF00006018.

Reese, M., Marquart, C., Malmali, M. et al. (2016). Performance of a small-scale Haber process. *Industrial & Engineering Chemistry Research* 55 (13): 3742–3750. https://doi.org/10.1021/acs.iecr.5b04909.

Reynolds, M.P. and Borlaug, N.E. (2006). Impacts of breeding on international collaborative wheat improvement. *Journal of Agricultural Science* 144: 3–17. https://doi.org/10.1017/S0021859606005867.

Robert, C., Bancal, M.O., Ney, B. et al. (2005). Wheat leaf photosynthesis loss due to leaf rust, with respect to lesion development and leaf nitrogen status. *New Phytologist* 165 (1): 227–241. https://doi.org/10.1111/j.1469-8137.2004.01237.x.

Roberts, E. (1847). On the management of wheat. *Journal of the Royal Agricultural Society of England* 8: 60–77.

Rodríguez, D., Keltjens, W.G., and Goudriaan, J. (1998). Plant leaf area expansion and assimilate production in wheat (*Triticum aestivum* L.) growing under low phosphorus conditions. *Plant and Soil* 200 (2): 227–240. https://doi.org/10.1023/A:1004310217694.

Rodríguez, D., Andrade, F.H., and Goudriaan, J. (1999). Effects of phosphorus nutrition on tiller emergence in wheat. *Plant and Soil* 209 (2): 283–295. https://doi.org/10.1023/A:1004690404870.

Rodríguez, D., Andrade, F.H., and Goudriaan, J. (2000). Does assimilate supply limit leaf expansion in wheat grown in the field under low phosphorus availability? *Field Crops Research* 67 (3): 227–238. https://doi.org/10.1016/S0378-4290(00)00098-8.

Roelfs, A.P. (1982). Effects of Barberry eradication. *Plant Disease* 66 (2): 177–181.

Roelfs, A.P. (1992). *Rust Diseases of Wheat: Concepts and Methods of Disease Management*. CIMMYT.

Royal Society (2020). *Ammonia: zero-carbon fertiliser, fuel and energy store*. The Royal Society. https://royalsociety.org/green-ammonia

Ruske, R.E., Gooding, M.J., and Jones, S.A. (2003a). The effects of triazole and strobilurin fungicide programmes on nitrogen uptake, partitioning, remobilization and grain N accumulation in winter wheat cultivars. *Journal of Agricultural Science* 140 (4): 395–407. https://doi.org/10.1017/S0021859603003228.

Ruske, R.E., Gooding, M.J., and Jones, S.A. (2003b). The effects of adding picoxystrobin, azoxystrobin and nitrogen to a triazole programme on disease control, flag leaf senescence, yield and grain quality of winter wheat. *Crop Protection* 22 (7): 975–987. https://doi.org/10.1016/S0261-2194(03)00113-3.

Ryan, P.R., Delhaize, E., and Jones, D.L. (2001). Function and mechanism of organic anion exudation from plant roots. *Annual Review of Plant Biology* 52 (1): 527–560. https://doi.org/10.1146/annurev.arplant.52.1.527.

Sadler, R. and Scott, K.J. (1974). Nitrogen assimilation and metabolism in barley leaves infected with the powdery mildew fungus. *Physiological Plant Pathology* 4 (2): 235–247. https://doi.org/10.1016/0048-4059(74)90012-5.

Salmon, S.E. and Cook, R.J. (1987). Effects of fungicides on the milling and baking quality of wheat. In: *Aspects of Applied Biology, 15, Cereal Quality*, 373–384. Association of Applied Biologists.

Salvagiotti, F., Castellarín, J.M., Miralles, D.J. et al. (2009). Sulfur fertilization improves nitrogen use efficiency in wheat by increasing nitrogen uptake. *Field Crops Research* 113 (2): 170–177. https://doi.org/10.1016/j.fcr.2009.05.003.

Saunders, D.G.O., Pretorius, Z.A., and Hovmøller, M.S. (2019). Tackling the re-emergence of wheat stem rust in Western Europe. *Communications in Biology* 2: 51. https://doi.org/10.1038/s42003-019-0294-9.

Saville, R.J., Gosman, N., Burt, C.J. et al. (2012). The 'Green Revolution' dwarfing genes play a role in disease resistance in Triticum aestivum and Hordeum vulgare. *Journal of Experimental Botany* 63 (3): 1271–1283. https://doi.org/10.1093/jxb/err350.

Schiere, J.B., Ibrahim, M.N.M., and Van Keulen, H. (2002). The role of livestock for sustainability in mixed farming: criteria and scenario studies under varying resource allocation. *Agriculture, Ecosystems & Environment* 90 (2): 139–153. https://doi.org/10.1016/S0167-8809(01)00176-1.

Schils, R., Olesen, J.E., Kersebaum, K.C. et al. (2018). Cereal yield gaps across Europe. *European Journal of Agronomy* 101: 109–120. https://doi.org/10.1016/j.eja.2018.09.003.

Scholz, R.W. and Wellmer, F.W. (2015). Losses and use efficiencies along the phosphorus cycle-part 2: understanding the concept of efficiency. *Resources, Conservation and Recycling* 105: 259–274. https://doi.org/10.1016/j.resconrec.2015.10.003.

Schwessinger, B. (2017). Fundamental wheat stripe rust research in the 21st century. *New Phytologist* 213 (4): 1625–1631. https://doi.org/10.1111/nph.14159.

Scott, R.K., Jaggard, K.W., and Sylvester-Bradley, R. (1994). Resource capture by arable crops. In: *Resource Capture by Crops* (eds. J.L. Monteith, R.K. Scott and M.H. Unsworth), 279–302. Nottingham University Press.

Serrago, R.A. and Miralles, D.J. (2014). Source limitations due to leaf rust (caused by *Puccinia triticina*) during grain filling in wheat. *Crop and Pasture Science* 65 (2): 185–193. https://doi.org/10.1071/CP13248.

Serrago, R.A., Carretero, R., Bancal, M.O. et al. (2009). Foliar diseases affect the eco-physiological attributes linked with yield and biomass in wheat (*Triticum aestivum* L.). *European Journal of Agronomy* 31 (4): 195–203. https://doi.org/10.1016/j.eja.2009.06.002.

Shaw, M.W., Bearchell, S.J., Fitt, B.D. et al. (2008). Long-term relationships between environment and abundance in wheat of Phaeosphaeria nodorum and Mycosphaerella graminicola. *New Phytologist* 177 (1): 229–238. https://doi.org/10.1111/j.1469-8137.2007.02236.x.

Shepon, A., Gildor, H., Labrador, L.J. et al. (2007). Global reactive nitrogen deposition from lightning NOx. *Journal of Geophysical Research: Atmospheres* 112 (D6): D06304. https://doi.org/10.1029/2006JD007458.

Shewry, P.R., Zhao, F.J., Gowa, G.B. et al. (2009). Sulphur nutrition differentially affects the distribution of asparagine in wheat grain. *Journal of Cereal Science* 50 (3): 407–409. https://doi.org/10.1016/j.jcs.2009.07.001.

Shindo, H. and Nishio, T. (2005). Immobilization and remineralization of N following addition of wheat straw into soil: determination of gross N transformation rates by 15N-ammonium isotope dilution technique. *Soil Biology and Biochemistry* 37 (3): 425–432. https://doi.org/10.1016/j.soilbio.2004.07.027.

Sidhu, H.S., Jat, M.L., Singh, Y. et al. (2019). Sub-surface drip fertigation with conservation agriculture in a rice-wheat system: a breakthrough for addressing water and nitrogen use efficiency. *Agricultural Water Management* 216: 273–283. https://doi.org/10.1016/j.agwat.2019.02.019.

Simkin, S.M., Allen, E.B., Bowman, W.D. et al. (2016). Conditional vulnerability of plant diversity to atmospheric nitrogen deposition across the United States. *Proceedings of the National Academy of Sciences* 113 (15): 4086–4091. https://doi.org/10.1073/pnas.1515241113.

Simón, M.R., Fleitas, M.C., Castro, A.C. et al. (2020). How foliar fungal diseases affect nitrogen dynamics, milling and end-use quality of wheat. *Frontiers in Plant Science 11*: 1568. https://doi.org/10.3389/fpls.2020.569401.

Singh, H. and Gill, H.S. (1987). Interacting effects of magnesium with nitrogen on the growth and yield of wheat. *Journal of Agricultural Science* 109 (2): 399–403. https://doi.org/10.1017/S0021859600080837.

Singh, R.P., Hodson, D.P., Huerta-Espino, J. et al. (2008). Will stem rust destroy the world's wheat crop? *Advances in Agronomy* 98: 271–309. https://doi.org/10.1016/S0065-2113(08)00205-8.

Singh, R.P., Hodson, D.P., Huerta-Espino, J. et al. (2011). The emergence of Ug99 races of the stem rust fungus is a threat to world wheat production. *Annual Review of Phytopathology* 49: 465–481. https://doi.org/10.1146/annurev-phyto-072910-095423.

Smil, V. (2004). *Enriching the Earth: Fritz Haber, Carl Bosch, and the Transformation of World Food Production*. MIT press.

Smith, C., Hill, A.K., and Torrente-Murciano, L. (2020). Current and future role of Haber–Bosch ammonia in a carbon-free energy landscape. *Energy & Environmental Science* 13 (2): 331–344. https://doi.org/10.1039/C9EE02873K.

Solomon, P.S., Wilson, T.G., Rybak, K. et al. (2006). Structural characterisation of the interaction between *Triticum aestivum* and the dothideomycete pathogen *Stagonospora nodorum*. *European Journal of Plant Pathology* 114 (3): 275–282. https://doi.org/10.1007/s10658-005-5768-6.

Speirs, S.D., Scott, B.J., Moody, P.W. et al. (2013). Soil phosphorus tests II: a comparison of soil test–crop response relationships for different soil tests and wheat. *Crop and Pasture Science* 64 (5): 469–479. https://doi.org/10.1071/CP13111.

Spink, J., Foulkes, M.J., Gay, A. et al. (2000a). *Reducing Winter Wheat Production Costs Through Crop Intelligence Information on Variety and Sowing Date, Rotational Position, and Canopy Management in Relation to Drought and Disease Control. HGCA Project Report, 235.* Agriculture and Horticulture Development Board.

Spink, J.H., Semere, T., Sparkes, D.L. et al. (2000b). Effect of sowing date on the optimum plant density of winter wheat. *Annals of Applied Biology* 137 (2): 179–188. https://doi.org/10.1111/j.1744-7348.2000.tb00049.x.

Stapper, M. and Fischer, R.A. (1990a). Genotype, sowing date and plant spacing influence on high-yielding irrigated wheat in southern New South Wales. I. Phasic development, canopy growth and spike production. *Australian Journal of Agricultural Research* 41 (6): 997–1019. https://doi.org/10.1071/AR9900997.

Stapper, M. and Fischer, R.A. (1990b). Genotype, sowing date and plant spacing influence on high-yielding irrigated wheat in southern New South Wales. II. Growth, yield and nitrogen use. *Australian Journal of Agricultural Research* 41 (6): 1021–1041. https://doi.org/10.1071/AR9901021.

Steiner, G., Geissler, B., Watson, I. et al. (2015). Efficiency developments in phosphate rock mining over the last three decades. *Resources, Conservation and Recycling* 105: 235–245. https://doi.org/10.1016/j.resconrec.2015.10.004.

Stukenbrock, E.H., Banke, S., and McDonald, B.A. (2006). Global migration patterns in the fungal wheat pathogen Phaeosphaeria nodorum. *Molecular Ecology* 15 (10): 2895–2904. https://doi.org/10.1111/j.1365-294X.2006.02986.x.

Swaminathan, M.S. (2006). An evergreen revolution. *Crop Science* 46: 2293–2303. https://doi.org/10.2135/cropsci2006.9999.

Swarbreck, S.M., Wang, M., Wang, Y. et al. (2019). A roadmap for lowering crop nitrogen requirement. *Trends in Plant Science* 24 (10): 892–904. https://doi.org/10.1016/j.tplants.2019.06.006.

Sweeney, D.W., Granade, G.V., Eversmeyer, M.G. et al. (2000). Phosphorus, potassium, chloride, and fungicide effects on wheat yield and leaf rust severity. *Journal of Plant Nutrition* 23 (9): 1267–1281. https://doi.org/10.1080/01904160009382099.

Sylvester-Bradley, R., and Clarke, S. (2009). Using grain N% as a signature for good N use. *Project Report No. 458*. Home-Grown Cereals Authority.

Sylvester-Bradley, R. and Kindred, D.R. (2009). Analysing nitrogen responses of cereals to prioritize routes to the improvement of nitrogen use efficiency. *Journal of Experimental Botany* 60 (7): 1939–1951. https://doi.org/10.1093/jxb/erp116.

Sylvester-Bradley, R. and Stokes, D.T. (2001). Fertiliser strategies to control the uptake and distribution of nitrogen in winter wheat. In: *Aspects of Applied Biology, 64, Wheat Quality* (eds. M.J. Gooding, S.A. Barton and G.P. Smith), 185–186. Association of Applied Biologists.

Sylvester-Bradley, R., Scott, R.K., Stokes, D.T. et al. (1997). The significance of crop canopies for N nutrition. In: *Aspects of Applied Biology 50 Optimising Cereal Inputs Part 1: Genetics and Nutrition* (eds. M.J. Gooding and P.R. Shewry), 103–116. Association of Applied Biologists.

Sylvester-Bradley, R., Stokes, D., and Scott, R. (2001). Dynamics of nitrogen capture without fertilizer: the baseline for fertilizing winter wheat in the UK. *Journal of Agricultural Science* 136 (1): 15–33. https://doi.org/10.1017/S0021859600008479.

Sylvester-Bradley, R., Kindred, D.R., Blake, J. et al. (2008). *Optimising Fertiliser Nitrogen for Modern Wheat and Barley Crops. Project Report No. 438*. Home-Grown Cereals Authority.

Sylvester-Bradley, R., Kindred, D.R., Wynn, S.C. et al. (2014). Efficiencies of nitrogen fertilizers for winter cereal production, with implications for greenhouse gas intensities of grain. *Journal of Agricultural Science* 152 (1): 3–22. https://doi.org/10.1017/S0021859612000810.

Takahashi, S. and Anwar, M.R. (2007). Wheat grain yield, phosphorus uptake and soil phosphorus fraction after 23 years of annual fertilizer application to an Andosol. *Field Crops Research* 101 (2): 160–171.

Tallaksen, J., Bauer, F., Hulteberg, C. et al. (2015). Nitrogen fertilizers manufactured using wind power: greenhouse gas and energy balance of community-scale ammonia production. *Journal of Cleaner Production* 107: 626–635. https://doi.org/10.1016/j.jclepro.2015.05.130.

Tao, Y. and Murphy, J.G. (2018). Evidence for the importance of semivolatile organic ammonium salts in ambient particulate matter. *Environmental Science & Technology* 53 (1): 108–116. https://doi.org/10.1021/acs.est.8b03800.

Tate, R.L. (1995). *Soil Microbiology*. Wiley.

Teng, W., Deng, Y., Chen, X.P. et al. (2013). Characterization of root response to phosphorus supply from morphology to gene analysis in field-grown wheat. *Journal of Experimental Botany* 64 (5): 1403–1411. https://doi.org/10.1093/jxb/ert023.

Thilakarathna, S.K., Hernandez-Ramirez, G., Puurveen, D. et al. (2020). Nitrous oxide emissions and nitrogen use efficiency in wheat: N fertilization timing and formulation, soil N, and weather effects. *Soil Science Society of America Journal* https://doi.org/10.1002/saj2.20145.

Thind, H.S., Kumar, A., and Vashistha, M. (2012). Establishment of threshold leaf colour greenness for need-based fertilizer nitrogen management in irrigated wheat (Triticum aestivum L.) using leaf colour chart. *Field Crops Research* 130: 109–119. https://doi.org/10.1016/j.fcr.2012.02.005.

Thompson, A.J., Gooding, M.J., and Davies, W.P. (1993). The effect of season and management on the grain yield and breadmaking quality of organically grown wheat cultivars. In: *Aspects of Applied Biology 36, Cereal Quality III* (eds. P.S. Kettlewell, J.R. Garstang, C.M. Duffus, et al.), 179–188. Association of Applied Biologists.

Thomsen, I.K. (2005). Crop N utilization and leaching losses as affected by time and method of application of farmyard manure. *European Journal of Agronomy* 22 (1): 1–9. https://doi.org/10.1016/j.eja.2003.10.008.

Tian, H., Xu, R., Canadell, J.G. et al. (2020). A comprehensive quantification of global nitrous oxide sources and sinks. *Nature* 586: 248–256. https://doi.org/10.1038/s41586-020-2780-0.

Tie, X., Zhang, R., Brasseur, G., and Lei, W. (2002). Global NOx production by lightning. *Journal of Atmospheric Chemistry* 43 (1): 61–74. https://doi.org/10.1023/A:1016145719608.

Tiedemann, A.V. (1996). Single and combined effects of nitrogen fertilization and ozone on fungal leaf diseases on wheat. *Zeitschrift fur Pflanzenkrankheiten und Pflanzenschutz* 103 (4): 409–419.

Tompkins, D.K., Fowler, D.B., and Wright, A.T. (1992). Foliar disease development in no-till winter wheat: influence of agronomic practices on powdery mildew development. *Canadian Journal of Plant Science* 72 (3): 965–972. https://doi.org/10.4141/cjps92-121.

Tompkins, D.K., Fowler, D.B., and Wright, A.T. (1993). Influence of agronomic practices on canopy microclimate and Septoria development in no-till winter wheat produced in the Parkland region of Saskatchewan. *Canadian Journal of Plant Science* 73 (1): 331–344. https://doi.org/10.4141/cjps93-050.

Trail, F. (2009). For blighted waves of grain: *Fusarium graminearum* in the postgenomics era. *Plant Physiology* 149 (1): 103–110. https://doi.org/10.1104/pp.108.129684.

Tränkner, M., Tavakol, E., and Jákli, B. (2018). Functioning of potassium and magnesium in photosynthesis, photosynthate translocation and photoprotection. *Physiologia Plantarum* 163 (3): 414–431. https://doi.org/10.1111/ppl.12747.

Tunå, P., Hulteberg, C., and Ahlgren, S. (2014). Techno-economic assessment of nonfossil ammonia production. *Environmental Progress & Sustainable Energy* 33 (4): 1290–1297. https://doi.org/10.1002/ep.11886.

Ulrich, A.E. and Frossard, E. (2014). On the history of a reoccurring concept: phosphorus scarcity. *Science of the Total Environment* 490: 694–707. https://doi.org/10.1016/j.scitotenv.2014.04.050.

UNEP-IFA (2001). *Environmental Aspects of Phosphate and Potash Mining*. United Nations Environment Programme, International Fertilizer Industry Association.

USGS (2020). Phosphate Rock Statistics and Information. Minerals Information Centre, U.S. Geological Survey. https://www.usgs.gov/centers/nmic/phosphate-rock-statistics-and-information

Vaidyanathan, L.V., Sylvester-Bradley, R., and Bloom, T.M. (1987). Effects of previous cropping, and applied nitrogen on grain nitrogen content in winter wheat. In: *Aspects of Applied Biology 15, Cereal Quality*, 227–239. Association of Applied Biologists.

Valle, S.R. and Calderini, D.F. (2010). Phyllochron and tillering of wheat in response to soil aluminum toxicity and phosphorus deficiency. *Crop and Pasture Science* 61 (11): 863–872. https://doi.org/10.1071/CP09310.

Van de Velde, K., Ruelens, P., Geuten, K. et al. (2017). Exploiting DELLA signaling in cereals. *Trends in Plant Science* 22: 880–893. https://doi.org/10.1016/j.tplants.2017.07.010.

Van Grinsven, H.J., Ward, M.H., Benjamin, N. et al. (2006). Does the evidence about health risks associated with nitrate ingestion warrant an increase of the nitrate standard for drinking water? *Environmental Health* 5 (1): 1–6. https://doi.org/10.1186/1476-069X-5-26.

Vance, C.P., Uhde-Stone, C., and Allan, D.L. (2003). Phosphorus acquisition and use: critical adaptations by plants for securing a nonrenewable resource. *New Phytologist* 157: 423–447. https://doi.org/10.1046/j.1469-8137.2003.00695.x.

Vanloqueren, G. and Baret, P.V. (2008). Why are ecological, low-input, multi-resistant wheat cultivars slow to develop commercially? A Belgian agricultural 'lock-in'case study. *Ecological Economics* 66 (2–3): 436–446. https://doi.org/10.1016/j.ecolecon.2007.10.007.

Verbruggen, N. and Hermans, C. (2013). Physiological and molecular responses to magnesium nutritional imbalance in plants. *Plant and Soil* 368 (1–2): 87–99. https://doi.org/10.1007/s11104-013-1589-0.

Vergauwen, D. and De Smet, I. (2017). From early farmers to Norman Borlaug - the making of modern wheat. *Current Biology* 27: R858–R862. https://doi.org/10.1016/j.cub.2017.06.061.

Wagner-Riddle, C., Congreves, K.A., Abalos, D. et al. (2017). Globally important nitrous oxide emissions from croplands induced by freeze–thaw cycles. *Nature Geoscience* 10 (4): 279–283. https://doi.org/10.1038/ngeo2907.

Wakelin, S.A., Warren, R.A., Harvey, P.R. et al. (2004). Phosphate solubilization by Penicillium spp. closely associated with wheat roots. *Biology and Fertility of Soils* 40 (1): 36–43. https://doi.org/10.1007/s00374-004-0750-6.

Walley, F., Yates, T., van Groenigen, J.W. et al. (2002). Relationships between soil nitrogen availability indices, yield, and nitrogen accumulation of wheat. *Soil Science Society of America Journal* 66 (5): 1549–1561. https://doi.org/10.2136/sssaj2002.1549.

Walters, D.R. and Bingham, I.J. (2007). Influence of nutrition on disease development caused by fungal pathogens: implications for plant disease control. *Annals of Applied Biology* 151 (3): 307–324. https://doi.org/10.1111/j.1744-7348.2007.00176.x.

Wang, Q., Zhou, F., Shang, Z. et al. (2020). Data-driven estimates of global nitrous oxide emissions from croplands. *National Science Review* 7 (2): 441–452. https://doi.org/10.1093/nsr/nwz087.

Watson, C.A., Fowlerf, S.M., and Wilman, D. (1993). Soil inorganic-N and nitrate leaching on organic farms. *Journal of Agricultural Science* 120 (3): 361–369. https://doi.org/10.1017/S0021859600076528.

Webb, J., Sylvester-Bradley, R., and Seeney, F.M. (1997). The effects of site and season on the fate of nitrogen residues from root crops grown on sandy soils. *Journal of Agricultural Science* 128 (4): 445–460. https://doi.org/10.1017/S002185969700436X.

Webb, J., Seeney, F.M., and Sylvester-Bradley, R. (1998). The response to fertilizer nitrogen of cereals grown on sandy soils. *Journal of Agricultural Science* 130 (3): 271–286. https://doi.org/10.1017/S0021859698005309.

Wegulo, S.N., Baenziger, P.S., Nopsa, J.H. et al. (2015). Management of Fusarium head blight of wheat and barley. *Crop Protection* 73: 100–107. https://doi.org/10.1016/j.cropro.2015.02.025.

Whaley, J.M., Sparkes, D.L., Foulkes, M.J. et al. (2000). The physiological response of winter wheat to reductions in plant density. *Annals of Applied Biology* 137 (2): 165–177. https://doi.org/10.1111/j.1744-7348.2000.tb00048.x.

White, E.M., Wilson, F.E.A., and Kettlewell, P.S. (1998). *Exploitation of varieties for UK cereal production. III. Varietal responses to soil fertilizer N availability. HGCA Project Report No. 174.* Agriculture and Horticulture Development Board.

WHO (2004). *Sulfate in Drinking-water: Background document for development of WHO Guidelines for Drinking-water Quality*. World Health Organisation. https://www.who.int/water_sanitation_health/dwq/chemicals/sulfate.pdf

WHO (2017). *Guidelines for Drinking-Water Quality* (4th ed.). World Health Organisation. https://www.who.int/publications/i/item/9789241549950

Wibberley, E.J. (1989). *Cereal Husbandry*. Farming Press.

Wiese, M.V. (1987). *Compendium of Wheat Diseases*, 2e. Minnesota: APS Press.

Williams, A.G., Audsley, E., and Sandars, D.L. (2010). Environmental burdens of producing bread wheat, oilseed rape and potatoes in England and Wales using simulation and system modelling. *The International Journal of Life Cycle Assessment* 15 (8): 855–868. https://doi.org/10.1007/s11367-010-0212-3.

Yamashita, M. and Fraaije, B. (2018). Non-target site SDHI resistance is present as standing genetic variation in field populations of *Zymoseptoria tritici*. *Pest Management Science* 74 (3): 672–681. https://doi.org/10.1002/ps.4761.

Yue, S., Meng, Q., Zhao, R. et al. (2012). Critical nitrogen dilution curve for optimizing nitrogen management of winter wheat production in the North China plain. *Agronomy Journal* 104 (2): 523–529. https://doi.org/10.2134/agronj2011.0258.

Zadoks, J.C., Chang, T.T., and Konzak, C.F. (1974). A decimal code for the growth stages of cereals. *Weed Research* 14: 415–421.

Zhang, X., Davidson, E.A., Mauzerall, D.L. et al. (2015). Managing nitrogen for sustainable development. *Nature* 528 (7580): 51–59. https://doi.org/10.1038/nature15743.

Zhao, F.J., McGrath, S.P., Crosland, A.R. et al. (1995). Changes in the sulphur status of British wheat grain in the last decade, and its geographical distribution. *Journal of the Science of Food and Agriculture* 68 (4): 507–514. https://doi.org/10.1002/jsfa.2740680415.

Zhao, F.J., Hawkesford, M.J., Warrilow, A.A. et al. (1996). Responses of two wheat varieties to sulphur addition and diagnosis of sulphur deficiency. *Plant and Soil* 181 (2): 317–327. https://doi.org/10.1007/BF00012066.

Zhao, F.J., Hawkesford, M.J., and McGrath, S.P. (1999). Sulphur assimilation and effects on yield and quality of wheat. *Journal of Cereal Science* 30 (1): 1–17. https://doi.org/10.1006/jcrs.1998.0241.

7

The Structure and Composition of the Wheat Grain

If you could understand a single grain of wheat you would die of wonder

Martin Luther

Although the wheat grain is generally assumed to be a seed, it is in fact a single-seeded fruit, termed a caryopsis. This means that the zygotic seed tissues are surrounded by maternal tissues, which form a protective outer layer. The grains are borne on a spike (ear), with the major axis (the rachis) bearing alternately arranged spikelets (Figure 7.1).

7.1 Grain Development

7.1.1 Fertilisation

The wheat plant is largely self-pollinated, although low levels of cross-pollination do occur with the frequency varying between genotypes. Also, because the pollen is heavy and short-lived, cross-fertilisation is generally only observed between plants growing close together. The female gametophyte is termed the embryo sac and comprises seven cells (Figure 7.2). Three cells are present at one end, comprising one egg cell and two synergids: these together form the egg apparatus. The central part of the embryo sac consists of a single large central cell which contains two polar nuclei, while three antipodal cells are present at the opposite end. Whereas the central cell is diploid, all the other cells are haploid.

The pollen germinates on the stigma and the pollen tube grows down the style to reach the embryo sac. The pollen tube contains two reproductive nuclei that are discharged into the embryo sac. One reproductive nucleus enters the egg cell and fuses with the egg nucleus, giving rise to the diploid embryo. The second pollen reproductive nucleus enters the central cell and fuses with the two female polar nuclei, giving rise to the triploid endosperm. Hence, the event is referred to as 'double fertilisation' and the two major tissues of the grain, the embryo and endosperm, have separate origins and are diploid and triploid, respectively. These tissues are surrounded by several tissues of maternal origin: the nucellus and nucellar epidermis, the inner integument and outer integuments, which

Wheat: Environment, Food and Health, First Edition. Mike J. Gooding and Peter R. Shewry.

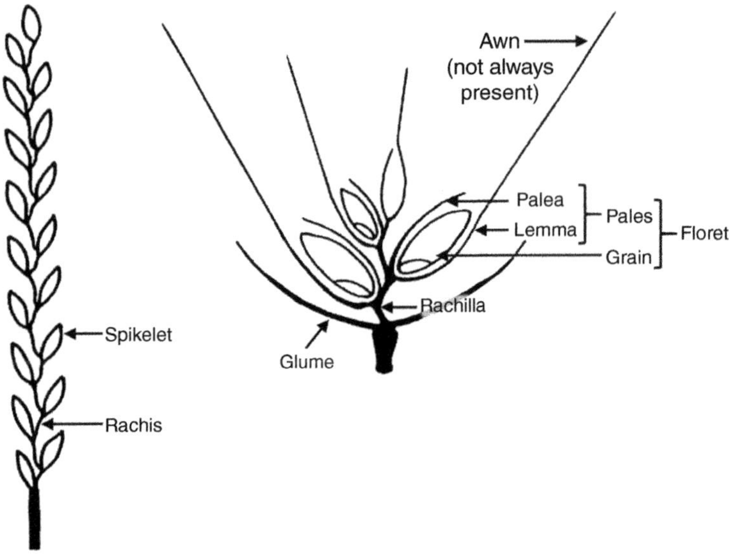

Figure 7.1 Structure of the wheat spike and spikelet. Left, spikelets arranged alternately on the main axis or rachis. Right, the structure of the mature spikelet. Its axis, or rachilla, bears (sterile) glumes at its base and florets arranged alternately along its length. In each floret, a grain is enclosed between two pales or fertile glumes. *Source:* Taken from Bechtel et al. (2009) with permission.

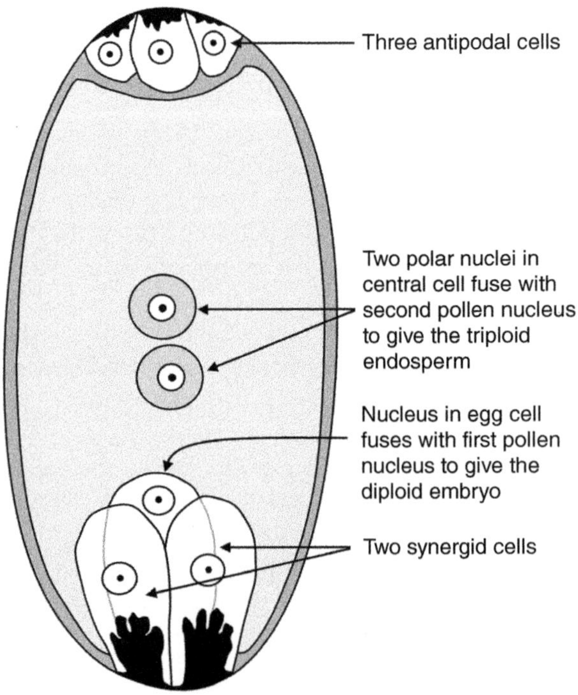

Figure 7.2 Diagram of the mature embryo sac of wheat, comprising seven cells. *Source:* Redrawn from Vijayaraghavan et al. (1988) with permission.

together form the testa (seed coat), and the pericarp which comprises the tube cells, cross cells, and outer pericarp cells in the mature grain. These structures are described in detail by Bechtel et al. (2009).

7.1.2 Post-fertilisation

The pericarp and nucellus account for most of the total caryopsis during the early stages of caryopsis development (Figure 7.3) but decline as the endosperm and embryo develop and the outer layers degenerate and die. Consequently, at maturity only the epidermis of the nucellus remains, consisting of a single layer of crushed cells between the aleurone and testa (seed coat).

The developmental changes in the three major tissues of the mature grain, the pericarp, endosperm, and embryo, are illustrated by the expression patterns of genes associated with these tissues, shown in Figure 7.4b, with transcripts associated with the pericarp declining after 10 days after anthesis, transcripts associated with the endosperm peaking between about 10 and 30 days, and embryo transcripts continuing to increase up to physiological maturity (40 days). These changes are important when interpreting the outcomes of studies of caryopsis development.

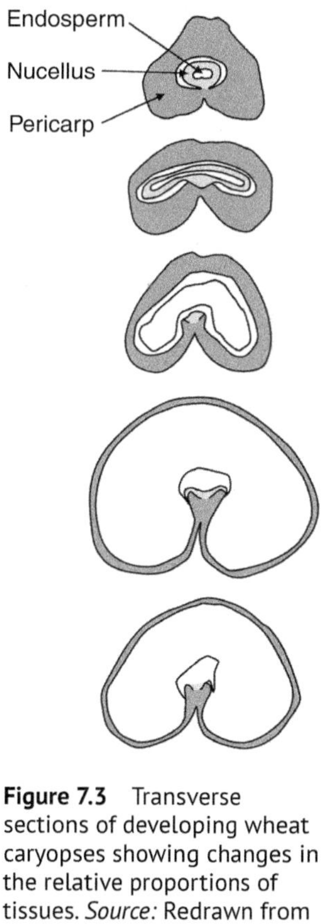

Figure 7.3 Transverse sections of developing wheat caryopses showing changes in the relative proportions of tissues. *Source:* Redrawn from Bechtel et al. (2009) with permission.

7.1.3 Endosperm Development

The formation of the triploid endosperm nucleus is followed by numerous mitotic divisions which are not accompanied by cell division, leading to the formation, after one or two days, of a single multinucleate endosperm cell with a peripheral zone of cytoplasm and a large central vacuole (syncytium). Cellularisation is then initiated by the growth of cell walls between the nuclei in the endosperm syncytium. Periclinal and anticlinal walls form between the dividing nuclei, resulting in the formation of several layers of cells. Finally, the cell walls growing from the dorsal and ventral sides of the grain meet, with cellularisation being complete by about four days after fertilization.

Differentiation of the endosperm cells starts at the beginning of the cellularisation stage when the peripheral layer of cells begins to differentiate into the aleurone layer. The cells inside the aleurone are highly vacuolated with thin walls and continue to divide up to 10 to 14 days after anthesis, depending on growth conditions, after which they continue to expand. This results in two distinct cell types, prismatic cells that radiate from the aleurone towards the centre of the grain and the central cells in the lobes of the grain, which are rounded or polygonal (Figure 7.5). By contrast, the aleurone cells remain meristematic and divide periclinally up to or beyond 14 days after anthesis (daa) (Bechtel et al. 2009). The cells immediately below the aleurone (termed sub-aleurone) are therefore considered to result from the division and redifferentiation of aleurone cells and form a third type of starchy endosperm cell, which differs in composition from the prismatic and central starchy endosperm cells.

The basic structure of the endosperm is therefore established by about 14 daa, when the synthesis of storage components (principally starch and proteins) for grain filling is initiated (Figure 7.4b, c). Grain filling is most active between about 14 and 35 daa in plants

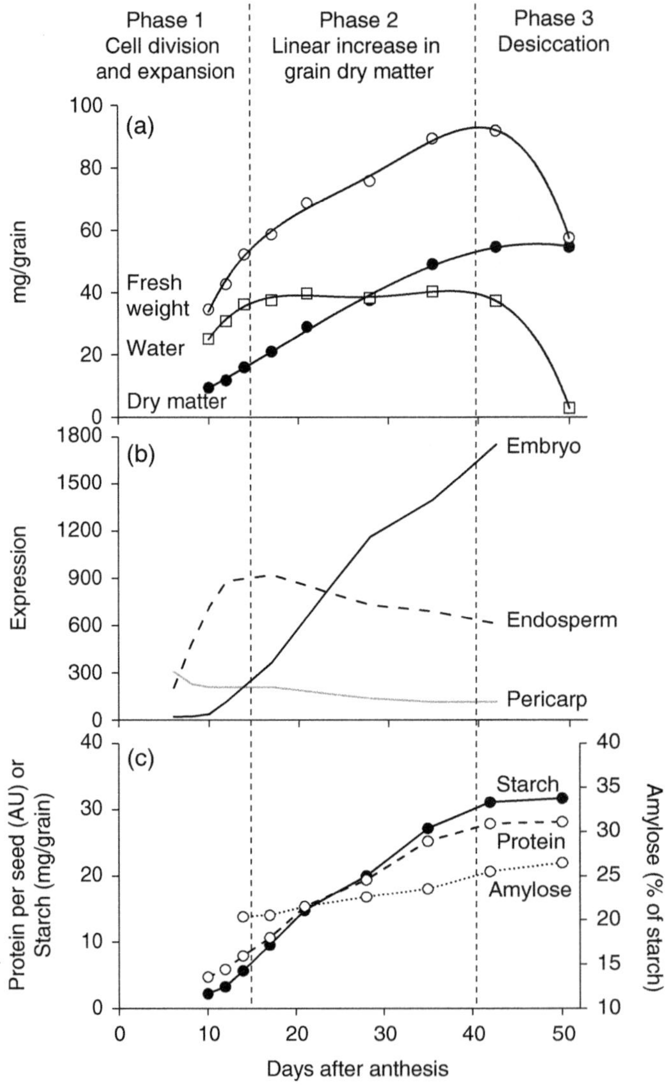

Figure 7.4 Development of grains of wheat cv. Hereward grown in the UK. (a) Changes in the quantities of water and dry matter in the grain. The three phases are delineated by the changes in water content as described in Figure 3.10; (b) expression patterns of transcripts associated with the pericarp, endosperm, and embryo; (c) accumulation of protein and starch, and changes in the amylose content of starch. *Source:* Redrawn from Shewry et al. (2009a) with permission.

grown under typical UK conditions, with the grain dry weight increasing by over twofold, and is essentially complete by 42 daa, after which fresh weight falls due to desiccation (Figure 7.4a). However, it should be noted that the precise durations of all the phases of development are affected by the environment, with development being faster at higher temperatures (Chapter 4).

Nutrients are supplied to the developing grain through a vascular bundle in the rachis close to the region of attachment of the grain (Figure 7.5). They are then transported into the developing endosperm through transfer cells, which have two origins. Maternal transfer cells are formed by the nucellar projection. This is derived from the nucellus, a tissue that is present between the integuments and embryo sac but largely degenerates during the early stages of grain development. These nucellar transfer cells transfer the nutrients into the endosperm cavity where they are reabsorbed by endosperm transfer cells, which are modified aleurone cells.

The starchy endosperm cells undergo endoreduplication (continued DNA synthesis without cells division) during grain development, leading to nuclei with up to ×8 the amount of DNA in a normal triploid nucleus (Chojecki et al. 1986). The functional significance of this is not known, but it could facilitate the high levels of storage product synthesis occurring in these cells. Finally, the starchy endosperm cells undergo programmed cell death, including degradation of nuclear DNA. Consequently, the starchy endosperm of the mature grain comprises dry dead cells.

7.1.4 Embryo Development

The fertilized egg cell divides to form a club-shaped embryo, which starts to differentiate about one week after anthesis to give the embryonic axis (comprising the plumule

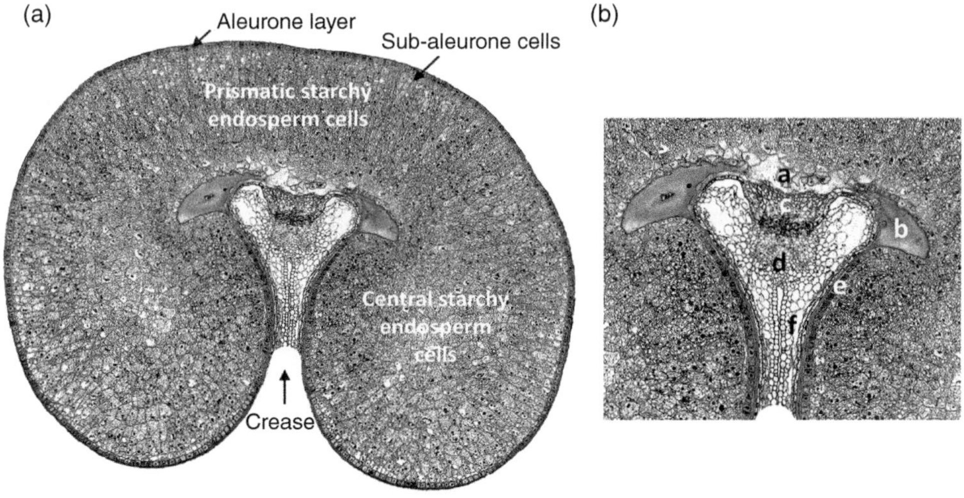

Figure 7.5 Cross-section of a developing wheat grain at about 28 days after anthesis. (a) Whole section. The outer layers of the grain are not present as they detach during sample preparation. (b) Enlargement of the transfer cell region. Labelling on part (b): a, endosperm transfer cells; b, endosperm cavity; c, nucellar projection transfer cells; d, vascular bundle; e, aleurone layer; f, seed coat. *Source:* Figure provided by Dr. Yongfang Wan (Rothamsted Research) and the Rothamsted Bioimaging Unit.

surrounded by the coleoptile, the mesocotyl and the radicle) and a single cotyledon called the scutellum. The mature embryo comprises about 3% of the grain dry weight, with approximately equal proportions of embryonic axis and scutellum (Barron et al. 2007).

7.2 Structure of the Mature Grain

The mature grain (Figure 7.6) has an elongated oval shape with a marked longitudinal crease on its ventral side, which is the site of entry of nutrients into the grain. The embryo is located on the dorsal side of the grain. A scar marks the region of attachment at the basal end of the grain with a bush of hairs at the apex.

The outer layers of the mature grain account for 7–8% of the grain dry weight. They comprise several layers of dry dead cells that are fused together and derived from maternal tissues: the nucellar epidermis, the inner and outer integuments, which form the testa (seed coat) and the tube cells, cross cells (which together form the inner pericarp) and outer pericarp, which together form the pericarp (Figure 7.6). At maturity the testa consists of several layers of crushed cells that are fused to the pericarp. Nevertheless, it is functionally important, contributing to the regulation of water uptake during late development and in the mature seed, and being the location of pigment in red grains.

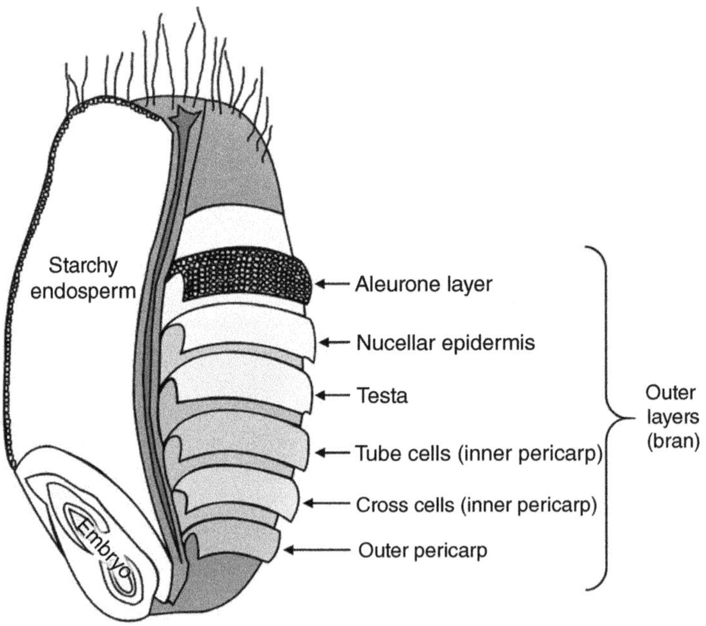

Figure 7.6 Schematic structure of the mature wheat grains showing the different tissues. *Source:* Taken from Hazard et al. (2020). Figure kindly provided by Dr. Brittany Hazard.

Table 7.1 Histological composition (% DM) of wheat mature grains from two wheat cultivars (Caphorn and Crousty).

	Caphorn	Crousty
Embryo	3.0	3.2
embryonic axis	1.5	1.7
scutellum	1.5	1.5
Endosperm	89.2	90.1
starchy endosperm	82.7	83.7
aleurone layer	6.5	6.4
Outer layers (nucellar epidermis, testa, pericarp)	7.8	6.7

Source: Taken from Barron et al. (2007) with permission.

The mature endosperm comprises two tissues. The outer cells form the aleurone layer. These cells only account for about 7% of the total endosperm, 6.5% of the whole grain (Table 7.1), and have thick walls which contribute 35–40% of their dry weight. The mature aleurone cells contain oil droplets and *aleurone grains*: these are protein bodies with *globoid* inclusions that are rich in minerals bound to phytate (see Chapter 9). The aleurone cells within the crease of the grain differentiate to form the endosperm transfer cells.

The starchy endosperm accounts for about 83% of the whole grain. It comprises thin-walled cells, which are differentiated into several types differing in their locations in the tissue and their compositions. The cells adjacent to the aleurone layer, which are called sub-aleurone cells, are small with high contents of protein (as high as 50%) and little or no starch. Prismatic cells radiate from the sub-aleurone towards the centre of the grain, where the central cells in the lobes of the grain are rounded or polygonal. There is, in general terms, a progressive increase in cell volume from the outer to inner parts of the grain, with an increased proportion of starch and decreased proportion of protein (as discussed in Section 7.4). This is illustrated in Figure 7.7, which shows a section from the lobe of a developing caryopsis at about 28 days after anthesis.

In contrast to the dead starchy endosperm cells, the aleurone cells remain alive in the mature grain and, in conjunction with the scutellum, secrete enzymes to digest the starchy endosperm during germination.

The embryo is situated at the point of attachment to the spikelet and accounts for about 3% of the grain dry weight (Table 7.1). It consists of a single cotyledon (the scutellum) which lies adjacent to the starchy endosperm and is joined by the hypocotyl to the embryonic axis (Figure 7.8). The embryonic axis comprises the shoot (plumule), which consists of several leaf primordia and the shoot apex, or dome, and the primary root (radicle), which are protected by the coleoptile and coleorhiza, respectively. The epiblast is a scale-like outgrowth attached on the opposite side of the embryo to the scutellum. Although it had been suggested to represent a rudimentary second cotyledon or modified leaf, Brown (1959)

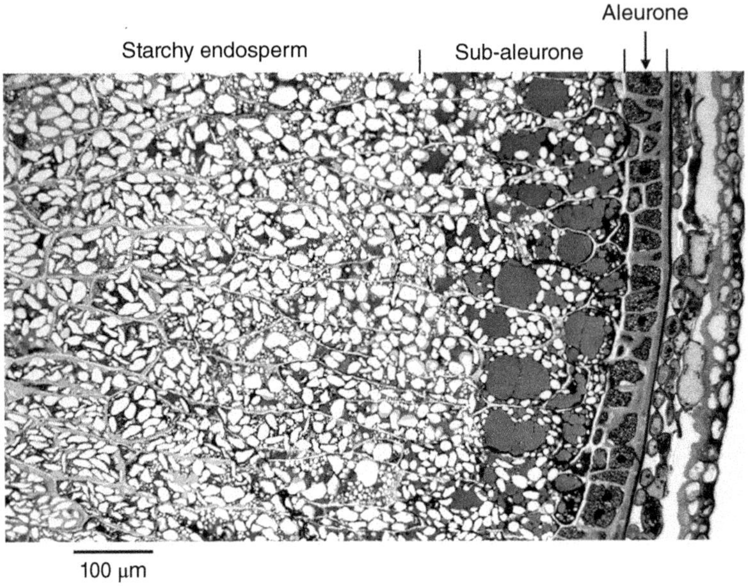

Figure 7.7 Light micrograph of a cross section from the cheek region of a developing wheat grain at about 28 days after anthesis, stained to show protein and labelled to show the aleurone, sub-aleurone and central starchy endosperm cells. *Source:* Figure kindly provided by Dr. Cristina Gritsch (Rothamsted Research) and Dr. Paola Tosi (University of Reading).

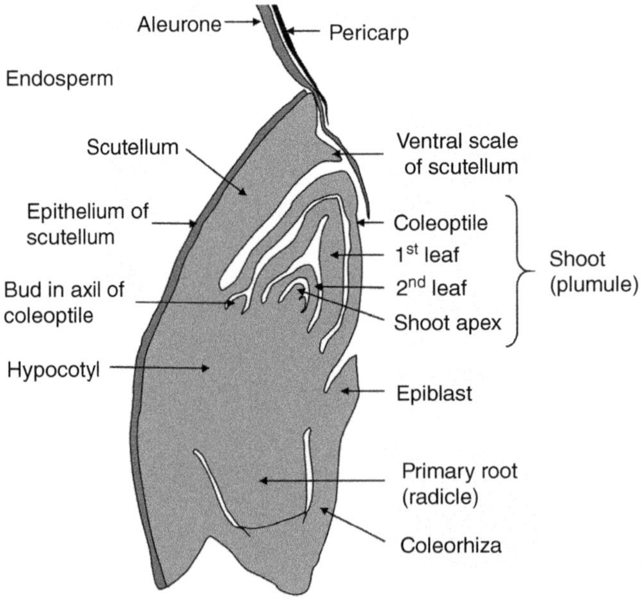

Figure 7.8 Diagrammatic structure of the embryo of a mature wheat grain. *Source:* Redrawn from Percival (1921).

concluded that it represented an outgrowth of the coleorhiza. Whereas the embryonic axis forms the seedling during germination, the scutellum is a storage tissue, providing nutrients for the seedling and also secreting enzymes to digest the starchy endosperm during germination. It therefore has a similar biological role to the aleurone layer, which it resembles in many respects, including the presence of similar storage components: globulin storage proteins, oil, and minerals.

7.3 Major Components of the Mature Grain

The contents of different components of the mature grain are summarized in Table 7.2. The *outer layers* correspond to the maternal tissues that protect the mature grain: the nucellar epidermis, testa, and pericarp. These account for just over half of the bran fraction produced by milling, the remainder being the aleurone layer. Hence values for bran and aleurone are also given in Table 7.2.

As implied by its name, the starchy endosperm is rich in starch, which comprises about 70–80% of its dry weight. It also contains 10–15% protein, mainly gluten storage proteins, about 4–5% dietary fibre (Hazard et al. 2020), and 2–3% lipid. However, the starchy endosperm only contains very low contents of other components, such as phytochemicals, minerals (determined as ash in Table 7.2), and vitamins.

The aleurone cells comprises 35–40% fibre, which is derived from the thick cell walls, about 20% of protein, and up to 10% lipid (oil). They are also rich in minerals, B vitamins, and phytochemicals, with little or no starch. The embryo is also rich in protein, oil, minerals, vitamins, and phytochemicals, with a clear distinction between the compositions of the scutellum (which stores protein, starch, and minerals) and the embryonic axis. However, most of the minerals in the aleurone and scutellum are bound to phytate (as discussed in Chapter 10).

The major component of the outer layers is fibre (cell wall polysaccharides and lignin), which accounts for about half of the dry weight (Table 7.2), but they also contain phytochemicals, notably phenolic components.

Table 7.2 Approximate composition (% DM) of major components in grain tissues.

	Whole grain	Starchy endosperm (white flour)	Aleurone layer	Outer layers (nucellar epidermis, testa and pericarp)	Bran (aleurone + outer layers)	Embryo (germ)
Protein	10–15	7–12	18	4	12–15	20–25
Starch	65–75	70–80	0	0	0	0
Fibre	10–15	4–5	35–40	45–50	40–50	15–20
Lipids	1.8	1–2	6–10	0.5–1	5–6	25–30
Ash	1–3	0.2–0.5	14–17	3	7–9	4–6

Source: Data from various sources.

7.3.1 Carbohydrates

7.3.1.1 Monosaccharides, Disaccharides, and Oligosaccharides

Wheat grain contains a range of sugars, including the monosaccharides D-glucose (0.03–0.09% of dry weight) and D-fructose (0.06–0.08%), the disaccharides sucrose (comprising glucose and fructose units) (0.54–1.55%) and maltose (two glucose units) (0.05–0.18%), and the trisaccharide raffinose (galactose, glucose, and fructose units) (0.19–0.68%) (Stone and Morell 2009).

However, the most abundant oligosaccharides in wheat grain are polymers of fructose, which are usually called fructans (reviewed by Grausgruber et al. 2020). Fructans comprise three or more fructose units, with some forms also having a single glucose unit. Wheat contains graminan-type (branched) fructans with both β-(2,1) and β-(2,6) linkages and an average degree of polymerization of 5–8. The fructan content of whole wheat grain varies from 0.84 to 1.85% (mean 1.28%). Fructans also differ in their concentrations in different tissues of the grain. Analysis of milling fractions shows higher contents in bran (outer layers including aleurone) (mean 7%) and shorts (a fraction comprising pieces of starchy endosperm attached to outer layers) (3.6%), and lower contents in the germ (embryo) (2.1%) and white flour (starchy endosperm) (1.5%) (Haska et al. 2008).

7.3.1.2 Starch

Starch is located almost entirely in the starchy endosperm. It consists of glucose molecules assembled into two polymers: amylose (~20–30% of the total) and amylopectin (~70–80% of the total). Amylose polymers are largely linear chains with α-1,4 linkages and masses of up to 8×10^5 while amylopectin molecules are larger (molecular mass 10^6–10^8), with α-1,4 linkages and α-1,6 linkages which form branch points (Figure 7.9).

Starch is deposited in granules in the amyloplasts (modified plastids) of the developing starchy endosperm cells. Starch granules have a highly organised structure with concentric amorphous lamellae that contain amylose polymers and branch points of amylopectin alternating with crystalline lamellae formed from amylopectin chains.

Wheat starch granules occur in two sizes. About 90% of the total starch is present in large A-granules, which are disc-shaped with an average diameter of about 25 μm. However, these only account for about 10% of the total number of granules, with 90% of the total (10% of the weight) being small B-granules which are more spherical with average diameters of about 5 μm. These two populations are readily observed in light micrographs of developing grain (Figure 7.10a) and scanning electron micrographs of freeze-fractured cells of mature grain (Figure 7.10b).

Although the proportions of amylose and amylopectin are similar in all cereals, they can vary due to natural mutations in biosynthetic genes. Most mutations in all organisms are recessive and hence are more readily observed in diploid species than in polyploids where mutations on two or more genomes may be required to observe a phenotype. However, they may also be observed in polyploid durum and bread wheats due to partial effects on phenotype.

Four types of enzyme are required for starch synthesis. Starch synthases (SSs) and granule-bound starch synthases (GBSSs) extend the glucose chains in amylopectin and amylose, respectively, by forming α-1,4 bonds. Starch branching enzymes (SBEs) introduce

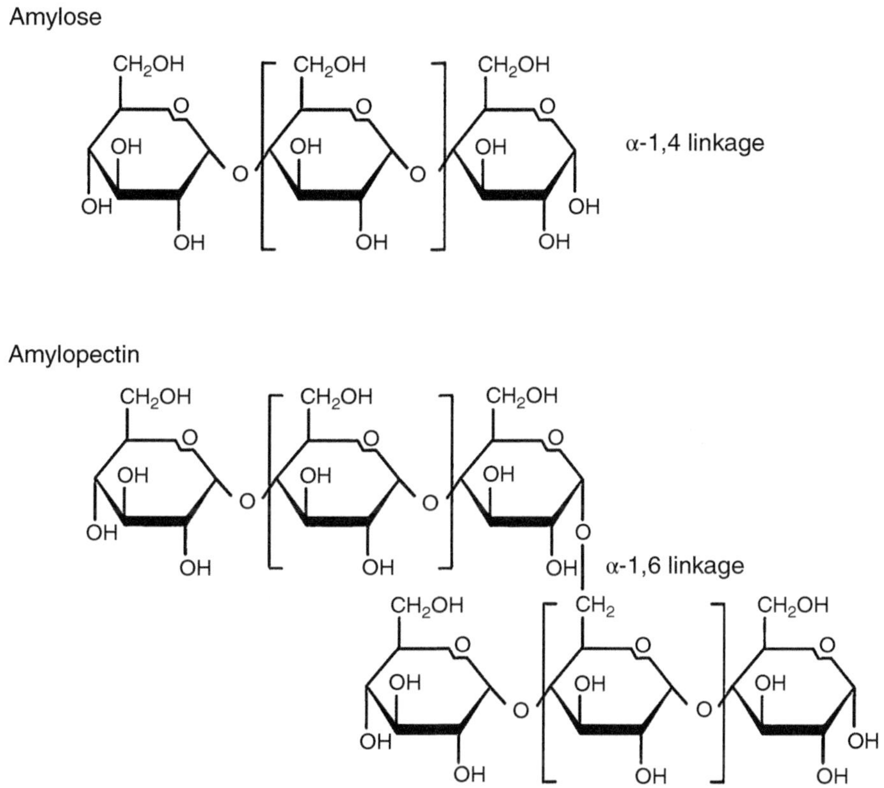

Figure 7.9 Schematic structures of starch polymers: amylose and amylopectin. *Source:* Modified from Shewry et al. (2020).

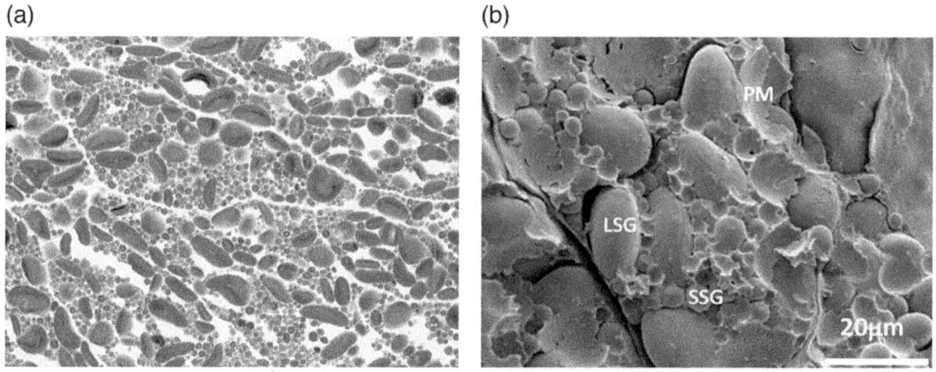

Figure 7.10 Light microscopy of a section of developing grain (a) and scanning electron microscopy of a freeze-fractured mature grain (b) showing large starch granules (LSG), small starch granules (SSG), and protein matrix (PM). *Source:* Figures provided by Dr Paola Tosi, Dr Cristina Gritsch, and Dr Louise Wells (Rothamsted Research).

α-1,6 branch points into amylopectin while starch debranching enzymes (DBEs) *fine tune* the structure by trimming the branches of the chains to create a structure that can crystallize to form the granule matrix.

Mutations in the GBSS enzymes that catalyse amylose biosynthesis lead to a *waxy* phenotype, with the starch comprising up to 100% amylopectin. However, the expression of the full waxy phenotype requires mutations in all three genomes of bread wheat and has not been observed in natural populations. By contrast, partial waxy phenotypes are quite common, with mutations in one, or less frequently two, of the genomes, resulting in reduced amylose content (Graybosch 1998). These naturally occurring GBSS mutations in the three genomes can also be combined by crossing to give the full waxy phenotypes. Waxy wheat is of interest to the food industry as waxy starch has higher viscosity and water-retention, which can be exploited in refrigerated and frozen foods and may slow the staling of bread.

Analysis of diploid cereals (such as barley and rice) has shown that an increased proportion of amylose in starch can result from reduced activity of SSs or SBEs (reviewed by Lafiandra et al. 2014). Naturally occurring mutations in these genes have therefore been identified in wheat by biochemical analysis and combined to give amylose levels of up to 40%, while even higher levels can be achieved by combining mutations induced by chemical mutagenesis (Lafiandra et al. 2014; Hazard et al. 2020).

High amylose starch has been of increasing interest in recent years as it is digested more slowly by humans and hence has health benefits in reducing the rate of release of glucose, which is a risk factor for type 2 diabetes. High amylose starch also differs from normal starch in its biochemical and biophysical properties but, in contrast to waxy starch, these properties make it less suitable for food processing and high amylose wheat lines are not currently used in large-scale commercial food production.

Starch accounts for 65–75% of the whole wheat grain and its pattern of accumulation during development essentially reflects the increase in grain dry weight, increasing rapidly from 14 daa until around 40 daa (Figure 7.4c). This is accompanied by an increase in the proportion of amylose, from about 20 to 26% in the study shown in Figure 7.4c (Shewry et al. 2009a).

7.3.1.3 Cell Wall Polysaccharides

The cell walls of the starchy endosperm and aleurone cells have similar compositions, comprising mainly polysaccharides. The widely accepted compositions of the two tissues are 70% and 65% arabinoxylan, 20% and 29% β-glucan, 7% and 2% glucomannan, and about 2% and 2% cellulose respectively (Stone and Morell 2009). However, a recent study has reported that the content of cellulose in the starchy endosperm may be higher and of β-glucan lower; this is discussed in Chapter 9. .

By contrast, the cell walls of the pericarp are more similar in composition to those of leaves and other vegetative tissues, being rich in cellulose (30%) and arabinoxylan (60%) but also containing about 12% lignin, a highly cross-linked phenolic polymer (reviewed by Stone and Morell 2009). The cell wall polysaccharides in wheat are often divided into soluble (water-extractable or WE) and insoluble (water-unextractable, WU) forms, which differ in their effects on human health and grain processing quality (Chapter 9).

Arabinoxylan (AX) is a polymer of two five-carbon sugars, arabinose, and xylose, and hence is often referred to as pentosan. The AX present in white flour has a simple structure, consisting of a backbone of β-D-xylopyranosyl (xylose) residues linked through (1,4) linkages. Some of these xylose residues are substituted with α-L-arabinofuranosyl (arabinose) residues at either one position (position 3) or at two positions (positions 2 and 3) (Figure 7.11).

Some arabinofuranosyl residues at position 3 of the xylose residues may themselves be substituted with ferulic acid (a phenolic acid) at the five position (as shown in Figure 7.11). Ferulate groups present on adjacent AX chains can also form cross-links, by oxidation to give dehydrodimers (diferulates). The extent of AX cross-linking is an important parameter as it affects the physiochemical properties (notably solubility and viscosity) of AX and hence the behaviour in food processing and the health benefits. The AX in starchy endosperm contain only low levels of ferulic acid, estimated to account for less than 0.5% (w/w) of WE-AX and less than 1% (w/w) of WU-AX.

The arabinoxylan present in the aleurone cells has a more complex structure than in white flour, with higher levels of ferulic acid and diferulic acids and additional esterification with *p*-coumaric acid and acetyl groups. The arabinoxylan in the pericarp is even more complex in structure, being highly branched with galactose and glucuronic acid residues. It is therefore sometimes termed glucuronoarabinoxylan. It also has high contents of ferulic acid, diferulic acids and, in addition, significant amounts of ferulic acid trimer.

The higher levels of complexity and substitution of the arabinoxylan in the bran tissues (aleurone and pericarp) account for the very low solubility of arabinoxylan in these tissues:

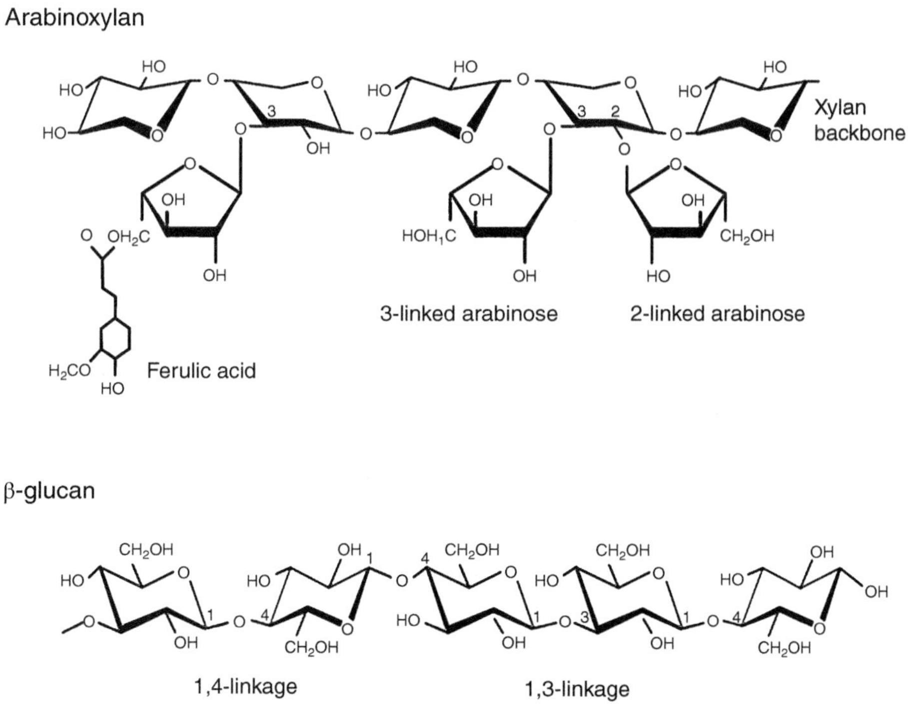

Figure 7.11 Schematic structures of the major cell wall polysaccharides present in the wheat grain: arabinoxylan and β-glucan. *Source:* Modified from Shewry et al. (2020).

whereas 5% or less of bran arabinoxylan is water-soluble, the proportion in white flour arabinoxylan varies from about 25 to 50% (Gebruers et al. 2008).

β-glucan has a simpler structure than AX, comprising only glucose residues joined by (1,3) and (1,4) linkages (Figure 7.11). Single (1,3) linkages are usually separated by two or three (1,4) linkages but longer stretches of (1,4) linked glucan (comprising over 10 units) have been reported for β-glucan in wheat bran. Such regions are sometimes referred to as *cellulose-like* and may form hydrogen bonds between adjacent glucan chains, as in cellulose. These intermolecular interactions are responsible for the insolubility of cellulose and may also account for the low water-solubility of wheat β-glucan (generally about 10–20% of the total).

The contents of cell wall polysaccharides vary between genotypes, from about 5.5 to 7.4% dry weight for arabinoxylan and 0.5 to 0.95% for β-glucan in whole grains (see discussion of dietary fibre in Chapter 9). Ever wider variation occurs in the arabinoxylan content of white flour, with a comparison of 150 lines showing that total arabinoxylan ranged from 1.35 to 2.75% dry wt (mean 1.93%) and water-extractable arabinoxylan from 0.3 to 1.4% dry wt (mean 0.5%) (Gebruers et al. 2008) (Figure 7.12).

The cell wall polysaccharides of the starchy endosperm cells show well-defined changes in composition and structure during grain development (Toole et al. 2009). These include an increase in the proportion of arabinoxylan compared to β-glucan, increased cross-linking of the arabinoxylan by the formation of diferulate cross-links, and remodelling of arabinoxylan structure with a decreased extent of arabinosylation (observed as a decreased proportion of disubstituted and increased proportion of monosubstituted xylose residues). These changes may reflect the biological requirements for the cell walls during the cell division and expansion, grain filling, and desiccation stages of grain development but will also affect the health benefits and processing properties of flours.

7.3.1.4 Arabinogalactan Peptide (AGP)

AGP accounts for up to 0.4% of the dry weight of white flour. It consists of a short (15 residue amino acid) peptide with three hydroxyproline residues, which are *o*-glycosylated

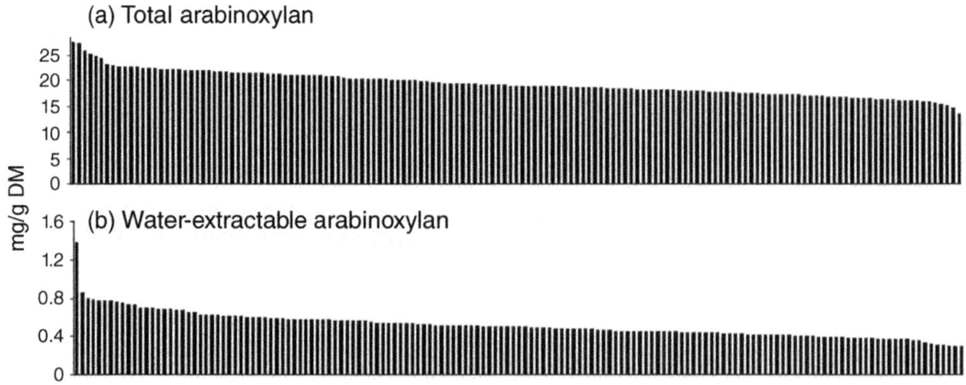

Figure 7.12 Concentrations of water-extractable and total arabinoxylan in 150 wheat genotypes grown on a single site, plotted in order of amount. *Source:* Data from study reported by Gebruers et al. (2008).

with branched arabinogalactan chains (Tryfora et al. 2010). The arabinogalactan is estimated to account for about 90% of the molecular mass and AGP was long considered to be present in the cell wall of the starchy endosperm. It is now considered to be located in the cytoplasm or vacuole of the wheat cell but does not appear to be essential for grain development and its biological function is not known (Wilkinson et al. 2017). However, it forms a significant proportion of the dietary fibre present in white flour (as discussed in Chapter 9).

7.3.2 Proteins

The content and properties of the proteins present in the mature wheat grain have major impacts on the end use quality of wheat as discussed in Chapters 1 and 8.

7.3.2.1 Grain Protein Content (GPC)

Total protein generally accounts for 10–15% of whole grain, 7–12% of white flour (Table 7.2). Farmers routinely adjust the protein content of their crop by the application of nitrogen fertiliser, but GPC is also under genetic control. Consequently, selective breeding has resulted in a difference of about 2% (dry weight basis) between the protein contents of high protein breadmaking cultivars and low protein cultivars bred for livestock feed when grown under the same conditions. Although GPC is clearly under multigenic control, major *high protein genes* have been identified and exploited in breeding programmes. Early studies identified high protein genes in lines derived from South America (Atlas 50 and Atlas 66) and India (Nap Hal), which were used extensively for breeding in North America. More recently attention has focused on high protein genes derived from wild emmer (see Chapter 1), which are often referred to as *dicoccoides* genes. This led to the identification of a major locus on chromosome 6B (*Gpc-B1*), which encodes a transcription factor that accelerates the senescence of vegetative tissues, resulting in increased mobilisation and transport of nitrogen and minerals into the grain (Uauy et al. 2006). Hence, the gene is associated with increased concentrations of iron and zinc in the grain as well as high protein (reviewed by Shewry 2007).

7.3.2.2 Grain Protein Deviation

There is a well-established negative relationship between grain yield and the concentration of protein in the mature wheat grain, allowing a simple negative regression line to be fitted (as shown in Figure 7.13). However, some wheat genotypes consistently deviate from this relationship, falling above or below the fitted line. This phenomenon is known as Grain Protein Deviation (GPD) (Monaghan et al. 2001) with positive GPD being of particular interest because it shows that the plant is able to transfer a high proportion of applied nitrogen into the grain. High GPD therefore reduces the need to apply nitrogen fertiliser and hence the financial and environmental costs of producing high protein wheat for breadmaking.

Comparisons of collections of genotypes grown in multiple environments (years and/or field sites) show consistent differences in GPD (Figure 7.14), with variation also being observed in the stability of the trait (indicated by the heights of the boxes in the plot). The heritability of GPD has been estimated as 0.44, with the largest component contributing to

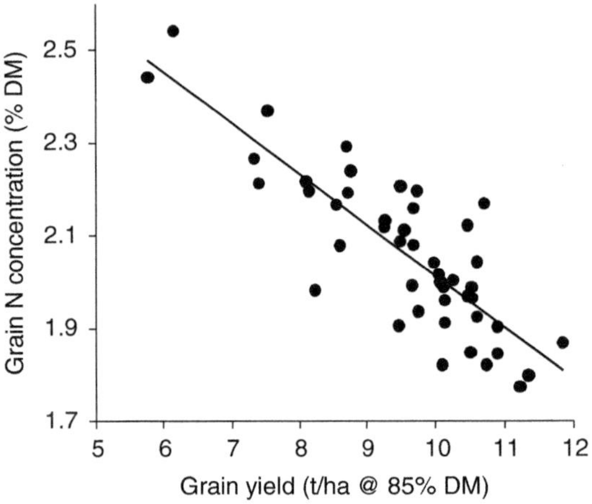

Figure 7.13 Mean grain N contents and yields for 47 wheat cultivars grown at Rothamsted Research between 2004 and 2012. *Source:* Data and figure kindly provided by Andrew Riche and Malcolm Hawkesford (Rothamsted Research).

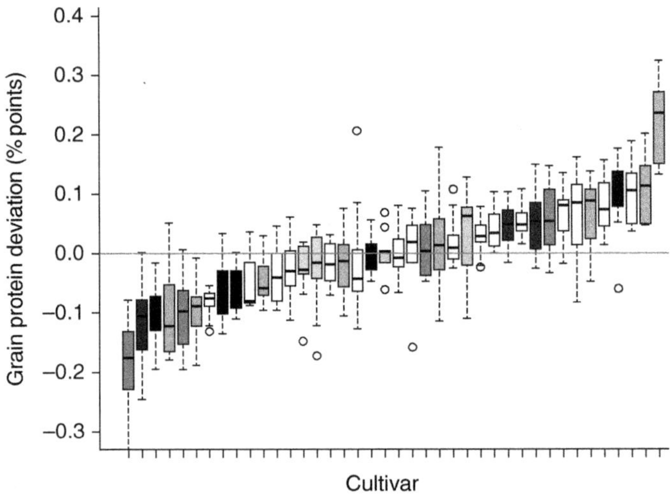

Figure 7.14 Boxplot of mean Grain Protein Deviation (GPD) for 40 wheat cultivars grown in 10 environments (five sites each in 2016 and 2017). *Source:* Redrawn from Mosleth et al. (2020).

the genetic variance being genotype (0.30), followed by the interactions between genotype and environment (i.e. different years and/or sites) (0.11) and a small contribution of the level of nitrogen fertilisation (Mosleth et al. 2020). This suggests that GDP is a multigenic trait but that selection is a viable target for breeders. However, the relatively low genetic component (0.30) suggests that a significant commitment will be required.

7.3.2.3 Essential Amino Acid Composition

The nutritional quality of proteins, whether for humans or livestock, is determined by their contents of essential amino acids (EAA). EAAs must be provided in the diet and if the amount of only one is limiting, the others will be broken down and excreted. The EAA composition of wheat is not a major concern for human nutrition, as most wheat-consuming populations also consume other sources of dietary protein. However, it is a major factor when formulating diets for livestock where cereal grain is blended with other components.

Ten amino acids are strictly essential for humans: lysine, isoleucine, leucine, phenylalanine, tyrosine, threonine, tryptophan, valine, histidine, and methionine, but cysteine is often included as it is synthesized from methionine, and combined values are used for tyrosine and phenylalanine (which can be interconverted). The requirements also differ with age, with higher levels being required by children than by adults, as they require amino acids for growth as well as maintenance.

In wheat the first limiting amino acid is lysine, followed by threonine, and then the sulphur-containing amino acids (cysteine and methionine). These deficiencies arise from the low levels of these EAAs in the gluten storage proteins present in the starchy endosperm tissue (see Section 7.3.2.5.1 and Table 7.3). Consequently, deficiencies are more severe in white flour than in whole grain. For example, Shewry (2007) quotes values for 2.8g lysine/100g protein in whole grain and 2.2g in white flour, compared with FAO recommendations of 6.6g for children and 1.6g for adults. The proportions of lysine and other EAA also decrease at higher grain protein contents, due to an increased proportion of gluten proteins (see below).

7.3.2.4 Wheat Grain Proteins

The study of wheat grain proteins has a long history, with wheat gluten being first described over 250 years ago. Since then, there has been a massive volume of research, and hence of publications, with a strong focus on their role in determining processing quality. However, in recent years attention has been increasingly focused in their role in human health, and in particular in triggering adverse reactions (including allergy and intolerance). These aspects are considered in Chapters 8 and 9, respectively.

The precise number of proteins in wheat grain has not been determined, but proteomic studies of white flour and whole grain have indicated up to 500 in the former and over 1000 in the latter. Wheat proteins have been classified on several criteria, with the oldest, and still most widely used, classification being on solubility.

The classification of proteins on solubility is based largely on the work of the great American protein chemist T.B. Osborne. Working between 1886 and 1928, Osborne published studies of seed proteins from 32 plant species including wheat. He developed a sequential extraction procedure, in which proteins are extracted with water (albumins), dilute salt solution (globulins), alcohol-water mixtures (60–70% v/v ethanol) (prolamins), and alkali (glutelins). The first two classes and names are widely accepted and used for proteins outside the plant kingdom (such as egg albumin and globulin antibodies), whereas the terms prolamin and glutelin are only used in cereal science.

Osborne coined the name prolamin to reflect the fact that the proteins had high contents of proline and amide nitrogen, which is now known to be derived from the amino acid glutamine. These early analyses were borne out by subsequent studies, which showed that wheat prolamins contain 30–50% glutamine and 15–30% proline. Osborne concluded that prolamins were restricted to the grains of cereals and grasses, and they have been given names based on the species of origin: gliadins in wheat, hordeins in barley, secalins in rye, avenins in oats, etc.

The glutelins have proved more problematic to define, as recognised by Osborne himself, because they comprise the mixture of proteins that remains after the albumins, globulins, and prolamins have been extracted. Nevertheless, Osborne considered that they were sufficiently well-characterised in wheat to be defined as a separate protein group and called them glutenins. We now know that the gliadins and glutenins of wheat actually form a single group of proteins, which are the major storage proteins in the starchy endosperm of the grain, with only small amounts of other storage protein types.

Other proteins present in the grain have structural, metabolic, and protective roles, including proteins in cell walls and membranes, enzymes, and inhibitors of hydrolytic enzymes.

7.3.2.5 Gluten Proteins: Gliadins and Glutenins

The term gluten is widely used in the press and media, but rarely defined for the non-expert. It can be defined as the cohesive visco-elastic mass that remains when a dough made from wheat flour is gently washed with water to remove soluble components, starch, and other particulate material. It is mostly protein, but also contains small amounts of starch, lipids, and other components which may be either entrapped or chemically bound. Similarly, although most of the proteins present in isolated gluten are gliadins and glutenins (i.e. gluten proteins), small amounts of other proteins are also present. Although gluten can be prepared as a cohesive mass, it is present in dough as a continuous network that extends throughout the dough. The importance of gluten in grain processing is discussed in Chapter 8.

Gluten is generally described as accounting for about 80% of the total grain proteins, but the precise proportion varies, being higher when the total grain protein content is high. This reflects the biological role of gluten proteins, which act as a store for nitrogen that is surplus to the requirements of the plant for growth and metabolism and hence are increased in amount when the surplus nitrogen increases. Glutenin generally accounts for about half of the total gluten proteins.

Gliadins are classically defined as prolamins and glutenins as glutelins, based on the fact that gliadins are soluble in alcohol-water mixtures whereas glutenins are insoluble. However, we now know that this difference in solubility does not reflect fundamental differences between the two groups of proteins, but results from the fact that the glutenins are large polymeric complexes stabilized by inter-chain disulphide bonds, whereas the gliadins are monomers. When the disulphide bonds stabilizing the glutenin polymers are reduced, the monomeric subunits are soluble in alcohol-water and have been shown to be closely related to the gliadins.

Both fractions, gliadins and glutenin subunits, comprise mixtures of proteins, which vary in their numbers and properties between genotypes, with a total number of between 50 and 100 in a single genotype. Both fractions are also divided into several protein sub-groups (Table 7.3).

7.3.2.5.1 Types and Properties of Gliadins and Glutenin Subunits

The earliest classifications of gliadins and glutenin subunits were based on electrophoretic separations. The gliadins were initially classified into four groups, called α-gliadins, β-gliadins, γ-gliadins, and ω-gliadins based on decreasing mobility when separated by

Table 7.3 Summary of the types and characteristics of wheat gluten proteins.

Gluten protein type	Polymers/ monomers	Molecular mass	% of total gluten fraction	Partial amino acid composition (mol %)
HMW prolamins				
HMW subunits	polymers	65–90 000	6–10	30–35 glutamine, 10–16 proline, 15–20 glycine, 0.5–1.5 cysteine, 0.7–1.4 lysine
S-rich prolamins				
α-gliadins, γ-gliadins	monomers	30–45 000	70–80	30–40 glutamine, 15–20 proline, 2–3 cysteine, < 1% lysine
B-type LMW subunits, C-type LMW subunits	polymers			
S-poor prolamins				
ω-gliadins	monomers	30–75 000	10–20	40–50 glutamine, 20–30 proline, 0–0.5 phenylalanine, 0–0.5 lysine, 0 cysteine (1 cysteine residue in D-type LMW subunits)
D-type LMW subunits	polymers			

Source: Taken from Shewry et al. (2019).

electrophoresis at low pH. Similarly, early studies showed that the glutenin subunits released by reduction of the disulphide bonds stabilizing the glutenin polymers could be separated by sodium dodecylsulphate electrophoresis (SDS-PAGE) on the basis of molecular mass into two major groups of components, called the low molecular weight (LMW) and high molecular weight (HMW) subunits (Figure 7.15). The LMW glutenin subunits are then further classified into three types: a major group called B-type and two minor groups called C- and D-types. Although these groups were initially defined based on electrophoretic separations, reflecting difference in charge at low pH (gliadins) and molecular mass (glutenin subunits), the classification has been largely supported by more detailed studies. However, an early modification to the classification, based on partial amino acid sequences, was to combine the α- and β-gliadins to form a single group which is sometimes called α-type gliadins.

Since this classification was established there has been a massive increase in our knowledge of the amino acid sequences of the proteins, based on the analysis of cloned cDNAs (derived from the RNA of expressed genes) and genomic DNA. For example, over 24 000 sequences corresponding to gluten proteins were present in the UniProt database in 2017, which corresponded to over 600 individual proteins (Bromilow et al. 2017).

These sequences confirm that the gliadins and glutenin subunits are related and show that they can be grouped into three families of prolamin, which also occur in related cereals (barley, rye) in the same grass Tribe (the Triticeae). This classification is summarised in

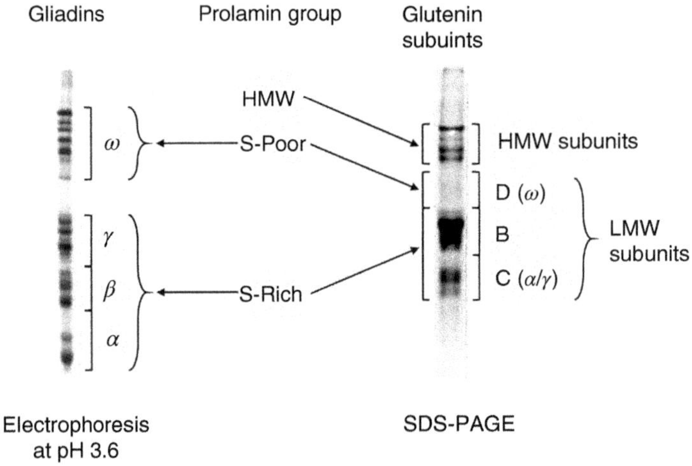

Figure 7.15 Separation of the major groups of gluten protein by electrophoresis. The gliadin fraction is separated by electrophoresis at low pH and the reduced glutenin subunits by SDS-PAGE. *Source:* Modified from Shewry et al. (1999) with permission.

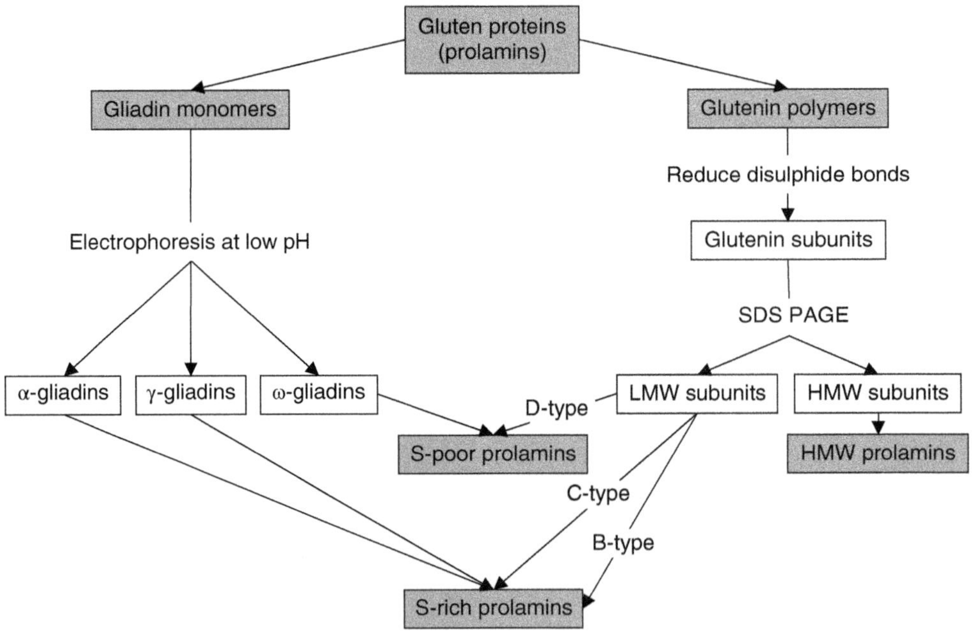

Figure 7.16 The classification of wheat gluten proteins.

Figure 7.16 and the characteristics of the groups in Table 7.3. The properties of the three groups can be briefly summarized as follows:

1) The high molecular weight (HMW) prolamins comprise the HMW subunits of glutenin and account for up to 10% of the gluten proteins. They are characterized by high contents of glutamine (30–35%) and glycine (15–20%) but relatively low contents of proline

(10–16%) compared with other gluten proteins. They also contain several (usually between 4 and 7) cysteine residues.

2) The sulphur-rich (S-rich) prolamins comprise the α-type gliadins, γ-gliadins, B-type LMW subunits, and C-type LMW subunits and account for 70–80% of the total gluten proteins. They are rich in cysteine residues (hence their name), which form intra-chain disulphide bonds (in α-type and γ-gliadins) or inter-chain and intra-chain disulphide bonds (in LMW subunits). Whereas the B-type LMW subunits form a distinct group within the S-rich prolamin family, the C-type LMW subunits are essentially forms of α-gliadins and γ-gliadins in which mutations have resulted in the presence of additional cysteine residues which form inter-chain disulphide bonds with other glutenin subunits.
3) The sulphur-poor prolamins comprise the ω-gliadins and D-type LMW subunits and account for 10–20% of the total gluten proteins. They have highly unusual amino acid compositions, with 40–50% glutamine, 20–30% proline and 8–9% phenylalanine. The ω-gliadins contain no cysteine residues while the D-type LMW subunits are mutant forms with single cysteine residues which allow their incorporation into the glutenin polymers. All have low contents of lysine, which is therefore the first limiting amino acid in the whole grain.

7.3.2.5.2 Sequences and Structures of Gluten Proteins

The sequences of wheat gluten proteins, and related proteins from barley and oats, show three features that allow them to be defined as a distinct protein family, and provide explanations for their properties (Figure 7.17):

1) Their amino acid sequences can be divided into distinct regions, or domains, which differ in their amino acid compositions.
2) One of these domains comprises repeated sequences of one or more short (generally 3 to 9 amino acid residue) peptide motifs.
3) With the exception of the ω-gliadins, they contain cysteine residues that form intra- and, in some cases, inter-chain disulphide bonds.

Sequence domains: The repetitive domains account for between 30 and 50% of the sequences of the S-rich prolamins and 75–85% of the sequences of the HMW subunits, and in both cases are flanked by non-repetitive domains. In the S-poor prolamins the repetitive domains account for almost the whole protein, with only short non-repetitive sequences at each end. This domain structure has several consequences for the properties of the proteins:

1) The peptide sequences present in the repetitive domains are rich in glutamine, proline and, depending on the protein group, either glycine and tyrosine (HMW prolamins) or phenylalanine (S-rich and S-poor prolamins) (Figure 7.17). Hence, they are responsible for the unusual amino acid compositions of the whole proteins.
2) Glutamine is generally regarded as a hydrophilic amino acid and hence glutamine-rich proteins are usually water-soluble. However, the regularly repeated glutamine residues in the repetitive domains of prolamins facilitate the formation of hydrogen bonds between the sequences present in adjacent proteins, resulting in insolubility in aqueous solvents.

The repetitive domains are therefore responsible for the unique features (solubility and amino acid composition) which allow prolamins to be defined as a group.

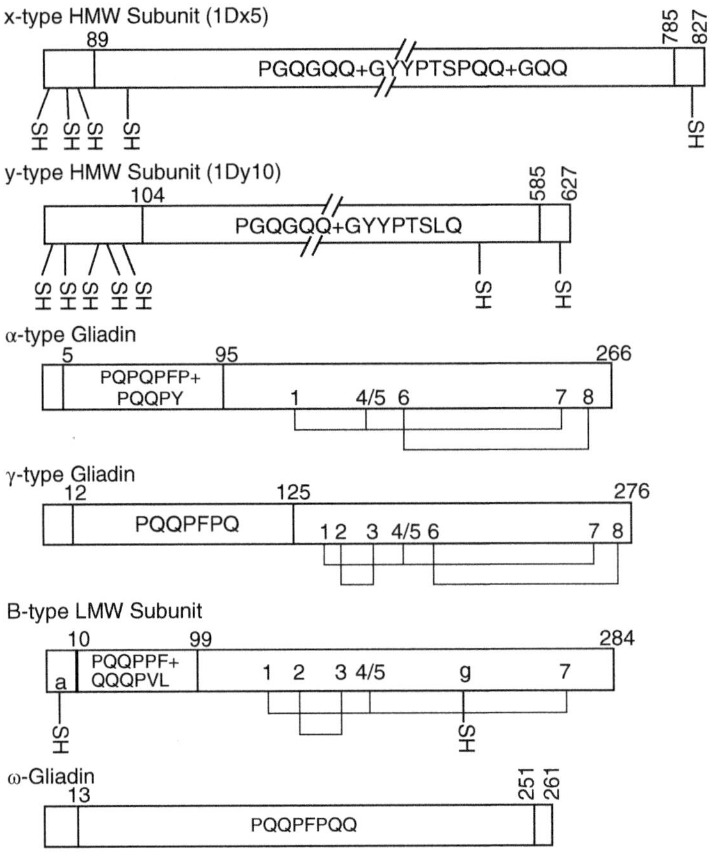

Figure 7.17 Schematic structures of typical wheat gluten proteins. The figure shows the division of prolamin sequences into domains and the consensus repeat motifs of the repetitive domains. Whereas single consensus repeat motifs can be defined for the α-, γ-, and ω-gliadins shown, the HMW subunits and B-type LMW subunit have tandem and/or interspersed repeats of two or three motifs. Detailed alignments of motifs in a range of gluten proteins are shown in Shewry et al. (2009a). Conserved cysteine residues in α- and γ-gliadins and B-type LMW subunits are numbered and established intra-chain disulphide bonds indicated. *Unpaired* cysteine residues in HMW and LMW subunits are indicated by SH. However, current evidence indicates that all of these are involved in the formation of disulphide bonds. Standard single letter abbreviations for amino acids are used: F, phenylalanine; G, glycine; P, proline; Q, glutamine; S, serine; T, threonine; Y, tyrosine. *Source:* Taken from Shewry et al. (2003) with permission.

Protein structures: The repetitive and non-repetitive domains of prolamins form different structures. The non-repetitive domains form globular structures that are rich in α-helix and similar to those formed by small water-soluble proteins (such as ATIs, Section 7.3.2.6.4). These domains also contain most (usually all) of the cysteine residues which form inter-molecular and intra-molecular disulphide bonds. By contrast, the repetitive domains may form unusual extended helical structures which reflect their repeated sequences. A possible model for this structure is shown in Figure 7.18, comprising β-reverse turns orientated to form a loose, spiral structure.

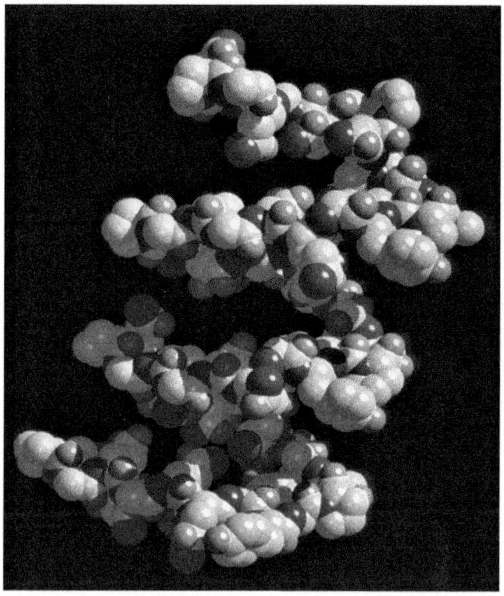

Figure 7.18 Space filling model for the repetitive domain of a wheat HMW subunit, showing β-reverse turns forming a spiral structure. *Source:* Taken from Parchment et al. (2001) with permission.

Disulphide structures: The α-gliadins contain six conserved cysteine residues, which form three intra-chain bonds. These are numbered 1, 4, 5, 6, 7, and 8 in the schematic in Figure 7.17. These same six cysteine residues and three disulphide bonds are also conserved in the γ-gliadins, which also contain two additional cysteine residues (numbered 2 and 3 in Figure 7.17) that form a fourth intra-chain bond. The B-type LMW subunits contain eight cysteine residues. Six of these are equivalent to residues 1, 2, 3, 4, 5, and 7 in the γ-gliadins and form three intra-chain bonds. The other two are *unpaired* cysteine residues and available to form inter-chain disulphide bonds. As discussed, the C-type LMW subunits are essentially α-gliadins and γ-gliadins with additional unpaired cysteine residues available to form inter-chain disulphide bonds. Most C-type LMW subunits contain only a single unpaired cysteine, and hence will terminate (rather than extend) chains of subunits in glutenin polymers.

The HMW subunits usually contain between 4 and 7 cysteine residues, most of which appear to form inter-chain disulphide bonds which stabilize the glutenin polymers. However, one intra-chain disulphide bond has also been reported. The role of the HMW subunits in the formation of glutenin polymers and their importance in determining processing quality is discussed in Chapter 8.

7.3.2.5.3 Polymorphism and Genetics of Gluten Proteins

As discussed previously, the gluten protein fraction comprises between about 50 and 100 proteins, which vary in number, amount, and sequences between genotypes (including commercial cultivars). This polymorphism arises from the presence of multigene families located at multiple genetic loci on the group 1 and 6 chromosomes (Figure 7.19). The simplest loci, called *Glu-1*, encode the HMW subunits of glutenin. These are located on the long arms of the group 1 chromosomes and comprise just two genes encoding HMW subunit proteins. The LMW subunits, γ-gliadins, and ω-gliadins are encoded by a series of

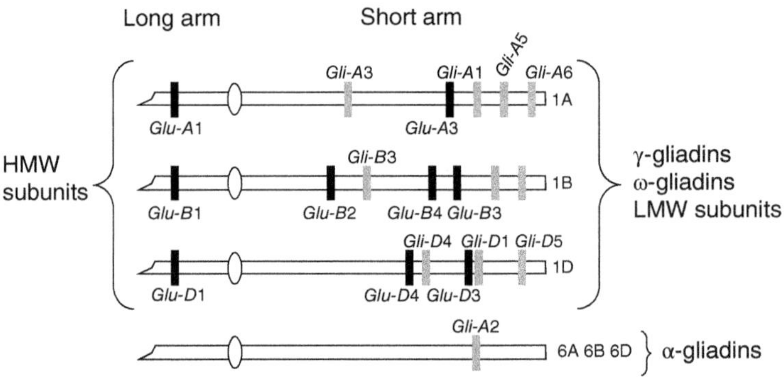

Figure 7.19 Summary of the location of loci encoding gluten proteins on the group 1 (1A, 1B, 1D) and group 6 (6A, 6B, 6D) chromosomes of wheat. Black bars are for glutenins, grey bars are for gliadins. *Source:* Taken from Shewry et al. (2003) with permission.

complex loci on the short arms of the group 1 chromosomes (*Gli-1, 3, 4, 5,* and *6,* which encode gliadins and *Glu-2, 3, 4,* which encode LMW subunits) while the α-type gliadins are encoded by the *Gli*-2 loci on the group 6 chromosomes. These are summarised in Figure 7.17 and discussed in detail by Shewry et al. (2003). Individual loci vary in complexity but may encode 20–30 individual α-type and γ-gliadin proteins in a single genotype.

Additional protein polymorphism may be generated by post-translational modification of the proteins, notably deamidation of glutamine residues, although this is probably limited in extent and modifications occurring *in planta* are difficult to distinguish from modifications occurring in extraction and analysis.

7.3.2.5.4 Gluten Protein Synthesis and Deposition

Although gluten protein transcripts may be detected in the developing starchy endosperm at 8–10 days after anthesis, significant amounts of protein are not detected until about 12 days, with rapid accumulation occurring between 14 and 35 days (Shewry et al. 2009a) (Figure 7.4b). Small differences in the profiles of gene expression and protein accumulation are observed between the major protein groups, and to a lesser extent within the groups, with the accumulation of the HMW subunits being slightly delayed compared to the other gluten protein groups (Shewry et al. 2009a).

Gluten proteins are termed *secretory proteins* because they are synthesized on ribosomes attached to the endoplasmic reticulum (ER) with an N-terminal signal peptide, which directs them into the lumen of the ER (where it is co-translationally cleaved).

Folding and the formation of intra-chain disulphide bonds probably occur immediately after the proteins enter the lumen of the ER, catalysed by molecular chaperones and by the enzyme protein disulphide isomerase. However, the mechanisms controlling the formation and specificity of the inter-chain disulphide bonds that stabilise the glutenin polymers are not known. A possible explanation is that the HMW subunits aggregate within the ER lumen, allowing the formation of disulphide bonds between adjacent chains and forming a basis for the attachment of LMW subunits.

Secretory proteins usually pass from the lumen of the ER into the Golgi apparatus where modifications (such as glycosylation) may take place and they are sorted and transported in vesicles to one of two destinations: the plasmalemma and apoplast (outside the cell) or the vacuole. Most of the gluten proteins are transported to the vacuole where they aggregate to form protein bodies. However, some gluten proteins are not transported to the Golgi apparatus but remain within the lumen of the ER where they aggregate to form a second population of protein bodies of ER origin. Although the segregation of gluten protein types into the two populations of protein bodies is not precise, it is clear that the ER-derived protein bodies are enriched in HMW subunits of glutenin, and hence in high molecular weight glutenin polymers, while the vacuolar protein bodies are enriched in gliadins. This distinction may result from the propensity of the proteins to aggregate, with the HMW glutenin polymers being less readily transported than the gliadins.

These two populations of protein bodies eventually merge to form a continuous matrix when the starchy endosperm cells die during the later stages of grain maturation, but the extent to which the contents of the two populations of protein bodies merge is not known (see Chapter 8).

The pathway and processes of gluten protein trafficking and deposition are discussed in detail by Tosi (2012) and summarized in Figure 7.20.

7.3.2.6 Other Proteins of the Prolamin Superfamily

Wheat grain contain a wide range of non-gluten proteins, many of which have effects on the processing properties and/or the health benefits of the grain, or contribute to defence against pests and pathogens. They include a number of small sulphur-rich proteins, which are related in their evolutionary origins to the non-repetitive domains of the prolamins, and are therefore classified with gluten proteins in the *prolamin superfamily* of plant proteins. Several of these are discussed below and their properties summarised in Table 7.4.

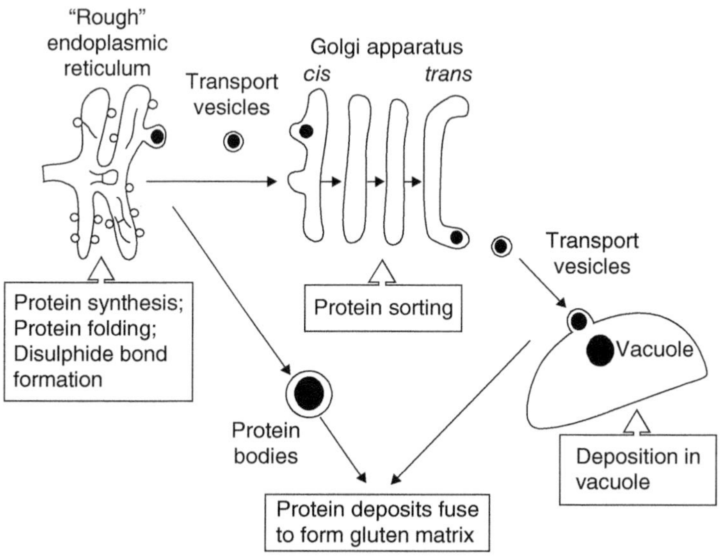

Figure 7.20 The synthesis, processing, trafficking, and deposition of gluten proteins in the secretory pathway of the starchy endosperm cells.

Table 7.4 Wheat grain proteins of the prolamin superfamily.

Protein group	Molecular mass	Characteristics	Abundance
Farinins	17 000–30 000	Correspond to avenin-like proteins	Not determined
Purinins (low molecular weight gliadins)	17 000–19 000	Possibly correspond to *ancestral* type of prolamin	Not determined
Puroindolines a and b	13 000	Tryptophan-rich loop region which may be involved in binding to starch granule surface. Major determinants of grain texture (Chapter 9).	0.029–0.060% dry wt of Pin a and 0.004–0.031% dry wt. of Pin b in wholemeal flour
Grain softness protein (GSP)	12 000	May contribute to grain texture (Chapter 9.).	Not determined
Non-specific lipid-transfer proteins (LTP)	9000 (LTP1) + 7000 (LTP2)	Bind and transport lipids *in vitro*. Concentrated in aleurone layer and embryo.	Not determined
α-amylase/trypsin inhibitors (ATIs)	12 000 to 16 000	Monomeric, dimeric, and tetrameric forms.	3–3.5% of flour proteins

Source: Adapted from Shewry (2019).

7.3.2.6.1 Farinins and Purinins

Farinins and purinins are closely related to gliadins but lack repeated sequences (Kasarda et al. 2013). Hence, they have been classed as globulins based on solubility. The farinins are most closely related in sequence to the prolamins (avenins) of oats and have also been called the avenin-like proteins. They comprise two groups of proteins, called a-type and b-type, which differ in that the b-type proteins contain a duplicated sequence of about 120 residues, resulting in a higher molecular mass (about 30 000 compared with 17 000). The purinins, which have also been called LMW gliadins, have masses of about 17 000–19 000 and are most closely related to the γ-gliadins in sequence. It has been suggested that they may resemble *ancestral* prolamin proteins, before they diverged by the development and amplification of the repetitive sequence domains.

7.3.2.6.2 Puroindolines

Two puroindoline (Pin) proteins are present in the starchy endosperm cells of wheat. Each comprises about 120 amino acids, including six conserved cysteine residues, with a mass of about 13 000. They differ from other wheat proteins in the presence of short sequences containing either three or five tryptophan residues and this is reflected in their name (tryptophan contains an indole ring). Pins are of particular interest in that they are the major determinants of grain texture (hardness) and are encoded by genes at the *Hardness* (*Ha*) locus on chromosome 5D (Chantret et al. 2005). Allelic variation in their expression and/or their amino acid sequences accounts for 60 to 80% of the variation in endosperm texture (hardness) in bread wheat (Bhave and Morris 2008). The biological function of Pins is unknown, and they have no known impact on human health. The *Ha* locus also contains a third gene, called *Gsp-1*, which encodes the grain softness protein (GSP) (Chapter 8) and the wheat arabinogalactan peptide (AGP) (see Section 7.3.1.4 and Chapter 9) (Chantret et al. 2005). The genes encoding Pin a and Pin b have been lost from the *Ha* loci on chromosomes 5A and 5B due to deletions, resulting in the ultrahard texture of tetraploid durum wheat. Four minor forms of Pin b have also been reported, encoded by genes on the long arms of the group 7 chromosomes (7A, 7B, 7D) (Wilkinson et al. 2008; Chen et al. 2010).

The structures of Pins and their roles in determining texture are discussed in detail in Chapter 8.

7.3.2.6.3 Lipid Transfer Proteins

The non-specific lipid transfer proteins (LTPs) are not restricted to seeds but occur in a range of plant tissues and species. They occur in two forms, called LTP1 (molecular mass of about 9000) and LTP2 (molecular mass of about 7000). They are related to other small sulphur-rich proteins of the prolamin superfamily and contain eight cysteine residues, which form four intra-chain disulphide bonds. They are able to bind phospholipids and transfer them from liposomes to mitochondria *in vitro*, but this activity lacks specificity and has not been demonstrated *in planta*. They are concentrated in the outer layers of plant tissues and this, together with their ability to inhibit fungal grown *in vitro*, suggests a role in plant defence. However, this role remains to be confirmed and they are of most interest in relation to health, LTPs being allergens in many plant foods (Marion et al. 2004) including wheat grain (Battais et al. 2005).

7.3.2.6.4 Amylase/Trypsin Inhibitors (ATIs)

Wheat grains contain several types of protein that inhibit hydrolytic enzymes, particularly inhibitors of α-amylase and proteases. Most of these proteins do not inhibit endogenous wheat enzymes, but enzymes in pests and pathogens. Hence, they are generally considered to be protective, forming part of a broad-spectrum defence system (Shewry and Lucas 1997). The major group of inhibitors in wheat are termed ATIs (cereal inhibitors of α-amylase and/or trypsin). These are water-soluble (albumin) proteins with masses between about 11 000 and 15 500 and are estimated to account for about 3% of the total grain proteins.

ATIs were first identified over 70 years ago and have been widely studied since. They occur in monomeric, dimeric, and tetrameric forms (Table 7.5). A single major monomeric form and two homodimeric forms are termed 0.28 and 0.19/0.53, respectively, based on their electrophoretic mobilities at alkaline pH. These have masses between about 13 000 and 13 500.

The tetrameric forms were initially extracted with a mixture of chloroform and methanol and hence called *CM proteins*. Each tetrameric form of ATI appears to consist of only three subunit types: one copy of either CM1 or CM2, one copy of either CM16 (which may be glycosylated) or CM17 and two copies of CM3 (which occurs in two forms). All of the unglycosylated subunits have masses similar to those of the monomeric and dimeric subunits except CM3 which has a mass of about 15 500. Finally, monomeric trypsin inhibitors (CMX) with masses between 11 400 and 11 500 have also been reported.

Table 7.5 Nomenclature, amylase inhibitory activity and abundance of wheat amylase/trypsin inhibitors.

Aggregation state	Amylase inhibitory activity	% total flour protein	Protein subunit		Widely used names
Monomeric	Human and insects	0.5			
			1	WMAI-1	0.28
Homodimer	Human and insects	1.0			
			1	WDAI-1	0.53
			2	WDAI-2	0.19
Tetramer	Human and insects	1.7			
			1	WTAI-CM1	CM1
			1	WTAI-CM2	CM2
			2	WTAI-CM16	CM16
			2	WTAI-CM16 glycosylated form	CM16
			2	WTAI-CM17	CM17
			3 (2 copies)	WTAI-CM3	CM3
Monomeric	No known activity	0.2			
			1	CMX1/2/3	CMX/ Trypsin inhibitors

Source: Modified from Geisslitz et al. (2021), based on Carbonero and Garcia-Olmedo (1999) with the addition of data on protein amount from Dupont et al. (2011) and Altenbach et al. (2011).

Proteomics studies have shown that at least some of the subunits occur in isoforms. For example, sequences corresponding to two monomeric, four dimeric, and six tetrameric subunits and three forms of CMX were identified in a single bread wheat cultivar (Altenbach et al. 2011).

Although the individual ATIs have low amino acid sequence homology (Figure 7.21A), they have highly conserved structures, consisting of four α-helices which are folded to form a right-handed superhelix stabilized by inter-chain disulphide bonds (Figure 7.21C). This structure is characteristic of members of the prolamin superfamily (see also puroindolines, Figure 8.2). Comparison of their amino acid sequences shows clear separation into four groups: the monomeric 0.28 and homodimeric 0.19/0.53 subunits; three forms of CMX, and the tetrameric subunits CM3/CM16/CM17 and CM1/CM2 (Figure 7.21b).

The monomeric, dimeric, and tetrameric forms are all able to inhibit amylases from humans (salivary and pancreatic enzymes) and amylases from Coleopteran (beetles) and Lepidopteran (caterpillar) insects (Table 7.6), but differ in their relative activities to amylases from the various sources (as discussed by Carbonero and Garcia-Olmedo 1999).

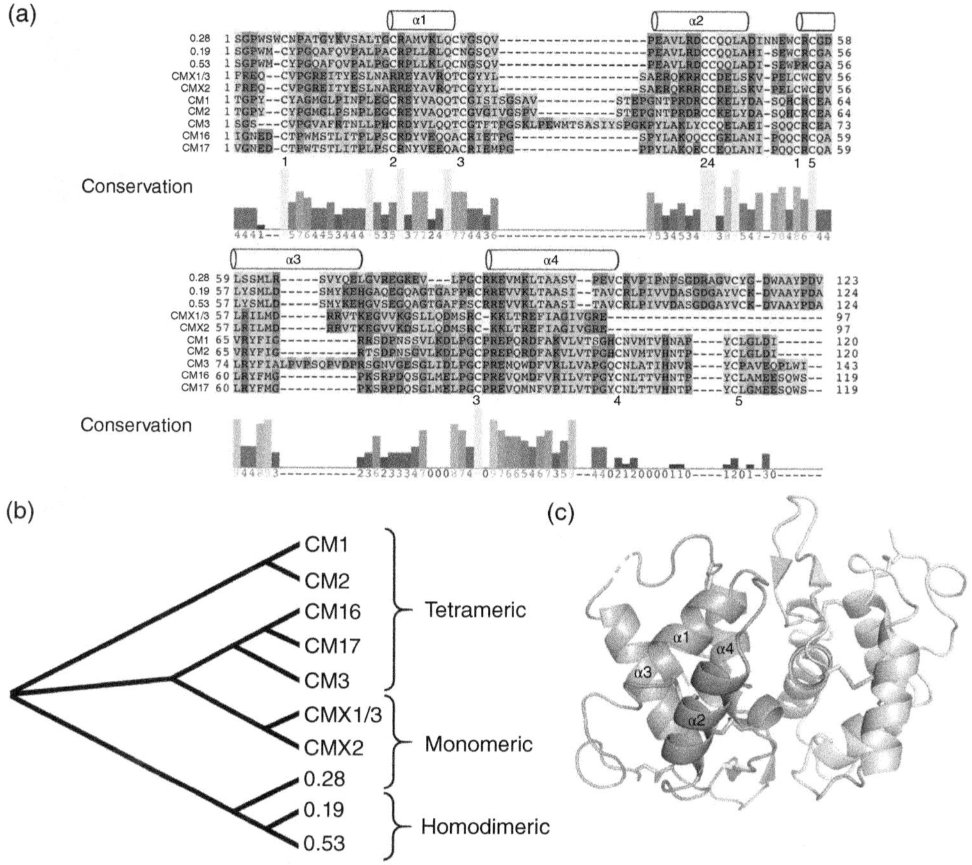

Figure 7.21 Sequences and structures of wheat ATIs. (a) Multiple sequence alignments of the mature proteins (after removal of signal peptides) showing the conservation of 10 cysteine residues which form five intra-chain disulphide bonds (numbered 1–5 below) and the sequences forming four α-helices (numbered α1-α4 above) based on the 3D structure of the 0.19 subunit (shown in c). Standard single letter notations for amino acids are used and the alignment is shaded to show amino acids with similar properties. (b) Phylogenetic tree based on alignment of amino acid sequences shown in (a). (c) Three-dimensional structure of the dimeric 0.19 subunit (PDB entry 1HSS; Oda et al. 1997). *Source:* Taken from Geisslitz et al. (2021) with permission of Professor Marie Sofie Møller (DTU, Lyngby).

The biological role of ATIs remains to be established and it is unlikely that their functions are essential for the plant. However, they are of great interest in relation to their role in human health, and particularly in triggering allergy, coeliac disease and non-coeliac wheat sensitivity (NCWS) (see Chapter 9).

7.3.2.7 Other Storage Proteins

The gluten proteins are restricted to the starchy endosperm cells and occur only in the seeds of cereals and other grass species. By contrast, the aleurone cells and scutellum contain deposits of a storage protein which falls into the family of vicilin-type storage proteins which are widely distributed in seeds of legumes and other dicotyledonous plants (reviewed by Casey and Domoney 1999). Vicilins are globulins with sedimentation coefficients ($S_{20.w}$) of about 7. They comprise three subunits with molecular masses of between 50 000 and 70 000 but may be post-translationally processed to give a wider range of subunit sizes. The 7S globulin of wheat has not been characterized in detail at the protein level but is reported to comprise subunits with molecular masses of about 40 000–55 000 (Shewry et al. 2009b).

Small amounts of storage proteins related to the legumin-type 11S storage globulins of dicotyledonous plants have also been reported in wheat and are named triticins. These are present in the starchy endosperm cells where they form inclusions in the prolamin-containing protein bodies (Shewry et al. 2009b).

7.3.2.8 Other Inhibitors and Putative Defensive Proteins

Wheat grain contains a range of other proteins, some of which may contribute broad defence against pest and pathogens. These are reviewed in detail by Shewry et al. (2009b) and Brijs et al. (2009) and are only briefly described here.

7.3.2.8.1 Purothionins

Purothionins were first characterized by workers in the 1940s who coined the name to reflect their origin and high content of sulphur. They comprise three components, α_1, α_2 and β, all of which consist of 45 amino acid residues including 8 cysteine residues that form 4 intra-chain disulphide bonds. They have been shown to be toxic *in vitro* to a wide range of organisms, including bacteria, yeast, filamentous fungi and cultured mammalian cells, with the ability to bind to and permeabilise membranes and inhibit RNA synthesis *in vitro*. However, there is no evidence that they are toxic to humans in wheat-based foods. Although they are generally considered to be defensive, isolated thionins also have redox properties indicating that they could contribute to redox metabolism *in vivo*.

7.3.2.8.2 Wheat Germ Agglutinin (WGA)

WGA is a lectin (carbohydrate-binding protein), which is able to bind N-acetylglucosamine as a monomer and in its polymeric form, chitin. It is only present in the embryo of the grain, particularly in the tissues that come into direct contact with the soil. WGA comprises three forms (isolectins) called A, B, and D, each consisting of two identical subunits of 171 amino acids. Each subunit is composed of four related domains, which can bind chitin and are often called *hevein domains* based on their similarity to hevein, a lectin from rubber latex. Chitin is present in the cell walls of fungi and the exoskeletons of insects, and WGA is active against a range of insects, but not fungi. There is no evidence for adverse effects of WGA in wheat-based foods on human health (van Buel and Brouns 2014).

7.3.2.8.3 Tritins (RIPs)

Tritins are ribosome-inactivating proteins (RIP) related to the well-characterised castor bean toxin ricin. They are able to irreversibly inactivate ribosomes by hydrolysis of a specific bond in the 26S ribosomal RNA. Wheat contains three forms of RIP, called tritin 1, 2, and 3, which do not affect protein synthesis in wheat, presumably due to compartmentalisation within the cell, but are active against pathogenic fungi.

7.3.2.8.4 Wheatwin1 (PR4)

Pathogenesis-related (PR) proteins are produced in plants in response to a range of stimuli, including infection, damage, and chemical elicitors. They comprise a number of families of structurally unrelated proteins with the PR4 group having antifungal activity. Four forms of PR4 protein (wheatwin) have been characterised, all comprising 125 amino acids with six cysteine residues forming three disulphide bonds, and all exhibiting antifungal properties.

7.3.2.8.5 Other Protease Inhibitors

In addition to ATIs, wheat contains several other types of protease inhibitor. Most are LMW proteins but an exception to this are the serpin serine protease inhibitors. These have masses of 40 000–45 000 and are major components of the albumin fraction (3 to 4 mg/g grain). They occur in a number of isoforms and may have regulatory functions within the plant.

The wheat subtilisin/chymotrypsin inhibitor (WASI) belongs to the potato 1 inhibitor family and comprises 72 amino acids with no cysteine residues and a high content of lysine (in common with the well-characterised barley homologue, BASI). It inhibits bacterial subtilisin and mammalian chymotrypsin and is presumably defensive. Other characterised inhibitors include a Bowman-Birk type endopeptidase inhibitor and cystatins (inhibitors of cysteine endopeptidases).

7.3.2.9 Xylanases and Xylanase Inhibitors

Endoxylanases (xylanases) hydrolyse the xylan backbone of arabinoxylans present in the grain cell walls and may contribute to the remodelling of arabinoxylan structure in the developing grain (see Section 7.3.1.3). Xylanases are synthesised during germination, but both endogenous xylanases and xylanases of microbial origin are also present in the mature grain (Dornez et al. 2006b). However, their content depends on the cultivar and weather conditions before harvest (Dornez et al. 2006a).

Xylanase activity may be difficult to measure in wheat flour because of high levels of endogenous xylanase inhibitors which consist of at least three families of proteins. At least seven isoforms of *Triticum aestivum* xylanase inhibitor (TAXI) exist. All have masses of about 40 000 with six disulphide bonds. Proteolytically processed forms also exist, comprising subunits of mass about 10 000 and 30 000 which remain linked by one of the disulphide bonds. The second group of xylanase inhibitors, called xylanase-inhibiting proteins (XIP), are related to chitinases present in other species. They also exist in multiple isoforms and have masses of about 30 000. Finally, the thaumatin-like xylanase inhibitors (TLXI) have masses of about 18 000 and belong to the thaumatin (sweet protein) family which are also called PR5 proteins.

The xylanase inhibitors may have roles in plant protection and in the regulation of xylanases during grain development. In addition, they may affect the functionality of wheat during processing as xylanases are widely used in breadmaking and other processes.

The reader is referred to Brijs et al. (2009) for a detailed review of these and other enzymes and inhibitors present in wheat.

7.3.3 Lipids

Lipids account for about 2.6% dry weight of wholegrain (Chung et al. 2009), although there is wide variation in the reported analyses of amount and composition, which probably reflects strong effects of genetics and environment as well as the methods used.

Grain lipids are classified into two classes: polar and non-polar, with the polar fraction being sub-divided into phospholipids (PL) and glycolipids (GL), and the non-polar comprising mainly triacylglycerols (TAG). The major PLs are phosphatidyl choline and lysophosphatidyl choline, a form with only one fatty acid esterified to the glycerol backbone (compared with two fatty acids in most PL), which is largely bound within the starch granules and not readily extracted. The GL comprise monogalactosyl diglyceride (MGDG) and digalactosyl diglyceride (DGDG).

TAGs are the major storage lipids in grain and the amounts reported range up to half or more of the total lipids in both wholegrain and white flour. TAGs are located in oil bodies that are abundant in the cells of the aleurone and scutellum but also occur in the starchy endosperm cells (González-Thuillier et al. 2021).

PLs and GLs are located in membranes, including the remnants of the amyloplast membranes (which is the sole known location of MGDG). MGDG and DGDG are of particular interest in relation to effects on breadmaking quality, as discussed in Chapter 8.

7.3.4 Minor Components: Minerals, Vitamins, and Phytochemicals

The mineral content of the grain is routinely determined as *ash* after combustion. Because minerals are concentrated in the bran fraction the ash content of wholemeal is generally about 1.3% DM compared with only 0.3% in the purest white flour fractions (derived from the starchy endosperm). Hence, ash content is used as measure of the purity of flour.

The major minerals in wholemeal and white flours are summarized in Table 7.6, showing that the degree of dilution in white flour compared to wholegrain ranges between x2 (calcium) and x9 (manganese).

Wheat grain also contains vitamins, notably B vitamins, and phytochemicals. These are discussed in relation to their impacts on the health of consumers in Chapter 9.

Table 7.6 Concentrations of selected minerals in wholemeal and white flour fractions.

Minerals	Wholemeal flour	White flour (starchy endosperm)	Ratio wholemeal: white flour
Ash (%)	1.7	0.51	3.4 : 1
Potassium (g/kg)	4.5	0.17	2.6 : 1
Calcium (g/kg)	0.30	0.15	2 : 1
Magnesium (g/kg)	1.51	0.24	3.6 : 1
Manganese (mg/kg)	50	5.50	9 : 1
Iron (mg/kg)	57	14	4.1 : 1
Copper (mg/kg)	5.6	2.3	2.4 : 1
Zinc (mg/kg)	38	9	4.2 : 1
Nickel (mg/kg)	0.35	0.12	2.9 : 1

Source: Values are recalculated from Varo et al. (1980) and expressed on a dry weight basis.

7.4 Gradients in Composition within the Starchy Endosperm

Differences in the compositions of wheat tissues (starchy endosperm, aleurone, embryo, and outer layers) are well understood and discussed above. By contrast, it is widely assumed that the starchy endosperm is a homogeneous tissue, and this is consistent with the aim of the miller to recover the whole tissue as white flour.

In fact, grain processors (including millers) and grain scientists have known for many years that this is not the case, at least for the two major components: starch and protein. In particular, the small sub-aleurone cells may contain 40% or more protein, compared to less than 10% in the central starchy endosperm cells, with few or no starch granules (Kent 1966) (see also Figure 7.7). Further studies have shown that radial gradients also exist in the content and composition of several other components, as summarised in Table 7.7.

Although the protein content of the starchy endosperm cells decreases from the outer layers to the centre, the protein quality for breadmaking actually increases, due to increases in the proportions of gluten proteins compared to total proteins, and of glutenins within the gluten protein fraction, especially HMW subunits of glutenin (He et al. 2013). Similarly, in addition to the increased total starch content, the proportions of large (A-type) granules and of amylose also increase from the outside of the grain while the temperature at which the granules gelatinise (an important factor in food processing, see Chapter 8) decreases.

Table 7.7 Summary of gradients in radial composition within the starchy endosperm of wheat grain.

Component	Gradient from outer to inner starchy endosperm
Protein:	
Total protein (% dry wt)	Decrease
Gluten proteins (% total protein)	Increase
Proportions of glutenins and large glutenin polymers (% gluten proteins)	Increase
Starch:	
Total starch (% dry wt)	Increase
A-type granules (% total granules)	Increase
% amylose	Increase
DSC onset temperature	Decrease
Dietary fibre:	
Total arabinoxylan (% dry wt)	Decrease
Arabinoxylose substitution (% disubstituted xylose residues)	Increase
Total β-glucan (% dry wt)	Decrease
Lipids:	
Total lipids (% dry wt)	Decrease
Unsaturated fatty acids (% total fatty acids)	Decrease

Source: Modified from Tosi et al. (2018). DSC onset temperature is the temperature at which starch begins to gelatinise, measured by differential scanning calorimetry.

Gradients also exist in other grain components, with decreases in the proportions of cell wall polysaccharides (arabinoxylan and β-glucan) and lipids (González-Thuillier et al. 2015) and changes in their compositions. These are summarised in Table 7.7.

Longitudinal gradients have been studied in less detail, but have been reported for cell wall polysaccharides, proteins, and amino acids (reviewed by Shewry et al. 2019). They are particularly marked in the proximal region (adjacent to the embryo) which may reflect interactions between the starchy endosperm and embryo.

The difference in composition between the sub-aleurone cells and those of the central starchy endosperm, particularly in their contents of starch and protein, may result from differences in developmental programming due to their different lineages, the sub-aleurone cells being derived from anticlinal divisions of the aleurone cells (Bechtel and Wilson 2003). However, the gradients in other components are not as clear cut and the explanations may differ. Saulnier et al. (2012) suggest that variation in AX structure modulates the hydration and permeability of cell walls and that variation in composition reflects the functional requirements of the cells (in the same way that the changes during development may reflect functional requirements). Hence, in this case, and perhaps also for other components such as lipids, the gradients may reflect biological rather than developmental differences between cell types.

The potential to exploit the gradients in composition within the starchy endosperm to produce flour fractions differing in composition and end use quality is discussed in Chapter 8.

References

Altenbach, S.B., Vensel, W.H., and Dupont, F.M. (2011). The spectrum of low molecular weight α-amylase/protease inhibitor genes expressed in the US bread wheat cultivar Butte 86. *BMC Research Notes* 4: 242. https://doi.org/10.1186/1756-0500-4-242.

Barron, C., Surget, A., and Rouau, X. (2007). Relative amounts of tissues in mature wheat (*Triticum aestivum* L.) grain and their carbohydrate and phenolic acid composition. *Journal of Cereal Science* 45: 88–96. https://doi.org/10.1016/j.jcs.2006.07.004.

Battais, F., Courcoux, P., Popineau, Y. et al. (2005). Food allergy to wheat: differences in immunoglobulin E-binding proteins as a function of age or symptoms. *Journal of Cereal Science* 42: 109–117. https://doi.org/10.1016/j.jcs.2005.01.004.

Bechtel, D.B. and Wilson, J.D. (2003). Amylose development and starch granule development in hard red winter wheat. *Cereal Chemistry* 80: 175–183. https://doi.org/10.1094/cchem.2003.80.2.175.

Bechtel, D.B., Abecassis, J., Shewry, P.R. et al. (2009). The development, structure and mechanical properties of the wheat grain. In: *Wheat: Chemistry and Technology*, 4e (eds. K. Khan and P.R. Shewry), 51–95. American Association of Cereal Chemists https://doi.org/10.1016/B978-1-891127-55-7.50010-0.

Bhave, M. and Morris, C.F. (2008). Molecular genetics of puroindolines and related genes: allelic diversity in wheat and other grasses. *Plant Molecular Biology* 66: 205–219. https://doi.org/10.1007/s11103-007-9263-7.

Brijs, K., Courtin, C.M., Goesart, H. et al. (2009). Enzymes and enzyme inhibitors endogenous to wheat. In: *Wheat: Chemistry and Technology*, 4e (eds. K. Khan and P.R. Shewry), 401–435. American Association of Cereal Chemists https://doi.org/10.1016/B978-1-891127-55-7.50018-5.

Bromilow, S., Gethings, L.A., Buckley, M. et al. (2017). A curated gluten protein sequence database to support development of proteomics methods for determination of gluten in gluten-free foods. *Journal of Proteomics* 163: 67–75. https://doi.org/10.1016/j.jprot.2017.03.026.

Brown, W.V. (1959). The epiblast and coleoptile of the grass embryo. *Bulletin of the Torrey Botanical Club* 86: 13–16. https://www.jstor.org/stable/2482656.

van Buel, V.J. and Brouns, F.J.P.H. (2014). Health effects of wheat lectins: a review. *Journal of Cereal Science* 59: 112–117. https://doi.org/10.1016/j.jcs.2014.01.010.

Carbonero, P. and Garcia-Olmedo, F. (1999). A multigene family of trypsin/α-amylase inhibitors from cereals. In: *Seed Proteins* (eds. P.R. Shewry and R. Casey), 617–633. Springer https://doi.org/10.1007/978-94-011-4431-5_26.

Casey, R. and Domoney, C. (1999). Pea globulins. In: *Seed Proteins* (eds. P.R. Shewry and R. Casey), 171–208. Kluwer Academic Publishers https://doi.org/10.1007/978-94-011-4431-5.

Chantret, N., Salse, J., Sabot, F. et al. (2005). Molecular basis of evolutionary events that shaped the hardness locus in diploid and polyploidy wheat species (Triticum and Aegilops). *The Plant Cell* 17: 1033–1045. https://doi.org/10.1105/tpc.104.029181.

Chen, F., Beecher, B., and Morris, C.F. (2010). Physical mapping and a new variant of *Puroindoline b-2* genes in wheat. *Theoretical and Applied Genetics* 120: 745–751. https://doi.org/10.1007/s00122-009-1195-y.

Chojecki, A.J.S., Bayliss, M.W., and Gale, M.D. (1986). Cell production and DNA accumulation in the wheat endosperm, and their association with grain weight. *Annals of Botany* 58: 809–811. https://academic.oup.com/aob/article/58/6/809/254382.

Chung, O.K., Ohm, J.B., Ram, M.S. et al. (2009). Wheat lipids. In: *Wheat: Chemistry and Technology*, 4e (eds. K. Khan and P.R. Shewry), 363–399. American Association of Cereal Chemists https://doi.org/10.1016/B978-1-891127-55-7.50017-3.

Dornez, E., Joye, I.J., Gebruers, K. et al. (2006a). Insight into variability of apparent endoxylanase and endoxylanase inhibitor levels in wheat kernels. *Journal of the Science of Food and Agriculture* 86: 1610–1617. https://doi.org/10.1002/jsfa.2505.

Dornez, E., Joye, I.J., Gebruers, K. et al. (2006b). Wheat kernel associated endoxylanases consist of a majority of microbial and a minority of wheat endogenous endoxylanases. *Journal of Agricultural and Food Chemistry* 54: 4028–4034. https://doi.org/10.1021/jf060129d.

Dupont, F.M., Vensel, W.H., Tanaka, C.K. et al. (2011). Deciphering the complexities of the wheat flour proteome using quantitative two-dimensional electrophoresis, three proteases and tandem mass spectrometry. *Proteome Science* 9: 10. https://doi.org/10.1186/1477-5956-9-10.

Gebruers, K., Dornez, E., Boros, D. et al. (2008). Variation in the content of dietary fiber and components thereof in wheats in the HEALTHGRAIN diversity screen. *Journal of Agricultural and Food Chemistry* 56: 9740–9749. https://doi.org/10.1021/jf800975w.

Geisslitz, S., Shewry, P.R., Brouns, F. et al. (2021). Wheat ATIs: characteristics and role in human disease. *Frontiers in Nutrition* 8: 667370. https://doi.org.10.3389/fnut.2021.667370.

González-Thuillier, I., Salt, L., Chope, G. et al. (2015). Distribution of lipids in the wheat grain determined by analysis of milling and pearling fractions. *Journal of Agricultural and Food Chemistry* 63: 10705–10716. https://doi.org/10.1021/acs.jafc.5b05289.

González-Thuillier, I., Pellny, T.K., Tosi, P. et al. (2021). Accumulation and deposition of triacylglycerols in the starchy endosperm of wheat grain. *Journal of Cereal Science* 98: 103167. https://doi.org/10.1016/j.jcs.2021.103167.

Grausgruber, H., Lovegrove, A., Shewry, P., and Békés, F. (2020). FODMAPs in wheat. In: *Wheat Quality for Improving Processing and Human Health* (eds. G. Igrejas, T. Ikeda and C. Guzmán), 515–532. Springer https://doi.org/10.1007/978-3-030-34163-3_21.

Graybosch, R.A. (1998). Waxy wheats: origin, properties, and prospects. *Trends in Food Science and Technology* 9: 135–142. https://doi.org/10.1016/S0924-2244(98)00034-X.

Haska, L., Nyman, M., and Andersson, R. (2008). Distribution and characterisation of fructan in wheat milling fractions. *Journal of Cereal Science* 48: 768–774. https://doi.org/10.1016/j.jcs.2008.05.002.

Hazard, B., Trafford, K., Lovegrove, A. et al. (2020). Improving the content and composition of starch and fibre in white wheat flour and bread. *Nature Food* 1: 475–480. https://doi.org/10.1038/s43016-020-0134-6.

He, J., Penson, S., Powers, S. et al. (2013). Spatial patterns of gluten protein and polymer distribution in wheat grain. *Journal of Agricultural and Food Chemistry* 61: 6207–6215. https://doi.org/10.1021/2Fjf401623d.

Kasarda, D.D., Adalsteins, E., Lew, E.J.-L. et al. (2013). Farinin: characterisation of a novel wheat endosperm protein belonging to the prolamin superfamily. *Journal of Agricultural and Food Chemistry* 61: 2407–2417. https://doi.org/10.1021/jf3053466.

Kent, N.L. (1966). Subaleurone endosperm cells of high protein content. *Cereal Chemistry* 43: 585–601.

Lafiandra, D.L., Riccardi, G., and Shewry, P.R. (2014). Improving cereal grain carbohydrates for diet and health. *Journal of Cereal Science* 59: 312–326. https://doi.org/10.1016/j.jcs.2014.01.001.

Marion, D., Douliez, J.-P., Gautier, M.-F. et al. (2004). Plant lipid transfer proteins: relationships between allergenicity, and structural, biological and technological properties. In: *Plant Food Allergens* (eds. E.N.C. Mills and P.R. Shewry), 57–86. Blackwell Science.

Monaghan, J.M., Snape, J.W., Chojecki, A.J.S. et al. (2001). The use of grain protein deviation for identifying wheat cultivars with high grain protein concentration and yield. *Euphytica* 122: 309–317. https://doi.org/10.1023/A:1012961703208.

Mosleth, E.F., Lillehammer, M., Pellny, T.K. et al. (2020). Genetic variation and heritability of grain protein deviation in European wheat genotypes. *Field Crops Research* 255: 107896. https://doi.org/10.1016/j.fcr.2020.107896.

Oda, Y., Matsunaga, T., Fukuyama, K. et al. (1997). Tertiary and quaternary structures of 0.19 α-amylase inhibitor from wheat kernel determined by X-ray analysis at 2.06 angstrom resolution. *Biochemistry* 36: 13503–13511. https://doi.org/10.1021/bi971307m.

Parchment, O., Shewry, P.R., Tatham, A.S. et al. (2001). Molecular modelling of unusual spiral structure in elastomeric seed protein. *Cereal Chemistry* 78: 658–662. https://doi.org/10.1094/CCHEM.2001.78.6.658.

Percival, J. (1921). *The Wheat Plant: A Monograph*. Duckworth and Co.

Saulnier, L., Guillon, F., and Chateigner-Boutin, A.-L. (2012). Cell wall deposition and metabolism in wheat grain. *Journal of Cereal Science* 56: 91–108. https://doi.org/10.1016/j.jcs.2012.02.010.

Shewry, P.R. (2007). Improving the protein content and composition of cereal grain. *Journal of Cereal Science* 46: 239–250. https://doi.org/10.1016/j.jcs.2007.06.006.

Shewry, P. (2019). What is gluten—why is it special? *Frontiers in Nutrition* 6: 101. https://doi.org/10.3389/fnut.2019.00101.

Shewry, P.R. and Lucas, J.A. (1997). Plant proteins that confer resistance to pests and pathogens. In: *Advances in Botanical Research*, vol. 26 (ed. J. Callow), 135–192. Academic Press https://www.jstor.org/stable/42796316.

Shewry, P.R., Tatham, A.S., and Halford, N. (1999). The prolamins of the Triticeae. In: *Seed Proteins* (eds. P.R. Shewry and R. Casey), 35–78. Kluwer Academic Publishers https://doi.org/10.1007/978-94-011-4431-5.

Shewry, P.R., Halford, N.G., and Lafiandra, D. (2003). The genetics of wheat gluten proteins. In: *Advances in Genetics*, vol. 49 (eds. J.C. Hall, J.C. Dunlap and T. Friedman), 111–184. Academic Press https://doi.org/10.1016/S0065-2660(03)01003-4.

Shewry, P.R., Underwood, C., Wan, Y. et al. (2009a). Storage product synthesis and accumulation in developing grains of wheat. *Journal of Cereal Science* 50: 106–112. https://doi.org/10.1016/2Fj.jcs.2009.03.009.

Shewry, P.R., D'Ovidio, R., Lafiandra, D. et al. (2009b). Wheat grain proteins. In: *Wheat: Chemistry and Technology*, 4e (eds. K. Khan and P.R. Shewry), 223–298. American Association of Cereal Chemists https://doi.org/10.1016/B978-1-891127-55-7.50015-X.

Shewry, P.R., Wan, Y., Hawkesford, M.J. et al. (2019). Spatial distribution of functional components in the starchy endosperm of wheat grains. *Journal of Cereal Science* 91: 102869. https://doi.org/10.1016/j.jcs.2019.102869.

Shewry, P.R., Hazard, B., Lovegrove, A. et al. (2020). Improving and starch fibre in wheat grain for human health. *The Biochemist* 42: 40–45. http://portlandpress.com/biochemist/article-pdf/42/4/40/891017/bio20200051.pdf.

Stone, B. and Morell, M.K. (2009). Carbohydrates. In: *Wheat: Chemistry and Technology*, 4e (eds. K. Khan and P.R. Shewry), 299–362. American Association of Cereal Chemists https://doi.org/10.1016/B978-1-891127-55-7.50016-1.

Toole, G.A., Barron, C., LeGall, G. et al. (2009). Remodelling of arabinoxylan in wheat (*Triticum aestivum*) endosperm cell walls during grain filling. *Planta* 229: 667–680. https://doi.org/10.1007/s00425-008-0865-3.

Tosi, P. (2012). Trafficking and deposition of prolamins in wheat. *Journal of Cereal Science* 56: 81–90. https://doi.org/10.1016/j.jcs.2012.02.004.

Tosi, P., He, J., Lovegrove, A. et al. (2018). Gradients in compositions in the starchy endosperm of wheat have implications for milling and processing. *Trends in Food Science and Technology* 82: 1–7. https://doi.org/10.1016/j.tifs.2018.09.027.

Tryfora, T., Liang, H.-C., Kotake, T. et al. (2010). Carbohydrate structural analysis of wheat arabinogalactan peptide. *Carbohydrate Research* 345: 2648–2656. https://doi.org/10.1016/j.carres.2010.09.018.

Uauy, C., Distelfeld, A., Fahima, T. et al. (2006). A NAC gene regulating senescence improves grain protein, zinc and iron content in wheat. *Science* 314: 1298–1301. https://doi.org/10.1126/science.1133649.

Varo, P., Nuurtamo, M., Saari, E. et al. (1980). Mineral element composition of Finnish foods. IV. Flours and bakery products. *Acta Agriculturae Scandinavica Supplementum (Sweden)* 22: 37–55.

Vijayaraghavan, M.R., Misra, G., and Saxena, P. (1988). Nutrient transfer during embryonic development of angiosperm plants. *Science Progress Oxford* 72: 467–480. https://www.jstor.org/stable/43421006.

Wilkinson, M., Wan, Y., Tosi, P. et al. (2008). Identification and genetic mapping of variant forms of puroindoline b expressed in developing wheat grain. *Journal of Cereal Science* 48 (3): 722–728. https://doi.org/10.1016/j.jcs.2008.03.007.

Wilkinson, M.D., Tosi, P., Lovegrove, A. et al. (2017). The Gsp-1 genes encode the wheat arabinogalactan peptide. *Journal of Cereal Science* 74: 155–164. https://doi.org/10.1016/j.jcs.2017.02.006.

8

Components and Mechanisms that Determine Grain Processing Properties

Of all foods bread is the most noble

Carl von Linné (Linnaeus)

The processing quality of wheat is of immense importance, determining the end use and value of the crop. It is therefore a major target for wheat breeders (together with yield), and not surprising that a massive volume of work has been carried out on the components that determine quality and their mechanisms of action.

Proteins usually only account for 9– to 16% dry matter (DM) of the mature grain but have been studied in detail because of their roles in the two most important properties which determine end use quality: grain hardness and dough viscoelasticity. The practical importance of these properties is discussed in Chapter 1, with the present chapter focusing on their underpinning mechanisms.

In addition to proteins, other components, including lipids, starch, and non-starch polysaccharides, may also have significant effects on processing quality, either by their interactions with proteins or independently. These components are therefore also considered.

8.1 Grain Hardness and Vitreousness

Although many factors may affect grain hardness as measured by particle size index (PSI), the most important is the degree of adhesion between the starch granules and the surrounding matrix proteins in the starchy endosperm cells. The starchy endosperm cells die during grain maturation and their contents become disrupted. This results in fusion of the gluten proteins present in discrete protein bodies to form a continuous matrix (as discussed in Section 7.3.2.5.4) and also allows these proteins to come into contact with the surfaces of starch granules. In soft wheats the binding of proteins to the starch granule surface is weak and the granules are readily released during milling. By contrast, the strong binding of proteins to the granule surface in hard wheats means that they are not readily released by milling, which requires more energy and results in more starch damage (SD). This difference is clearly shown in electron micrographs of the surfaces of starchy endosperm cells after freeze fracture of hard and soft grains (Figure 8.1).

Wheat: Environment, Food and Health, First Edition. Mike J. Gooding and Peter R. Shewry.

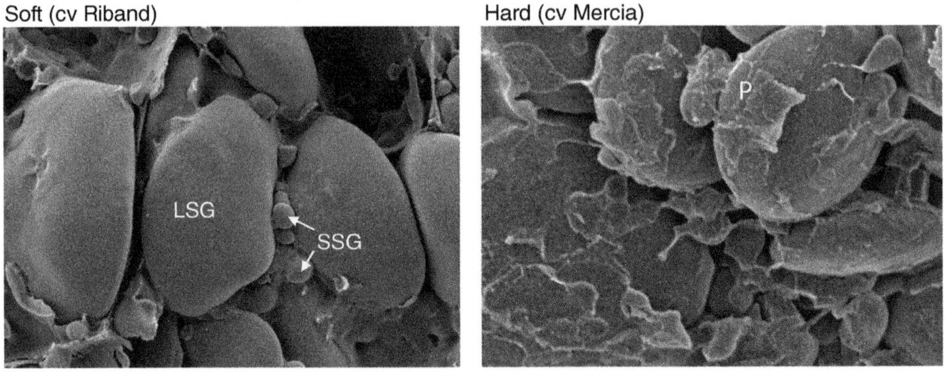

Figure 8.1 Comparison of the surfaces of starchy endosperm cells of soft and hard types of wheat after freeze fracture. Freeze fracture results in breakage of the cells exposing the large starch granules (LSG, diameter about 25 μm) and small starch granules (SSG). The presence of protein (P) adhering to the granule surfaces is clearly seen in the hard type only. *Source:* Taken from Chen et al. (2005) with permission.

8.1.1 Friabilin and Puroindolines

Greenwell and Schofield (1986) prepared starch granules from hard and soft types of wheat using standard water-washing procedures and then extracted the proteins present on the granule surface. This showed that a group of proteins with masses of about 15 000 were present on the surface of the starch granules in soft but not hard wheats. They called these proteins *friabilin*, to reflect their association with grain friability, and suggested that they formed a non-stick layer, which prevented the binding of the gluten proteins in the matrix to the surfaces of the granules. Friabilin comprises several components, with the major ones being called puroindolines (Pins) a and b. These two proteins are encoded by genes (designated *Pina-D1* and *Pinb-D1*) at the hardness (*Ha*) locus, which is located on chromosome 5D of hexaploid bread wheat and accounts for about 75% of the variation in hardness between bread wheat lines in the UK (Turner et al. 2004). Although *Ha* loci are also present on chromosomes 5A and 5B, the *Pina-D1* and *Pinb-D1* genes have been lost by deletion, resulting in the ultrahard texture of tetraploid durum wheat.

As discussed in Chapter 7, Pin a and Pin b comprise about 120 amino acids, including six conserved cysteine residues, with masses of about 13 000. However, they differ from other wheat proteins in the presence of short sequences containing either three (in Pin b) or five (in Pin a) tryptophan residues. Although the three-dimensional structures of Pins are not known, they are thought to be similar to those of related small sulphur-rich seed proteins, such as the amylase/trypsin inhibitors (ATIs) and non-specific lipid transfer proteins (see Chapter 7), which are tightly folded with four α-helices to form a right-handed superhelix stabilised by inter-chain disulphide bonds (three in Pins). The tryptophan-rich region probably forms a loop between two of the helices, which is exposed on the outer surface of the protein. This is illustrated by the hypothetical structure shown in Figure 8.2.

The identification of an association between the presence of Pins on the surface of starch granules and grain texture resulted in the demonstration that hardness can result from mutations which affected either the ability of the Pin proteins to bind to starch or their

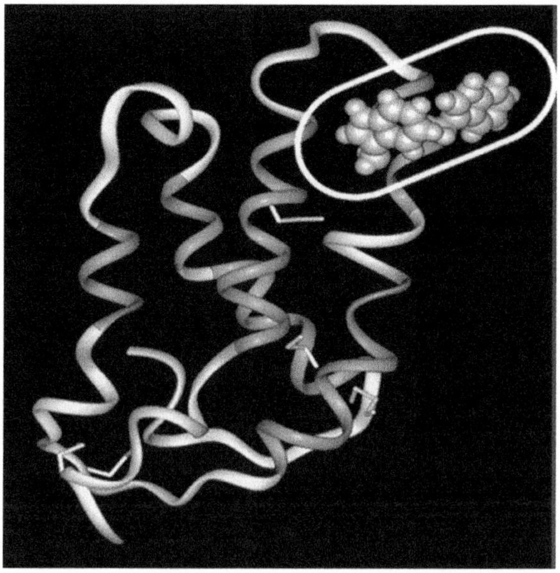

Figure 8.2 Possible molecular structure for wheat puroindoline b, showing the protein backbone as a ribbon and the tryptophan residues in space filling presentation (with the atoms as spheres) (circled). *Source:* Model kindly provided by Dr. F. Beaudoin (Rothamsted Research).

expression. Grain softness results from the presence of the wild type alleles of the genes, which are called *Pina-D1a* and *Pinb-D1a*. The most common mutations which result in hardness are the *Pina-D1* gene deletion (null allele *Pina-D1b*) and the *Pinb-D1b* allele, which encodes a protein with a single amino acid substitution (the replacement of glycine at position 6 with serine). Five other point mutations in *Pinb-D1* that result in single amino acid substitutions or premature stop codons (alleles *Pinb-D1c-g)* were initially described (Morris 2002) but analysis of a wide range of germplasm has identified an increasing number of rare alleles including lines with mutations at both genes. These mutations may confer subtle differences in texture. Bhave and Morris (2008) and Morris et al. (2021) provide reviews of this increasingly complex topic.

Despite our extensive knowledge of variation in the sequences of Pin proteins, their mechanism of action remains poorly understood. It is most likely that the tryptophan-rich sequence interacts with the starch polymers at the granule surface (Wall et al. 2010). This is consistent with our knowledge of other carbohydrate-binding proteins and would account for the fact that the property appears to be unique among wheat grain proteins. However, this does not explain why the binding of Pins has such a major impact on the adhesion of other proteins; this may result from a further interaction with lipids, as discussed in Section 8.1.3.

8.1.2 Other Proteins that Affect Grain Hardness

Pins only account for about 75% of the variation in grain hardness and it is therefore of interest to identify other proteins, or other components, which also have significant effects. Two proteins that make small but significant contributions have been identified, both of which are related to Pins. Willkinson et al. (2008) identified transcripts encoding variant forms of Pin b, which lacked one of the tryptophan residues. Transcript levels are low, about a tenth of those of the major form of Pin b, but one variant was mapped to a locus on chromosome 7A that was linked to a minor hardness locus, which accounted for about 10% of the hardness measured using the Perten SKCS.

Grain softness protein (GSP) was initially identified as a component of the friabilin group of proteins present on the surface of starch granules from soft genotypes. It is encoded by *Gsp-1* genes which are also at the *Ha* locus. The gene transcript actually encodes a pre-protein, with the first 15 residues being removed by proteolysis and then glycosylated to form arabinogalactan peptide (AGP). This forms part of the grain fibre fraction and is discussed in Chapters 7 and 9. The remaining part of the protein, which comprises 113 residues in the major form, is GSP. However, whereas the genes encoding Pin a and Pin b are absent from the A and B genomes of bread wheat due to gene deletions, the *Gsp-1* genes are present on all three genomes (Chantret et al. 2005). Despite its name, most studies have failed to demonstrate a direct relationship between GSP and grain texture between genotypes. However, when the expression of *Gsp-1* was downregulated in transgenic wheat, a small increase in hardness was observed (by about 10 units measured with the Perten Single Kernel Characterisation System) (Wilkinson et al. 2017). Hence, both minor Pins and GSP may contribute to the determination of grain texture.

8.1.3 Role of Lipids

Both the total content of polar lipids in flour and the amount of polar lipids present on the surface of starch granules are inversely correlated with grain hardness (Konopka et al. 2005), being lower in lines which are null for Pin a or have mutant forms of Pin b (Finnie et al. 2010). Earlier work also mapped two loci for free polar lipid content, one of which (*Flp*-1) is located close to or at the *Ha* locus (Morrison et al. 1989).

Pins have been shown to bind to polar lipids *in vitro* and spectroscopic studies show that the tryptophan residues are involved in this binding (Dubreil et al. 1997; Clifton et al. 2007). However, the binding of lipids to the tryptophan residues has not been demonstrated *in vivo* and it is possible that these residues bind to starch (as discussed in Section 8.1.1). It is therefore more likely that polar lipids bind to the outer exposed surface of the Pin proteins, forming anti-stick patches that prevent interactions with the matrix proteins.

8.1.4 When and How Is Grain Softness Established?

It is likely, therefore, that grain softness is determined by a three-way interaction between starch, Pins, and polar lipids. If so, it is important to consider how and when these interactions are established. Pins are *secretory proteins* (see Chapter 7, Section 7.3.2.5.4), being synthesised on the rough endoplasmic reticulum and transported through the lumen, Golgi apparatus, and Golgi-derived vesicles into the vacuole. By contrast, starch is synthesised in the amyloplast, a specialised type of plastid, which is surrounded by a double membrane. Hence, the Pins and starch can only come into contact during the later stages of grain maturation when the starchy endosperm cells dry, the organelle membranes are disrupted, and the cell contents merge. It is therefore likely that the Pins migrate to the starch granule surface by diffusion during the late stage of development, with the granules acting as an affinity matrix to establish a concentration gradient. The polar lipids are presumably derived from the remains of membranes, particularly those surrounding the amyloplasts and vacuolar protein bodies. It is also of interest to speculate on relationships between the Pin concentration in the cell, the total surface area of the starch granules in the cell, and the surface coverage required to confer softness; these have not been established.

8.1.5 Vitreousness

The description of a grain as vitreous implies that the endosperm has a glassy, translucent appearance rather than a mealy or floury nature. The term is used particularly for durum wheat, where it correlates with semolina yields. The difference in appearance of floury and vitreous grain has been suggested to result from the density of the protein matrix, and whether it has included air bubbles. Vitreousness is largely determined by the environment and the level is greater in high-protein grain and in samples exposed to high temperatures during grain development (Kindred et al. 2008). Grains can also be piebald, with both vitreous and floury areas. In such grain the vitreous areas are higher in protein than the mealy areas (McDermott and Pace 1960).

8.2 Dough Viscoelasticity

8.2.1 Wheat Gluten and Dough Viscoelasticity

Gluten is classically defined as the largely proteinaceous mass that remains when a dough made from wheat flour and water is gently washed in an excess of water or dilute salt solution to remove most of the starch and soluble material (Wrigley and Bietz 1988). It comprises about 75–80% protein on a DM basis, depending on how well it has been washed, with small amounts of starch (entrapped in the gluten network) and lipids (which bind to and plasticise the proteins). The major protein components are gliadins and glutenins (Chapter 7), with small amounts of other proteins, such as ATIs (Chapter 7). Gluten is unique to wheat: although related proteins from other cereals (barley and rye) are sometimes also referred to as gluten proteins, flours from these cereals cannot be used to make doughs that are equivalent to wheat dough.

Isolated gluten has a unique combination of physical properties which is generally described as *visco-elasticity*. In simple terms this means that the cohesive mass extends when stretched but also exhibits some elastic recoil, and the balance of these properties (extensibility and elasticity) is crucial for the behaviour of dough in food systems.

Gluten is not present in dough as a cohesive mass but as a continuous network that extends throughout the dough. The formation of the network starts during grain maturation, when the starchy endosperm cells die and the gluten proteins, which are deposited in protein bodies, fuse to form a continuous proteinaceous matrix surrounding the starch granules (Figure 8.3a). This network can be revealed in individual flour particles if the starch granules are removed by enzyme digestion (Figure 8.3b). Mixing the flour with water therefore brings together the individual networks in the particles to form a single continuous network.

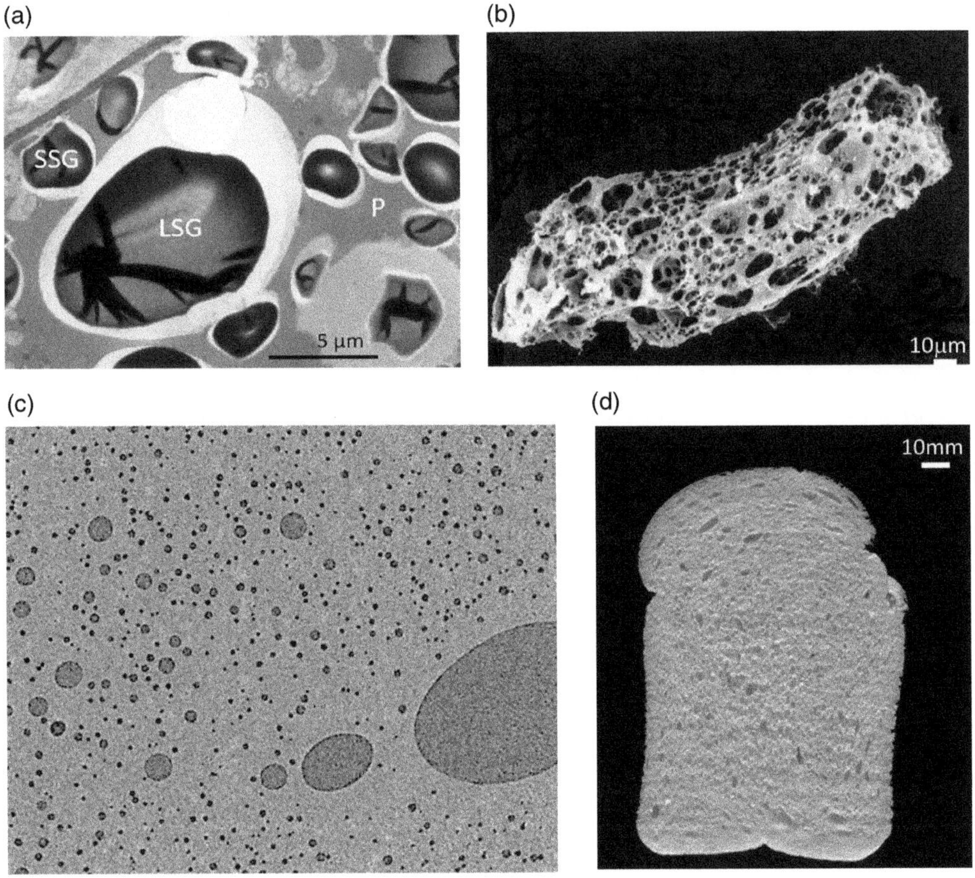

Figure 8.3 Gluten forms networks in the grain and in dough and bread. The wheat gluten proteins form a continuous network in the sub-aleurone and other starchy endosperm cells of the mature wheat grain (a). This network can be revealed when a flour particle is digested to remove starch (b). The network entraps gas to form bubbles in bread, as demonstrated for a gluten : starch blend (20 : 80) by synchrotron x-ray microtomography (c). Denaturation of the protein network during baking leads to the open crumb structure of the loaf (d). *Source:* Part (a) taken from Shewry et al. (1995) with permission. Figure originally provided by Dr. M. Parker. Part (b) taken from Amend and Beauvais (1995) with publisher's permission; Part (c), kindly provided by Professor Filiz Koksel, University of Manitoba; Part (d), kindly provided by Dr. Mervin Poole, Heygates Ltd., UK.

The gluten network provides the cohesion required for making products such as noodles and biscuits, and also stabilises the gas bubbles that are formed during mixing (Figure 8.3c), allowing them to be expanded by carbon dioxide generated during proofing without coalescence and resulting in over 70% of the final loaf volume being made of gas cells (Figure 8.3d).

The importance of wheat gluten in determining processing quality has resulted in a massive literature extending over more than 250 years. Despite this, the precise details are still not fully understood. Nevertheless, several points can be noted. Firstly, although protein content is important, the quality of the protein varies and is therefore also important. Poor gluten quality can be partially, but not completely, compensated for by higher protein content. Secondly, the properties of gluten, and the differences in quality between different types of wheat depend on the contributions of both the gliadins and glutenins. The glutenin subunits form large three-dimensional polymers stabilised by inter-chain disulphide bonds that interact with gliadins and with other glutenin polymers by non-covalent forces, particularly hydrogen bonds. In broad terms, the glutenin polymers confer elasticity while the gliadins contribute particularly to viscosity and extensibility. Hence, the ratio of gliadins:gutenins relates to differences in quality.

Although gluten quality is affected by environmental factors, there is strong genetic control, which has allowed us to gain a more detailed insight into mechanisms and allowed breeders to select for gluten quality in their breeding programmes. In this respect, the HMW (high molecular weight) subunits of glutenin (Chapter 7) are particularly important.

8.2.2 HMW Subunits, Dough Strength, and Breadmaking Quality

The HMW subunits generally account for about 8–10% of the total gluten proteins. Nevertheless, they appear to be largely responsible for conferring elasticity to gluten and dough. There are two types of evidence for this. Firstly, biochemical analyses show that the HMW subunits are concentrated in HMW glutenin polymers, and that the proportions of these polymers correlate positively with dough strength. Secondly, genetic studies show that allelic variation in HMW subunit composition is associated with variation in dough strength (Payne 1987).

The HMW subunits are encoded by a small multigene family, with two genes each located at the *Glu-1* loci on the long arms of the group 1 chromosomes (Figure 7.19). These genes encode two types of HMW subunit called x-type and y-type, but not all of the genes are expressed in all genotypes. Thus, all modern bread wheat cultivars express the two genes on chromosome 1D (encoding 1Dx and 1Dy subunits) and one gene on chromosome 1B (encoding a 1Bx subunit). Some cultivars also express the second gene on chromosome 1B (encoding 1By subunit) and/or one gene on chromosome 1A (encoding a 1Ax subunit). The second gene on chromosome 1A (encoding a 1Ay subunit) is not expressed in commercial bread wheats. This variation in the number of expressed HMW subunits (from 3–5) is associated with quantitative differences in HMW subunit amount, with each expressed subunit accounting for about 2% of the total gluten proteins. Cultivars expressing more subunits, particular 1Ax subunits, generally have higher gluten elasticity.

Further variation in elasticity is associated with allelic variation in expressed subunits. In common with other gluten proteins (Chapter 7), the HMW subunits show extensive polymorphism between genotypes. This is readily observed as differences in mobility when total grain proteins are separated by sodium dodecylsulphate electrophoresis (SDS-PAGE) (Figure 8.4). The allelic forms vary in their prevalence in modern cultivars, with two pairs of subunits encoded by chromosome 1D (1Dx2 + 1Dy12 and 1Dx5 + 1Dy10) being most widely present in European wheats. Genetic analyses have shown that subunits 1Dx5 + 1Dy10 are associated with good quality while subunits 1Dx2 + 1Dy12 are associated with poor quality. Further studies allowed widely occurring subunits encoded by all three genomes to be assigned quality scores (Payne et al. 1987). SDS-PAGE has therefore been widely used by wheat breeders to select for combinations of quality subunits in their breeding programmes for breadmaking cultivars.

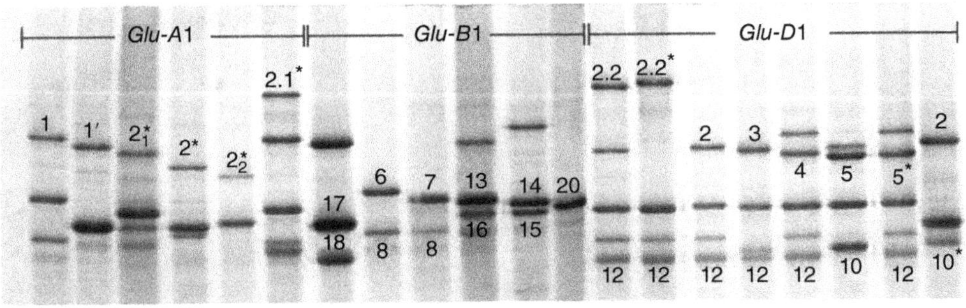

Figure 8.4 Allelic variation in HMW subunits encoded by the Glu-A1, Glu-B1 and Glu-D1 loci, shown by SDS-PAGE. *Source:* Taken from Shewry et al. (2003a) with permission.

8.2.3 Effects of Other Gluten Proteins on Dough Quality

Breeding for good quality HMW subunits has had a major impact on wheat grown in Western Europe because the dough strength of European wheat is often low compared to the stronger wheats that had been widely imported to Europe from North America before the increased use of homegrown wheat which occurred in the later part of the twentieth century. In light of this success, similar attempts have been made to identify quality-related alleles of the second group of glutenin subunits, the LMW (low molecular weight) subunits. This has led to the identification of allelic proteins and shown some relationships between alleles and quality, but this variation has not been exploited by breeders. There are probably three reasons for this. Firstly, the level of genetic complexity of the LMW subunits is much greater than that of the HMW subunits, with multiple loci on the group 1 chromosomes (Figure 7.19). Secondly, the effects of individual alleles are smaller than those associated with HMW subunit alleles. Finally, the analytical systems required to identify the alleles are more sophisticated than SDS-PAGE and not appropriate for routine use in breeding programmes.

Finally, although genetic improvement has largely focused on increasing dough strength, it should be emphasised that a balance of gluten properties are required for processing quality, which means that the amount and composition of the gliadins are also important. Overemphasis on strength can in fact lead to the production of over-strong types of wheats, which are not suitable for modern processing systems.

8.2.4 Molecular Basis for the Role of the HMW Subunits in Gluten Structure and Properties

The demonstration of the importance of the HMW subunits in determining dough strength has led to a range of studies designed to explain their role in the structure and elastomeric properties of gluten. These include analyses of the structures and biophysical properties of isolated proteins and of polymer fractions, including the mapping of disulphide bonds. The picture is still incomplete but can be summarised as follows (and illustrated in Figure 8.5). A more detailed explanation is provided by Shewry et al. (2003b).

The HMW subunits are considered to form the backbone of gluten, being assembled head-to-tail by disulphide bonds between cysteine residues in the N- and C-terminal domains.

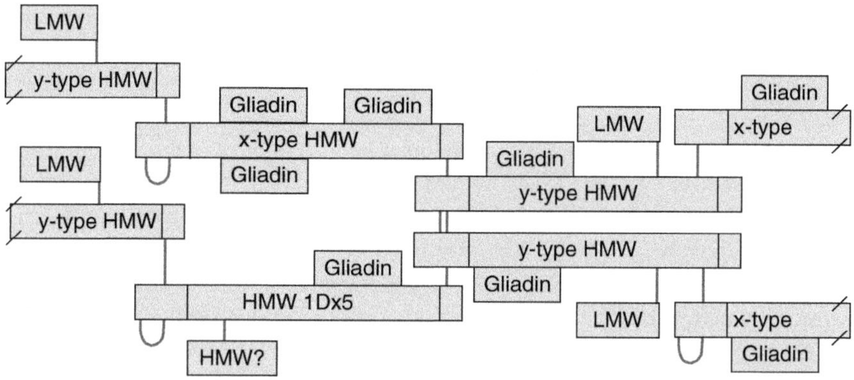

Figure 8.5 Model to explain the role of the HMW subunits of glutenin in determining the structure of wheat glutenin polymers, based on the identification of interchain disulphide bonds (shown as vertical lines). A single interchain bond has been identified in the N-terminal domain of x-type subunits. HMW subunit 1Dx5 is also proposed to form an addition cross-link to another HMW subunit based on the effects of overexpression on gluten rheology. The HMW subunit polymers also form disulphide bonds with LMW subunits while gliadins interact by non-covalent forces. Based on Shewry et al. (2003a), which gives a more detailed explanation.

There is also evidence that x-type and y-type subunits may alternate in this structure, forming x–y dimers within the polymers. The HMW polymers also form interchain disulphide bonds with LMW subunits while the gliadins interact by non-covalent forces, notably hydrogen bonds, formed between the glutamine residues present in the repetitive domains.

There is no doubt that the assembly of the HMW subunits into large polymers is essential for dough elasticity, as this property is lost if the disulphide bonds are reduced to release the individual subunits. However, non-covalent hydrogen bonds formed between glutamine residues are also crucial. A simplified explanation for the role of HMW subunits is that the structure of hydrated gluten is stabilised by both disulphide bonds between individual subunits and by hydrogen bonds. Hydrogen bonds are formed both within the repetitive domains of individual subunits, where they stabilise the loose spiral structure, and between the repetitive domains of adjacent subunits. These hydrogen bonds are in equilibrium in hydrated gluten and this equilibrium is disturbed on stretching, resulting in their reformation on release of the stress (i.e. elastic recoil). However, some slippage between polymers also occurs, allowing extensibility. The increased resistance that is observed when dough is mixed therefore results from optimisation of the number and distribution of hydrogen bonds, and the decreased resistance that occurs during over-mixing to the breakdown of the optimised structure. Although this model focuses on the role of the HMW subunits, it also allows for the effects of LMW subunits and gliadins, via their contributions to polymer structure and hydrogen-bonding patterns.

8.3 Functional Properties of Starch

Starch accounts for about 70% of the grain dry weight and 80% of white flour. It is therefore inevitable that it will affect the functional properties. However, whereas seed proteins vary widely in their structure and properties, starch is much more uniform across species.

Hence, the account below focuses on the role of starch in wheat processing, rather than properties which are generic for all starches.

As discussed in Chapter 7, wheat starch consists of glucose molecules assembled into two polymers. Amylose (~20–30% of the total) is largely linear chains with α-1,4 linkages and molecular masses of up to 8×10^5, while amylopectin (~70–80% of the total) comprises larger (molecular mass 10^6–10^8) branched polymers with both α-1,4 linkages and α-1,6 branch points. These polymers are deposited in granules that have a highly organised structure with concentric amorphous lamellae that contain amylose polymers and branch points of amylopectin alternating with crystalline lamellae formed from amylopectin chains. Intact granules can be readily identified when viewed in polarised light as they exhibit birefringence (in the form of a Maltese cross), which reflects the presence of an ordered structure. The functional properties of starch are determined by the structures of the individual polymers and their organisation.

8.3.1 Starch Gelatinisation

Intact starch granules are permeable to water, and hydration at room temperature results in a small increase in volume, by about 5%. This process is fully reversible. However, heating in excess water results in granule swelling and starch gelatinisation. This is a progressive loss of structure and is irreversible above a certain temperature (called the gelatinisation temperature). This change is endothermic and can be measured by thermal analysis, with differential scanning calorimetry (DSC) being widely used. Starches from different cereals vary in their gelatinisation temperatures, but in wheat gelatinisation occurs between 51 °C and 60 °C, with 50% of the granules having lost their birefringence by 53 °C.

Gelatinisation is associated with increased leaching of amylose from the granules, increased viscosity of aqueous suspensions and increased susceptibility to enzyme digestion. In practice the increase in viscosity on heating can be monitored using a recording viscometer, such as the Brabender Amylograph. Starch gelatinisation and viscosity are particularly important for the use of wheat flour in sauces and batters and as an ingredient in processed foods.

Delcour and Hoseney (2010) provide an excellent introduction to cereal starches and grain processing.

8.3.2 Starch Damage

Whereas intact starch granules are relatively inert, this is not the case when they are damaged. Damaged starch absorbs 3–4 times its mass of water, compared to about 0.5 times for intact starch. Hence damaged starch is a major contributor to the water absorption (WA) of flour, as discussed in Chapter 1. Damaged starch is also much more susceptible to amylolysis, either during processing or in the human gastrointestinal tract, with the enzyme forming cavities inside the granule. At the molecular level, starch damage leads to loss of structural organisation and increased solubility of individual polymers, notably amylose. In functional terms, damage affects the gelatinisation properties and the viscosity of starch pastes.

8.3.3 Starch Retrogradation and Staling

The gelatinisation of starch during cooking results in the loss of structural organisation with starch pastes forming a gel on cooling. If starch-rich foods are allowed to stand, the starch starts to crystallise, a process known as retrogradation. This is accelerated at low temperatures (such as refrigeration) and is largely due initially to the crystallisation of amylose, followed by the linear (amylose-like) fraction of amylopectin. Starch retrogradation contributes to the staling of bread and other baked goods and to the firmness of cold pasta. Although staling is a major issue for food retailers it may have health benefits, retrograded starch being resistant to digestion in the human gastrointestinal tract and defined as RS3 (Chapter 9). Hence, whereas the storage of bread in the refrigerator increases staling, the storage of other wheat-based foods can lead to a lower glycaemic index.

8.3.4 Waxy and High Amylose Starches

Waxy wheat is of interest to the food industry as waxy starch has higher viscosity and water retention, which can be exploited in refrigerated and frozen foods and may slow the staling of bread. High amylose starch also differs from normal starch in its biochemical and biophysical properties but, in contrast to waxy starch, these properties make it less suitable for food processing and high amylose wheat lines are not currently used in large-scale commercial food production.

8.4 Other Functional Components

8.4.1 Arabinoxylan

Arabinoxylans (AX) have a higher water-binding capacity than other grain components, absorbing about 10 times their own weight of water. This contributes to the water absorption (WA) of flour (Chapter 1 and below). It also affects the distribution of water in food systems and may result in stickiness (insoluble AX) and high viscosity of aqueous systems (soluble AX). In addition, AX has a major impact on the quality of wheat flour for breadmaking, affecting a range of quality parameters. These effects have been reviewed in detail by Courtin and Delcour (2002), which provides a basis for the brief discussion here.

Soluble and insoluble AX have positive and negative effects, respectively, on loaf volume, one of the most important quality traits. The positive effect of soluble AX is suggested to result from increasing the stability of the liquid films around gas bubbles, resulting in a more homogeneous crumb structure as well as greater loaf volume. It is also possible that they form a weak network, which reinforces the gluten network. By contrast, insoluble AX appear to destabilise dough structure, possibly by physical disruption of the gluten network and bubble structure. Endoxylanase enzymes may therefore be used as processing aids to convert insoluble arabinoxylan into large soluble polymers. However, the same enzymes may also degrade the native soluble polymers, leading to detrimental effects. It is therefore necessary to optimise the enzyme specificity and reaction conditions to maximise the generation of large soluble AX and minimise their further breakdown. Optimisation of AX size and solubility leads to increased dough stability, a greater rise in the oven, and loaves with a higher volume and a finer, softer, and more homogeneous texture.

8.4.2 Functional Properties of Lipids in Dough

Lipids are relatively minor components, accounting for about 2% of white flour. Nevertheless, they have significant effects on processing quality, affecting the volume and the crumb structure of bread (Pareyt et al. 2011). This may partially result from their interactions with gluten, as they bind to and plasticise the gluten network. However, their main impact is probably due to effects on the structure and stability of the gas bubbles that are formed during mixing and then expanded by carbon dioxide generated during proofing.

The most active lipids in this respect are the glycolipids (GL) monogalactosyl diglyceride (MGDG) and digalactosyl diglyceride (DGDG), which are highly surface-active and increase the elasticity of the liquid film at the gas bubble interface. Improved quality for breadmaking can be achieved by exploiting genetic variation in the amounts of GLs (Min et al. 2020) and by using specific lipases to generate surface-active lyso-forms (Melis et al. 2019).

8.4.3 Water Absorption: Effects of Starch, Protein, and Fibre

The WA of flour is determined by the amounts and compositions of the three major grain components: starch, protein, and cell wall polysaccharides (fibre). Because starch accounts for about 80% of white flour, it is considered to have the greatest impact on WA. Intact starch granules only absorb about 0.5 times their dry weight of water, but this increases to 3–4 times when the starch is damaged (Kent and Evers 1994).

The two starch polymers, amylose and amylopectin (see Chapter 7), have different WA capacities, and the swelling capacity of starch granules decreases as the amylose content increases. The large A-type starch granules of wheat are richer in amylose than the small B-type granules and the WA will therefore vary with differences in granule size distribution and composition. These parameters may be affected by the genotype and the environment. Mutant high amylose starch also has a lower WA than high amylopectin starch which is described as 'sticky' or 'waxy'.

Protein accounts for about 11–12% of the dry weight of white flour used for breadmaking and has a relatively low water-absorption capacity (about 1.8 times dry wt.). It therefore contributes less than starch to the WA of flour. Although differences in the composition and properties of the grain proteins could affect the WA of flour, such as the ratio of gluten proteins to soluble proteins, this variation is unlikely to have a major impact.

The contributions of starch and protein to WA are reflected in the widely used Farrand equation (Farrand 1969), which predicts WA based on % protein content (P), starch damage (SD) in Farrand Units, and % moisture content (MC):

$$WA = 68.26 + 0.878P + 0.334SD - 1.97MC \tag{8.1}$$

However, the Farrand equation does not explain all the variation in WA, and there has been some debate about the components that are responsible for the variation that is not accounted for (up to about 25%). The most likely candidates are cell wall polysaccharides (dietary fibre), particularly arabinoxylan.

Although AX is a minor component of white flour, it has very high water-absorption capacity, absorbing about 10 times the dry weight for the water-insoluble fraction and 11 times for the water-soluble fraction. β-glucan only accounts for about 20% of the total cell wall polysaccharides in wheat, but it is the major cell wall polysaccharide in barley and oats, where it forms highly viscous solutions (Lazaridou and Biliaderis 2007). It could therefore also contribute to WA in wheat flour. Lovegrove et al. (2020) showed that both AX and β-glucan contribute to the variation in WA in wheats grown in the UK.

8.5 Effects of Crop Nutrition and Environmental Factors on Grain Composition and Quality

As discussed in Chapter 1, the vast majority of wheat grown in the world is used to prepare foods for human consumption. Hence, processing quality is of paramount importance for breeders, farmers, and processors.

Processing quality is under strong genetic control, as discussed, but is also significantly affected by the environment, including nutrient availability and the climate (temperature and precipitation).

8.5.1 Nitrogen Fertilisation

Nitrogen is the major input which determines crop yield and nitrogen application is usually optimized to obtain the most economical yields (which are usually below the maximum intrinsic yield). However, historical increases in yield achieved by wheat breeders have resulted in overall dilution of grain protein (Chapter 10) and farmers may need to apply additional nitrogen to achieve the protein content required by processors. This is illustrated in Figure 8.6, which shows the yields and grain protein contents of wheat grown in the Broadbalk long-term experiment at Rothamsted Research. Nitrogen fertiliser is the

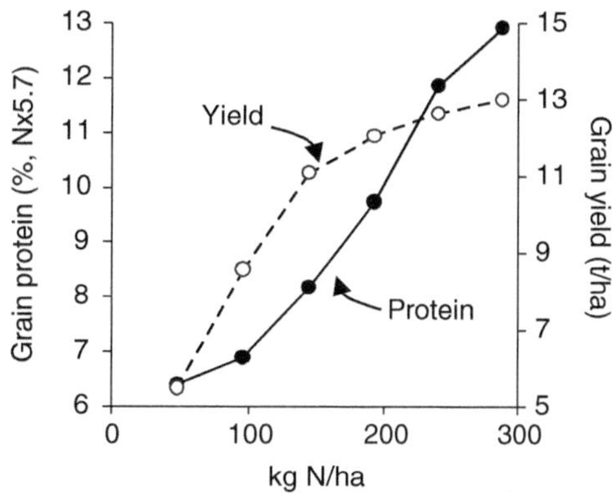

Figure 8.6 Relationship between application of nitrogen fertiliser, grain yield, and grain protein content for the breadmaking cultivar Crusoe grown as first wheat in the Broadbalk experiment at Rothamsted Research in 2014. Yield is expressed on an 85% dry matter basis and protein content on a dry weight basis. We thank the Lawes Agricultural Trust and Rothamsted Research for data from the e-RA database. The Rothamsted Long-term Experiments National Capability (LTE-NCG) is supported by the UK Biotechnology and Biological Sciences Research Council and the Lawes Agricultural Trust.

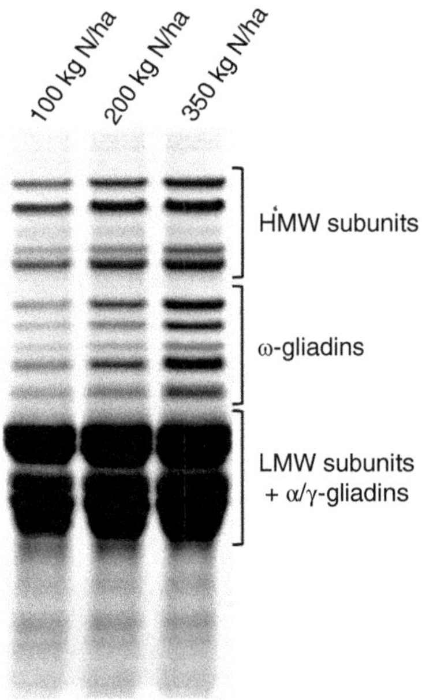

Figure 8.7 Sodium dodecyl-sulphate polyacrylamide gel electrophoresis (SDS-PAGE) of wheat grain proteins, showing increases in the proportions of ω-gliadins under conditions of high nitrogen fertilization. *Source:* Taken from the study reported by Wan et al. (2014) and kindly provided by Dr. Yongfang Wan.

major production cost in many countries and hence the requirement for additional nitrogen fertiliser may have a significant impact on the price of breadmaking wheat.

Nitrogen fertilisation results not only in increased total grain protein but also in an increased proportion of gluten proteins, which effectively act as a sink for nitrogen that is not required for growth and metabolism. Although there is variation between the results reported from different studies, most show that increased grain protein content resulting from nitrogen fertilisation is associated with an increased proportion of gliadin relative to glutenin, resulting in decreased dough strength and increased extensibility. Specific increases in the proportions of ω-gliadins are also often observed (Figure 8.7), which could result from limitation in sulphur availability (if sulphur fertiliser is not also applied), as discussed in Section 8.5.2.

Nitrogen fertiliser is often applied during grain filling in order to increase the grain protein content for breadmaking. However, this may not lead to increased protein quality. In fact, it appears that late nitrogen application results in a decreased proportion of large insoluble glutenin polymers, and hence lower grain quality, than when nitrogen is applied early in grain development (Johansson et al. 2004).

8.5.2 Sulphur Availability

The second mineral nutrient which significantly affects grain quality is sulphur. Sulphur is an essential nutrient for plants, being required for the synthesis of the sulphur-containing amino acids (cysteine and methionine) present in proteins. Cysteine is particularly important for wheat grain quality, as it forms disulphide bonds within and between proteins, including the interchain disulphide bonds which stabilise the glutenin polymers (as discussed). Hence, sulphur deficiency affects gluten protein composition and processing quality.

It has been estimated that the total sulphur uptake by winter wheat grown in the UK varies between 15 and 25 kg/ha, with uptake below 15 kg/ha being deficient. However, the requirement for sulphur fertilisation was not apparent until about 30 years ago because crops received adequate sulphur from atmospheric deposition (burning fossil fuels) and the use of sulphur in fertilisers (such as ammonium sulphate and single superphosphate). Changes in fertilisers and reduction of emissions can result in rapid changes in sulphur status. For example, a survey of samples of wheat grown across the UK showed that the concentration of sulphur in the grain decreased from a mean of 1.72 mg/g in 1982 and 1983 to a mean of 1.35 mg/g in 1992 and 1993, with 26% of the samples grown in 1993 having less than 1.2 mg/g and therefore considered to be S deficient (Zhao et al. 1999). Because the requirements for nitrogen and sulphur are inter-related, deficiency has also been defined based on the ratio of N : S in the grain, with values above 17 : 1 considered to be deficient.

Sulphur deficiency results in changes in the composition of the gluten proteins, notably increased proportions of the ω-gliadins, which lack cysteine residues. The ratio of HMW subunits to LMW subunits in glutenin may also be increased, resulting in a higher proportion of HMW glutenin polymers, but the total amount of glutenins is decreased due to reduced accumulation of LMW subunits. These changes result in increased dough strength and reduced loaf volume when used for breadmaking (Zhao et al. 1999; Naeem and MacRitchie 2003).

An unfortunate and unintended consequence of fertilisation with sulphur is reduced contents of two essential micronutrients, selenium and molybdenum (Stroud et al. 2010). This probably results from competition for transporters involved in nutrient uptake and transport within the plant. The consequences of this for selenium and human health are discussed in Chapter 9.

8.5.2.1 Sulphur Nutrition, Asparagine Content, and Acrylamide Formation

Cysteine and methionine are required for the synthesis of grain proteins (with the exception of ω-gliadins) and hence sulphur deficiency results in decreased protein synthesis and the accumulation of free amino acids, in particular of asparagine. Asparagine is an amino acid amide, with a carbon : nitrogen ratio of 2 : 1. Hence, it is an efficient form in which to store nitrogen. Asparagine is also relatively metabolically inert, being synthesized and broken down by single enzymes (asparagine synthase and asparaginase, respectively). It is therefore an ideal compound for storing nitrogen.

The interest in asparagine relates not to its role in plant metabolism but to its contribution to the formation of acrylamide in processed foods. Acrylamide is a neurotoxin and potential carcinogen that is formed in foods by a Maillard reaction between asparagine and reducing sugars (such as glucose), with asparagine generally being the limiting factor in wheat products. Because asparagine is concentrated in bran the formation of acrylamide is a greater problem in products made from wholemeal than in white flour products (Raffan et al. 2020).

The simplest strategy to minimize asparagine in wheat products is to ensure that crops are grown with adequate sulphur and farmers in the UK are advised to apply 20 kgS/ha for wheat production. Acrylamide formation in processed foods can also be reduced by using the enzyme asparaginase, produced by fermentation, to break down free asparagine during processing.

Genetic variation in the content of free asparagine has been reported, but there are strong effects of environment and it has not so far been possible to select for low free asparagine in breeding programmes.

8.5.3 Temperature and Water Availability

Effects of environmental factors on grain composition and quality have been widely studied, focusing on temperature, water availability, and carbon dioxide concentration, particularly in the context of climate change.

In general terms, grain processing quality is higher when the temperature during grain development is higher, provided the temperature does not exceed about 30 °C when heat shock effects occur (Johansson et al. 2013). Warm and dry conditions are associated with faster rates, but shorter durations of grain filling with DM, N, and S (Table 8.1). The earlier curtailment of grain filling often results in smaller, more shrivelled grain, of low mean grain weight. The amounts of N and S accumulated are also reduced, but not usually to the same extent as that of DM (and starch), so the concentrations of N and S in the grain are higher in hot/dry conditions (Tables 8.1 and 8.2). If the grain are milled to produce flour with the same level of bran contamination (as indicated by flour colour), the flour extraction rates are reduced from the more shrivelled grain. Nonetheless, the beneficial effects of hot/dry conditions during grain filling on protein concentration persist in the white flour that is extracted (Table 8.2).

The temperature during grain maturation is also positively correlated with glutenin polymer size, a positive indicator of breadmaking quality, up to 30 °C, above which the correlation is negative. Hence, protein content, protein quality, and loaf volumes increase at temperatures up to 30 °C (Table 8.2).

High temperature during grain maturation is often associated with drought stress, which may also contribute to higher protein quality, possibly by increasing the formation of hydrogen bonds between proteins. Genotypes also vary in the impact of temperature and

Table 8.1 The effect of environmental conditions during phase 2 (Figure 7.4) on the filling of wheat grain with dry matter (DM), nitrogen (N), and sulphur (S). Mean of seven genotypes.

	Duration[a] (d)		Max. rate[a] (μg/grain/d)		Final weight (mg/grain)	
	Cool/wet[b]	Hot/dry[b]	Cool/wet	Hot/dry	Cool/wet	Hot/dry
DM	45.7	28.1	1300	1800	40.2	27.2
N	47.9	32.7	27.1	41.8	0.944	0.776
S	47.1	31.0	1.72	2.75	0.0594	0.0486

[a] Duration and maximum rate of grain filling are derived from logistic fits to the weight per grain *v* time data as described by Zahedi and Jenner (2003).
[b] Cool/wet is full irrigation and ambient temperatures; Hot/dry is 5–10 °C above ambient (but below 28 °C) and water availability restricted to close to the wilting point.
Source: Data from Wan et al. (2009).

Table 8.2 The effect of environmental conditions during phase 2 (Figure 7.4) on grain composition and flour and baking quality. Mean of seven genotypes.

	Cool/wet[a]	Hot/dry[a]
Grain		
Dry matter yield (t/ha)	8.9	6.2
Nitrogen yield (kg/ha)	220	186
Sulphur yield (kg/ha)	13.5	11.4
N:S ratio	16.3	16.3
N concentration (% DM)	2.46	2.99
S concentration (% DM)	0.151	0.184
Flour		
Extraction rate (%)	75.6	71.7
Flour whiteness score	32.8	32.4
Flour protein (% DM)	12.5	14.9
Bread		
Crumb texture score	8.7	8.3
Loaf volume (ml)	1234	1341

[a] Cool/wet is full irrigation and ambient temperatures; Hot/dry is 5–10 °C above ambient (but below 28 °C) and water availability restricted to close to the wilting point.
Source: Data from Wan et al. (2009).

water availability on grain development and processing quality, with genotype x environment interactions (Wan et al. 2009; Gooding 2017).

Increases in temperature above 30 °C result in progressive decreases in grain protein content and also negatively affect protein quality, resulting in increased proportions of gliadins and small soluble glutenin polymers, and decreased proportions of large insoluble glutenin polymers. This results in weaker dough and poorer quality for breadmaking.

8.5.4 Carbon Dioxide Concentration

Long-term increases in atmospheric carbon dioxide are of concern for food security. Increased CO_2 concentrations result in higher grain yields, by increasing the fixation of carbon by photosynthesis. This inevitably results in dilution of other grain components, including protein concentration, which is predicted to decrease by over 8%, the precise magnitude of the effect varying with the experimental set up (see meta-analysis by Broberg et al. 2018).

Several studies have reported detrimental effects of elevated CO_2 on grain processing quality, with lower peak resistance during dough mixing and lower loaf volume on baking (Broberg et al. 2018). Fewer studies of effects on protein composition have been reported, but Wieser et al. (2008) reported decreases in gliadins and glutenins, with ω-gliadins and HMW subunits of glutenin being affected more than other types of gluten protein. It is not known whether these effects can be ameliorated by increased application of fertiliser (nitrogen and sulphur), or whether there is sufficient genetic variation in the response to be exploited by wheat breeders.

References

Amend, T. and Beauvais, F. (1995). Der mechanismus der Teigbildung: Vorstoß in den molekularen Strukturbereich. *Getreide Mehl Und Brot* 49: 359–362.

Bhave, M. and Morris, C.F. (2008). Molecular genetics of puroindolines and related genes: allelic diversity in wheat and other grasses. *Plant Molecular Biology* 66: 205–219. https://doi.org/10.1007/s11103-007-9263-7.

Broberg, M.C., Hogy, P., and Pleijel, H. (2018). CO_2-induced changes in grain composition: meta-analysis and response functions. *Agronomy* 7: 32. https://doi.org/10:3390/agronomy7020032.

Chantret, N., Salse, J., Sabot, F. et al. (2005). Molecular basis of evolutionary events that shaped the hardness locus in diploid and polyploidy wheat species (Triticum and Aegilops). *The Plant Cell* 17: 1033–1045. https://doi.org/10.1105/tpc.104.029181.

Chen, M., Wilkinson, M., Tosi, P. et al. (2005). Novel puroindoline and grain softness protein alleles in *Aegilops* species with the C, D, S, M and U genomes. *Theoretical and Applied Genetics* 111: 1159–1166. https://doi.org/10.1007/s00122-005-0047-7.

Clifton, L.A., Green, R.J., and Frazier, R.A. (2007). Puroindoline-b mutations control the lipid binding interactions in mixed puroindoline-a:puroindoline-b systems. *Biochemistry* 46: 13929–13937. https://doi.org/10.1021/bi701680w.

Courtin, C. and Delcour, J. (2002). Arabinoxylans and endoxylanases in wheat flour bread-making. *Journal of Cereal Science* 35: 225–243. https://doi.org/10.1006/jcrs.2001.0433.

Delcour, J.A. and Hoseney, R.C. (2010). *Principles of Cereal Science and Technology*, 3e. AACC International.

Dubreil, L., Compoint, J.P., and Marion, D. (1997). The interaction of puroindolines with wheat polar lipids determines their foaming properties. *Journal of Agriculture and Food Chemistry* 45: 108–116. https://doi.org/10.1021/jf960402j.

Farrand, E.A. (1969). Starch damage and alpha-amylase as bases for mathematical models relating to flour water- absorption. *Cereal Chemistry* 46: 103–116.

Finnie, S.M., Jeannotte, R., Morris, C.F. et al. (2010). Variation in polar lipid composition among near-isogenic wheat lines possessing different puroindoline haplotypes. *Journal of Cereal Science* 51: 66–72. https://doi.org/10.1016/j.jcs.2009.09.006.

Gooding, M.J. (2017). The effects of growth environment and agronomy on grain quality. In: *Cereal Grains: Assessing and Managing Quality*, 2e (eds. C. Wrigley, I. Batey and D. Miskelly), 493–512. Woodhead Publishing.

Greenwell, P. and Schofield, J.D. (1986). A starch granule protein associated with endosperm softness in wheat. *Cereal Chemistry* 63: 379–380. https://doi.org/10.1006/jcrs.2000.0312.

Johansson, E., Prieto-Linde, M.L., and Svensson, G. (2004). Influence of nitrogen application rate and timing on grain protein composition and gluten strength in Swedish wheat. *Journal of Plant Nutrition and Soil Science* 167: 345–350. https://doi.org/10.1002/jpln.200320332.

Johansson, E., Malik, A.H., Hussain, A. et al. (2013). Wheat gluten polymer structures: the impact of genotype, environment and processing on their functionality in various applications. *Cereal Chemistry* 90: 367–376. https://doi.org/10.1094/CCHEM-08-12-0105-FI.

Kent, N.L. and Evers, A.D. (1994). *Kent's Technology of Cereals*, 4e. Woodhead Publishing.

Kindred, D.R., Verhoeven, T., Weightman, R. et al. (2008). Effects of variety and fertiliser nitrogen on alcohol yield, grain yield, starch and protein content, and protein composition of winter wheat. *Journal of Cereal Science* 48: 46–57. https://doi.org/10.1016/j.jcs.2007.07.010.

Konopka, I., Rotkiewicz, D., and Tańska, M. (2005). Wheat endosperm hardness. Part II. Relationships to content and composition of flour lipids. *European Food Research and Technology* 220: 20–24. https://doi.org/10.1007/s00217-004-1038-7.

Lazaridou, A. and Biliaderis, C.G. (2007). Molecular aspects of β-glucan functionality: physical properties, technological applications and physiological effects. *Journal of Cereal Science* 46: 101–118. https://doi.org/10.1016/j.jcs.2007.05.003.

Lovegrove, A., Wood, A.J., Hassall, K. et al. (2020). The contribution of fibre components to water absorption of wheat grown in the UK. *Cereal Chemistry* 97: 940–948. https://doi.org/10.1002/cche.10316.

McDermott, E.E. and Pace, J. (1960). Comparison of the amino-acid composition of the protein in flour and endosperm from different types of wheat, with particular reference to variation in lysine content. *Journal of the Science of Food and Agriculture* 11: 109–115. https://doi.org/10.1002/jsfa.2740110207.

Melis, S., Meza Morales, W.R., and Delcour, J.A. (2019). Lipases in wheat flour bread making: importance of an appropriate balance between wheat endogenous lipids and their enzymatically released hydrolysis products. *Food Chemistry* 298: 125002. https://doi.org/10.1016/j.foodchem.2019.125002.

Min, B., Salt, L., Wilde, P. et al. (2020). Genetic variation in wheat grain quality is associated with differences in the galactolipid content of flour and the gas bubble properties of dough liquor. *Food Chemistry: X* 6: 100093. https://doi.org/10.1016/j.fochx.2020.100093.

Morris, C.F. (2002). Puroindolines: the molecular genetic basis of wheat grain hardness. *Plant Molecular Biology* 48: 633–647. https://doi.org/10.1023/A:1014837431178.

Morris, C.F., Luna, J., and Caffe-Treml, M. (2021). The Vromindolines of cv. Hayden oat (*Avena sativa* L.) – a review of the Poeae and Triticeae indolines and a suggested system for harmonization of nomenclature. *Journal of Cereal Science* 97: 103135. https://doi.org/10.1016/j.jcs.2020.103135.

Morrison, W.R., Law, C.N., Wylie, L.J. et al. (1989). The effect of group M. 5 chromosomes on the free polar lipids and breadmaking quality of wheat. *Journal of Cereal Science* 9: 41–51. https://doi.org/10.1016/S0733-5210(89)80020-7.

Naeem, H.A. and MacRitchie, F. (2003). Effects of Sulphur nutrition on agronomy and quality attributes of wheat. In: *Sulphur in Plants* (eds. Y.P. Abrol and A. Ahmad), 305–322. Kluwer Academic Publishers https://link.springer.com/content/pdf/10.1007%2F978-94-017-0289-8_17.pdf.

Pareyt, B., Finnie, S.M., Putseys, J.A. et al. (2011). Lipids in bread making: sources, interactions, and impact on bread quality. *Journal of Cereal Science* 54: 266–279. https://doi.org/10.1016/j.jcs.2011.08.011.

Payne, P.I. (1987). Genetics of wheat storage proteins and the effect of allelic variation on breadmaking quality. *Annual Review of Plant Physiology* 38: 141–153. https://doi.org/10.1146/annurev.pp.38.060187.001041.

Payne, P.I., Nightingale, M.A., Krattiger, A.F. et al. (1987). The relationship between HMW glutenin subunit composition and the breadmaking quality of British grown wheat varieties. *Journal of the Science of Food and Agriculture* 40: 51–65. https://doi.org/10.1002/jsfa.2740400108.

Raffan, S., Oddy, J., and Halford, N.G. (2020). The sulphur response in wheat grain and its implications for acrylamide formation and food safety. *International Journal of Molecular Sciences* 21: 3786. https://doi.org/10.3390/ijms21113876.

Shewry, P.R., Tatham, A.S., Barro, F. et al. (1995). Biotechnology of breadmaking: unravelling and manipulating the multi-protein gluten complex. *Biotechnology* 13: 1185–1190. https://doi.org/10.1038/nbt1195-1185.

Shewry, P.R., Halford, N.G., and Lafiandra, D. (2003a). The genetics of wheat gluten proteins. In: *Advances in Genetics*, vol. 49 (eds. J.C. Hall, J.C. Dunlap and T. Friedman), 111–184. Academic Press.

Shewry, P.R., Halford, N.G., Tatham, A.S. et al. (2003b). The high molecular weight subunits of wheat glutenin and their role in determining wheat processing properties. *Advances in Food and Nutrition Research* 45: 221–302. https://doi.org/10.1016/S1043-4526(03)45006-7.

Stroud, J.L., Zhao, F.J., Buchner, P. et al. (2010). Impacts of sulphur nutrition on selenium and molybdenum concentrations in wheat grain. *Journal of Cereal Science* 52: 111–113. https://doi.org/10.1016/j.jcs.2010.03.011.

Turner, A.S., Bradburne, R.P., Fish, L. et al. (2004). New quantitative trait loci influencing grain texture and protein content in bread wheat. *Journal of Cereal Science* 40: 51–60. https://doi.org/10.1016/j.jcs.2004.03.001.

Wall, M.L., Wheeler, H.L., Huebsch, M.P. et al. (2010). The tryptophan-rich domain of puroindoline is directly associated with the starch granule surface as judged by trypric shaving and mass spectrometry. *Journal of Cereal Science* 52: 115–120. https://doi.org/10.1016/j.jcs.2010.04.002.

Wan, Y., Underwood, C., Toole, G. et al. (2009). A novel transcriptomic approach to identify candidate genes for grain quality traits in wheat. *Plant Biotechnology Journal* 7 (5): 401–410. https://doi.org/10.1111/j.1467-7652.2009.00405.x.

Wan, Y., Sanchis Gritsch, C., Hawkesford, M.J. et al. (2014). Effects of nitrogen nutrition on the synthesis and deposition of the ω-gliadins of wheat. *Annals of Botany* 113: 607–615. https://doi.org/10.1093/aob/mct291.

Wieser, H., Manderscheid, R., Erbs, M. et al. (2008). Effects of elevated atmospheric CO_2 concentrations on the quantitative protein composition of wheat grain. *Journal of Agricultural and Food Chemistry* 56: 6531–6535. https://doi.org/10.1021/jf8008603.

Wilkinson, M.D., Tosi, P., Lovegrove, A. et al. (2017). The *Gsp-1* genes encode the wheat arabinogalactan peptide. *Journal of Cereal Science* 74: 155–164. https://doi.org/10.1016/j.jcs.2017.02.006.

Willkinson, M., Wan, Y., Tosi, P. et al. (2008). Identification and genetic mapping of variant forms of purolindoline b expressed in developing wheat grain. *Journal of Cereal Science* 48: 722–728. https://doi.org/10.1016/j.jcs.2008.03.007.

Wrigley, C.W. and Bietz, J.A. (1988). Proteins and amino acids. In: *Wheat Chemistry and Technology*, 3e (ed. Y. Pomeranz), 159–275. American Association of Cereal Chemists.

Zahedi, M. and Jenner, C.F. (2003). Analysis of effects in wheat of high temperature on grain filling attributes estimated from mathematical models of grain filling. *Journal of Agricultural Science* 141 (2): 203–212.

Zhao, F.J., Hawkesford, M.J., and McGrath, S.P. (1999). Sulphur assimilation and effects on yield and quality of wheat. *Journal of Cereal Science* 30: 1–17. https://doi.org/10.1006/jcrs.1998.0241.

9

The Role of Wheat in Diet and Health

If the English people only knew the immense value of wholemeal bread as a preserver of health they would make it a regular article of diet and never eat any other.... No home is complete unless it contains a wholemeal loaf

Thomas Allinson

9.1 Contribution of Wheat to the Human Diet

The wheat grain has been a staple food in temperate countries for millennia, accounting for between about 10% and over 50% of total calories in wheat-consuming countries. However, it also contributes a range of other components that are essential for nutrition and health. Because wheat is highly pervasive in food systems, and is often present in processed foods with other cereals, precise data for the contributions of wheat to diets are not readily available. However, we do have data on the consumption of bread, which is the major wheat-based food in many countries. For example, data from the UK National Diet and Nutrition Survey (NDNS) shows that bread contributes 10–12% of the daily intake of protein, between 10% and 20% of the daily intakes of minerals and B vitamins (B1 thiamine, B3 niacin, B9 folate, iron, zinc, copper, magnesium, selenium and calcium), and about 20% of the daily fibre intake (Table 9.1). Furthermore, these contributions vary little with age or gender. It should also be noted that these values are for the UK, which has a mixed diet: they will be much higher in countries in which wheat forms a more important part of the diet.

9.2 Dietary Fibre

9.2.1 Proposed and Approved Benefits of Dietary Fibre

Dietary fibre comprises 'carbohydrate oligomers and polymers that are not digested and absorbed in the small intestine and are partially or completely fermented in the colon. It also includes associated components such as lignin' (CGA 2021). Dietary fibre, and particularly fibre in wholegrain cereals, has been shown to have a number of health benefits,

Wheat: Environment, Food and Health, First Edition. Mike J. Gooding and Peter R. Shewry.

Table 9.1 Percentage contributions of bread to average daily intake of some essential nutrients in the UK, separated by age and gender.

	Children aged 4–18		Adults aged 19–64		Adults aged 65+	
	Boys	**Girls**	**Men**	**Women**	**Men**	**Women**
Energy	11	11	12	10	13	12
Protein	12	10	11	11	12	9
Carbohydrates	17	16	21	18	20	17
Non-Starch Polysaccharides (fibre)	18	17	21	18	21	18
Thiamine (B1)	15	15	16	15	17	14
Riboflavin (B2)	4	4	5	4	5	4
Niacin (B3) (equivalents)	12	11	11	10	13	12
Pyridoxine (B6)	5	5	5	4	4	4
Folates (B9)	12	11	12	12	13	12
Iron	16	15	16	15	17	15
Calcium	16	16	19	15	19	14
Magnesium	13	11	13	13	14	14
Sodium	18	18	19	18	20	18
Zinc	11	11	12	12	13	11
Copper	16	15	15	14	15	14
Selenium	14	13	13	12	13	12

Source: (Bates et al. 2014a,b)

including lowering blood pressure and serum cholesterol, improving insulin sensitivity (and hence reducing the risk of type 2 diabetes) and reducing the incidence of certain types of cancer, notably colorectal cancer (Aune et al. 2011, 2013, 2016; Reynolds et al. 2019), but also breast cancer in pre-menopausal women (Cade et al. 2007), small intestinal cancer (Schatzkin et al. 2008), and pancreatic cancer (Chan et al. 2007) (Table 9.2).

The mechanisms for these benefits are undoubtedly complex (Section 9.2.3) but include increasing faecal bulk and reducing intestinal transit time, binding cholesterol and carcinogens, reducing the rate of digestion and glucose release in the small intestine, and fermentation in the colon to produce short chain fatty acids (SCFA), which have health benefits. Soluble and insoluble forms of fibre may also differ in their benefits and mechanisms (Section 9.2.3).

Three health claims for wheat fibre have been approved by the European Food Safety Authority (EFSA) (EU 2006; Table 9.2). These are one claim for arabinoxylan (reduction of post-prandial glucose) (EFSA Panel on Dietetic Products, Nutrition and Allergies [NDA], 2011) and two claims for bran fibre: increased faecal bulk and reduction in intestinal transit time (EFSA Panel on Dietetic Products, Nutrition and Allergies [NDA], 2010). In

Table 9.2 Summary of proposed and established health benefits of dietary fibre components present in wheat grain and other cereals.

Component and proposed health benefit for cereals or other foods	Health claim supported by the European Food Safety Authority for food labelling
Dietary fibre	
Reduces post-prandial glycaemic response (and risk of type 2 diabetes)	No
Reduces intestinal transit time	Yes
Increases faecal bulk	Yes
Reduces cholesterol (and risk of coronary heart disease)	No
Reduces risk of colorectal cancer	No
Reduces risk of other cancers (breast, small intestine, pancreas)	No
Reduces risk of stroke	No
Prebiotic effects	No
Stimulates immune responses	No
Arabinoxylan	
Reduces post-prandial glycaemic response (and risk of type 2 diabetes)	Yes
β-glucan	
Reduces post-prandial glycaemic response (and risk of type 2 diabetes)	Yes (barley and oats)
Reduces cholesterol (and risk of coronary heart disease)	Yes (barley and oats)
Resistant starch	
Reduces post-prandial glycaemic response	Yes (all sources)
Other benefits as part of dietary fibre above	No
Fructans	
Prebiotic effects	No
Promotes calcium and iron absorption	No

addition, relevant accepted health claims for β-glucans from oats and barley and for resistant starch (RS) (which is considered to be a fibre component) from other foods are also noted in Table 9.2. Approved health claims are important for the commercial development of nutritionally enhanced foods, as they allow the products to be marketed on the basis of the claims. However, the level of scientific evidence that is required is very stringent, hence some widely accepted benefits of fibre (such as reduced risk of colorectal cancer) are not supported by approved health claims and therefore cannot be used in promoting fibre-rich foods.

9.2.2 Wheat Grain Fibre

Whole wheat grains are relatively rich in fibre, which accounts for 11.5–15.5% of the dry weight, as illustrated by the comparative analyses of 129 wheat lines reported by Andersson et al. (2013) (Table 9.3). The major dietary fibre components in whole grain are cell wall polysaccharides (arabinoxylan, β-glucan, and cellulose), lignin (a phenolic polymer present in cell walls), and fructo-oligosaccharides (fructans). Additionally, arabinogalactan peptide (AGP), present only in the starchy endosperm, also contributes to dietary fibre (Chapter 7).

White flour contains less fibre than whole grain. Lignin is not present and about 70% of the total cell wall polysaccharides is arabinoxylan (about 1.5–2.5% DM [dry matter]) (Mares and Stone 1973). The second major component is usually stated as β-glucan, accounting for about 20% of the total cell wall polysaccharides (about 0.5% DM) (as shown in Table 9.3). However, Gartuala et al. (2018) recently reported that cellulose may account for up to 20% of wheat starchy endosperm cell walls (approx. 0.5% DM), with a much lower proportion of β-glucan (5–6% cell wall polysaccharides). This disparity is still not resolved but it is likely that these two components together account for 20–25% of the starchy endosperm cell walls.

Combining the data for cell wall polysaccharides, fructans (about 1.5% DM), and AGP (up to 0.4% DM) suggests that mean total fibre content of white flour is about 4.5% DM, with bread and other processed foods containing an additional 1% DM fibre from RS (Table 9.4). However, whereas the fibre in bran is largely insoluble, a high proportion of the fibre in white flour is soluble, including between 20 and 50% of the total arabinoxylan (compared with only 2–5% in bran) (Chapter 8) (Gebruers et al. 2008), a proportion of the β-glucan (probably 10–15%), and all the fructans and AGP (Table 9.4).

9.2.2.1 Cell Wall Polysaccharides

Because arabinoxylan accounts for 50% or more of the total fibre in white flour it has been the main target for increasing flour fibre content. This can be achieved by exploiting genetic variation, as the arabinoxylan content varies by over twofold (from 1.35 to 2.75%) between

Table 9.3 Contents (% DM) of total dietary fibre and dietary fibre components in whole grain of 129 winter wheat varieties (based on data in Andersson et al. 2013). The values do not include resistant starch, which accounts for about 1.2% of total starch (about 1% DM) in bread made from white flour (Murphy et al. 2008) (about 0.8% DM of bread made from wholemeal).

	Range	Mean
Total dietary fibre	11.5–15.5	13.4
Klason lignin	0.74–2.03	1.33
Arabinoxylan	5.53–7.42	6.49
Cellulose	1.67–3.05	2.11
β-Glucan	0.51–0.96	0.73
Fructan	0.84–1.85	1.28

Table 9.4 Dietary fibre components (% DM) in white flour of wheat. Values are based on literature discussed above and in Chapter 8. In addition to the components listed, white flour also contains cellulose (insoluble) and glucomannan (solubility not known) (see Chapter 8). The values do not include resistant starch, which accounts for about 1.2% of total starch (about 1% DM) in bread made from white flour (Murphy et al. 2008).

	Total	Soluble
Arabinoxylan	1.35–2.75 (mean = 1.93)	0.3–1.4 (mean = 0.51)
β-glucan	0.25–0.63 (mean = 0.44)	≈0.05
Fructans	≈1.5	≈1.5
Arabinogalactan peptide	≈0.4	≈0.4

genotypes (Figure 7.12) (Gebruers et al. 2008). Although the content of arabinoxylan is under complex multigenic control, several major quantitative trait loci (QTL) have been identified, including major loci on chromosomes 1B and 6B, allowing the establishment of molecular markers for selection by breeders (Lovegrove et al. 2020; Ibba et al., 2021). The concentration of β-glucan in 49 white flour samples varies to a similar extent, from 0.25 to 0.63% DM (mean 0.44% DM) (Andersson et al. 1992), but robust QTLs have not been identified.

9.2.2.2 Fructans

The fructan content in whole grain has been reported to range from under 1% DM to over 3% DM with moderate to high heritability (Grausgruber et al. 2020). However, QTL (Chapter 10) and genes controlling fructan content have not been identified.

9.2.2.3 Resistant Starch

RS can be defined as 'the starch and products of starch digestion that are not absorbed in the small intestine of healthy individuals' (Englyst and Cummings 1985). Consequently, it passes into the colon where it forms part of the fibre fraction and is fermented by the colonic microflora. RS is categorised into several types, two of which are relevant to wheat.

Type 2 RS (RS2) is starch that is inaccessible to enzymes due to its conformation and includes high amylose starches (Section 9.2.2.4), while type 3 (RS3) is formed when foods are cooked and cooled and results from starch crystallisation (retrogradation). Starch retrogradation is well understood and the benefits of eating starchy foods after cooking and cooling (such as pasta) have been communicated to consumers. However, starch retrogradation can also have detrimental effects on food texture and acceptability, being one of the processes that contribute to staling of bread (see Chapter 8).

9.2.2.4 High Amylose Starch

Increasing the amylose content of starch (RS2) can be achieved by the modification of starch synthesis, particularly downregulation of the starch syntheses and starch branching enzymes that catalyse the synthesis of amylopectin. The combination of naturally occurring mutations in starch synthase IIa enzymes on the A and B genomes of tetraploid pasta (durum) wheat result in an amylose content of over 40%, compared with about 25% in control lines, while amylose contents of over 70% have been achieved by reducing the expression of starch branching enzymes in bread wheat, using induced mutations or transgenesis (Chapter 10) (reviewed by Lafiandra et al. 2014).

This research has led to the commercial availability of high amylose types of wheat but there has so far been little interest in exploiting this material from mainstream wheat processors. This is because the mutations are associated with adverse effects on yield, and hence cost, and on processing quality. Processing quality is particularly important for wheat and any changes in the starch fraction, which represents 80% of white flour, are likely to affect product quality. For example, increased amylose affects the hydration and gelatinisation of starch during processing and may also reduce the shelf life of baked products due to more rapid staling (see also Chapter 8).

9.2.3 Mechanism of Action of Dietary Fibre

A widely used sub-division of fibre is into soluble and insoluble forms, as these may have different functionalities. In wheat, lignin, cellulose, RS, and proportions of AX and β-glucan are insoluble, while fructans, AGP, and proportions of AX and β-glucan are soluble. However, solubility depends on the conditions in the human gastrointestinal (GI) tract and Gill et al. (2021) suggest that it is important to consider the balance of three physicochemical properties: solubility, viscosity, and fermentability. Insoluble, non-viscous fibres (such as cellulose and lignin) are either slowly or not fermented in the colon and have benefits in increasing intestinal transit time and stool bulking; soluble non-viscous components such as fructans are rapidly fermented to SCFA and have prebiotic effects in the colon; while soluble viscous fibres (such as some AX and β-glucan components) affect the uptake of nutrients in the small intestine as well as being fermented to SCFA in the colon. The beneficial effects of fibre in the small intestine include reducing the rate of starch breakdown, and hence the glycaemic response, and lowering serum cholesterol. Gill et al. (2021) state that 'viscosity remains the accepted model for the cholesterol-lowering capacity of β-glucan', through the prevention of reabsorption of bile salts. However, whether this is the sole, or even major, reason for the beneficial effects of wheat fibre in the small intestine needs to be determined.

Several mechanisms may therefore contribute to the beneficial effects of AX (and to a lesser extent β-glucan) in the GI tract.

9.2.3.1 Role of Food Structure and Breakdown

AX, β-glucan, cellulose, and glucomannan are intrinsic components of the cell walls of the starchy endosperm cells and there is an increasing volume of evidence that the porosity of cell walls controls the rate of digestion of the cell contents, including the digestion of starch and release of glucose (Edwards et al., 2015, 2021; Li et al., 2021). The extent to which this structural barrier is retained in processed foods (such as white bread) has not been established, but support is provided by Gouseti et al. (2019), who showed a lower rate of glucose release from high fibre than conventional white bread when digested in a model duodenal system.

9.2.3.2 Role of Luminal Viscosity

Increasing luminal viscosity is considered to have beneficial effects on glucose homeostasis and appetite control, by giving a feeling of fullness and delaying gastric emptying. However, effects of viscosity in the small intestine are less clear, being subject to dilution by digestive secretions and being difficult to measure *in vivo* (reviewed by Muller et al. 2018; Gill et al. 2021).

9.2.3.3 Role of Prebiotic Activity

The beneficial effects of fermentation of fibre in the colon have led to the concept of prebiotics. These are 'non-digestible (by the host) food ingredients that have a beneficial effect through their selective metabolism in the intestinal tract' (Gibson et al. 2004). The major fermentation products of fibre are SCFAs, notably acetic acid, butyric acid, and propionic acid. These SCFAs have accepted benefits for cardiovascular and metabolic health (Chambers et al. 2018) and for reducing the risk of colorectal cancer (Scheppach et al. 1995). However, the SCFAs differ in their spectra of activity; for example, butyric acid is particularly effective in reducing the risk of colorectal cancer (Russo et al. 1999).

Prebiotic activity is generally accepted as including increases in the populations of beneficial bacteria, such as *Bifidobacterium* species and/or *Lactobacillus* species, and is often measured *in vitro* by determining the effects of isolated components on bacterial populations and SCFAs released by faecal cultures. Such studies have demonstrated that soluble arabinoxylan and AGP fractions from wheat are effective prebiotics, with activities that compare well with inulin, a fructan from Jerusalem artichoke that is used as a standard in such studies (Hughes et al. 2007; Harris et al. 2019).

Both AX and β-glucan are highly fermentable (Gill et al. 2021) and it is therefore probable that most, if not all, of the AX and β-glucan in bread and other products made from white flour are fermented in the colon. However, it is likely that insoluble forms are fermented more slowly than soluble forms, with fermentation extending into the distal colon. This should have benefits in preventing cancer as the beneficial effects of the butyrate produced may reduce the deleterious effects of fermentation of proteins in the distal colon (which can lead to the production of potentially toxic metabolites such as ammonia, amines, phenols, and sulphides) (Windey et al. 2012).

9.3 Micronutrients and Phytochemicals

Wheat grain contains a range of micronutrients, including minerals and vitamins that are essential for health, and phytochemicals, which may contribute health benefits.

9.3.1 Iron and Zinc

Both iron (Fe) and zinc (Zn) are essential micronutrients for plants. Iron is essential for a range of biological processes, as a cofactor for enzymes and as a central component of electron transport chains. It contributes to chlorophyll synthesis, photosynthesis, respiration, the control of redox balance, and the regulation of development (Rout and Sahoo 2015).

However, it is also toxic within the cell and hence plants have mechanisms to bind excess iron to reduce toxicity, including binding to the chelator nicotianamine (Schmidt et al. 2020). Similarly, zinc is an essential cofactor for a number of plant enzymes, is essential for chloroplast development and function, and is an integral part of the Zn finger family of transcription factors that regulate cell growth and differentiation (Sharma et al. 2013).

Deficiencies of Fe and Zn in humans are estimated to affect about 2 billion people globally, making them a primary target for global health initiatives. The WHO has estimated that 43% of children and 29% of women of reproductive age globally have anaemia, half of which results from iron deficiency (WHO 2015), while zinc deficiency is estimated to affect about 155 million children globally (WHO 2013).

Iron and zinc deficiencies are widespread in developing countries but also occur in the developed world: iron deficiency anaemia is estimated to affect 5–6% of women of child-bearing age in the UK. Although red meat and liver are the best sources of dietary Fe, most of the iron in our diet comes from plant-based foods, particularly cereals, which provide about half of the total iron in the UK diet. The amounts of Fe and Zn in wheat grain are strongly affected by the amounts of available minerals in soils, but in broad terms both range from about 20–50 mg/kg DM.

Many countries have mandatory fortification of Fe in flours from wheat, maize, and rice (FFI 2019), with white wheat flour milled in the UK being fortified with inorganic forms of Fe to 16.5 mg/kg (equivalent to levels in wheat flour at an 80% extraction rate). Voluntary fortification of wheat flour with Zn is also practiced in several countries.

Both Fe and Zn are concentrated in the embryo and aleurone layer of the grain, but their relative distributions between these two tissues differ, with Fe being more concentrated in the aleurone layer and Zn in the embryo (particularly in the embryonic axis). This is illustrated in Figure 9.1. In common with other minerals (including calcium and magnesium), the Fe and Zn in the aleurone cells and in the scutellum of the embryo are largely present as phytates in phytin globoids. These are clearly seen in the images in Figure 9.1. Phytates are complexes with phytic acid (inositol hexakisphosphate), which has a cyclic structure with six phosphate groups that can bind metal ions. The form of the zinc in the wheat embryonic axis has not been determined, but it is presumed to include zinc cofactors in enzymes.

The localisation of Fe and Zn as phytates in the aleurone and embryo has two consequences for human health. Firstly, their concentrations are depleted in white flour compared to wholemeal, by about fourfold. Secondly, because phytates have low solubility the bioavailability of Fe and Zn in wholegrain wheat is low, although probably higher for Zn (about 25%) than for Fe (about 10%) (Bouis and Welch 2010). The higher bioavailability of Zn could relate to the presence of non-phytate bound forms in the embryonic axis.

Comparisons of grain samples from older and modern types of wheat have shown significant decreases in the concentrations of Fe, Zn, and most other minerals (including magnesium but not apparently calcium and selenium), in the grain of modern semi-dwarf types compared to traditional tall wheats (Figure 6.15). This is discussed further in Chapter 10. Increases in the concentrations of Fe and Zn in the grain can be achieved by their application in fertilizers, a strategy known as agronomic biofortification (Velu et al. 2014). Increases in the content of Zn in whole grain have also been achieved by

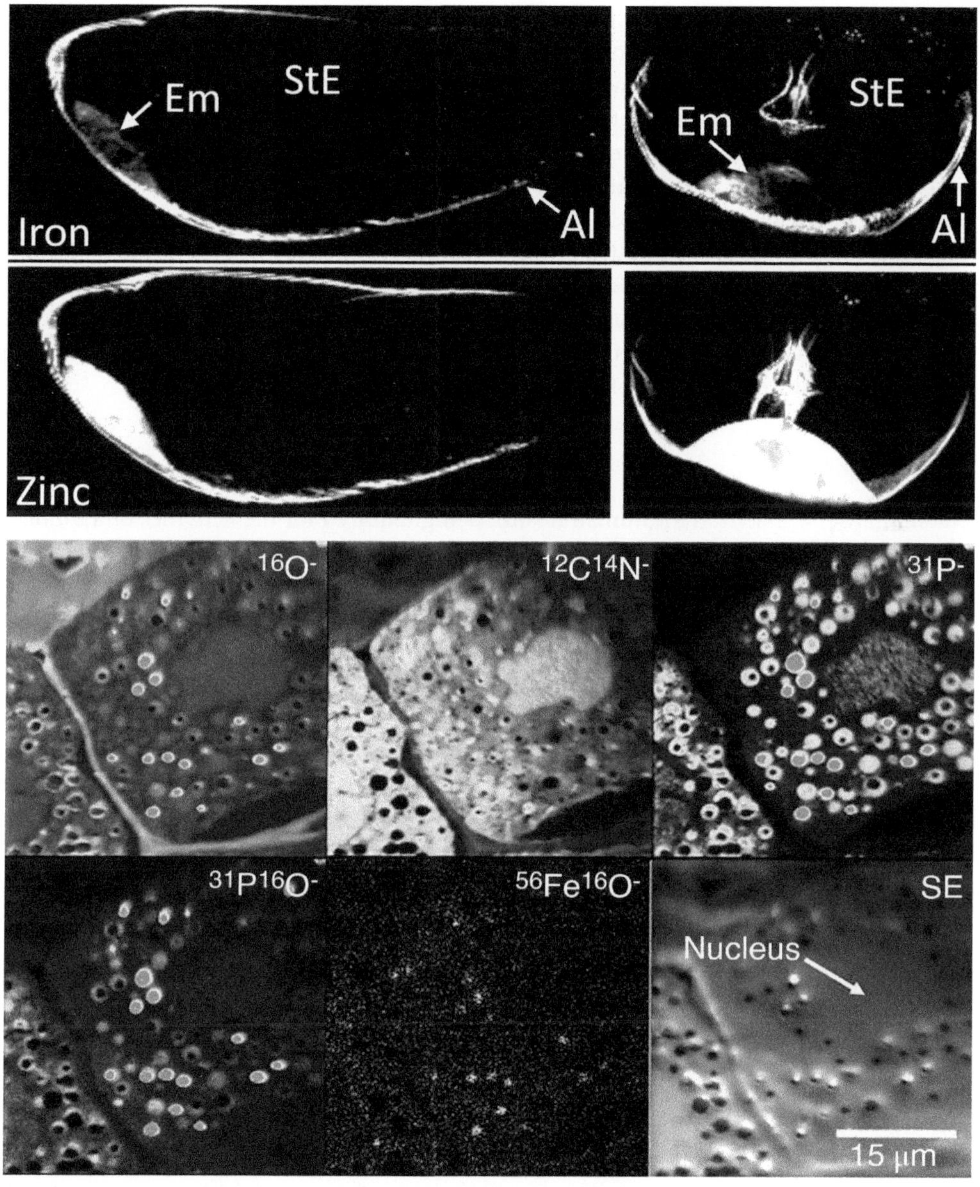

Figure 9.1 The locations of iron and zinc in wheat grain. Top: Heat map representation of the distribution of iron and zinc in longitudinal (left) and transverse (right) sections of wheat grain, revealed by X-ray fluorescence. White indicates high signal intensity. A, aleurone layer; E, embryo; StE, starchy endosperm. Taken from Neal et al. (2013) with permission, figure kindly provided by Dr. Andy Neal (Rothamsted Research). Bottom: NanoSIMS images of an aleurone cell of an immature wheat grain showing localisation of $^{56}Fe^{16}O^{-}$ to the phytin globoids (as indicated by the $^{31}P^{16}O^{-}$ image). White indicates high signal intensity. SE is secondary electron. *Source:* Taken from Moore et al. (2012) with permission, figure kindly provided by Dr. Katie Moore (University of Manchester).

marker-assisted selection (see Chapter 10) for genetic variation in total grain Zn (Velu et al. 2018a). Lines from this study have been released to farmers in South Asia (Velu et al. 2018b) and their impact on the Zn status of populations is being studied by dietary intervention in Pakistan (Lowe et al. 2020).

Genetic selection has not so far led to stable increases in Fe in wheat, but the concentrations of both minerals in the starchy endosperm tissue can be increased by transgenesis, expressing genes which result either in the sequestration of the minerals in the vacuoles or their chelation by components such as nicotianamine (which is involved in mineral transport in the plant) (reviewed by Balk et al. 2018) (Section 10.7.2.3). The concentrations of minerals in white flour are increased in these lines and the minerals have high bioavailability. However, the approach has not been used to produce improved cultivars for human consumption due to regulation of the use of transgenesis (Section 10.6.2).

9.3.2 Selenium

Selenium (Se) is essential for human health, being required for the activities of a number of enzymes. Wheat is an important source of Se in the diet, but the concentration in the grain is largely determined by selenium availability in soil. Consequently, the geographical source of wheat has a major effect on the Se status of consumers. For example, the increased use of home-grown wheat for breadmaking that occurred in the UK starting in the 1970s, replacing wheat imported from North America, resulted in a significant decrease in the intake of Se in the UK diet, from about 60 to 30–40 μg/day. However, the concentration of selenium in grain of modern wheats does not appear to have declined when compared to older types, with historical changes in concentration in UK wheats being negatively related to the availability of sulphur from fertilisation, atmospheric deposition, and environmental pollution (Fan et al. 2008).

There is little genetic variation in the content of Se in wheat grain grown under low Se conditions but agronomic biofortification using Se-containing fertilisers is effective and widely used in some countries. However, unlike most minerals, Se is not taken up using dedicated transporters, but by sulphur transporters, competing with the uptake of sulphur. Because sulphur is itself deficient for the production of breadmaking wheat in some countries (such as the UK) and therefore routinely applied as a fertiliser, the competing requirements for Se and sulphur must be considered when designing strategies for biofortification.

9.3.3 B Vitamins

The B vitamin complex was initially considered to be a single compound but is now known to comprise eight water-soluble components, which often occur together in the same foods but are not related in their structures, biosynthetic pathways, or physiological roles.

Wheat is a significant dietary source of B vitamins, providing over 10% of the UK daily intakes of thiamine (B1), niacin (B3), and folates (B9), and about 5% of the daily intakes of pyridoxine (B6) and riboflavin (B2) (Table 9.1). There has been considerable debate on whether to fortify white flour with folates (B9) to reduce the incidence of neural tube

Table 9.5 Contents (μg/g) of B vitamins in wheat grain fractions (compiled from various sources by Piironen et al. 2009) and in wholegrain fractions from the EU HEALTHGRAIN study.

Fraction	Thiamine (B1)	Riboflavin (B2)	Niacin (B3)	Pyridoxine (B6)	Biotin (B7)	Folate (B9)
Bran	5.2–8.9	3.2–5.8	136–296	7.3–16	0.2–0.5	0.8–2.6
Germ	15–20	5.0–10	38–68	4.9–33	0.2–0.3	1.9–5.2
White flour	0.8–2.6	0.3–0.6	7.0–15	0.4–2.8	0.01–0.03	0.2–0.3
Wholegrain[a]	7.2-11.7	0.86-1.07	0.69–1.18[b]	1.64-2.21	–	0.47–0.70

[a] Mean values for 26 cultivars grown in 4 (vitamins B1,2,3,6) and 6 (folates) environments.
[b] Bioavailable form only determined.
Source: (Kariluoto et al. 2010; Shewry et al. 2011).

defects during foetal development. The benefits in this respect are clear and mandatory fortification is carried out in over 60 countries including the USA and UK. It is not yet mandatory in the EU, with a debate focused on whether fortification will result in risks for certain individuals.

Table 9.5 summarises the contents of B vitamins in various grain fractions, showing that all are concentrated in the bran and germ. It also includes data for whole grain samples from the EU HEALTHGRAIN project, in which 26 cultivars were grown in four environments (thiamine, riboflavin, niacin, pyridoxine) and six environments (folates).

Much of the niacin in grains is not bioavailable to humans due to binding to other food components, including fibre. The values determined for bioavailable niacin in the HEALTHGRAIN study therefore indicate that only about 10–15% of the niacin in the wholegrain is bioavailable.

The data from the multi-environment trials also allowed the 'broad sense' heritability of the content of B vitamins to be calculated, showing that the genotype accounted for between 7% (bioavailable niacin) and 31% (thiamine) of the variance, the remainder being accounted for by the environment or genotype × environment interactions. It is therefore not surprising that there has been no success in increasing the contents of B vitamins in wheat by breeding (Shewry et al. 2011).

9.3.4 Phytochemicals

Although the term phytochemical is widely used, it is not readily defined. In broad terms it is applied to compounds present in plants that are not essential but may have benefits in terms of the plants' defence against pests and pathogens and/or abiotic stresses, and also have positive or negative effects on human health. They are immensely variable in plants, but most can be classified into two groups based on their biosynthetic pathways: phenolics and terpenoids.

9.3.4.1 Phenolics

Phenolic compounds contain at least one aromatic ring bearing at least one hydroxyl group, and are the most abundant and diverse group of phytochemicals in plants.

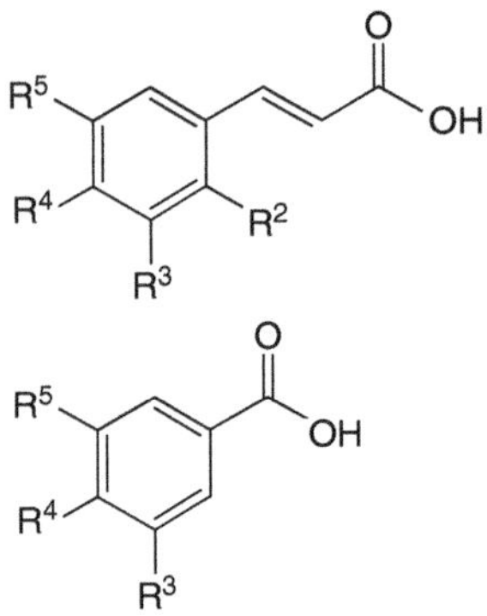

Cinnamic acid derivatives

R^2=OH, o-coumaric acid R^3=R^4=OH, caffeic acid

R^3=OH, m-coumaric acid R^3=OCH_3, R^4=OH, ferulic acid

R^4=OH, p-coumaric acid R^3=R^5=OCH_3, R^4=OH, sinapic acid

Benzoic acid derivatives

R^3=OH, m-hydroxybenzoic acid R^3=OCH_3, R^4=OH, vanillic acid

R^4=OH, p-hydroxybenzoic acid R^3=R^5=OCH_3, R^4=OH, syringic acid

R^{3-} R^4=OH, protocatechuic acid

Figure 9.2 Chemical structures of major types of phenolic acid present in wheat grain.
Source: Taken from Shewry (2014).

Table 9.6 Concentrations (μg/g DM) of phenolic acids and alkylresorcinols in wholegrain of 150 lines of wheat grown on a single site.

Component	Range	Mean
Total phenolic acids	326–1171	657
Total phenolic acids	171–190	180
Free phenolic acids	3–30	10.6
Conjugated phenolic acids	76–297	162.5
Bound phenolic acids	208–878	484.9
Bound ferulic acid	162–721	367.4
Alkylresorcinols	241–677	432

Source: Data from Li et al. (2008) and Andersson et al. (2008).

The major group are the phenolic acids (PAs), which have an organic carboxylic function and fall into two groups derived from cinnamic acid or benzoic acid (Figure 9.2). PAs occur in three forms, either as free compounds, as soluble conjugates bound to low molecular weight compounds such as sugars and sterols, and as bound forms, which are linked to cell wall polysaccharides by ester bonds. The contents of total PAs in whole wheat grain vary from about 300 to over 1000 μg/g DM (Table 9.6) but the concentration is much lower in white flour, below 200 μg/g DM (reviewed by Shewry and Hey 2015). Although there is wide variation in the proportions of individual PAs between different genotypes, the bound fraction generally accounts for over 75% of the total PA fraction, with ferulic acid accounting for about 75% of the bound fraction (Table 9.6). Hence, ferulic acid is generally the most abundant phytochemical in the grain.

Other phenolics present in wheat grain include flavonoids, which include the anthocyanins present in the outer layers of coloured grain types, alkylresorcinols, which are phenolic lipids, and lignans, which have phytoestrogen activity (see Shewry and Hey (2015) for further details).

All phenolics are present at low concentrations in white flour and enriched in bran and germ fractions. Alkylresorcinols occur only in the grains of wheat and related species and have therefore been proposed as biomarkers to measure the consumption of wholegrain in diets.

9.3.4.2 Terpenoids

The second major group of phytochemicals in wheat are terpenoids, which are derived from five carbon isoprene units. Terpenoids can comprise up to eight or more isoprene units and most form several ring structures with oxygens containing functional groups. Like phenolics, they are highly diverse in structure.

The major group of terpenoids in wheat are phytosterols, comprising a tetracyclic ring with a hydroxyl group and a flexible side chain (Figure 9.3). The most abundant sterol in wheat is β-sitosterol, which accounts for about half or more of the total fraction. Significant amounts of saturated forms of sterols (called stanols) are also present, accounting for between about 10 and 30% of the total fraction while proportions of both sterols and stanol are modified, either by esterification to a fatty acid or PA to form sterol esters, or β-linked to a carbohydrate to form a sterol glycoside, with the latter also sometimes being acylated. For example, β-sitosterol is present in approximately equal amounts in all four forms: free, esterified, glycosides, and acylated glycosides. Unlike most phytochemicals, sterols have as essential biolological role as components of plant membranes. This may account for the fact that their concentrations vary less than those of other phytochemicals: by × 1.43 (Table 9.7) compared with up to × 10 for free PAs (Table 9.6).

The second most abundant terpenoids in wheat are tocols, which comprise a chromanol ring with a C16 phytol side chain, which can be either saturated (tocopherols) or contain three double bonds (tocotrienols) (Figure 9.3). Tocopherols and tocotrienols exist in four forms, which differ in the positions of methyl groups on the chromanol ring and are called α-, β-, γ-, and δ-forms. They are less abundant than sterols, with a mean of 49.9 μg/g DM for the 150 lines in Table 9.7 compared with a mean of 844 μg/g DM for total sterols.

Carotenoids are isoprenoid pigments derived from long polyene chains of 35–40 carbons. They occur in two major forms: the oxygen-containing xanthophylls (including lutein and zeaxanthin) and the unoxygenated carotenes (including α-carotene and β-carotene). The levels of carotenoids are generally low in bread wheat grain due to selection for white flour during breeding: ranges for β-carotene, zeaxanthin, and lutein of 0.10–0.21, 0.20–0.39, and 0.82–1.14 μg/g DM, respectively, have been reported for six samples of American soft wheat (Moore et al. 2005). Higher contents of carotenoids are present in durum wheat, where yellow flour is a quality trait for pasta making, and in the related emmer wheat (see also Chapter 10) (Shewry and Hey 2015).

In common with phenolics, terpenoids are concentrated in the bran and germ fractions and depleted in white flour.

9.3.5 Betaine and Choline

Betaine (more correctly called glycine betaine [N, N,Ntrimethylglycine]) and choline are both able to donate methyl groups in biochemical reactions and are therefore often referred to as *methyl donors*. Choline is an essential nutrient that is required for the synthesis of

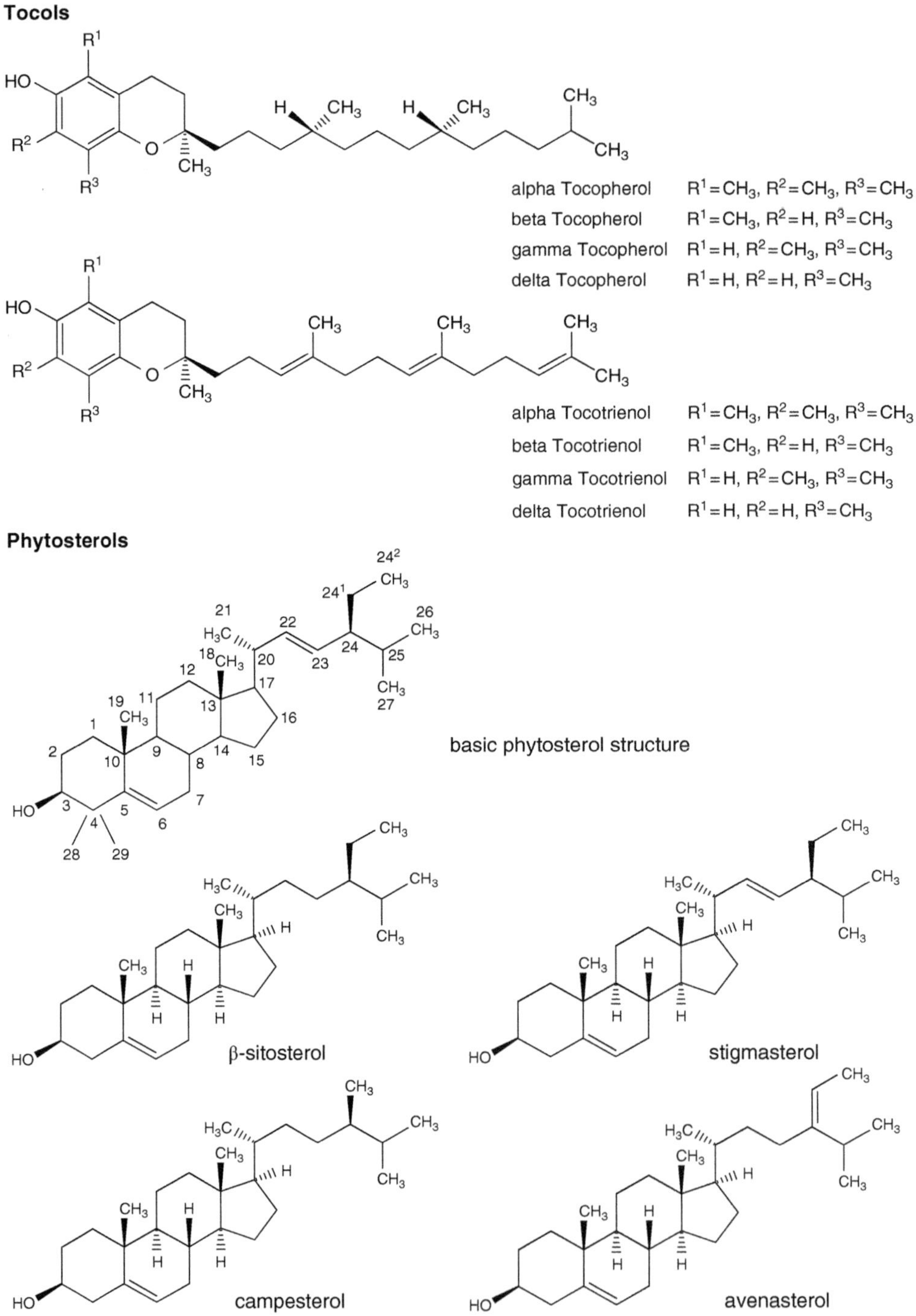

Figure 9.3 Chemical structures of major types of terpenoid compounds present in wheat grain (tocols and phytosterols). *Source:* Taken from Shewry (2014).

acetylcholine (a neurotransmitter), phospholipids, sphingolipids, and other components. It is also a precursor of betaine, which is widely used as a dietary supplement for livestock.

Wheat is one of the richest known sources of betaine in the diet, with the betaine concentration generally being about 10 times higher than that of choline. Table 9.8 shows mean values of 1600 μg/g DM of betaine and 110 μg/g DM of choline in wholemeal flours of 150 wheat lines. Both compounds are concentrated in the bran. Betaine has a well-established role as an osmoregulant, in salt-tolerant (halophytic) and drought-stressed plants (Chen and Murata 2002), and in other organisms. However, it is not known whether the amount present in grain changes in response to drought or other stresses.

9.3.6 Health Benefits of Phytochemicals

Phytochemicals in wheat (particularly wholegrain wheat) and in other plant-based foods have been proposed to have a number of health benefits, some of which are listed in Table 9.9. These suggestions are based on a range of studies, including comparisons with related components in other foods, biochemical analyses, *in vitro* studies using cell cultures, and dietary interventions in humans.

Antioxidant activity is readily determined and is the most widely studied property. However, the *in vivo* significance of this activity is debatable and it has not been accepted

Table 9.7 Concentrations (μg/g DM) of terpenoids in wholegrain of 150 lines of wheat grown on a single site.

Component	Range	Mean
Total tocols	27.6–79.7	49.9
Tocopherols	12.3–33.2	19.9
α-Tocopherol (vitamin E)	9.1–19.9	13.6
Tocotrienols	12.5–52.0	30.02
Total sterols (including stanols)	670–959	844
% stanols	11–29%	23.9%

Source: Data from Piironen et al. (2009) and Lampi et al. (2008).

Table 9.8 Concentrations (μg/g DM) of choline and betaine in wholegrain of 150 lines of wheat grown on a single site.

Component	Range	Mean
Betaine	970–2940	1596
Choline	76–135	110

Source: Data from Corol et al. 2012.

Table 9.9 Some proposed and established health benefits of phytochemical components present in wheat grain and other cereals.

Component and proposed health benefit for cereals or other foods	Health claim supported by the European Food Safety Authority for food labelling
Betaine	
Normal homocysteine metabolism (reduced risk of coronary heart disease)	Yes
Choline	
Normal homocysteine metabolism (reduced risk of coronary heart disease)	No
Phenolic acids	
Improves vascular function	No
Antitumour properties	No
'Antioxidant'	No
Alkylresorcinols	
Antimicrobial	No
'Antioxidant'	No
Lignans	
Phytoestrogen	No
Antitumour	No
Antimicrobial	No
Sterols, stanols and derivatives	
Reduces serum cholesterol and risk of coronary heart disease	Yes
Anti-cancer effects	No
Tocols	
Vitamin E activity	Yes
Prevention of neurodegeneration	No
Induction of immune responses	No
Anti-cancer effects	No
Reduces serum cholesterol and risk of coronary heart disease	No
'Antioxidant'	No

as a basis for health claims (EU 2006). PAs are the major antioxidants in wheat grain and the total antioxidant activity of extracts is usually correlated with the contents of PAs (or ferulic acid as the major component). Although antioxidant activity has been suggested to contribute to health benefits (reviewed by Shahidi and Ambigaipalan 2015), alternative mechanisms have been proposed, mediated by the interactions of PAs with cellular signalling pathways that maintain cell function under both normal and pathological conditions

(Vauzour et al. 2010). Direct effects of ferulic acid on cardiovascular function have also been demonstrated following the consumption of bread treated with feruloyl esterase enzyme to release the bound ferulic acid (Turner et al. 2021). The proposed health benefits of phenolics are largely based on short-term dietary interventions and *in vitro* studies. However, it is likely that PAs, and to a lesser extent other less-abundant phenolic compounds, could contribute to the long-term health benefits accepted for wholegrain foods.

EFSA has approved several health claims relating to the ability of plant sterols, stanols, and stanol esters to lower blood cholesterol levels or maintain healthy levels, recognising that high cholesterol is a risk factor in the development of coronary heart disease (e.g. EFSA Panel on Dietary Products, Nutrition and Allergies [NDA], 2011b).

Although all tocols are sometimes referred to as vitamin E, only α-tocopherol is currently considered to possess vitamin E activity (EFSA Panel on Dietetic Products, Nutrition, and Allergies 2015). Some carotenoids, including carotenes, are converted to vitamin A (retinol) in mammals and sometimes referred to as provitamin A. However, the major carotenoid in wheat is lutein, which does not have provitamin A activity.

A high level of homocysteine in the blood (hyperhomocysteinemia) is a major risk factor in cardiovascular disease. Homocysteine is produced by the demethylation of methionine and can either be remethylated to methionine, metabolized to give cysteine, or converted to S-adenosylhomocysteine. The remethylation of homocysteine to form methionine requires the addition of a methyl group which may be provided by betaine. This activity forms the basis for a health claim for betaine, which has been accepted by EFSA (EFSA Panel on Dietary Products, Nutrition and Allergies [NDS] 2011a). However, the intake required is substantially above that present in wheat products: 0.5 g/portion, contributing to a daily intake of 1.5 g/day.

9.3.7 Environmental Effects on the Concentrations of Phytochemicals and Minerals

In addition to genetic variation, the contents of bioactive components are also affected by environmental factors, and by genotype × environment interactions. These effects are most obvious for phytochemicals where they may be related to their role in response to environmental stress. For example, the contents of free and bound PAs, sterols, tocols, betaine, and folates were all positively correlated with the mean temperature during grain development when 26 cultivars were grown in 6 environments, with the contents of free and bound PAs, betaine, and folates also being negatively correlated with total precipitation between heading and harvest (Shewry et al. 2010; Corol et al. 2012). These strong environmental effects, together with the interactions of the genotype with the environment, result in low heritability of most phytochemicals, which limits the ability of plant breeders to select for genetic variation.

Of greater long-term concern are the potential impacts of climate change, and in particular of increases in the concentration of CO_2. Analysis of wheat grown under field conditions with levels of CO_2 predicted for the mid-twenty-first century showed decreased concentrations of Fe and Zn (Myers et al. 2014) and meta-analysis of the results from a number of studies have confirmed this observation and showed that the concentration of magnesium also decreased (Brogberg et al. 2017). The effects cannot be explained by simple yield dilution, and effects on uptake and transport mechanisms may contribute. However, there is significant variation between genotypes, which may allow the development of new cultivars adapted to the higher concentration of CO_2 predicted for the future.

9.4 Adverse Effects of Wheat on Health

9.4.1 Wheat as Part of the Western Diet

The per capita consumption of wheat and its contribution to the total intake of calories have declined with increased prosperity in most developed countries over the past century. In addition, there has been a marked decrease in consumption of wheat over the past decade in some countries, particularly in Europe and North America, due to perceived negative effects of wheat or gluten on health. To a large extent this trend has been driven by social and popular media, including high-profile popular books (Davis 2019). However, it also reflects increasing consumer awareness of the importance of diet in long-term health, and concern about the changes in food production systems and processing methods that have occurred with industrialisation and urbanisation.

This has led to the misconception that wheat contributes to increases in obesity and diet-related chronic diseases. This argument is based on the fact that most wheat-based foods are made from white flour, which is highly refined with low contents of fibre and other 'beneficial' components. It is therefore regarded as part of the 'Western diet' (highly refined foods rich in sugar, rapidly digested carbohydrates, and fat), which is itself associated with increases in the prevalence of obesity and type 2 diabetes. This conclusion is not supported by publications in peer-reviewed journals, and critiques have been published by Jones (2012) and Brouns et al. (2013). The conclusion from scientific studies is that wheat products, including white bread, do not contribute directly to the development of these diseases and should be consumed as part of a balanced diet (which should include other sources of protein).

However, wheat can trigger several types of adverse reaction in susceptible individuals.

9.4.2 Allergy and Intolerance to Wheat

Adverse reactions to wheat can be broadly divided into two well-defined types – allergy and intolerance – and several less well understood conditions (referred to as 'others' in Figure 9.4).

9.4.3 Allergy

Classical allergic responses are mediated by immunoglobulin E (IgE) antibodies (as opposed to the IgG antibodies produced against infections) that bind to the allergen and then trigger a response by basophils (mast cells). The etiology is understood and diagnosis is straightforward.

Wheat is frequently listed as one of the eight major food allergies (with milk, eggs, crustaceans, shellfish, tree nuts, peanuts, and soybean) However, classical (IgE-mediated)

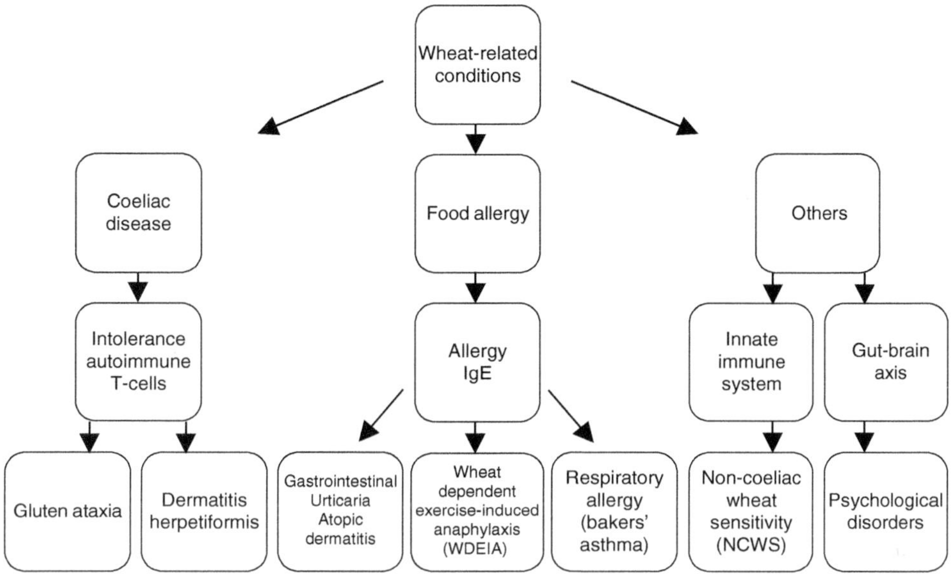

Figure 9.4 Summary of the types of wheat-related disorders in humans. *Source:* Modified from Sapone et al. (2012).

allergy to wheat actually has a very low prevalence and this listing reflects the wider prevalence of intolerances (see Section 9.4.4).

The prevalence of wheat allergy is higher in children than adults, with a detailed meta-analysis showing a mean prevalence of 0.25% (Zuidmeer et al. 2008). However, children tend to grow out of wheat allergy and the prevalence in adults is lower (below 0.2%). Symptoms of wheat allergy include atopic dermatitis, urticaria, and respiratory and gastro-intestinal symptoms. A range of proteins have been implicated, including a number of enzyme inhibitors, lipid transfer proteins, enzymes, and gluten proteins (Tatham and Shewry 2008; Cianferoni 2016). The inhibitors include the α-amylase/trypsin inhibitors (ATIs), which are also implicated in other adverse reactions to wheat (see Section 9.4.5 and Geisslitz et al. 2021).

The most severe form of food allergy to wheat is 'wheat-dependent exercise-induced anaphylaxis' (WDEIA), where the major allergen is ω-5 gliadin and the allergic response is triggered by physical exercise after the ingestion of food (Morita et al. 2003). WDEIA has been studied most widely in Japan, where the prevalence in adults is about 0.2% (Morita et al. 2012).

Allergy to wheat also occurs via inhalation of flour, contributing to the development of Bakers' asthma. This is a major occupational disease, estimated to affect between 4 and 25% of bakers worldwide (Brisman 2002). Although several allergenic foods have been identified, including cereal and non-cereal flours, nuts, and moulds, wheat flour is the dominant allergen (Brisman 2002). Several wheat allergenic proteins have been identified in Bakers' asthma, including α/β-gliadins, various enzymes, and lipid transfer proteins (Section 7.3.2.6). However, the dominant allergens appear to be ATIs (Geisslitz et al. 2021), with 38% of wheat-positive bakers having antibodies to at least one form of ATI (Sander et al. 2011).

9.4.4 Coeliac Disease and Related Intolerances

Coeliac disease is an autoimmune reaction mediated by an inflammatory T cell response. Coeliac disease affects the small intestine, resulting in flattening of the villi which facilitate the absorption of food, and leading to malabsorption, weight loss, fatigue, abdominal pain, vomiting, and diarrhoea. Consequently, it often results in nutrient deficiencies, including Fe deficiency anaemia and folate deficiency. However, individuals may also be asymptomatic or present only mild symptoms.

Coeliac disease has been recognised since Roman times, and its relationship to wheat gluten since the 1950s. However, the molecular basis has only been understood for about two decades. This understanding has facilitated more accurate diagnosis, including the recognition that the disease develops more commonly in adults than in children. This improved diagnosis, together with greater awareness by physicians and patients, has contributed to the perception that the prevalence has increased greatly in past decades. Available evidence suggests that the prevalence has increased in some countries, but probably to the same extent as increases in other autoimmune diseases. The current prevalence in Europe and other populations of European origin is about 1%.

Dermatitis herpetiformis is a chronic skin condition that occurs in about 10–15% of coeliac patients, while coeliac disease may also be associated with neurological conditions, notably gluten ataxia. The latter is reported to account for 15% of all ataxias, but the prevalence in the whole population has not been reported (Brouns et al. 2019).

Coeliac disease is triggered in susceptible individuals by the ingestion of wheat gluten proteins and related proteins from other cereals: barley, rye and possibly (at least for some individuals) oats. Susceptible individuals have the HLA-DQ2 or HLA-DQ8 haplotypes, but these haplotypes account for 25–40% of the total population and only a small proportion of individuals develop the disease.

The triggering of coeliac disease in susceptible individuals results from the presence of repetitive sequences, which are rich in proline and glutamine, in gluten proteins (Chapter 8). The presence of proline in proteins limits their digestibility due to hindrance of digestive proteases. Consequently, the repetitive sequences are incompletely digested in the human gastrointestinal tract. This results in the presence of undigested glutamine-rich peptides, which enter the lamina propria (mucous membrane) of the small intestine where they are deamidated by tissue transglutaminase: this converts glutamine to glutamate residues, increasing their affinity for HLA DQ2/DQ8 receptors. An inflammatory T4 cell response to the DQ2/DQ8 bound peptides then occurs, based on the recognition of specific short peptide sequences which are usually referred to as epitopes. Almost 40 epitopes have been identified in gluten and related proteins from barley, rye, and oats. All comprise nine amino acids, with three-quarters of the total being associated with HLA DQ2 (Sollid et al. 2020).

Gluten proteins vary widely in their sequences, and the contents of coeliac disease epitopes vary both between and within different gluten protein groups. In particular, the α-gliadins are rich in epitopes. Differences may also occur in the abundances of epitopes in proteins encoded by the homoeologous genomes of bread wheat and ancient wheats. This variation therefore provides an opportunity to identify and combine gluten protein variants with low coeliac toxicity (Gilissen et al. 2014; Shewry and Tatham 2016).

Eliminating coeliac toxicity has also been a target for genetic engineering, including gene editing (Chapter 10). However, the production of coeliac-safe wheat is unlikely to be achieved without the elimination of a substantial proportion of the total gluten proteins, which could have catastrophic effects on the processing properties.

9.4.5 Other Adverse Reactions to Wheat

Wheat allergy and coeliac disease are well-understood conditions with established diagnostic criteria. This does not apply to wheat sensitivity, which has only been recognised for the last decade.

The range of reported symptoms varies widely, including gastrointestinal symptoms, tiredness, headache, dermatitis, pains in muscles and joints, depression, anxiety, and anaemia. This new condition was initially defined as 'non-coeliac gluten sensitivity' (NCGS) (Sapone et al. 2012). However, because the role of gluten has not been established it is now more widely referred to as non-coeliac wheat sensitivity (NCWS). There is also significant overlap between the symptoms reported for NCWS and those of irritable bowel syndrome (IBS) (see Section 9.4.6) and coeliac disease (Borghini et al. 2017).

The etiology of NCGS/NCWS is not understood but is likely to feature a mixture of factors including the stimulation of the innate immune system. A range of causative substances have been suggested, including gluten proteins and fermentable oligosaccharides (Fermentable Oligo-, Di-, Monosaccharides And Polyols [FODMAPs]). However, current attention is focused on the role of ATIs, which have been shown to induce an innate immune response via activation of the TRL4 receptors. (Schuppan and Gisbert-Schuppan 2019).

The role of ATIs in triggering NCWS, coeliac disease, and other adverse reactions in wheat is an active area of research with a rapidly increasing number of publications. However, definitive studies are largely lacking, due mainly to the problems in obtaining sufficient amounts of pure ATI preparations (Geisslitz et al. 2021).

The diagnosis of NCWS remains difficult and most patients are self-diagnosed. Hence, the prevalence is difficult to determine and even the most scientifically based estimates vary widely (from less than 1 to 10% of the population) (Brouns et al. 2019).

In addition to immune-mediated responses, wheat/gluten has also been proposed to lead to mental-psychological conditions via effects on the gut–brain axis (Brouns et al. 2019). However, these conditions are even more difficult to define than NCWS and their prevalence, if they exist at all, has not been determined.

9.4.6 FODMAPs and Gastro-Intestinal Disorders

FODMAPs are defined as small fermentable carbohydrates present in foods, with the main FODMAPs in wheat grain being fructans and raffinose (Chapter 7). They are of interest because it has been shown that diets low in FODMAPs improve the management of IBS and inflammatory bowel disease (Crohn's disease and ulcerative colitis) by reducing fermentation and the production of gas (which leads to discomfort) in the colon (Gibson and Shepherd 2010; Barrett and Gibson 2012).

However, fructans may have benefits for healthy patients, as their fermentation has prebiotic effects, increasing the populations of beneficial bacteria and leading to the production of SCFA. Hence, manipulating the content of fructans in the grain may have detrimental or beneficial effects on the health of consumers, depending on their status.

9.4.7 Bloating

A common perception is that bread causes bloating, a feeling of discomfort and abdominal swelling after consumption. However, a systematic review of the literature found no evidence that bread, including bread produced by the Chorleywood Breadmaking Process (CBP) (see Section 9.5.3), caused bloating or other gastrointestinal symptoms (Weichselbaum 2012).

9.5 Producing Healthier Wheat Products by Processing

Grain processing has significant impacts on the composition of wheat products, including the contents of health-promoting components. This approach has the advantage over genetic improvement for food processors in that it is possible to use conventional types of wheat rather than pay a premium for novel types, thereby reducing costs and widening availability. It is therefore being widely studied by grain scientists and exploited by processors. Readers are also referred to an excellent review article, which provides more detailed discussions of the effect of processing on wheat grain composition in relation to health (Kucek et al. 2015).

9.5.1 Debranning

Debranning of the wheat grain can be achieved by friction (peeling), abrasion (pearling), or a combination of the two methods. It is now routinely used before conventional roller milling in some commercial mills, although it adds to the cost and the advantages for flour quality are not clearly established.

Removal of the outer 3–4% of the grain will remove the outer pericarp, which is rich in insoluble fibre with low contents of ferulic acid and minerals and give flour with a higher content of bioactive components than white flour produced by conventional milling (Hemery et al. 2007). For example, removal of 3.5% of the grain dry weight by peeling reduced the content of total sterols from 708 to 669 μg/d DM, while removal of a further 3% of the grain by pearling reduced the total sterol content to 621 μg/d DM. However, these decreases were small compared to the effects of milling: flours milled from the peeled and peeled/pearled grains to 76% extraction rate (based on 100% wholegrain) contained 344 and 361 μg/g DM total sterols, respectively (Nurmi et al. 2012). Debranning can also be used to remove contaminants such as bacteria, mycotoxins, pesticides residues and heavy metals (Laca et al. 2006) (see Section 9.6.4).

9.5.2 Flour Particle Size

The bioavailability of vitamins, minerals, and phytochemicals in the aleurone and other bran layers can be increased by reducing the particle size by ultrafine grinding (Hemery et al. 2007; Aslam et al. 2018). The rate of starch digestion, and hence post-prandial glycaemia, is also influenced by the flour particle size, with coarsely ground particles from durum wheat being digested less rapidly than fine particles (Edwards et al. 2015, 2021). Hence, ultrafine flours from the aleurone are advantageous for health whereas the opposite applies to starchy endosperm flours.

9.5.3 Fermentation

A high proportion of wheat is consumed as bread, which includes unleavened and leavened types. Leavened breads are dominant over much of the world, including Europe and the Americas, with unleavened flat breads being more widely consumed in North Africa, West Asia, and South Asia (and in other parts of the world by communities and in foods originating from these regions) (Chapter 1).

The oldest type of leavened bread is considered to be sourdough, which probably dates back over 5000 years and originated by contamination of unfermented dough products with natural microorganisms. In sourdough the fermentation results from the presence of wild yeasts and lactobacilli, which are usually added in a starter (a small piece of dough that is carried over from batch to batch). Sourdough has a characteristic flavour, imparted by the lactobacilli, and has been increasing in popularity in recent years, after being largely restricted to low-volume artisan breads (and the famous San Francisco sourdough bread in the USA).

In non-sourdough breads the fermentation is carried out by bakers' yeast (*Saccharomyces cerevisiae*). Although bakers' yeast is readily available in pure form, some artisan processes still use starters from previous batches and the length of fermentation can vary widely including long overnight systems. In the UK and other English-speaking countries most of the high-volume 'factory' bread is now produced using the CBP, or processes related to this. In the CBP the mixing and dough development are carried out in a single operation, using a high concentration of yeast, high energy, and control of the headspace pressure during mixing. After this only short intermediate and final proving periods are required.

A full discussion of breadmaking processes is outside the scope of this chapter, as is a debate as to the sensory and organoleptic quality of the end products. However, it is relevant to note that breadmaking processes differ widely in the time required for fermentation and in the microorganisms involved, which can have impacts on the composition and properties (beneficial or negative) of the final products for human health.

Sourdough has been reported to have significant potential for improving the health benefits of cereal products, including reducing the digestibility of starch (and hence decreasing postprandial glycaemia) and increasing the bioavailability of minerals by degradation of phytate. Positive and negative effects of sourdough fermentation on the amounts of bioactive components have also been reported, depending on the components and the type of process (Katina et al. 2005), with one possible positive effect being the release of bound ferulic acid.

However, the main interest in exploiting sourdough for improving health benefits has been to reduce coeliac toxicity by proteolysis of the gluten proteins during fermentation. Although breads made after sourdough fermentation have been reported to either lack or have reduced contents of coeliac-toxic sequences, their safety for coeliac patients has not been established and, to the best of our knowledge, they are still not widely recommended for coeliac patients by clinicians. In addition, a long fermentation period (24 hours) is required to maximise the reduction in coeliac-toxic sequences, compared with an optimum for breadmaking of four hours, which poses a challenge for producing acceptable products (Cabrera-Chavez and Calderon de la Barca 2010).

IBS is the most widespread gastrointestinal disorder in many countries and is strongly affected by diet (and in particular the presence of FODMAPs), as discussed. Sourdough fermentation could be advantageous for patients suffering from IBS by reducing the content of fructans and other FODMAPs (Longin et al. 2020), but this has not been proven by dietary interventions.

9.5.4 Sprouting

Although bean sprouts (mung bean and soybean) and germinated brown rice (GBR) are widely consumed, particularly in Asia, germinated wheat is not widely consumed. Germinated wheat has been shown to contain increased amounts of bioactive components, including B vitamins (folates, niacin), phenolic compounds, and tocols (see, for example, Koehler et al. 2007). Although sprouted grain cannot be used for making conventional products such as pasta and noodles, it may be used for products such as weaning foods (Hussain and Uddin 2012).

9.5.5 Enzyme Treatments

The benefits associated with fermentation and sprouting result from the action of enzymes produced by microorganisms (fermentation) or the germinating wheat grain (sprouting). Benefits can also result from treatment with exogenous enzymes. Such enzymes, which are usually produced by fermentation with filamentous fungi, are currently used as processing aids and include α-amylase and endoxylanase. Endoxylanases may have secondary benefits for health, by increasing soluble fibre. Turner et al. (2021) also showed that feruloyl esterase present in a mixture of enzymes developed as an aid for brewing could be used to release ferulic acid during the production of flatbread, leading to improved cardiovascular function in a controlled dietary intervention trial. Asparaginase may also be used to hydrolyse asparagine in flours, reducing the formation of acrylamide by the Maillard reaction during processing.

9.5.6 Conclusions: Processing for Improved Health Benefits

It is clear that innovative processing can be used to increase the health benefits of wheat, by increasing the contents and bioavailability of some beneficial components (minerals, vitamins, phytochemicals, fibre) and reducing adverse effects (for example, reducing FODMAPs and the rate of glucose release). However, several factors must be considered if these benefits are to be realised. Firstly, they should not result in increased cost compared to conventional products, as this will restrict their availability to lower-income groups. Secondly, consumers are increasingly opting for products that have minimal processing, and particularly reduced contents of additives. Hence, new marketing strategies may be required. Finally, most of the processes discussed here will affect the functional properties of the flour and hence product quality and consumer acceptability. They may therefore require innovative processing strategies. Despite the challenges, however, processing currently offers the most realistic option for improving the health benefits of wheat-based foods in the short to medium term.

9.6 Fungal Toxins in Wheat

Although they are not natural components of wheat grains, toxins produced during fungal infections of grain are major health challenges globally.

9.6.1 Ergot

Ergots are the sclerotia of *Claviceps purpurea*, which infects rye and triticale as well as wheat and occurs in all major wheat-growing areas of the world. The sclerotium is a hard purple-black mass which develops in the ear in place of a grain. When the sclerotia are shed or 'sown' onto the soil they overwinter and germinate in spring to produce perithecia, which release ascospores. These are carried on the wind and penetrate the ovaries of susceptible florets. After a few days conidial spores are formed in honeydew, which can cause secondary infections following splash, wind, or insect dispersal. The infected ovaries enlarge and are converted into sclerotia.

Ergot contains indole alkaloids, notably ergotamine, ergosine, and ergotoxine, which cause serious illness, and sometimes death in humans and livestock. All ergot alkaloids contain a lysergic acid structure and the hallucinogenic drug LSD (lysergic acid diethylamide) was developed as part of a research project on ergot alkaloids. Convulsive effects of ergot contamination are well documented historically, and the consumption of infected rye grain has been proposed (but not proven) as the origin of the effects ascribed to witchcraft in Salem (Massachusetts) in the 1690s.

9.6.2 Fusarium Mycotoxins

Mycotoxins are secondary metabolites that are produced during fungal infections of food crops and can have both chronic and acute effects on human health. They occur in a range of crops and foods and can be the result from infection of the crop in the field or of the harvested food during transport and storage.

Mycotoxin production in wheat most commonly results from infection of the crop in the field by fusarium head blight (FHB), which is caused by infection of developing heads with the *Fusarium graminearum* species complex (FGSC) or other *Fusarium* species (notably *Fusarium avenaceum*, *Fusarium culmorum*, and *Fusarium poae*) (Torres et al. 2019). Infection occurs most readily at anthesis when airborne spores germinate in flowering spikelets (Wang et al. 2018). Symptoms are usually first seen as small brownish spots at the base or middle of the glume. These then spread in all directions and a pinkish fungal growth may be seen along the edge of the glumes or at the base of the spikelet. *Fusarium* infected grains shrink and become grey/brown with a floury discoloured interior (Parry et al. 1995).

Fusarium infection results in changes in grain composition and reduced processing quality (Bechtel et al. 1985, Szunics et al. 1987, Boyacioglu and Hettiarachchy 1995). However, the major concern is not yield loss, which is reported to run into billions of dollars annually, but effects on human health. *Fusarium* spp. produce a range of toxins, the major types being trichothecenes and zearalenone. Although fumonisins are produced by species infecting maize they are of little importance in wheat and will not be considered here.

Trichothecenes are a large group of structurally related sesquiterpenes produced by a number of genera of fungi including *Fusarium*. All contain a common 12,13-epoxytricothecene skeleton and an olefinic bond with various side chain substitutions that provide a basis for their classification into types. Infection of wheat grains by *Fusarium* species may result in the presence of type A trichothecenes such as the T-2 and HT-2 toxins and type B trichothecenes, notably deoxynivalenol (DON) which is probably the most common mycotoxin, and nivalenol (NIV) which often occurs with DON.

Zearalenone (ZON) has a highly complex structure and is produced by *F. graminearum*. Hence, it often occurs with DON and other *Fusarium* mycotoxins.

Trichothecenes are potent inhibitors of eukaryotic protein synthesis, particularly by effects on peptidyl transferase activity, and this is probably the basis for a wide range of effects on the health of humans and livestock (Bennett and Klitch 2003). Although type A trichothecenes are more toxic than type B, the latter (particularly DON) are the most widely consumed in grains, causing reduced feed intake, diarrhoea, and vomiting. Hence DON is sometimes referred to as vomitoxin. ZON has potent activity as an oestrogen but low toxicity. The reader is referred to the comprehensive review of Bennett and Klitch (2003) for further details of the chemistry and biological activity of these compounds.

The EU has established maximum levels (MLs) for DON and ZON in foodstuffs, which are summarised for wheat in Table 9.10 (EU 2020).

Table 9.10 Maximum levels (μg/g) of the mycotoxins deoxynivalenol (DON) and zearalenone (ZON) permitted in wheat and wheat-based foods for human consumption by the European Union.

	DON	ZON
Unprocessed wheat apart from durum wheat	1250	100
Unprocessed durum wheat	1750	100
Wheat intended for direct human consumption, wheat flour, bran and germ as end product marketed for direct human consumption, with the exception of foods for infants and young children	750	75
Dry pasta	750	
Bread, pastries, biscuits, snacks, cereal bars	500	50
Processed wheat-based foods and foods for infants and young children	200	20

Source: Adapted from EU (2020).

9.6.3 Mycotoxins from Storage Fungi

Fusarium infection and associated mycotoxin production initiated in the field can develop further in store (Magan et al. 2010). This development is particularly likely when grain is stored wet. For example, in UK conditions, grain stored above 14.5% moisture content is at risk of the development of further fungal contamination, infestation with insects, and excessive grain respiration. This activity is associated with significant increases in temperature, further increases in biological activity, and significant increases in mycotoxin contamination (Magan et al. 2010). Several fungal groups are associated with grain spoilage in store, and some of these can also release mycotoxins. Most notable examples include *Aspergillus flavus*, which together with some other species of the same genus, release aflatoxins. Aflatoxins are powerful human carcinogens that can also affect liver function. They cause adverse effects in various animals, especially chickens (Magan et al. 2011). Maximum limits of aflatoxin in wheat-based foods in the EU are 4 and 0.1 μg/kg for adults and infants, respectively. Additionally, *Penicillium* spp. and *Aspergillus* spp. can release ochratoxin A (OTA) (El Khoury and Atoui 2010), which has been shown to be affect kidney function and to be carcinogenic in laboratory animals and pigs (Magan et al. 2011). EU limits for OTA in wheat-based foods are 3 μg/kg for adults and 0.5 for infants (EU 2020).

9.6.4 Removing Fungal Toxins by Processing

The most effective way to reduce mycotoxin contamination is, of course, to avoid infection of the crop by good husbandry and of the harvested grain by appropriate storage and transport conditions. Nevertheless, grain processors do need to deal with infected grain samples and have developed strategies to reduce the levels in the products below the EU thresholds. These strategies have been reviewed in detail by Schaarschmidt and Fauhl-Hassek (2018) and are only discussed briefly here.

The simplest and most obvious first step is cleaning of the grain before processing. This may involve sieving and separation on density, but also scouring of the outside of the grain. Cleaning has been reported to reduce the level of DON by between 10% and over 80%, with lower ranges (based on smaller numbers of studies) for other mycotoxins (Schaarschmidt and Fauhl-Hassek 2018). Sieving to remove small shrunken grain resulting from FHB certainly contributes to the effectiveness of cleaning, but it does not exclude larger infected grain. These may be amenable to removal by optical sorting based on differences in grain colour, although this has only been carried out at the laboratory level.

The effectiveness of removing mycotoxins by milling depends on their location in the grain, which reflects the colonisation pattern of the pathogens. In general, the concentration of mycotoxins is higher in the grain outer layers and hence the concentration in white flour may be reduced by between 10% and 50% compared to whole grain (Schaarschmidt and Fauhl-Hassek 2018). However, this varies between types of wheat and is also affected by the environment. Conversely, the concentration in the bran is generally higher than in whole grain and the use of commercial debranning before milling may therefore reduce the levels in the milling fractions. The use of wet processing to fractionate wheat flour to produce starch and *vital gluten* may also be effective in removing water-soluble components such as DON and NIV, but less water-soluble components (such as ochratoxin) may be concentrated in the gluten fraction which is widely used in making bread and other foods.

A large amount of research has been carried on the effects of *secondary processing* of flour on mycotoxin levels, including the effects of dough improvers, fermentation using yeast and sourdough systems, baking, pasta making, frying, and extrusion cooking (Schaarschmidt and Fauhl-Hassek 2018). It is perhaps not surprising in view of the range of methods and conditions that have been used that the results also vary considerably. However, in the case of breadmaking, which is the most widely used process, Schaarschmidt and Fauhl-Hassek (2018) estimated that, on average, the DON content of cleaned wheat may be reduced by a factor of 0.4 during the whole process (from 1250 to 500 µg/kg), with contributions from milling, fermentation, and baking.

The fact that reducing mycotoxin contamination of processed foods remains a challenge clearly suggests that avoiding infection in the field (Chapter 6) and in stored grain should be the priority rather than relying on technology to deliver safe products.

References

Andersson, A.A.M., Kamal-Eldin, A., Fras, A. et al. (2008). Alkylresorcinols in wheat varieties in the HEALTHGRAIN diversity screen. *Journal of Agricultural and Food Chemistry* 56: 9722–9725. https://doi.org/10.1021/jf8011344.

Andersson, A., Westerlund, E., Tilly, A.-C. et al. (1992). Natural variations in the chemical composition of white flour. *Journal of Cereal Science* 17: 183–199. https://doi.org/10.1006/jcrs.1993.1018.

Andersson, A.A.M., Andersson, R., Piironen, V. et al. (2013). Contents of dietary fibre components and their relation to associated bioactive components in whole grain wheat samples from the HEALTHGRAIN diversity screen. *Food Chemistry* 136: 1243–1248. https://doi.org/10.1016/j.foodchem.2012.09.074.

Aslam, M.F., Ellis, P.R., Berry, S.E. et al. (2018). Enhancing mineral bioavailability from cereals: current strategies and future perspectives. *Nutrition Bulletin* 43: 184–188. https://doi.org/10.1111/nbu.12324.

Aune, D., Chan, D.S.M., Lau, R. et al. (2011). Dietary fibre, whole grains, and risk of colorectal cancer: systematic review and dose-response meta-analysis of prospective studies. *British Medical Journal* 343: d6617. https://doi.org/10.1136/bmj.d6617.

Aune, D., Norat, T., Romundstad, P. et al. (2013). Whole grain and refined grain consumption and the risk of type 2 diabetes: a systematic review and dose-response meta-analysis of cohort studies. *European Journal of Epidemiology* 28: 845–858. https://doi.org/10.1007/s10654-013-9852-5.

Aune, D., Keum, N., Giovannucci, E. et al. (2016). Whole grain consumption and risk of cardiovascular disease, cancer, and all cause and cause specific mortality: systematic review and dose-response meta-analysis of prospective studies. *British Medical Journal 353*: i2716. https://doi.org/10.1136/bmj.i2716.

Balk, J., Connorton, J.M., Wan, Y. et al. (2018). Improving wheat as a source of iron and zinc for global nutrition. *Nutrition Bulletin* 44: 53–59. https://doi.org/10.1111/nbu.12361.

Barrett, J.S. and Gibson, P.R. (2012). Fermentable oligosaccharides, disaccharides, monosaccharides and polyols (FODMAPs) and nonallergic food intolerance: FODMAPs or food chemicals. *Therapeutic Advances in Gastroenterology* 5: 261–268. https://doi.org/10.1177/1756283X11436241.

Bates, B., Lennox, A., Prentice, A. et al. (Eds.) (2014a). *National Diet and Nutrition Survey: Results from Years 1–4 (combined) of the Rolling Programme (2008/2009–2011/2012). Executive Summary*. Public Health England. https://assets.publishing.service.gov.uk/government/uploads/system/uploads/attachment_data/file/594360/NDNS_Y1_to_4_UK_report_executive_summary_revised_February_2017.pdf

Bates, B., Lennox, A., Prentice, A. et al. (Eds.) (2014b). *National Diet and Nutrition Survey: Results from Years 1–4 (combined) of the Rolling Programme (2008/2009–2011/2012)*. Public Health England. https://assets.publishing.service.gov.uk/government/uploads/system/uploads/attachment_data/file/216484/dh_128550.pdf

Bechtel, D.B., Keleikau, L.A., Gaines, R.L. et al. (1985). The effects of *Fusarium graminearum* infection on wheat kernals. *Cereal Chemistry* 62: 191–197.

Bennett, J.W. and Klitch, M. (2003). Mycotoxins. *Clinical Microbiology Reviews* 16: 497–516. https://doi.org/10.1128/CMR.16.3.497–516.2003.

Bouis, H.E. and Welch, R.M. (2010). Biofortification-a sustainable agricultural strategy for reducing micronutrient malnutrition in the global south. *Crop Science* 50: S20–S32. https://doi.org/10.2135/cropsci2009.09.0531.

Borghini, R., Donato, G., Alvaro, D. et al. (2017). New insights in IBS-like disorders: Pandora's box has been opened; a review. *Gastroenterology and Hepatology from Bed to Bench* 10: 79–89. https://www.ncbi.nlm.nih.gov/pmc/articles/PMC5495893.

Boyacioglu, D. and Hettiarachchy, N.S. (1995). Changes in some biochemical components of wheat grain that was infected with *Fusarium graminearum*. *Journal of Cereal Science* 21: 57–62. https://doi.org/10.1016/S0733-5210(95)80008-5.

Brisman, J. (2002). Baker's asthma. *Occupational and Environmental Medicine* 59: 498–502. https://doi.org/10.1136/oem.59.7.498.

Brogberg, M., Hogy, P., and Pleijey, H. (2017). Co_2-induced changes in wheat grain composition: meta-analysis and response functions. *Agronomy* 2017: 7020032. https://doi.org/10.3390/agronomy7020032.

Brouns, F.J.P.H., van Buul, V.J., and Shewry, P.R. (2013). Does wheat make us fat and sick? *Journal of Cereal Science* 58: 209–215. https://doi.org/10.1016/j.jcs.2013.06.002.

Brouns, F., van Rooy, G., Shewry, P. et al. (2019). Adverse reactions to wheat or wheat components. *Comprehensive Reviews in Food Science and Food Safety* 18: 1437–1452. https://doi.org/10.1111/1541-4337.12475.

Cabrera-Chavez, F. and Calderon de la Barca, A.M. (2010). Trends in wheat technology and modification of gluten proteins for dietary treatment of coeliac disease patients. *Journal of Cereal Science* 52: 337–341. https://doi.org/10.1016/j.jcs.2010.06.020.

Cade, J.E., Burley, V.J., Greenwood, D.C. et al. (2007). Dietary fibre and risk of breast cancer in the UK Women's Cohort Study. *International Journal of Epidemiology* 36: 431–438. https://doi.org/10.1093/ije/dyl295.

CGA. (2021). Dietary Fibre. Cereal and Grains Association. https://www.cerealsgrains.org/initiatives/definitions/Pages/DietaryFiber.aspx

Chambers, E.S., Preston, T., Frost, and Morrison, D.J. (2018). Role of gut microbiota-generated short-chain fatty acids in metabolic and cardiovascular health. *Current Nutrition Reports* 7: 198–206. https://doi.org/10.1007/s13668-018-0248-8.

Chan, J.M., Wang, F., and Holly, E.A. (2007). Whole grains and risk of pancreatic cancer in a large population-based case-control study in the San Francisco Bay area, California. *American Journal of Epidemiology* 166: 1174–1185. https://doi.org/10.1093/aje/kwm194.

Chen, T.H.H. and Murata, N. (2002). Enhancement of tolerance of abiotic stress by metabolic engineering of betaines and other compatible solutes. *Current Opinion in Plant Biology* 5: 250–257. https://doi.org/10.1016/S1369-5266(02)00255-8.

Cianferoni, A. (2016). Wheat allergy: diagnosis and management. *Journal of Asthma and Allergy* 9: 13–25. https://doi.org/10.2147/JAA.S81550.

Corol, D.I., Ravel, C., Raksegi, M. et al. (2012). Effects of genotype and environment on the contents of betaine, choline, and trigonelline in cereal grains. *Journal of Agricultural and Food Chemistry* 60: 5471–5481. https://doi.org/10.1021/jf3008794.

Davis, W.R. (2019). *Wheat Belly: Lose the Wheat, Lose the Weight, and Find Your Path Back to Health.* Rodale Books.

Edwards, C.H., Grundy, M.M.L., Grassby, T. et al. (2015). Manipulation of starch bioaccessibility in wheat endosperm to regulate starch digestion, postprandial glycemia, insulinemia, and gut hormone responses: a randomized controlled trial in healthy ileostomy participants. *The American Journal of Clinical Nutrition* 102: 791–800. https://doi.org/10.3945/ajcn.114.106203.

Edwards, C.H., Ryden, P., Mandalari, G. et al. (2021). Structure-function studies of chickpea and durum wheat uncover mechanisms by which cell wall properties influence starch bioaccessibility. *Nature Food* 2: 118–126. https://doi.org/10.1038/s43016-021-00230-y.

EFSA Panel on Dietary Products, Nutrition and Allergies (NDA) (2010). Scientific opinion on the substantiation of health claims related to wheat bran fibre and increase in faecal bulk (ID 3066), reduction in intestinal transit time (ID 828, 839, 3067, 4699) and contribution to the maintenance or achievement of a normal body weight (ID 829) pursuant to article 13(1) of regulation (EC) no 1924/2006. *EFSA Journal 8*: 8017. https://doi.org/10.2903/j.efsa.2010.1817.

EFSA Panel on Dietary Products, Nutrition and Allergies (NDA) (2011a). Scientific opinion on the substantiation of health claims related to betaine and contribution to normal homocysteine metabolism (ID 4325) pursuant to article 13(1) of regulation (EC) no 1924/2006. *EFSA Journal* 9: 2052. https://doi.org/10.2903/j.efsa.2011.2052.

EFSA Panel on Dietary Products, Nutrition and Allergies (NDA) (2011b). Scientific opinion on health claims already evaluated (ID 215, 568, 674, 712, 1398, 1633, 1974, 4191, 4192, 4193, 4236, 4335, 4698, 4704) pursuant to article 13(1) of regulation (EC) no 1924/2006. *EFSA Journal 9*: 2203. https://doi.org/10.2903/j.efsa.2011.2203.

EFSA Panel on Dietetic Products, Nutrition and Allergies (NDA) (2011). Scientific opinion on the substantiation of health claims related to arabinoxylan produced from wheat endosperm and reduction of post-prandial glycaemic responses (ID 830) pursuant to article 13(1) of regulation (EC) no 1924/2006. *EFSA Journal* 9: 2205. https://doi.org/10.2903/j.efsa.2011.2205.

EFSA Panel on Dietetic Products, Nutrition and Allergies (NDA) (2015). Scientific opinion on dietary reference values for vitamin e as α-tocopherol. *EFSA Journal* 13: 4149–4221. https://doi.org/10.2903/j.efsa.2015.4149.

El Khoury, A. and Atoui, A. (2010). Ochratoxin a: general overview and actual molecular status. *Toxins* 2 (4): 461–493. https://doi.org/10.3390/toxins2040461.

Englyst, H.N. and Cummings, J.H. (1985). Digestion of the polysaccharides of some cereal foods in the human small intestine. *American Journal of Clinical Nutrition* 42: 778–787. https://doi.org/10.1093/ajcn/42.5.778.

EU (2006). Regulation (EC) no 1924/2006 of the European Parliament and of the council of 20 December 2006. On nutrition and health claims made on foods. *Official Journal of the European Union*: L404/9. http://data.europa.eu/eli/reg/2006/1924/oj.

EU (2020). Commission Regulation (EC) no 1881/2006 of 19 December 2006 setting maximum levels for certain contaminants in foodstuffs. *Official Journal of the European Union*: L364/5. http://data.europa.eu/eli/reg/2006/1881/2020-10-14.

Fan, M.-S., Zhao, F.-J., Poulton, P. et al. (2008). Historical changes in the concentrations of selenium in soil and wheat grain from the Broadbalk experiment over the last 160 years. *Science of the Total Environment* 389: 532–538. https://doi.org/10.1016/j.scitotenv.2007.08.024.

FFI (2019). Annual Report 2019: Plan, Implement, Monitor: *Food Fortification for a Healthier Future*. Food Fortification Initiative. https://www.ffinetwork.org

Gartuala, G., Dhital, S., Netzel, G. et al. (2018). Quantitive structural organisation model for wheat endosperm cell walls: cellulose as an important constituent. *Carbohydrate Polymers* 196: 199–208. https://doi.org/10.1016/j.carbpol2018.05.041.

Gebruers, K., Dornez, E., Boros, D. et al. (2008). Variation in the content of dietary fiber and components thereof in wheats in the HEALTHGRAIN diversity screen. *Journal of Agricultural and Food Chemistry* 56: 9740–9749. https://doi.org/10.1021/jf800975w.

Geisslitz, S., Shewry, P.R., Brouns, F. et al. (2021). Wheat ATIs: characteristics and role in human disease. *Frontiers in Nutrition* Submitted.

Gibson, G., Probert, H., van Loo, J. et al. (2004). Dietary modulation of the human colonic microbiota: updating the concept of prebiotics. *Nutrition Research Reviews* 17: 259–275. https://doi.org/10.1079/NRR200479.

Gibson, P.R. and Shepherd, S.J. (2010). Evidence-based dietary management of functional gastrointestinal symptoms: the FODMAP approach. *Journal of Gastroenterology and Hepatology* 25: 252–258. https://doi.org/10.1111/j.1440-1746.2009.06149.x.

Gilissen, L.J.W.J., van der Meer, I.M., and Smulders, M.J.M. (2014). Reducing the incidence of allergy and intolerance to cereals. *Journal of Cereal Science* 59: 337–353. https://doi.org/10.1016/j.jcs.2014.01.005.

Gill, S.K., Rossi, M., Bajka, B., and Whelan, K. (2021). Dietary fibre in gastrointestinal health and disease. *Nature Reviews Gastroenterology & Hepatology* 18: 101–116. https://doi.org/10.1038/s41575-020-00375-4.

Grausgruber, H., Lovegrove, A., Shewry, P. et al. (2020). FODMAPs in wheat. In: *Wheat Quality for Improving Processing and Human Health* (eds. G. Igrejas, T. Ikeda and C. Guzmán), 515–532. Springer https://doi.org/10.1007/978-3-030-34163-3_21.

Gouseti, O., Lovegrove, A., Kosik, O. et al. (2019). *Journal of Agricultural and Food Chemistry* 67: 8419–8424. https://doi.org/10.1021/acs.jafc.9b01847.

Harris, S., Powers, S., Monteagudo-Mera, A. et al. (2019). Determination of the prebiotic activity of wheat arabinogalactan peptide (AGP) using batch culture fermentation. *European Journal of Nutrition* 59: 297–307. https://doi.org/10.1007/s00394-019-01908-7.

Hemery, Y., Rouau, X., Lullien-Pellerin, V. et al. (2007). Dry processes to develop wheat fractions and products with enhanced nutritional quality. *Journal of Cereal Science* 46: 327–347. https://doi.org/10.1016/j.jcs.2007.09.008.

Hughes, S., Shewry, P., Li, L. et al. (2007). in vitro fermentation by human fecal microflora of wheat arabinoxylans. *Journal of Agricultural and Food Chemistry* 55: 4589–4595. https://doi.org/10.1021/jf070293g.

Hussain, I. and Uddin, M.B. (2012). Optimization effect of germination on functional properties of wheat flour by response surface methodology. *International Research Journal of Plant Science* 3: 31–37.

Ibba, M.I., Juliana, P., Hernandez-Espinosa, N. et al. (2021). Genome-wide association analysis for arabinoxylan content in common wheat (*T. eastivum)* flour. *Journal of Cereal Science* 98: 103166. https://doi.org/10.1016/j.jcs.2021.103166.

Jones, J. (2012). Wheat belly—an analysis of selected statements and basic theses from the book. *Cereal Foods World* 57 (4): 177–189. https://doi.org/10.1094/CFW-57-4-0177.

Kariluoto, S., Edelmann, M., and Piironen, V. (2010). Effect of environment on folate contents in wheat genotypes. *Journal of Agricultural and Food Chemistry* 58: 9324–9331. https://doi.org/10.1021/jf100251j.

Katina, K., Arendt, E., Liukkonen, K.H. et al. (2005). Potential of sourdough for healthier cereal products. *Trends in Food Science and Technology* 16: 104–112. https://doi.org/10.1016/j.tifs.2004.03.008.

Koehler, P., Hartmann, G., Wieser, H. et al. (2007). Changes in folates, dietary fibre, and proteins in wheat as affected by germination. *Journal of Agricultural and Food Chemistry* 55: 4678–4683. https://doi.org/10.1021/jf0633037.

Kucek, L.K., Veenstra, L.D., Amnuaycheewa, P. et al. (2015). A grounded guide to gluten: how modern genotypes and processing impact wheat sensitivity. *Comprehensive Reviews in Food Science and Safety* 14: 285–302. https://doi.org/10.1111/1541-4337.12129.

Laca, A., Mousia, Z., Diaz, M. et al. (2006). Distribution of microbial contamination within cereal grains. *Journal of Food Engineering* 72: 332–338. https://doi.org/10.1016/j.jfoodeng.2004.12.012.

Lafiandra, D., Riccardi, G., and Shewry, P.R. (2014). Improving cereal grain carbohydrates for diet and health. *Journal of Cereal Science* 59: 312–326. https://doi.org/10.1016/j.jcs.2014.01.001.

Lampi, A.-M., Nurmi, T., Ollilainen, V. et al. (2008). Tocopherols and tocotrienols in wheat genotypes in the HEALTHGRAIN diversity screen. *Journal of Agriculture and Food Chemistry* 56: 9716–9721. https://doi.org/10.1021/jf801092a.

Li, H.-T., Chen, S.-Q., Bui, A.T. et al. (2021). Natural "capsule" in food plants: cell wall porosity controls starch digestion and fermentation. *Food Hydrocolloids* 117: 1066557. https://doi.org/10.1016/j.foodhyd.2021.106657.

Li, L., Shewry, P.R., and Ward, J.L. (2008). Phenolic acids in wheat varieties in the HEALTHGRAIN Diversity Screen. *Journal of Agriculture and Food Chemistry* 56: 9732–9739. https://doi.org/10.1021/jf801069s.

Longin, C.F.H., Beck, H., Gutler, A. et al. (2020). Influence of wheat variety and dough preparation on FODMAP content in yeast-leavened wheat breads. *Journal of Cereal Science* 95: 103021. https://doi.org/10.1016/j.jcs.2015.01.003 0733-5210.

Lovegrove, A., Wingen, L.U., Plummer, A. et al. (2020). Identification of a major QTL and associated molecular marker for high arabinoxylan fibre in white wheat flour. *PLoS One* 15 (2): e0227826. https://doi.org/10.1371/journal.pone.0227826.

Lowe, N.M., Zaman, M., Moran, V.H. et al. (2020). Biofortification of wheat with zinc for eliminating deficiency in Pakistan: study protocol for a cluster-randomised doulbe-blind, controlled effectiveness study (BIZIFED2). *BMJ Open* 10: e039231. https://doi.org/10.1136/bmjopen-2020-039231.

Magan, N., Aldred, D., Mylona, K. et al. (2010). Limiting mycotoxins in stored wheat. *Food Additives and Contaminants* 27 (5): 644–650. https://doi.org/10.1080/19440040903514523.

Magan, N., Medina, A., and Aldred, D. (2011). Possible climate-change effects on mycotoxin contamination of food crops pre-and postharvest. *Plant Pathology* 60 (1): 150–163. https://doi.org/10.1111/j.1365-3059.2010.02412.x.

Mares, D.J. and Stone, B.A. (1973). Studies on wheat endosperm I. chemical composition and ultrastructure of the cell walls. *Australian Journal of Biological Sciences* 26: 793–812. https://doi.org/10.1071/BI9730793.

Morita, E., Matsuo, H., Mihara, S. et al. (2003). Fast ω-5 gliadin, is a major allergen in wheat-dependent exercise-induced anaphylaxis. *Journal of Dermatological Science* 33: 99–104. https://doi.org/10.1016/S0923-1811(03)00156-7.

Morita, E., Chinuki, Y., Takahashi, H. et al. (2012). Prevalence of wheat allergy in Japanese adults. *Allergology International* 61: 101–105. https://doi.org/10.2332/allergolint.11-OA-0345.

Moore, J., Hao, Z., Zhou, K. et al. (2005). Carotenoid, tocopherol, phenolic acid and antioxidant properties of Maryland-grown soft wheat. *Journal of Agricultural and Food Chemistry* 53: 6649–6657. https://doi.org/10.1021/jf050481b.

Moore, K.L., Zhao, F.-J., Gritsch, C. et al. (2012). Localisation of iron in wheat grain using high resolution secondary ion mass spectrometry. *Journal of Cereal Science* 55: 183–187. https://doi.org/10.1016/j.jcs.2011.11.005.

Muller, M., Canfora, E.E., and Blaak, E.E. (2018). Gastrointestinal transit time, glucose homeostasis and metabolic health: modulation by dietary fibres. *Nutrients* 10: 275. https://doi.org/10.3390/nu10030275.

Murphy, M.M., Douglass, J.S., and Birkett, A. (2008). Resistant starch intakes in the United States. *Journal of the American Dietetic Association* 108: 67–78. https://doi.org/10.1016/j.jada.2007.10.012.

Myers, S.S., Zanobetti, A., Kloog, I. et al. (2014). Increasing CO2 threatens human nutrition. *Nature* 510: 139–142. https://doi.org/10.1038/nature13179.

Neal, A.L., Geraki, K., Borg, S. et al. (2013). Iron and zinc complexation in wild-type and ferritin-expressing wheat grain: implications for mineral transport into developing grain. *Journal of Bioinorganic Chemistry* 18: 557–570. https://doi.org/10.1007/s00775-013-1000-x.

Nurmi, T., Lampi, A.-M., Nystrom, L. et al. (2012). Distribution and composition of phytosterols and steryl ferulates in wheat grain and bran fractions. *Journal of Cereal Science* 56: 379–388. https://doi.org/10.1016/j.jcs.2012.04.010.

Parry, D.W., Jenkinson, P., and McLeod, L. (1995). Fusarium ear blight (scab) in small grain cereals—a review. *Plant Pathology* 44: 207–238. https://doi.org/10.1111/j.1365-3059.1995.tb02773.x.

Piironen, V., Lampi, A.-M., Ekholm, P. et al. (2009). Micronutrients and phytochemicals in wheat grain. In: *Wheat: Chemistry and Technology*, 4e (eds. K. Khan and P.R. Shewry), 179–222. American Association of Cereal Chemists.

Reynolds, A., Mann, J., Cummings, J. et al. (2019). Carbohydrate quality and human health: a series of systematic reviews and meta-analyses. *Lancet* 393: 434–445. https://doi.org/10.1016/S0140-6736(18)31808-9.

Rout, G.Y. and Sahoo, S. (2015). Role of iron in plant growth and metabolism. *Reviews in Agricultural Science* 3: 1–24. https://doi.org/10.7831/ras.3.1.

Russo, G.L., Pietra, V.D., and Mercurio, C. (1999). Protective effects of butyric acid in colon cancer. In: *Advances in Nutrition and Cancer 2* (eds. V. Zappia, F.D. Ragione, A. Barbarisi, et al.), 131–147. Springer https://doi.org/10.1007/978-1-4757-3230-6_12.

Sander, I., Rozynek, P., Rihs, H.P. et al. (2011). Multiple wheat flour allergens and cross-reactive carbohydrate determinants bind IgE in baker's asthma. *Allergy* 66: 1208–1215. https://doi.org/10.1111/j.1398-9995.2011.02636.x.

Sapone, A., Bai, J.C., Ciacci, C. et al. (2012). Spectrum of gluten-related disorders: consensus on new nomenclature and classification. *BMC Medicine* 10: 13. https://doi.org/10.1186/1741-7015-10-13.

Schaarschmidt, S. and Fauhl-Hassek, C. (2018). The fate of mycotoxins during the processing of wheat for human consumption. *Comprehensive Reviews in Food Science and Food Safety* 17: 556–593. https://doi.org/10.1111/1541-4337.12338.

Schatzkin, A., Park, Y., Leitzmann, M.F. et al. (2008). Prospective study of dietary fiber, whole grain foods, and small intestinal cancer. *Gastroenterology* 135: 1163–1167. https://doi.org/10.1053/j.gastro.2008.07.015.

Schmidt, W., Thomine, S., and Buckhout, T.J. (2020). Editorial: iron nutrition and interactions in plants. *Frontiers in Plant Science* 10: 1670. https://doi.org/10.3389/fpls.2019.01670.

Schuppan, D. and Gisbert-Schuppan, K. (2019). *Wheat Syndromes. How Wheat, Gluten and ATI Cause Inflammation, IBS and Autoimmune Diseases*. Springer https://doi.org/10.1007/978-3-030-19023-1.

Shahidi, F. and Ambigaipalan, P. (2015). Phenolics and polyphenolics in foods, beverages and spices: antioxidant activity and health effects – a review. *Journal of Functional Foods* 18: 820–897. https://doi.org/10.1016/j.jff.2015.06.018.

Sharma, A., Patni, B., Shankhdar, D. et al. (2013). Zinc- an indispensable micronutrient. *Physiology and Molecular Biology of Plants* 19: 11–20. https://doi.org/10.1007/s12298-012-0139-1.

Scheppach, W., Bartram, H.P., and Richter, F. (1995). Role of short-chain fatty acids in the prevention of colorectal cancer. *European Journal of Cancer* 31A: 1077–1080. https://doi.org/10.1016/0959-8049(95)00165-f.

Shewry, P.R. (2014). Minor components of the barley grain: minerals, lipids, terpenoids, phenolics and vitamins. In: *Barley, Chemistry and Technology*, 2e (eds. P.R. Shewry and S.E. Ullrich), 169–192. American Association of Cereal Chemists https://doi.org/10.1016/B978-1-891127-79-3.50007-X.

Shewry, P.R. and Hey, S. (2015). The contribution of wheat to human diet and health. *Food and Energy Security* 4: 178–202. https://doi.org/10.1002/fes3.64.

Shewry, P.R. and Tatham, A.S. (2016). Improving wheat to remove coeliac epitopes but retain functionality. *Journal of Cereal Science* 67: 12–21. https://doi.org/10.1016/j.jcs.2015.06.005.

Shewry, P.R., Piironen, V., Lampi, A.-M. et al. (2010). The HEALTHGRAIN wheat diversity screen: effects of genotype and environment on phytochemicals and dietary fiber components. *Journal of Agricultural and Food Chemistry* 58: 9291–9298. https://doi.org/10.1021/jf100039b.

Shewry, P.R., Van Schaik, F., Ravel, C. et al. (2011). Genotype and environment effects on the contents of vitamins B1, B2, B3 and B6 in wheat grain. *Journal of Agricultural and Food Chemistry* 59: 10564–10571. https://doi.org/10.1021/jf202762b.

Sollid, L.M., Tye-Din, J.A., Qiao, S.-W. et al. (2020). Update 2020: nomenclature and listing of celiac-disease relevant gluten epitopes recognised by $CD4^+$ T cells. *Immunogenetics* 72: 85–88. https://doi.org/10.1007/s00251-019-01141-w.

Szunics, L., Szunics, L., and Stehli, L. (1987). Microorganisms isolated from wheat and damage caused by them. In: *Cereal Quality, Aspects of Applied Biology 15*, 283–287. Association of Applied Biologists.

Tatham, A.S. and Shewry, P.R. (2008). Allergy to wheat and related cereals. *Clinical and Experimental Allergy* 38: 1712–1726. https://doi.org/10.1111/j.1365-2222.2008.03101.x.

Torres, A.M., Palacios, S.A., Yerkovich, N. et al. (2019). Fusarium head blight and mycotoxins in wheat: prevention and control strategies across the food chain. *World Mycotoxin Journal* 12: 333–355. https://www.wageningenacademic.com/doi/pdf/10.3920/WMJ2019.2438.

Turner, A.L., Michaelson, L.V., Shewry, P.R. et al. (2021). Increased bioavailability of phenolic acids and enhanced vascular function following intake of feruloyl esterase-processed high fibre bread: a randomized, controlled, single blind, crossover human intervention trial. *Clinical Nutrition* 40 (3): 788–795. https://doi.org/10.1016/j.clnu.2020.07.026.

Vauzour, D.E.J.H., George, T.W., Corona, G. et al. (2010). Moderate champagne consumption promotes an acute improvement in acute endothelial-independent vascular function in healthy human volunteers. *British Journal of Nutrition* 103: 1168–1178. https://doi.org/10.1017/S0007114509992959.

Velu, G., Ortiz-Monasterio, I., Cakmak, I. et al. (2014). Biofortification strategies to increase grain zinc and iron concentrations in wheat. *Journal of Cereal Science* 59: 365–372. https://doi.org/10.1016/j.jcs.2013.09.001.

Velu, G., Singh, R.P., Crespo-Herrera, L. et al. (2018a). Genetic dissection of grain zinc concentration in spring wheat for mainstreaming biofortification in CIMMYT wheat breeding. *Scientific Reports* 8: 13526. https://www.nature.com/articles/s41598-018-31951-z.

Velu, G., Singh, R., Balasubramanian, A. et al. (2018b). Reaching out to farmers with high zinc wheat cultivars through public-private partnerships – an experience from eastern-Gangetic plains of India. *Advances in Food Technology and Nutritional Sciences* 1: 73–75. https://doi.org/10.17140/AFTNSOJ-1-112.

Wang, L., Li, Q., Liu, Z. et al. (2018). Integrated transcriptome and hormone profiling highlight the role of multiple phytohormone pathways in wheat resistance against fusarium head blight. *PLoS One* 13 (11): e0207036. https://doi.org/10.1371/journal.pone.0207036.

Weichselbaum, E. (2012). Does bread cause bloating? *Nutrition Bulletin* 37: 30–36. https://doi.org/10.1111/j.1467-3010.2011.01943.x.

Windey, K., De Preter, V., and Verbeke, K. (2012). Relevance of protein fermentation to gut health. *Molecular Nutrition & Food Research* 56: 184–196. https://doi.org/10.1002/mnfr.201100542.

WHO (2013). *World Health Report: Research for Universal Health Coverage*. World Health Organization. https://www.who.int/whr/2013/report/en
WHO (2015). The Global Prevalence of Anaemia in 2011. World Health Organization. https://www.who.int/nutrition/publications/micronutrients/global_prevalence_anaemia_2011/en
Zuidmeer, L., Goldhahn, K., Rona, R.J. et al. (2008). The prevalence of plant food allergies: a systematic review. *Journal of Allergy and Clinical Immunology* 121: 1210–1218. https://doi.org/10.1016/j.jaci.2008.02.019.

10

Wheat Genetics and Improvement

. . ..whoever could make two ears of corn, or two blades of grass, to grow upon a spot of ground where only one grew before, would deserve better of mankind, and do more essential service to his country, than the whole race of politicians put together.

Jonathan Swift, 'Gulliver's Travels'

10.1 Genetic Background to Wheat Breeding

Modern plant breeding started at the end of the nineteenth century with the rediscovery of the work of Gregor Mendel. Although his original work was carried out in the 1860s, it was not widely known until 1900, when it was rediscovered independently by three plant geneticists who were studying the inheritance of traits in plant hybrids: Hugo DeVries, Carl Correns, and Erich von Tschermak. It was subsequently popularised in the UK by William Bateson at the University of Cambridge. Bateson became Director of the John Innes Horticultural Institution (then at Merton Park, Surrey) in 1910, where he initiated the strong tradition in crop genetics that provided the foundation for the modern John Innes Centre in Norwich.

The traits studied by Mendel and his followers were controlled by small numbers of genes with strong dominant alleles. Hence, the analysis of such traits is often referred to as 'Mendelian genetics'. However, many traits do not show strong allelic effects, but show continuous variation, such as plant height, biomass, and yield. These are termed 'quantitative traits' and are usually controlled by multiple genes. The importance of quantitative traits led to the development of new mathematical treatments, initially by Sir Ronald Fisher, who analysed data from the Long Term Experiments at Rothamsted Experimental Station (now Rothamsted Research) and J.B.S. Haldane, and developed further by Sir Kenneth Mather and John L. Jinks at the University of Birmingham (Mather and Jinks 1982). The application of these approaches allows the mapping of quantitative trait loci (QTLs), which are regions of the chromosome that contribute to the trait of interest.

Wheat: Environment, Food and Health, First Edition. Mike J. Gooding and Peter R. Shewry.

The mapping and exploitation of QTLs have been facilitated by the application of molecular markers, as discussed here.

Classical plant breeding, based on Mendelian and quantitative genetics, has been immensely successful in terms of increasing the yield and quality of wheat, particularly in the high yielding intensive production systems used in Western Europe, and remains the mainstay of wheat breeding. For example, Mackay et al. (2011) analysed historical datasets from the UK National and Recommended List trials and calculated that about 88% of the gain in yield of winter wheat between 1981 and 2007, from about 6 to 8 t/ha, was attributable to genetic improvement. Similarly, Laidig et al. (2017) reported yield gains of 24% in official German variety trials during 1983–2014, while the yields of Canadian western red spring wheat were estimated to be increasing by 15 kg/ha per year in 1970 and by 23 kg/ha per year in 1995 (Hucl et al. 2015). However, in the past few decades classical plant breeding has been complemented by novel technologies, based on information from molecular genetics studies.

Nevertheless, despite the increasing application of novel technologies wheat breeding remains a long process, taking up to seven years from the first crosses to improved cultivars reaching the marketplace. This is partly due to the need for multiple generations of plants to be grown in field trials, a process which may be speeded up by growing alternative generations in the northern (e.g. UK) and southern (e.g. New Zealand) hemispheres, known as 'shuttle breeding'. In addition, several years of field trials are required to provide data for cultivar registration and assignment to quality classes.

Some of the approaches and technologies that are most widely used in wheat genetics and breeding are described in the next sections.

10.2 Technologies for Wheat Genetics and Breeding

10.2.1 Aneuploid Lines

Although they are rarely used now, aneuploid lines were the major resource used by wheat cytogeneticists for over half a century, following their development by Dr. Ernie Sears and colleagues at the University of Missouri (Sears 1954). Aneuploids are defined as 'individuals with a changed number of chromosomes, chromosome arms, or centromeres, or with arms so arranged that normal pairing and disjunction cannot occur at meiosis' (Sears 1975). Such drastic changes in the chromosome complement of cells would be lethal in diploid species but may be accepted in bread wheat and other polyploids, due to the buffering provided by related genes on the three genomes.

A range of aneuploid lines were generated in bread wheat, almost all in the Chinese Spring genotype, which was therefore subsequently adopted as the international reference genotype for wheat. They include lines with zero (nullisomic), one (monosomic), and four (tetrasomic) copies of all individual chromosomes, with nullisomics and tetrasomics often being combined to improve the stability and vigour of the lines. This improvement results from the ability of the homoeologous (homologous but non-pairing) chromosomes of the three genomes to partially compensate for each other. These are therefore called compensating nulli/tetrasomics. For example, the nulli1A/tetra1D line has no 1A chromosomes but four 1D chromosomes. Other aneuploids were developed with only single chromosome arms (telocentrics) and with two identical arms (isochromosomes).

It was therefore relatively easy to assign genes to chromosomes, and often also to chromosome arms, by analysing the appropriate series of aneuploids. This approach was widely used for mapping genes controlling grain composition, including the assignment of wheat gluten protein genes to chromosomes (Wrigley and Shepherd 1973).

10.2.2 Doubled Haploid Lines, Recombinant Inbred Lines, and Near-Isogenic Lines

Doubled haploid (DH) and recombinant inbred (RI) lines are widely used in wheat genetics and breeding as they are essentially stable between generations. Both are homozygous populations that preserve the variation present in the F2 generation of a cross, but they are produced in very different ways (Figure 10.1).

RILs are produced by self-pollinating F2 lines repeatedly for at least six generations, by which time they are about 97% homozygous. This requires a minimum of about four years, due to the generation time of wheat. By contrast, DHLs are produced in a single step, by generating haploids from the F1 plants. There are two ways to do this. In the first method, developing microspores (pollen) from anthers taken from the F1 plant are cultured *in vitro*, giving rise to callus, which can be regenerated to give haploid plants. Chromosome doubling can occur spontaneously but is usually induced by treatment of the haploid plants with colchicine. In the second and more widely used method, haploid plants are generated by pollination of the F1 wheat plant with maize pollen. The maize chromosomes are excluded from the developing embryos, which are therefore haploid. The developing

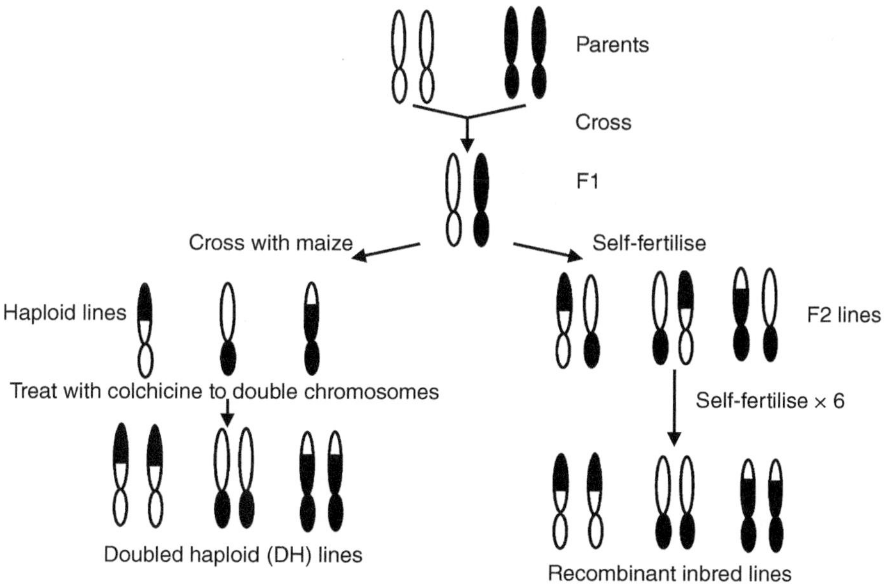

Figure 10.1 Scheme for the production of doubled haploid (DH) and recombinant inbred (RI) lines of wheat.

embryos are then rescued and grown in culture to produce haploid plants, which are treated with colchicine to double the chromosome number (Niu et al. 2014). Hence, DH technology allows the production of homozygous plants in a single generation.

Near-isogenic lines (NILs) can be developed to compare the effects of allelic variation in a single gene, by removing the effects of other genes in the lines which could affect the expression of the trait. This is achieved by transferring the two alleles to the same genetic background, by back crossing with the same parental line (or with a third line) at least six times and selecting for the alleles of interest. Hence, the lines should have the same genetic backgrounds and differ only in the alleles of interest.

10.2.3 Marker-Assisted Selection (MAS)

The development of molecular markers has revolutionised plant breeding, particularly for quantitative traits (i.e. traits controlled by multiple genes), and for traits which are difficult or expensive to identify using conventional assays. The concept was developed over 30 years ago by Steven Tanksley at Cornell University (Ithaca, USA), working on tomatoes. The technology was initially based on the hydridisation of DNA sequences from (or close to) the gene of interest to genomic DNA digested with restriction enzymes but has developed rapidly over the intervening period, with most marker systems now being based on identifying differences in single nucleotides of DNA (single nucleotide polymorphisms, SNPs) from sequence comparisons. SNPs can then be identified in populations of plants using high throughput systems, such as high-density arrays (which may allow the simultaneous detection of over 800 000 SNPs) (Winfield et al. 2016) or competitive allele-specific polymerase chain reaction (PCR) analyses (Allen et al. 2011). However, it should be noted that most molecular markers are not located in the genes that determine the traits but in the same region of DNA. Hence, they are referred to as 'imperfect', as opposed to 'perfect' markers, which are specific for the actual genes. Nevertheless, the use of high-density marker systems has greatly improved the resolution of QTL analysis, often allowing the identification of only a small number of candidate genes for traits, allowing them to be functionally characterised using approaches such as mutagenesis, transgenesis, and gene editing.

10.2.4 Genome-Wide Association Genetics (GWAS) and Genomic Selection (GS)

The increased efficiency and reduced cost of high-density genotyping mean that it is now possible to genotype whole populations in wheat breeding programmes and to apply this information in genome-wide association studies (GWAS) and genomic selection (GS).

The attraction of GWAS is that it is not usually necessary to construct bespoke populations to identify major alleles for traits, although this may be required if the alleles occur only rarely. The concept is very simple: phenotypic variation in a selection of genotypes is compared with marker data to identify associations between the two (Korte and Farlow 2013). In practice it is more challenging, with several requirements: the genotypes must be selected to avoid bias due to relatedness; the number of genotypes must be sufficient to give statistically significant data; and the genotypes must be grown in multiple replicated field experiments to minimise effects of environment and genotype × environment interactions.

Although GS also exploits high-density genotyping, it differs from GWAS in that the marker data are used as predictors of performance rather than to identify loci that determine individual traits. In practical breeding, an entire breeding programme can be used as a database, using statistical methods to predict the performance of lines that have not been assessed in the field (Jannink et al. 2010). As for GWAS, the database should include phenotyping carried out over several years in order to build a robust statistical model and, because major genes will be more accurately predicted than those with smaller effects, the larger the calibration database the better the prediction accuracy. GS should allow polygenic traits, such as yield, to be increased by selecting multiple markers, rather than by identifying and combining QTLs. Bentley et al. (2014) provide an excellent discussion of the application of these approaches to wheat improvement.

10.2.5 Intermated Populations (MAGIC and NAM)

Much of the variation exploited in conventional plant breeding is generated by reassorting alleles in the two parents. This reassortment forms the basis for populations generated by multi-parent advanced generation intercrosses (MAGIC), in which multiple parental lines are crossed in all combinations over several generations to generate a population with high genotypic and phenotypic diversity. For example, NIAB (Cambridge, UK) has developed strategies to construct MAGIC populations by crossing between 4 and 26 parents and constructed 2 populations, from 8 elite currently grown cultivars, and from 16 UK wheat cultivars listed for production in the UK between 1935 and 2004 (Ladejobi et al. 2016). The provision of such lines, together with high-density molecular marker and phenotypic data, is an important resource for both wheat scientists and commercial breeders (Huang et al. 2015).

Nested association mapping (NAM) is a related approach, in which a diverse range of genotypes are all crossed with the same reference line. The F1 plants are then self-fertilised to generate families of RILs that are genotyped with high-density markers. NAM combines the advantages of linkage analysis and association mapping, giving high resolution of the variation between the selected genotypes and that generated by recombination in the crosses.

Simon Griffiths and colleagues at the John Innes Centre (Norwich, UK) have developed a NAM panel comprising RILs from crosses between landraces from the Watkins collection (see Section 10.3.2) and the spring wheat Paragon to exploit landrace diversity (Wingen et al. 2017), while Kidane et al. (2018) described an Ethiopian NAM population of durum wheat, comprising 50 related biparental populations.

10.2.6 Hybrid Wheat

F1 hybrids derived from crosses between selected inbred lines often show significant increases in yield compared to the parental lines, a phenomenon known as hybrid vigour. Consequently, many major commercial crops are dominated by hybrid cultivars. This includes maize, where essentially all commercial production in the USA and many other countries uses hybrid seed, but also crops such as soybean and sugar beet. The production of hybrid rice is also increasing rapidly, particularly of the widely consumed *indica* form but also of *japonica* rice (Zheng et al. 2020). By contrast, hybrid wheat has been much more difficult to develop and market. There are two reasons for this.

Firstly, all of the major hybrid crops are diploid, and hybrid vigour is greater in diploids than in polyploids: this may reflect the fact that polyploids, such as wheat, already have a level of heterozygosity due to the presence of related genes on the separate genomes. Some wheat hybrids can produce higher yields than either parent by combining, e.g. high levels of light interception, radiation-use efficiency, and harvest index, from the different parents (Kindred and Gooding 2005). However, the yield gains are less than in the diploid crops, and in any case may be outpaced by yield improvements in conventionally bred wheat, as the selection of parents with good hybridizing ability takes time and restricts choice.

Secondly, it is necessary to have an efficient system for hybrid production, which is more difficult for inbreeding crops (such as wheat) than for outbreeding crops (such as maize). In commercial terms, hybrid crops are only sustainable if the profit from increased yields outweighs the higher cost of seed production.

Hybrid crops also have clear advantages for the seed production industry, as it is necessary for farmers to purchase seeds every year, rather than replant seed from their previous harvest. All of the major hybrid crops have also been engineered to express genes conferring resistance to herbicides, allowing companies to market a package of hybrid transgenic seeds and appropriate agrochemicals. However, it important to note that farmers are not obliged to purchase hybrid seeds but choose to because they give higher yields and profits.

The production of hybrid wheat has had several false starts, but is now becoming commercially viable, particularly in China (Gupta et al. 2019), but also in Western Europe. This has resulted from the development of systems to ensure male sterility in the female parent to allow the hybrid to be produced but to restore male fertility in the hybrid plant to allow fertilization and grain production (Tucker et al. 2017; Gupta et al. 2019). Although the yield gains are still not clear, part of the increased interest is due to the suggestion (which is still unproved) that they may be more stable to fluctuating weather conditions, an important consideration in an increasingly variable climate.

10.2.7 Perennial Wheat

Perennial crops should have better capture and more efficient use of resources (light, nutrients, water) than annual crops, due to increased persistency and extent of canopy and roots throughout the year; and improved retention and recycling of nutrients within the crop. Eliminating the need for soil disturbance to establish annual crops should also reduce losses of nutrients (particularly nitrate) and organic matter, greenhouse gas emissions, the risk of soil erosion, and damage to soil fauna and structure (Glover et al. 2010). Maintaining a greater autumn and winter crop cover and stubble should also increase competition against weeds and provide habitat and food for some farmland birds (Butler et al. 2010).

The concept of perennial wheat has been attractive for decades but long-term breeding programmes in the USSR and the USA were initially disappointing (Cox et al. 2006). A perennial is a plant that keeps meristem regeneration ahead of senescence (Thomas et al. 2000). In the grasses, perenniality is achieved through tillering, stolon, and rhizome production, traits that are present in some wild relatives of wheat such as *Thinopyrum* (wheatgrass), *Leymus* (lyme grass), and *Agropyron* (couch grass). All of these genera have been crossed with *Triticum*, but recent attention has focused on *Thinopyrum* spp., which have been crossed with *Triticum* spp. to produce perennial breeding lines (Cox et al. 2006; Curwen-McAdams and Jones 2017; Cui et al. 2018).

However, the long-term viability of perennial wheats is still not established, with possible susceptibility to frost damage and pathogen infections. It is perhaps more realistic to consider perennial wheat as an alternative pasture crop, providing grains and forage to livestock, rather than as an alternative to annual wheat: it will certainly never be competitive in the intensive wheat production systems in many countries in terms of yield.

10.3 Sources and Exploitation of Genetic Diversity

It is clear that plant breeding has led to lower diversity in modern bread wheats than in wheats grown in historical times. This is of increasing concern to breeders, and several approaches are being used to identify and exploit wider sources of variation.

10.3.1 Gene Banks

Much of the current variation in wheat and related species is captured in collections in gene banks. There are some 80 gene banks in the world, which together hold about 800 000 accessions. The most important is held by CIMMYT (International Maize and Wheat Improvement Centre, Mexico), with over 150 000 accessions, but substantial collections are also held at the USDA-NSGC (Aberdeen, Idaho, USA), the Vavilov Institute (Russia), ICARDA (Lebanon), NBPR (India), and Instituto del Germoplasma (Bari, Italy). These collections comprise three types of material: modern and obsolete cultivars, landraces (see Section 10.3.2) and wild relatives, with the vast majority of the accessions being modern cultivars.

10.3.2 Land Races

It is widely assumed that plant breeding is a modern activity. In fact, it has been carried out for millennia, by farmers selecting the best seeds (in terms of yield and perhaps other traits) from their crops to grow on in future years. This selection led to the development of landraces, which were distinct populations (rather than pure cultivars) adapted to local climatic conditions and processing requirements. Such landraces are still grown in some countries, such as Turkey, where they are valued for their high quality for making traditional foods.

Landraces are regarded as an important resource for crop improvement as they represent a 'snapshot' of the variation available when they were collected. Hence, they are well represented in gene banks.

The most famous collection of wheat landraces is held at the N I Vavilov Institute of Plant Industry in St. Petersburg (Russia). This originated from the collections made by Nicolai Vavilov (1887–1943), who was one of the first scientists to realise the importance of capturing the genetic diversity present in crops and wild relatives. Collecting expeditions to more than 50 countries led to over 250 000 accessions of crop plants and related species, with the most active period of collection being between 1917 and 1929. The gene bank currently holds almost 350 000 accessions, including 50 000 accessions of *Triticum*, Triticale, and *Aegilops* species (Dzyubenko 2018).

A second collection of wheat landraces was made during the late 1920s and early 1930s by A. E. Watkins at the School of Agriculture, Cambridge University, UK. Following Vavilov's lead, he used connections with the London Board of Trade to obtain samples of wheat from markets in several countries, leading to a collection of over 7000 accessions. Many of these accessions were lost due to poor storage during the second World War and the current collection is now 826 accessions of bread wheat. Analysis with molecular markers allowed the identification of a core set of 119 accessions, which capture 96–98% of the diversity in the whole collection (Wingen et al. 2014). This core set is being widely used by wheat researchers to identify novel variation in traits for cultivar improvement.

10.3.3 Wild Relatives

Introgression of variation from crosses with wild relatives in the field has contributed to the current diversity of bread wheat (Przewieslik-Allen et al. 2021) and is increasingly used as a source of variation by wheat geneticists and breeders.

Related species can be divided into gene pools (GP) based on their ease of crossing with wheat (Hao et al. 2020). No wild forms of hexaploid bread wheat exist but several variety groups are recognised, notably var. *spelta* (spelt), which is hulled, and the free-threshing var. *sphaerococcum* (Indian shot wheat) and var. *compactum* (club wheat). These are all readily crossed with bread wheat and hence form the primary gene pool (GP-1).

The secondary gene pool (GP-2) comprises wild and cultivated forms of tetraploid and diploid *Triticum* species that share one or more genomes with wheat and *Aegilops tauschii*, which is the origin of the D genome. These species are readily crossed with bread wheat but the hybrids show low fertility. Nevertheless, recombination between the homoeologous genomes can be achieved.

The tertiary gene pool (GP-3) is species which do not share a genome with wheat, comprising barley, rye, and related grasses (such as the genera *Aegilops*, *Agropyron*, *Haynaldia*, and *Thinopyrum*). These genera are much more difficult to introgress and the hybrid embryos often fail to develop on the plant and need to be rescued and cultured *in vitro*.

Introgression of genes from related chromosomes is usually achieved by homologous recombination, exploiting the *Ph* loci, particularly *Ph1* which is located on chromosome 5B. The homology between the three genomes of bread wheat is sufficiently high for the related chromosomes to come together at meiosis. This would lead to genetic instability and is prevented by the *Ph* genes, which effectively stabilise the polyploid by ensuring that meiosis results in regular segregation of the chromosomes. Mutations at *Ph* loci allow this pairing to occur, allowing recombination between the genomes. These mutants are therefore exploited to allow the transfer of DNA in the hybrids formed between relatives and cultivated wheats.

A major problem with introgression is 'genetic drag', due to adverse effects of genes closely linked to the gene of interest on traits such as yield. In many cases these can be eliminated by recombination, using marker-assisted selection.

Introgression has been widely used as a source of genes for resistance to pathogens, in particular from *Aegilops* species. Kishii (2019) lists resistance genes for eight fungal pathogens, one virus and six pests (two nematodes, three insects, and one mite), which have been identified in *Aegilops* species. The pathogens include stem, leaf and stripe rusts, mildew, eyespot, and Septoria, with work dating back to 1970.

One of the most widely used introgressions is high grain protein content from wild emmer (*Triticum turgidum* var. *dicoccoides*). High protein lines of wild emmer were identified in Israel in the 1970s and the *GPC-B1* (grain protein content) locus mapped to chromosome 1B using addition lines of *T. turgidum* chromosomes into durum wheat. It was subsequently mapped to a single Mendelian locus and the gene shown to encode a protein belonging to the well-characterised NAC-domain transcription factor family (Uauy et al. 2006). It has since been shown that *GPC-B1* regulates senescence and the redistribution of nitrogen and minerals from the senescing leaves to the developing grain. Hence, *GPC-B1* lines have high contents of iron and zinc in addition to high protein. Because *GPC-B1* essentially accelerates crop senescence grain yield may be reduced, but this effect can be reduced by breeding.

The active *GPC-B1* allele is widespread in wild and cultivated genotypes of emmer, but most cultivars of bread and durum wheat have inactive alleles (resulting from mutations or deletion). It is difficult to quantify the impact of *dicoccoides* high protein genes in commercial breeding, as breeding strategies are rarely reported in scientific publications. Tabbita et al. (2017) list a number of *GPC-B1* cultivars bred in the US and Canada, but the high protein trait from *dicoccoides* has also been extensively used in other countries, for example in breeding high-quality breadmaking wheats for the UK.

Further evidence for the use of wild emmer introgressions into modern commercial bread wheat cultivars is a non-glaucous (i.e. not grey-green) trait. This is associated with delayed senescence and a striking green colour of the canopy. The gene has been mapped to chromosome 2B of wild emmer and is associated with the wax inhibitor gene *Iw1*, which reduces the β-diketone component of the plant wax (Simmonds et al. 2008; Adamski et al. 2013; Clarke et al. 2017). The trait has been present in several commercially successful cultivars in the UK and has been linked to yield stability (Simmonds et al. 2008). Introgression from wild emmer may also contribute to greater rooting at depth of some elite lines (e.g. Clarke et al. 2017).

10.3.4 Rye Translocations

The challenge of removing genetic drag is illustrated by the 1BL/1RS translocation, in which the short arm of wheat chromosome 1B is replaced by the short arm of chromosome 1R of rye. This translocation was introduced in the 1920s from rye and triticale, and by the 1990s was widely distributed in high yielding cultivars, including cultivars developed in Europe, in North and South America, and by CIMMYT for the developing world. The translocation was widely adopted because it was associated with higher yields, which probably resulted from resistance to pathogens. The rye segment carries genes for resistance to fungal pathogens, notably *Lr26* for leaf rust, *Yr9* for yellow rust, *Sr31* for stem rust, and *Pm8* for powdery mildew resistance. However, in most countries these resistances were subsequently overcome by new strains of pathogen. A 1DL/1RS translocation has also been produced by scientists working in Australia but has been used as a research tool rather than for cultivar development.

The rye translocation was used most successfully in high yielding feed wheats because it also results in dough stickiness and poor processing quality. This effect probably results from the replacement of the gluten proteins encoded by genes on chromosome 1B (gliadins and low molecular weight subunits of glutenin) with secalins encoded by the short arm of chromosome 1R. The secalins are related to gliadins but have greater solubility in water while 1RS does not encode glutenin subunits. These changes may therefore contribute to greater stickiness. However, rye also has a higher content of arabinoxylan fibre than wheat, which has also been suggested to contribute to stickiness (see Graybosch 2001 for a detailed discussion).

Despite considerable effort it has not been possible to separate the benefits of the 1BL/1RS translocation from the deficits and the use has decreased over the past two decades; for example, it is not present in any cultivars currently grown in the UK.

10.3.5 Synthetic Wheats

As discussed in Section 1.3.4, the D genome of bread wheat displays less diversity than the A and B genomes. Further diversity can therefore be introduced by constructing new, synthetic hexaploids. These are generally produced by crossing durum wheat (*T. turgidum* var. *durum*) with *Ae. tauschii*, with the latter usually being the male parent. The resulting embryos are triploid and they are therefore cultured and the young plants treated with colchicine to double their chromosome number. This procedure was pioneered at CIMMYT in the late 1980s and they have used synthetic wheats in their breeding programmes. However, considerable breeding efforts are required to remove deleterious traits derived from the *Ae. tauschii* parent. It should also be noted that the A and B genomes in synthetic wheats are derived from durum wheat and therefore differ from those in bread wheat.

Nevertheless, synthetic wheats have been widely used over the past two decades, particularly as sources of resistance to biotic and abiotic stresses, leading to the release of cultivars in countries ranging from Spain to China (Li et al. 2018). *Ae. tauschii* is also a rich source of genes for resistance to pest and pathogens (Kishii 2019).

10.3.6 Ancient Wheats

The term 'ancient wheats' is usually applied to types of wheat that were widely cultivated in antiquity but are today only grown on small areas. However, it cannot be assumed that the currently grown genotypes are identical to wheats cultivated in ancient times. This is because the genotypes grown commercially today have been subjected to genetic improvement by breeders, and modern spelts may have been derived from crosses with bread wheats to improve their yields and agronomic performance. The best known and most widely studied ancient wheats are diploid einkorn (*Triticum monococcum* var. *monococcum*, genome AA), tetraploid emmer (*T. turgidum* var. *dicoccum*, genomes AABB), and hexaploid spelt (*Triticum aestivum* var. *spelta*, genomes AABBDD). Spelt, emmer, and most forms of einkorn differ from bread and durum wheats in being hulled (i.e. the glumes remain tightly closed over the grain and are not removed by threshing). They therefore need to be mechanically dehulled before milling or other forms of processing.

Spelt and modern bread wheat are both forms of *T. aestivum* and spelt breeders may use bread wheat as a parent in breeding programmes, as a source of traits for improved agronomic performance.

Similarly, emmer and modern durum wheats are forms of *T. turgidum*. Although other old forms of *T. aestivum* and *Triticum durum* are cultivated to a limited extent (Chapter 1), the only type that is often included as an 'ancient wheat' is the free threshing tetraploid Khorasan wheat (*T. turgidum* var. *turanicum*). *T. turgidum* also includes two trademarked types of wheat, Kamut® and Graziella Ra®.

Kamut seed was originally obtained from Egypt in the late 1940s, described as 'mummy wheat' from an Egyptian tomb. It is highly unlikely that the seed had come from a tomb, and it is more probable that the grain was derived from the diversity of crops grown by farmers in Egypt or Asia Minor (Michalcová et al. 2014). Kamut is a registered trademark of Kamut International Ltd. and is only grown on certified organic farms. Similarly, the tetraploid Italian wheat Graziella Ra is also purported to be derived from an Egyptian tomb (Gregorini et al. 2009). Comparative analyses show that Graziella Ra and Kamut are related but distinct (Colomba and Gregorini 2011).

There has been a resurgence of interest in these species in recent years due to their organoleptic properties and perceived health benefits. Notably, the cultivation of spelt has increased significantly in Germany and Austria.

10.3.7 Tritordeum: A Novel Cereal Derived from Wheat

Wheat has been used as a basis for developing two novel cereal species, by hybridization with rye (Triticale) and wild perennial barley (Tritordeum). Triticale is a well-established crop with an extensive literature and will not be discussed here. However, Tritordeum is a more recent development and will be briefly described.

Tritordeum was developed by Antonio Martin and colleagues in Cordoba (Spain) in the late 1970s, by hybridising hexaploid bread and tetraploid durum wheats with the perennial South American barley species *Hordeum chilense*. This led to the production of octoploid (genomes $H^{ch}H^{ch}AABBDD$) and hexaploid (genomes $H^{ch}H^{ch}AABB$) forms, with the latter being developed commercially (Martin et al. 1999). Tritordeum is adapted to similar environments as durum wheat. It has good agronomic performance and is also reported to have a lower environmental impact with higher nitrogen uptake and water-use efficiency (Martin et al. 1999). It also has higher quality for breadmaking than durum wheat, with good organoleptic properties and high consumer acceptability. Tritordeum has high contents of the carotenoid lutein, fivefold greater than durum wheat (Atienza et al. 2007), and has also been reported to contain lower numbers of coeliac-toxic gluten protein sequences than bread wheat (Vaquero et al. 2018).

Tritordeum is attracting considerable interest as an alternative cereal with putative environmental and health benefits, similar to ancient wheats. However, as with ancient wheats, more long-term studies are required to establish the extent of these benefits.

10.3.8 Mutagenesis and TILLING

Mutants have been exploited by plant breeders for many years, who initially used naturally occurring mutants such as waxy starch (Chapter 7). These naturally occurring mutations have been supplemented by mutations induced by chemical and physical mutagens.

The use of ionising radiation to induce mutations for plant breeding dates from the 1920s but was promoted by the Food and Agricultural Organisation(FAO) of the United Nations and the International Atomic Energy Agency (IAEA) after the Second World War as a peaceful use of nuclear energy. Although the early studies used X-rays, gamma-rays from ^{60}Co and other sources have been increasingly used since 1960. Other physical mutagens that have been used in plant breeding include neutrons, β-particles, α-particles, protons, and ion beams.

All physical mutagens are hazardous and require specialist facilities. The mutagens differ in mass, energy and penetration and can result in chromosome breakage leading to deletions and rearrangements. Although such rearrangements are usually lethal in diploid species they may be tolerated in polyploids such as bread and durum wheats and ionising radiation has therefore been used to develop resources for gene mapping. The ability of polyploids to 'buffer' the effects of mutations means that it is possible to generate lines with very high mutation frequencies, up to 50 000 mutations per plant. This allows mutations in all wheat genes to be captured in a small population of plants.

Although physical mutagens are still used in wheat genetics (Shaw et al. 2013), most modern studies use chemical mutagens, which do not require specialist facilities and have milder effects. They are also less hazardous to use, although this does not mean that they are safe: they are just as likely to induce mutations in scientists as they are in crops! The most widely used chemical mutagens in cereals are ethyl methane sulphonate (EMS), which is an alkylating agent, and sodium azide. Both react with the bases in DNA leading mainly to point mutations. Most of the mutants produced will be silent, being present outside expressed genes or affecting the third bases of codons (which often accept more than one base). Most induced mutations are recessive and hence only affect the phenotype when homozygous.

Because bread wheat is hexaploid, mutations on one, or even two, of the genomes may have little or no effect on the phenotype, with mutations on all three genomes being required to have an effect. This is illustrated by waxy wheat, which results from mutations in granule bound starch synthase (GBSS), the single enzyme that catalyses amylose synthesis. Partial waxy types result from mutations on one or two genomes but the full waxy (zero amylose) phenotype is only obtained by combining three mutations (Graybosch 1998) (Chapter 7). This requirement to identify and combine mutations on two or three genomes accounts for the fact that mutagenesis has had little impact on wheat breeding compared to other crops (reviewed by Oladosu et al. 2016).

The exploitation of mutations for gene discovery and crop improvement was revolutionized in 2000 by the development of TILLING (targeting induced local lesions in genomes) (McCallum et al. 2000). TILLING is based on high throughput screening of DNA samples from thousands of individuals for SNPs, which can be naturally occurring or induced (i.e. mutations). As initially established the screening was carried out on DNA from pools of individuals using PCR analysis. One of the first applications in crops was to develop wheat with waxy starch, which was achieved by screening a library of EMS-treated plants to identify mutations in the GBSS genes on the A and D genomes. These mutations were then combined with a naturally occurring mutation on the B genome to give a full waxy phenotype (Slade et al. 2005).

The development of rapid and low-cost genome sequences allowed Krasileva et al. (2017) to establish a novel genetic resource for wheat genetics and breeding. They established libraries of EMS mutants, comprising 1535 lines of the durum wheat cultivar Kronos and 1200 lines of the bread wheat cultivar Cadenza and used modern sequencing technologies to identify more than 10 million mutations in protein-coding regions. The mutation frequencies corresponded to averages of 2705 mutations per lines of Kronos and 5351 per line of Cadenza. The mutant lines are maintained as public collections and the sequence data are readily accessible to all researchers. Hence, researchers can request lines containing mutations in their genes of interest for either scientific study or crop improvement. However, the exploitation is still not trivial, as the effects of the mutations on the expression of the target genes, or on the functions of the gene products, need to be established, the background mutations removed by crossing, and mutations in the three genomes combined and selected together in breeding programmes. Nevertheless, this resource is revolutionising the exploitation of mutations in wheat genetics and breeding.

10.4 Impact of Breeding on Genetic Diversity in Wheat

As discussed above, modern wheat breeding has been immensely successful in increasing the yield and total production of wheat, particularly in the high yielding intensive production systems less limited by water and excessive temperatures.

However, detailed comparisons of genomic sequences show that both domestication of bread wheat and subsequent genetic improvement have resulted in decreased genetic diversity (Pont et al. 2019). Pont et al. (2019) also showed that the reduction in diversity occurred in two waves, during the development of early cultivars from landraces, and then during the Green Revolution (Chapter 6), and that the reduction was greater in the A and B genomes than in the D genome.

There is widespread concern that this erosion of genetic diversity in wheat will limit the ability of breeders to sustain the increases in wheat yield that are required to feed the growing global population. It has therefore led to international programmes to conserve genetic diversity and increased exploitation of early cultivars (often referred to as Heritage cultivars), landraces, ancient wheats, and related grass species (as discussed).

A further concern is that intensive selection for high yield (i.e. starch accumulation) and processing quality (determined by gluten proteins) has led to reduced contents of grain components that contribute to diet and health, including non-gluten proteins, minerals, vitamins, and beneficial phytochemicals (Morris and Sands 2006). Furthermore, there is concern that increases in the content of gluten and changes in the gluten protein composition may have contributed to increases in adverse reactions to the consumption of wheat-based foods.

The evidence for such effects is therefore discussed in the next sections.

10.4.1 Minerals

Iron and zinc are essential mineral micronutrients which are deficient in many populations and wheat is an important dietary source of both (Chapter 9). There is also broad agreement that the contents of both minerals are lower in modern wheats than in older cultivars.

The evidence comes from two types of study: comparative analyses of new and old cultivars grown together in replicated field trials and analysis of archived samples from historic field trials. These are illustrated by two studies carried out at Rothamsted Research.

Firstly, analyses of European lines released between 1838 and 2012 show clear decreases in the contents of both iron and zinc (Figure 10.2) (Shewry et al. 2016). This decline is particularly marked after about 1960, which agrees with analyses of wheat samples archived from the Broadbalk long-term wheat experiment at Rothamsted Research. Analyses of 362 grain samples from 16 cultivars grown between 1843 and 2000 showed decreased contents of several minerals (zinc, iron, copper, magnesium) despite the mineral concentrations in the soil having increased or remained constant (Fan et al. 2008).

The greatest decreases in iron and zinc have occurred since the introduction of semi-dwarf cultivars as part of the green revolution in the 1967s. The grain accounts for a higher proportion of the total biomass in these wheats, resulting in higher yields. Hence, the decreased concentrations of minerals may be partially due to yield dilution (i.e. to increased starch accumulation). However, decreases in mineral concentrations are also observed under growth conditions in which the yield is not increased (Fan et al. 2008), suggesting that the dwarfing genes used to reduce plant height may have other effects.

10.4.2 Protein Content

When compared at the same level of nitrogen availability, it is clear that modern, high yielding wheats have lower grain protein concentrations than their predecessors. The protein content has

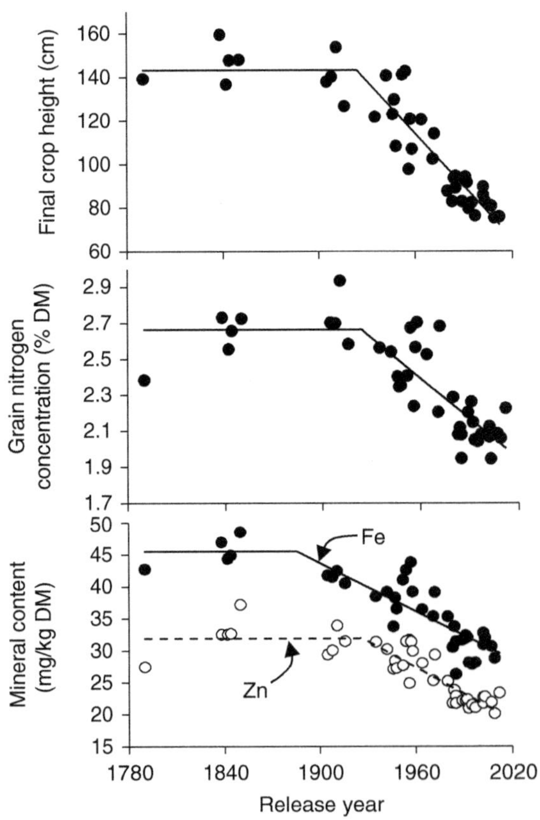

Figure 10.2 Crop heights and concentrations of nitrogen, iron (Fe), and zinc (Zn) in grains of wheat cultivars grown commercially in the UK between 1790 and 2012, based on analyses of replicate field trials grown at Rothamsted in 2014 and 2015. *Source:* Redrawn from Shewry et al. (2016) with permission.

clearly declined in wheats grown in the UK (Smith and Gooding 1999; Shewry et al. 2016) (Figure 10.2) and analysis of historical data suggests that the protein content of wheats grown in the Northern Plains of the US has also decreased (Kasarda 2013). Similarly, Pronin et al. (2020) reported a decrease in crude protein content of wheats grown widely in Germany between 1891 and 2020. This decrease in crude protein can probably be ascribed to yield dilution, as any increase in starch deposition will inevitably dilute the grain protein.

One consequence of the lower protein content is that modern high yielding wheats may not meet the protein content required for breadmaking when grown at levels of nitrogen fertiliser based purely on the yield response. Hence, many farmers in the UK may apply an additional 50 kg N/ha or more to increase the protein content of breadmaking wheat.

10.4.3 Wheat Proteins that Trigger Adverse Reactions

A major concern is that changes in the amount or composition of gluten proteins may lead to increases in the ability of wheat-based foods to trigger coeliac disease, which currently occurs in about 1% of the population (Chapter 9). Coeliac disease is a well-understood autoimmune response that is triggered in susceptible individuals by short peptide sequences present in gluten proteins. Almost 40 such epitopes have been identified (Sollid et al. 2020), which differ in their distribution between individual gluten proteins. In particular, coeliac epitopes are generally more abundant in gliadins than in glutenin subunits, with α-gliadins being particularly coeliac-toxic. Hence, changes in gluten protein composition could result in higher or lower coeliac toxicity.

Our current knowledge indicates that modern wheats are not more likely to be coeliac-toxic due to changes in gluten protein composition. Selection for breadmaking quality generally results in more elastic dough, and dough elasticity is strongly related to the content of high molecular weight glutenin polymers rather than the gliadins. This is demonstrated by a comparison of Austrian wheats dating from between 1790 and 2012, which showed a decreased ratio of gliadins to glutenins, but no change in the amounts of α-gliadins (Call et al. 2020). Similarly, Pronin et al. (2021) reported an increase in the proportion of glutenins within the gluten fractions of 60 German wheats grown from the nineteenth to twenty-first centuries, but no changes were reported in the immunoreactive potential of the same samples based on quantitative analysis of four coeliac-active peptides. Comparison of gluten proteins from 53 modern cultivars and 19 landraces with a monoclonal antibody that recognises a number of widely distributed coeliac-toxic sequences also showed no increases in coeliac epitopes in the modern cultivars (Ribeiro et al. 2016)

10.4.4 Dietary Fibre

Dietary fibre accounts between 11.5 and 15.5% of the grain dry weight, and wheat is a major source of fibre in the human diet. Most of the fibre is present in the outer layers of the grain and there is no evidence that the content has decreased with breeding (despite that fact that increased seed size would be expected to result in dilution) (Shewry et al. 2011). However, the contents of the two major fibre components in white flour, arabinoxylan and β-glucan, have increased in wheat cultivars grown in the UK over the last 200 years (Figure 10.3). These differences are sufficient to have impacts on health, as white bread provides about 10% of the total daily intake of dietary fibre in the UK.

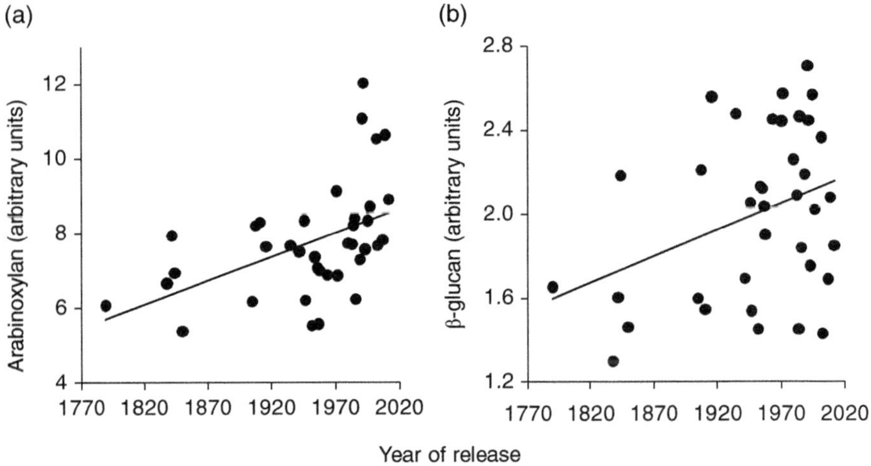

Figure 10.3 The concentrations of (a) arabinoxylan and (b) β-glucan in white flour of wheat cultivars gown commercially in the UK between 1790 and 2012. Data are expressed in units determined by HPLC analysis of oligosaccharides released by enzyme digestion. Data are means of samples from three replicate plots grown for each of three years. The lines shown are statistically significant ($p = 0.002$ and 0.032 for a and b, respectively) explaining 21% (a) and 10% (b) of the variation. *Source:* Redrawn from Lovegrove et al. (2020) with permission.

By contrast, a comparison of eight modern and seven older cultivars of Italian durum wheat showed no differences in the content of arabinoxylan in semolina or wholemeal flour, but significantly higher contents of β-glucan in both fractions from the modern cultivars (De Santis et al. 2018).

10.4.5 Other Components

The effects of intensive breeding on other components of the grain are less clear. Analyses of 146 bread wheat cultivars released between 1911 and 2004 and grown on a single site showed no statistically significant relationship between the cultivar release date and the contents of a range of phytochemicals (sterols, stanols, tocols, alkylresorcinols, and phenolic acids) in wholegrain. However, comparisons of smaller numbers of older and modern Italian cultivars have shown some differences in either the amount or profile of phenolic components. These are reviewed by Spisni et al. (2018), who note the need for further studies, and that the differences may not be relevant to human health in the context of the wider Mediterranean diet.

10.5 Are Ancient Wheats More Healthy than Modern Wheats?

An increasing number of studies of the compositions and 'health benefits' of ancient wheats have been published (reviewed by Shewry and Hey 2015; Shewry 2018). However, the results are sometimes difficult to interpret due to the designs of the studies.

It is crucial that comparisons between wheat species and types are carried out on comparable samples: this means that the grain should be grown together in randomised replicated field trials with the same agronomic treatments. The conditions for harvesting, storage, milling, and processing also need to be the same. However, this is rarely the case. This is partly due to practical considerations: ancient wheats are often tall in stature and not suited to the high nitrogen applications used in modern bread wheat production. Hence, they are usually grown in low input or organic systems (with organic systems being required for Kamut). Similarly, they are often processed using traditional stone milling and long fermentation or soughdough breadmaking systems. Consequently, comparisons with modern wheats and with wheat products from grain grown and processed using modern systems may not be valid. Taking these factors into account, it is not possible to conclude that ancient wheats have greater health benefits than modern wheat species (Shewry 2018).

10.5.1 Wheat Proteins that Trigger Adverse Reactions

As with 'Heritage' cultivars, much of the current interest in ancient wheats relates to the possibility that they may less likely to trigger adverse reactions (allergy, coeliac disease, and non-coeliac wheat sensitivity (NCWS) (Chapter 9).

Although spelt and bread wheat are separate variety groups of the same species, a recent proteomic comparison of 15 genotypes of each type showed differences in protein composition, with 84 and 193 proteins being unique for spelt and bread wheat, respectively (out of a total of 30050 identified [Afzal et al. 2020]). However, the implications of these differences for human health are not clear. The authors of the study addressed this by calculating an allergen index, based on proteins reported to trigger allergy (serpin protease inhibitors, ATIs, and wheat germ agglutinin). This showed wide variation between the individual genotypes with no significant differences between the two types of wheat. Similarly, several studies have shown that bread wheat and spelt do not differ significantly in the abundances of coeliac-stimulatory epitopes (Ribeiro et al. 2016; Prandi et al. 2017; Escarnot et al. 2018).

Several studies have compared the amounts and compositions of amylase/trypsin inhibitors (ATIs), which have been implicated in coeliac disease and NCWS (Chapter 9), between wheat species (Rogniaux et al. 2015; Geisslitz et al. 2018; Call et al. 2020). These show that diploid einkorn contains either very low concentrations or no ATIs, while the amounts in emmer, spelt, and bread wheat are essentially similar. Differences in the proportions of ATI isoforms have also been reported, but these are not consistent between studies and their relevance to health impacts has not been established.

10.5.2 Other Components

Comparisons of the compositions of ancient and modern wheats show only limited differences in grain composition (Shewry and Hey 2015; Shewry et al. 2017). However, two grain components are worth noting.

Firstly, although ancient wheats, particularly spelt, are often assumed to have a higher content of dietary fibre than bread wheat, this is not true. In fact, a number of comparative analyses of bread wheat and spelt have shown generally lower contents of total dietary fibre and individual fibre components in spelt (Shewry and Hey 2015).

Secondly, wheat species vary widely in their contents of carotenoids, and in particular of the major carotenoid lutein. Selection for white flour has led to low contents of carotenoids in bread wheat while selection for yellow colour of pasta has resulted in higher contents in durum wheat (by about twofold). The concentrations of carotenoids are also significantly higher in ancient wheats (including Kamut/Khorasan wheat) than in bread wheat, particularly in diploid einkorn, which has about five times the concentration of lutein compared with bread wheat (Shewry and Hey 2015). Although lutein is not converted into Vitamin A (retinol), the higher contents could still contribute to improved health in some populations.

It is not clear whether modern and ancient wheats differ in other components, due to a lack of detailed studies and the strong confounding effects of the environment on the composition of the grain and of processing on the compositions of products. For example, although Kamut/Khorasan wheat has been reported to contain higher levels of selenium, this may result from cultivation in organic systems, particularly in North America where the higher contents of selenium in soils result in higher concentrations in the wheat that is grown. The literature is reviewed in detail by Shewry and Hey (2015), Shewry (2018), Dinu et al. (2018) and Spisni et al. (2018).

The current conclusion, therefore, is that ancient wheats do not differ significantly from modern wheats in their effects on human health. Nevertheless, there is no doubt that baked products made from ancient wheat have high consumer acceptability and that demand is increasing. This may be due to specific flavour components. However, it may also result from the traditional processing methods, and the perception that they are more 'natural and green' than modern wheats.

10.6 Wheat Biotechnology

Wheat genetics and improvement have been facilitated in recent years by the developments of new genomic resources and technologies, with a major advance being the publication of the wheat genome assembly (RefSeq) from the reference genotype Chinese Spring (IWGSC 2018). This assembly listed and ordered the genes present on the three genomes providing a basis for gene identification and characterisation. This was followed in 2020 by the report of reference-level genome sequences of 10 wheat genotypes, generating a 'pangenome' resource to facilitate the analysis of genetic diversity (Walkowiak et al. 2020). However, a vastly greater number of genotypes are currently being sequenced, including landraces (the Watkins collection) as well as commercial cultivars. Publicly available sequences can be accessed in online databases, notably GrainGenes (https://wheat.pw.usda.gov/GG3/node/825), which is supported by the U.S. Department of Agriculture-Agricultural Research Service in Berkeley (California, USA).

10.6.1 Genetic Transformation

Although wheat was the last of the major cereal crops to be transformed, following rice and maize, the technology still dates back almost 30 years. Genetic transformation requires the development of technologies for DNA delivery and for the selection and regeneration of transgenic plants. It is also necessary to understand the control of gene expression in order to ensure the required pattern and level of expression of the transgene.

10.6.1.1 DNA Delivery

The development of methods to transform cereals started in the early 1980s, using a range of methods to transfer DNA into wheat cells or tissues which were subsequently regenerated into transformed plants (reviewed by Barcelo et al. 2001). Most of the approaches failed to give reliable transformation at acceptable efficiency but one became widely used.

This system is 'particle bombardment' or 'biolistics', in which DNA is coated onto the surface of metal particles (usually gold) and physically shot into the cells of a tissue that can be regenerated into a plant. Early systems used a shotgun cartridge to propel the particles, with a firearms licence being required, but by the early 1990s a commercial system had been developed using a burst of helium gas. The most appropriate tissue for regeneration is the scutellum of the immature embryo, which is hand-dissected from developing grain. Hence, the system is labour-intensive and the efficiency low (generally less than 10% of embryos bombarded giving rise to transformed plants). Nevertheless, particle bombardment was the method of choice for cereal transformation in academic and commercial laboratories in the 1990s and was used to produce the first transgenic wheat plants (Vasil et al. 1992).

Particle bombardment results in random insertions of foreign DNA into the genome, with multiple copies generally being inserted (which can be linked in concatenates in the same insert or as separate insets). Hence, the transgenic lines are difficult to characterise at the genomic level and not suitable for commercial development (which requires detailed characterisation of insertion sites). Nevertheless, it has been widely used for academic studies.

However, the need for higher transformation efficiencies and cleaner insertion patterns led to the development of delivery systems based on *Agrobacterium tumefaciens,* a soil bacterium that naturally infects many plant species. Infection of plants by *Agrobacterium* results in the transfer of a small piece of DNA (copied from the tumour-inducing [Ti] plasmid) into the genome of the host plant and this DNA can be engineered to transfer foreign DNA. *Agrobacterium*-mediated transformation rapidly became the standard system for transformation of dicotyledonous plants. However, many monocotyledonous plants, including wheat, are outside of the normal host range for *Agrobacterium*. Nevertheless, systems for the application of *Agrobacterium*-mediated transformation to cereals were developed in the 1990s, with the first transformation of wheat being reported by the Monsanto Company (Cheng et al. 1997). However, the most widely used system was developed by workers at the Japan Tobacco company and gives efficiencies of over 50% in some genotypes such as the American spring wheat Fielder (Ishida et al. 2015). This system, called PureWheat, is widely used under licence.

10.6.1.2 Selection of Transformed Plants

The production of transformed plants depends on the ability to regenerate whole plants from the transformed tissues. Most wheat tissues, such as leaves, cannot be regenerated into whole plants but two tissues can be regenerated via somatic embryogenesis to give whole plants: these are the scutella of immature embryos and immature inflorescences. However, only a small number of cells in the tissues are transformed and regeneration would therefore result in a chimaera with a mixture of transformed and non-transformed cells. Hence a selectable marker is used to allow the non-transformed cells to be killed and only the transformed cells to be regenerated. The most commonly used selectable marker genes are the *bar/pat* and *epsps* genes, which confer the ability to survive on media containing phosphinothricin and glyphosate (Round Up) herbicides, respectively, or the *npt*II gene, which confers resistance to a range of aminoglycoside antibiotics. Regeneration of the transformed tissues in the presence of these agents therefore results in the selection and regeneration of the transformed cells only.

The selectable marker genes can be linked to the gene of interest for transformation, leading to co-insertion, or present on a separate DNA fragment (for bombardment) or Ti plasmid leading to insertion at separate sites. The second option is preferred as it allows the selectable maker to be eliminated by selection after regeneration.

Transgenes are normally dominant in expression and present on only one chromosome (heterozygous). Hence, they show normal Mendelian segregation when the transgenic plants are self-fertilised. However, they may be inserted at multiple sites, which may also (in the case of bombardment) contain inserts of the selectable marker. Hence, the first priority is to obtain stable homozygous lines, which is usually achieved by several rounds of self-fertilisation and the selection of transgenic progeny. Doubled haploid technology (see Section 10.2.2) can also be used to produce homozygotes in a single step. During these processes the presence and numbers of transgenes are monitored by Southern (DNA) blotting or PCR.

In academic studies it is usual to select control plants from the 'null segregants' (i.e. the progeny which lack the transgenes), as these have originated from the same transformation and regeneration procedures. This is because transformation and regeneration can result in somaclonal variation in the progeny, which may occur as a response to stress during the procedures.

10.6.1.3 Targeting and Regulating Gene Expression

Expression of genes throughout the plant can be achieved using constitutive promoters, but it is more usual to direct expression to tissues and stages of development using specific promoters. Many transgenic studies of wheat have focused on manipulating the composition of the starchy endosperm to affect aspects of grain quality, with the most widely used promoter being the starchy endosperm-specific promoter of the HMW subunit 1Dx5 gene (Lamacchia et al. 2001).

Transgenesis can also be used to both up- and downregulate the expression of endogenous genes. Upregulation is often achieved by using the endogenous gene promoter, although a different promoter (such as the 1Dx5 promoter above) can be used. Downregulation is usually achieved by RNAi technology, in which the transgene constructs are designed to encode hairpin-shaped RNA molecules that generate small interfering RNAs (RNAi) (reviewed by Agrawal et al. 2003). This technology has the potential to downregulate the levels of harmful components such as allergens.

10.6.1.4 Gene Editing

Gene editing allows the targeted manipulation of specific sequences in the plant genome without the introduction of foreign DNA. Several systems have been developed but one of these, Clustered Regularly Interspaced Palindromic Repeats (CRISPR/Cas9) has become widely adopted (Gao 2015), including for wheat (Zhang et al. 2016; Shewry and Jones 2020;

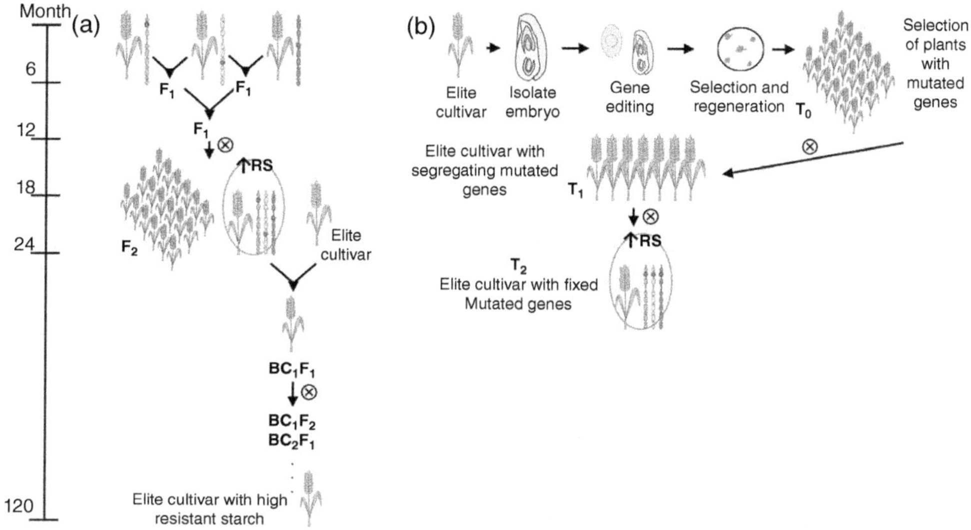

Figure 10.4 Predicted timeline for the development of an elite cultivar with increased resistant starch using TILLING versus gene editing. (a) Crossing scheme to combine TILLING mutations in wheat homoeologues that confer elevated levels of resistant starch followed by multiple rounds of backcrossing and selection to an elite cultivar. Panel b: Introduction of mutations in wheat homoeologues that confer elevated levels of resistant starch directly into an elite cultivar. F = filial generation, T = transgenic generation, BC = backcross. *Source:* Taken from Hazard et al. (2020) with permission.

Li et al. 2021). This provides a clear time advantage for developing improved cultivars of wheat compared to exploiting mutations by TILLING. This is illustrated for developing resistant starch in wheat in Figure 10.4.

10.6.2 Regulation, Impact and Consumer Acceptance of Genetic Transformation and Genome Editing in Wheat and Other Crops

The debate about the risks and benefits of transgenic crops has continued since their first introduction in the early 1990s. It is clearly a highly complex topic including geopolitical (globalisation), scientific (food safety and environmental impact), and ethical (is it right) issues, with consumer acceptance being a major consideration. We will therefore only consider it briefly here.

The growth of transgenic (widely referred to as genetically manipulated, GM) crops is regulated in all countries and has been approved for crops such as maize, soybean, and cotton in many countries. Consequently, GM types of these crops are widely grown. By contrast, the commercial uptake of transgenic wheat has been slow, with the first approval being granted only in the autumn of 2020. This was for growth in Argentina of drought-resistant wheat developed using the HB4 technology (discussed in Section 10.7.1.3) by Bioceres Crop Solutions Corp.

The failure to develop commercial GM wheat, even in countries where other GM crops are widely grown and accepted, may be due partly to economic factors: GM crops are expensive to develop, with high costs for regulation and approval. Consequently, the technology has been used mainly for hybrid crops, in which the profits are more readily realised due to the need to purchase basic seed every year; although the production of hybrid wheat is slowly increasing, it is still a very small part of the total production. A further barrier to the marketing of GM wheat may be resistance from consumers due to its central position in culture and religion.

It has been suggested that gene editing can be used to 'side-step' the tight regulation of GM crops that exists in many countries, including countries in the EU that have a common regulatory system. This is because it does not necessarily result in the introduction of foreign DNA into the crop genome. However, the current position in the EU is that edited crops will be subject to the same regulation as GM crops. This implies that the technology used for gene editing is considered, as well as the composition of the product, in defining the level of regulation that is required.

Although it is not possible to assess the impact of GM wheat we are able to draw broad conclusions based on almost 30 years of research on the impact of other transgenic crops. These show a number of clear benefits, in terms of improved yield, reduced cost of production, and reduced environmental impact. For example, Klumper and Qaim (2014) carried out a detailed meta-analysis of data from 147 studies. This showed that the use of chemical pesticides was reduced by 37%, crop yields increased by 22%, and farmer profits increased by 68%. The two major GM traits studied were insect resistance and herbicide resistance and the yield gains and pesticide reductions were greater for insect resistance than for herbicide resistance (by 24.9% compared with 9.3%). Similarly, Finger et al. (2011) concluded that the adoption of GM crops led to economic benefits to farmers, but that the magnitude of the benefits varied across crops, traits, countries, and regions, noting that the adoption of insect-resistant cotton had been particularly beneficial in developing countries.

Despite early concerns about the safety of GM technology, no cases of adverse effects on human health have been substantiated and it is now generally accepted that the GM crops accepted by regulatory authorities are safe for human consumption. In fact, there is evidence of benefits to farm workers and consumers in reducing the levels of pesticide use, and hence pesticide poisoning, and of toxins present in grain (Smyth 2020). In particular, a meta-analysis of field trials of insect-resistant maize showed that the contents of mycotoxins (see Sections 9.6.2) were reduced by about 30%, probably due to reduced fungal colonisation of damaged grains (Pellegrino et al. 2018).

GM also provides opportunities to develop biofortified crops to reduce the incidence of deficiencies in micronutrients and vitamins (Malik and Maqbool 2020) but it is too early to assess the impact of these on the health of populations.

10.7 Applications of Biotechnology to Wheat Improvement

In broad terms, crop traits can be divided into two classes: input and output. Input traits relate to the efficiency and cost of production, and include resistance to pests, pathogens, and herbicides. Output traits include aspects of crop composition and quality and include clear benefits to consumers as well as to producers and processors.

10.7.1 Input Traits

Increasing the production of wheat is crucial for global food security. Furthermore, this needs to be achieved against a background of increasing urbanisation (resulting in decreases in available land), environmental degradation, and climate change. It is therefore a priority for both national and international funding agencies. Biotechnological approaches are attractive because they offer the possibility to make *quantum leaps* in traits, rather than the small incremental improvements made by conventional plant breeding. However, few studies have been carried out under field conditions and only one technology (the HB4 technology for drought resistance) has been exploited in commercial cultivars. They are therefore only briefly summarised in the following sections.

10.7.1.1 Potential Yield

Grain yields are limited globally by abiotic (drought, heat) and biotic (pest and pathogen) stresses, and work on these is discussed here. However, it is also important to increase the intrinsic yield (i.e. the yield when grown under ideal conditions). There is a long-running debate on whether grain yield is limited by the source (i.e. the amount of carbon fixed by the canopy and transported to the developing grain) or the sink capacity (the demand from the developing grain) and both of these processes have been explored using transgenesis. Work on the former has focused on increasing the fixation of carbon by photosynthesis, by improving the canopy structure, the capture and conversion of light energy, and in particular the efficiency of photosynthesis (which is an inefficient process due to the loss of fixed carbon by photorespiration) (Ort et al. 2015). Sink capacity can potentially be increased by increasing grain size or grain number, although there is often a trade-off between these two traits in that higher seed weight is associated with lower seed number (see, for example, Griffiths et al. 2015). Furthermore, it is likely that the maximum seed size is established early in the development of the caryopsis, before the start of grain filling.

Calderini et al. (2020) therefore targeted cell expansion during early development, by expressing a gene encoding α-expansin in the developing grain using the promoter from the puroindoline b gene (conferring expression in the starchy endosperm, aleurone layer and pericarp). Expansins are proteins located in the plant cell wall where they appear to facilitate expansion by loosening non-covalent bonds between cellulose and other cell wall polysaccharides. The best-performing transgenic line was reported to have increases of 12.3% in grain weight and 11.3% in yield when compared to the control line in field experiments.

These results therefore indicate that it is possible to uncouple effects on mean grain weight and grain number.

10.7.1.2 Improving Nitrogen-Use Efficiency (NUE)

Improving nitrogen-use efficiency (NUE) is a major target for improvement, due to the financial and sustainability issues discussed in Chapter 6. Logical targets for manipulation are genes encoding transporters responsible for nitrogen uptake and transport, and enzymes involved in nitrogen assimilation and metabolism (McAllister et al. 2012). A number of these have been manipulated in transgenic plants but only one so far has given consistently promising results. This is the *AlaAT* gene encoding alanine amino acid transferase, which catalyses the reversible conversion of glutamate and pyruvate to alanine and α-ketoglutarate (Good et al. 2007). Over-expression of *AlaAT* has been reported to improve NUE when expressed in rice, barley, and wheat (Tiong et al. 2021). It has been evaluated in large-scale field experiments of rice (Tiong et al. 2021) but not yet developed into commercial cultivars.

10.7.1.3 Resistance to Abiotic Stresses

Several studies have focused on using transgenesis to improve resistance to abiotic stress, notably the efficiency of water use and tolerance to drought (Khan et al. 2019), but also to increase resistances to heat stress and salinity. However, only one of these has so far been translated into a commercial application. This is the HB4 technology that was used to develop the first commercially released GM wheat. HB4 is a modified version of the homeodomain-leucine zipper (HD-zip) transcription factor (TF) *HaHB4* from sunflower (*Helianthus annuus*) and results in reduced sensitivity to the plant hormone ethylene and delayed senescence (Gonzalez et al. 2019; Martignago et al. 2020). The technology is stated to result in 20% increases in wheat yields when grown in seasons affected by drought.

10.7.1.4 Resistance to Pests and Pathogens

A range of studies have reported on using transgenesis to confer resistances to pests and pathogens, including insect pests of stored grain. A number of plant proteins have been shown to inhibit the growth of pests and pathogens under laboratory conditions (Shewry and Lucas 1997; Selitrennikoff 2001). Several of these proteins have conferred resistance when expressed in transgenic plants. Similarly, resistance has been demonstrated under controlled conditions for plants engineered using other approaches, such as using host-induced gene silencing (HIST), in which the crop is engineered to express RNAi to silence genes in pests and pathogens (Borisjuk et al. 2019). However, despite the importance of pests and pathogens in limiting wheat yields and major investment in research on the topic, biotechnology has failed to deliver practical improvements in resistances in commercial wheat production.

This is perhaps not surprising as resistances are complex traits with strong genotype × environment interactions. Consequently, manipulating single genes will only have limited effects and achieving stable long-term improvements is still some way off.

10.7.2 Output Traits: Grain Quality

10.7.2.1 Dough Strength

Grain quality is an attractive target for the application of GM technology because of the established relationships between grain proteins (particularly gluten proteins) and quality traits. It is therefore not surprising that the first application of GM to wheat was to increase dough strength by inserting additional genes for HMW subunits of glutenin. The relationship between these proteins and dough strength is well-established (Chapter 9) and several genes are available. The first study inserted a hybrid gene, which encoded a protein comprising parts of subunits 1Dx5 and 1Dy10. This strategy was adopted to facilitate the identification of the protein among the endogenous HMW subunits (Blechl and Anderson 1996). This was followed by a number of studies in which genes encoding single subunits, notably the quality-related subunits 1Ax1, 1Dx5 and 1Dy10, and combinations of subunits were inserted (reviewed by Shewry and Jones 2020). A number of field trials were also conducted on the material, allowing the substantial equivalence of the GM and non-GM lines to be addressed (Shewry et al. 2007).

Dough strength can be determined as the increase in resistance which occurs as the gluten network is formed during dough mixing. Several types of dough mixer have been designed to measure this (see also Chapter 1), with the Mixograph being most widely used in studies of transgenic wheats. The Mixograph measures energy input during mixing and allows the calculation of a number of parameters, the most important of which are peak resistance (in arbitrary units), mixing time (in seconds), and resistance breakdown (i.e. stability to overmixing, also measured in arbitrary units). Strong doughs generally have higher peak resistances and longer mixing times than weak doughs. The application of the Mixograph to transgenic wheat is illustrated in Figure 10.5, which shows the effects of the expression of a single transgene (encoding subunit 1Ax1) in a control line expressing only two HMW subunits (1Bx17 + 1By18).

These transgenic studies have broadly confirmed our knowledge of the relationships between HMW subunit amount and composition in dough strength, in that the expression of additional HMW subunits at levels similar to those of the endogenous proteins results in

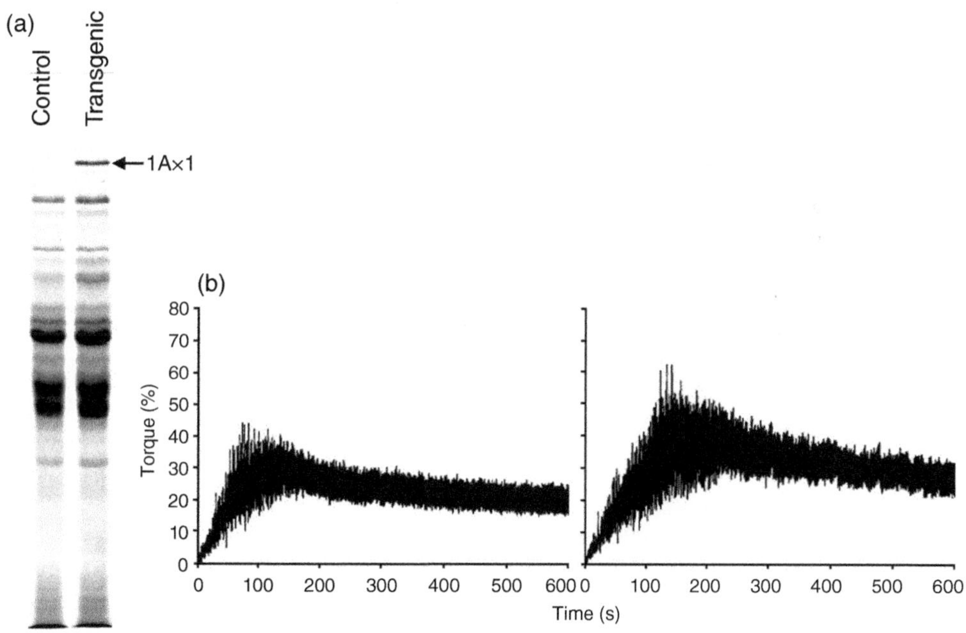

Figure 10.5 Expression of a transgene encoding HMW subunit 1Ax1 results in increased dough strength. Analyses of (a) protein compositions by sodium dodecylsulphate polyacrylamide gel electrophoresis (SDS-PAGE) and (b) dough mixing curves determined by Mixograph from a control line (L88-31) expressing two HMW subunits and a transgenic line (B102-1-2) expressing a transgenes encoding an additional HMW subunit (1Ax1 subunit). Torque indicates dough resistance. *Source:* Mixographs are taken from Popineau et al. (2001) with permission.

increased dough strength (Figure 10.5). However, when higher levels of protein are expressed, the dough becomes overstrong and fails to absorb water and mix correctly, resulting in a lower loaf volume.

Although these studies provide a good demonstration of the use of transgenesis to explore the relationship between grain composition and processing properties they have not translated into practical applications. This is because the naturally occurring variation in dough strength is sufficient to satisfy the requirements of most breeders and processors.

Several studies have also been reported the transgenic manipulation of other gluten proteins, with expression of a LMW subunit of glutenin in durum wheat (Tosi et al. 2005) and of avenin-like b proteins (farinins) in bread wheat (Ma et al. 2013), resulting in improved mixing properties.

10.7.2.2 Grain Texture

The dominant effects of puroindolines (Pins) on determining grain texture means that the trait can readily be modified by transgenesis. However, as with dough strength, this is of little practical interest to breeders because texture can readily be modified by conventional breeding. Nevertheless, Pins have been expressed in both durum and bread wheats (Wiley et al. 2007). Effects on dough properties were demonstrated by Ma et al. (2013), who showed that whereas the expression of Pin a alone in durum wheat resulted in detrimental effects on dough extensibility, co-expression of Pin a with HMW subunit 1Ax1 resulted in improvements with dough strength and dough resistance compared to lines expressing the single transgenes.

10.7.2.3 Increasing Mineral Micronutrients

Iron and zinc are depleted in the starchy endosperm but concentrated in the aleurone cells and scutellum of the wheat grain, where they are largely bound to phytates (Section 9.3.1). Consequently, the concentrations of iron and zinc in white (starchy endosperm) flour are lower than those required for human health, while the higher concentrations in the bran have low bioavailability. Furthermore, the higher contents of minerals achieved by genetic biofortification are largely present in the same tissues and forms and hence are likely to have similar bioavailability (Section 9.3.1). Transgenesis has therefore been used to divert higher levels of minerals into the starchy endosperm, to increase the concentrations of bioavailable minerals in products made from white flour, which are most widely consumed globally (Balk et al. 2018). The concept is to entrap the minerals as they are transported across the starchy endosperm cells from the crease to the aleurone layer and embryo and two complementary strategies have been used to achieve this.

The first strategy to be adopted was to bind minerals in the starchy endosperm cells by overexpression of a gene encoding the enzyme nicotianamine synthase (NAS). This is because nicotianamine plays key roles in the chelation and transport of metals, including Fe and Zn, in higher plants (von Wiren et al. 1999) and is also the precursor for the biosynthesis of 20-deoxymugenic acid (DMA), a phytosiderophore which also chelates Fe and Zn. There is also evidence that these two compounds, NA and DMA, are the main chelators of Fe in white flour of wheat (Eagling et al. 2014). This strategy was initially validated in rice, where expression of NAS led to 2.2-fold and 2.9-fold increases in the concentrations of zinc and iron, respectively, with both minerals showing high bioavailability in mice

(Lee et al. 2009). Similar success has since been achieved by overexpressing NAS in the wheat grain (Zheng et al. 2010; Johnson et al. 2011; Singh et al. 2017; Beasley et al. 2019).

The advantage of overexpressing NAS is that it leads to increases in iron, zinc, and possibly also other minerals that are bound by NA and DMA. By contrast, the second strategy, to concentrate minerals into the starchy endosperm cells by overexpressing metal transporters, is specific for single minerals. Connorton et al. (2017) showed that the expression of a wheat Vacuolar Iron Transporter (*TaVIT2*) in the starchy endosperm of wheat using the starchy endosperm-specific HMW subunit 1Dx5 promoter resulted in a twofold increase in the iron content of the white flour fraction, while Menguer et al. (2018) showed similar accumulation of zinc in barley by overexpression of the barley Metal Tolerance Protein 1 (*HvMTP1*). However, the minerals may not be uniformly distributed within the starchy endosperm. For example, Beasley et al. (2019) showed that whereas iron was distributed throughout the starchy endosperm, zinc was concentrated in the region surrounding the crease when the rice *OsNAS2* gene was expressed in wheat using a constitutive promoter. Similarly, Connorton et al. (2017) showed that iron was concentrated in the starchy endosperm surrounding the crease when the HMW subunit promoter was used to drive *TaVIT2*.

The coexpression of genes to concentrate and bind minerals is currently being explored and it is likely to lead to higher concentrations and more evenly distributed accumulation of minerals. However, the adoption of the biofortified lines will depend on the acceptability of the technology. Strategies to increase minerals in wheat are reviewed by Balk et al. (2018).

10.7.2.4 Reducing Adverse Effects

Most recent interest has focused not on adding gluten proteins but reducing their expression to alleviate adverse effects of wheat on susceptible individuals. The relationship between wheat gluten proteins and coeliac disease is well established and coeliac-toxic epitopes have been identified in all gluten protein types (see Section 10.5.1 and Chapter 9) (Sollid et al. 2020). However, α-gliadin genes are particularly coeliac-active and early studies were directed at reducing their amount using RNA-mediated downregulation. In these lines α-gliadins were reduced by over 60% compared to the control cultivar but compensatory increases in other proteins (albumins, globulins, other gliadins, and LMW subunits) occurred (Becker et al. 2012). Following this, Barro and coworkers (Gil-Humanes et al. 2010) generated two series of lines in which either γ-gliadins or all gliadins (α/β, γ, and ω) and LMW subunits were downregulated. Analyses of these wheat lines using cell lines derived from intestinal cells of patients with coeliac disease showed almost complete suppression of disease-related epitopes (Gil-Humanes et al. 2010). More recently, the same researchers used the CRISPR/Cas9 gene editing technology to target conserved regions adjacent to the coding sequence of the coeliac active 33-mer peptide in the α2-gliadin genes (Sanchez-Leon et al. 2018). Two single-guide RNAs were designed to induce mutations in specific gliadin genes, leading to an 85% reduction in immunoreactivity of mutant lines. This study therefore showed that CRISPR/Cas9 technology can be used to precisely and efficiently reduce the amount of coeliac epitopes. However, since coeliac epitopes have been identified in all types of gluten protein (Sollid et al. 2020) their complete elimination is a formidable challenge.

Although the prevalence of allergy to wheat consumption is low, wheat-dependent exercise-induced anaphylaxis (WDEIA) can lead to anaphylaxis and death (Chapter 9). The ω5-gliadins have been identified as the most important trigger for WDEIA and have therefore been silenced using RNAi technology (Altenbach et al. 2014).

Amylase/trypsin inhibitors (ATIs) are well-established allergens (Chapter 9) and have been implicated in triggering coeliac disease and non-coeliac wheat sensitivity. They are therefore currently one of the most active topics in wheat research. RNAi suppression has been used to produce transgenic lines of bread wheat silenced for three major forms of ATI (CM3, CM16 and 0.28), resulting in lower reactions with sera from patients with allergy to wheat (Kalunke et al. 2020). Gene editing using CRISPR-Cas9 has also been used to produce both small mutations and large deletions in CM3 and CM16 ATI genes in durum wheat (Camerlengo et al. 2020).

A further topic of interest is to the reduce the potential to form acrylamide in wheat-based foods during processing, by reducing the concentration of asparagine, the amino acid precursor. This has been achieved in potato by reducing the expression of asparagine synthase (reviewed by Raffan and Halford 2019) and this approach is also being used in wheat.

10.7.2.5 'Improving' Grain Polysaccharides

Reducing the rate and extent of starch digestion in the gastrointestinal tract can lead to reduced blood glucose levels and hence improved health impacts of wheat-based foods (as discussed in Chapter 9). The digestion of starch mainly occurs in the upper gastrointestinal tract and the rate at which this occurs in wheat-based foods is affected by a number of factors, including the structure and properties of starch. In particular, amylose is generally digested more slowly than amylopectin (reviewed by Hazard et al. 2020) and several strategies have been used to develop high amylose wheat starch. Although amylose content can be increased form 25–30% in normal plants to 40% by combining mutations (see Chapter 9), levels of up to 80% have been achieved by downregulation of genes encoding the branching enzymes responsible for amylopectin synthesis, using either RNAi-mediated suppression (Regina et al. 2006) or TILLING (Hazard et al. 2012; Botticella et al. 2018).

However, although high amylose starch may have health benefits there is little commercial interest in developing high amylose commercial cultivars. There are two reasons for this. Firstly, changes in starch composition are associated with decreases in yield, which would result in higher costs for processors and consumers. However, a more important concern is the impact on grain processing quality: starch accounts for about 80% of wheat flour and changes in its composition and physical properties therefore affect the processing properties and end-use quality (Chapter 9). These side effects are of particular concern for delivering health benefits through wheat improvement, as wheat is a high-volume commodity used mostly for producing low-cost staple foods. Hence, cost and processing quality are of paramount importance.

Arabinoxylan fibre only accounts for between 5.5 and 7.5% of whole wheat grain and has established health benefits (see Chapter 9). However, it also has a negative impact on the quality of grain used for livestock feed and for distilling, due to the high viscosity of aqueous extracts, which results mainly from the soluble arabinoxylan in the starchy endosperm (white flour) fraction. The amount of soluble arabinoxylan and the viscosity of aqueous extracts have been reduced to about half by combining TILLING mutants in the TaIRX9b genes, which encode a xylan synthase enzyme (that catalysis the synthesis of the backbone of the arabinoxylan polymer) (Pellny et al. 2020). This trait could be particularly valuable for distilling wheat providing it can be combined with acceptable grain yields.

References

Adamski, N.M., Bush, M.S., Simmonds, J. et al. (2013). The inhibitor of wax 1 locus (Iw1) prevents formation of β-and OH-β-diketones in wheat cuticular waxes and maps to a sub-cM interval on chromosome arm 2BS. *The Plant Journal* 74: 989–1002. https://doi.org/10.1111/tpj.12185.

Afzal, M., Pfannstiel, J., Zimmermann, J. et al. (2020). High resolution proteomics reveals differences in the proteome of spelt and bread wheat representing targets for research on wheat sensitivities. *Scientific Reports* 10: 14677. https://doi.org/10.1038/s41598-020-71712-5.

Agrawal, N., Dasaradhi, P.V.N., Mohmmed, A. et al. (2003). RNA interference: biology, mechanism, and applications. *Microbiology and Molecular Biology Reviews* 67: 657–685. https://doi.org/10.1128/MMBR.67.4.657-685.2003.

Allen, A.M., Barker, G.L.A., Berry, S.T. et al. (2011). Transcript-specific single-nucleotide polymorphism discovery and linkage analysis in hexaploidy bread wheat (*Triticum aestivum* L.). *Plant Biotech Journal* 9: 1086–1099. https://doi.org/10.1111/j.1467-7652.2011.00628.x.

Altenbach, S.B., Tanaka, C.K., and Seabourn, B.W. (2014). Silencing of omega-5 gliadins in transgenic wheat eliminates a major source of environmental variability and improves dough mixing properties of flour. *BMC Plant Biology* 14: 393. https://doi.org/10.1186/s12870-014-0393-1.

Atienza, S.G., Ballesteros, J., Martin, A. et al. (2007). Genetic variability of carotenoid concentration and degree of esterification among Tritordeum (x *Tritordeum* Ascherson et Graebner). *Journal of Agricultural and Food Chemistry* 55: 4244–4251. https://doi.org/10.1021/jf070342p.

Balk, J., Connorton, J.M., Wan, Y. et al. (2018). Improving wheat as a source of iron and zinc for global nutrition. *Nutrition Bulletin* 44: 53–59. https://doi.org/10.1111/nbu.12361.

Barcelo, P., Rasco-Gaunt, S., Thorpe, C. et al. (2001). Transformation and gene expression. In: *Advances in Botanical Research Incorporating Advances in Plant Pathology*, vol. 34 (eds. P.R. Shewry, P.A. Lazzeri and K.J. Edwards), 59–126. Academic Press.

Beasley, J.T., Bonneau, J.P., Sánchez-Palacios, J.T. et al. (2019). Metabolic engineering of bread wheat improves grain iron concentration and bioavailability. *Plant Biotechnology Journal* 17: 1514–1526. https://doi.org/10.1111/pbi.13074.

Becker, D., Wieser, H., Koehler, P. et al. (2012). Protein composition and techno-functional properties of transgenic wheat with reduced α-gliadin content obtained by RNA interference. *Journal of Applied Botany and Food Quality* 85: 23–33.

Bentley, A., Scutari, M., Gosman, N. et al. (2014). Applying association mapping and genomic selection to the dissection of key traits in elite European wheat. *Theoretical and Applied Genetics* 127: 2619–2633. https://doi.org/10.1007/s00122-014-2403-y.

Blechl, A.E. and Anderson, O.D. (1996). Expression of a novel high-molecular-weight glutenin subunit gene in transgenic wheat. *Nature Biotechnology* 14: 875–879. https://doi.org/10.1038/nbt0796-875.

Borisjuk, N., Kishenko, O., and Eliby, S. (2019). Genetic modification for wheat improvement: from transgenesis to genome editing. *BioMed Research International*: 6216304. https://doi.org/10.1155/2019/6216304.

Botticella, E., Sestili, F., Sparla, F. et al. (2018). Combining mutations at genes encoding key enzymes involved in starch synthesis affects the amylose content, carbohydrate allocation and hardness in the wheat grain. *Plant Biotechnology Journal* 16: 1723–1734. https://doi.org/10.1111/pbi.12908.

Butler, S.J., Mattison, E.H.A., Glithero, N.J. et al. (2010). Resource availability and the persistence of seed-eating bird populations in agricultural landscapes: a mechanistic modelling approach. *Journal of Applied Ecology* 47: 67–75. https://doi.org/10.1111/j.1365-2664.2009.01750.x.

Calderini, D.F., Castillo, F.M., Arenas-M, A. et al. (2020). Overcoming the trade-off between grain weight and number in wheat by the ectopic expression of expansin in developing seeds leads to increased yield potential. *New Phytologist* 230: 629–640. https://doi.org/10.1111/nph.17048.

Call, L., Kapeller, M., Grausgruber, H. et al. (2020). Effects of species and breeding on wheat protein composition. *Journal of Cereal Science* 93: 102974. https://doi.org/10.1016/j.jcs.2020.102974.

Camerlengo, F., Frittelli, A., Sparks, C. et al. (2020). CRISPR-Cas9 multiplex editing of the α-amylase/trypsin inhibitor genes to reduce allergen proteins in durum wheat. *Frontiers in Sustainable Food Systems* 4: 104. https://doi.org/10.3389/fsufs.2020.00104.

Cheng, M., Fry, J.E., Pang, S. et al. (1997). Genetic transformation of wheat mediated by *Agrobacterium tumefaciens*. *Plant Physiology* 115: 971–980. https://doi.org/10.1104/pp.115.3.971.

Clarke, C.K., Gregory, P.J., Lukac, M. et al. (2017). Quantifying rooting at depth in a wheat doubled haploid population with introgression from wild emmer. *Annals of Botany* 120 (3): 457–470. https://doi.org/10.1093/aob/mcx068.

Colomba, M.S. and Gregorini, A. (2011). Genetic diversity analysis of the durum wheat Graziella Ra, *Triticum turgidum* L. subsp. *durum* (Desf.) Husn. (Poales, Poaceae). *Biodiversity Journal* 2: 73–84.

Connorton, J.M., Jones, E.R., Rodriguez-Ramiro, I. et al. (2017). Wheat vacuolar iron transporter TaVIT2 transports Fe and Mn and is effective for biofortification. *Plant Physiology* 174: 2434–2444. https://doi.org/10.1104/pp.17.00672.

Cox, T.S., Glover, J.D., Van Tassel, C.M. et al. (2006). Prospects for developing perennial grain crops. *BioScience* 56: 649–659. https://doi.org/10.1641/0006-3568(2006)56[649:PFDPGC]2.0.CO;2.

Cui, L., Ren, Y., Murray, T.D. et al. (2018). Development of perennial wheat through hybridization between wheat and wheatgrasses: a review. *Engineering* 4: 507–513. https://doi.org/10.1016/j.eng.2018.07.003.

Curwen-McAdams, C. and Jones, S.S. (2017). Breeding perennial grain crops based on wheat. *Crop Science* 57: 1172–1188. https://doi.org/10.2135/cropsci2016.10.0869.

De Santis, M.A., Kosik, O., Passmore, D. et al. (2018). Comparison of dietary fibre composition of old and modern wheat (*Triticum turgidum* ssp. *durum*) genotypes. *Food Chemistry* 244: 304–310. https://doi.org/10.1016/j.foodchem.2017.09.143.

Dinu, M., Whittaker, A., Pagliali, G. et al. (2018). Ancient wheat species and human health: biochemical and clinical implications. *Journal of Nutritional Biochemistry* 52: 1–9. https://doi.org/10.1016/j.jnutbio.2017.09.001.

Dzyubenko, N.I. (2018). Vavilov's collection of worldwide crop genetic resources in the 21st century. *Biopreservation and Biobanking* 15: 377–383. https://doi.org/10.1089/2Fbio.2018.0045.

Eagling, T., Neal, A.L., McGrath, S.P. et al. (2014). Distribution and speciation of iron and zinc in grain of two wheat genotypes. *Journal of Agricultural and Food Chemistry* 62: 708–716. https://doi.org/10.1021/jf403331p.

Escarnot, E., Gofflot, S., Sinnaeve, G. et al. (2018). Reactivity of gluten proteins from spelt and bread wheat accessions towards A1 and G12 antibodies in the framework of celiac disease. *Food Chemistry* 268: 522–532. https://doi.org/10.1016/j.foodchem.2018.06.094.

Fan, M.-S., Zhao, F.-J., Fairweather-Tait, S. et al. (2008). Evidence of decreasing mineral density in wheat grain over the last 160 years. *Journal of Trace Elements in Medicine and Biology* 22: 315–324. https://doi.org/10.1016/j.jtemb.2008.07.002.

Finger, R., El Benni, N., Kaphengst, T. et al. (2011). A meta analysis on farm-level costs and benefits of GM crops. *Sustainability* 3: 743–762. https://doi.org/10.3390/su3050743.

Gao, C. (2015). Genome editing in crops: from bench to field. *National Science Review* 2: 13–15. https://doi.org/10.1093/nsr/nwu054.

Geisslitz, S., Ludwig, C., Scherf, K.A. et al. (2018). Targeted LC–MS/MS reveals similar contents of α-amylase/trypsin-inhibitors as putative triggers of nonceliac gluten sensitivity in all wheat species except einkorn. *Journal of Agricultural and Food Chemistry* 66: 12395–12403. https://doi.org/10.1021/acs.jafc.8b04411.

Gil-Humanes, J., Piston, F., Tollefsen, S. et al. (2010). Effective shutdown in the expression of celiac disease-related wheat gliadin T-cell epitopes by RNA interference. *Proceedings of the National Academy of Sciences, USA* 107: 17023–17028. https://doi.org/10.1073/pnas.1007773107.

Glover, J.D., Reganold, J.P., Bell, L.W. et al. (2010). Increased food and ecosystem security via perennial grains. *Science* 328: 1638–1639. https://doi.org/10.1126/science.1188761.

Gonzalez, F.B., Capella, M., Ribichich, K.F. et al. (2019). Field-grown transgenic wheat expressing the sunflower gene *HaHB4* significantly outyields the wild type. *Journal of Experimental Botany* 70: 1669–1681. https://doi.org/10.1093/jxb/erz037.

Good, A.G., Johnson, S.J., De Pauw, M. et al. (2007). Engineering nitrogen use efficiency with alanine aminotransferase. *Canadian Journal of Botany* 85: 252–262. https://doi.org/10.1139/B07-019.

Graybosch, R.A. (1998). Waxy wheats: origin, properties, and prospects. *Trends in Food Science and Technology* 9: 135–142. https://doi.org/10.1016/S0924-2244(98)00034-X.

Graybosch, R.A. (2001). Uneasy unions: quality effects of rye chromatin transfers to wheat. *Journal of Cereal Science* 33: 3–16. https://doi.org/10.1006/jcrs.2000.0336.

Gregorini, A., Colomba, M., Ellis, H.J. et al. (2009). Immunogenicity characterization of two ancient wheat α-gliadin peptides related to coeliac disease. *Nutrients* 1 (2): 276–290. https://doi.org/10.3390/nu1020276.

Griffiths, S., Wingen, L., Pietragalla, J. et al. (2015). Genetic dissection of grain size and grain number trade-offs in CIMMYT wheat germplasm. *PLoS One* 10: e0118847. https://doi.org/10.1371/journal.pone.0118847.

Gupta, K.M., Balyan, H.S., Gahlaut, V. et al. (2019). Hybrid wheat: past, present and future. *Theoretical and Applied Genetics* 132: 2463–2483. https://doi.org/10.1007/s00122-019-03397-y.

Hao, M., Zhang, L., Ning, S. et al. (2020). The resurgence of introgression breeding, as exemplified in wheat improvement. *Frontiers in Plant Science* 11: 252. https://doi.org/10.3389/fpls.2020.00252.

Hazard, B., Zhang, X., Colasuonno, P. et al. (2012). Induced mutations in the starch branching enzyme II (*SBEII*) genes increase amylose and resistant starch content in durum wheat. *Crop Science* 52: 1754–1766. https://doi.org/10.2135/cropsci2012.02.0126.

Hazard, B., Trafford, K., Lovegrove, A. et al. (2020). Strategies to improve wheat for human health. *Nature Food* 1: 475–480. https://doi.org/10.1038/s43016-020-0134-6.

Huang, E., Verbyla, K.L., Verbyla, A.P. et al. (2015). MAGIC populations in crops: current status and future prospects. *Theoretical and Applied Genetics* 128: 999–1017. https://doi.org/10.1007/s00122-015-2506-0.

Hucl, P., Briggs, C., Graf, R.J. et al. (2015). Genetic gains in agronomic and selected end-use quality traits over a century of plant breeding of Canada western red spring wheat. *Cereal Chemistry* 92: 537–543. https://doi.org/10.1094/CCHEM-02-15-0029-R.

Ishida, Y., Hiei, Y., and Komari, T. (2015). High efficiency wheat transformation mediated by agrobacterium tumefaciens. In: *Advances in Wheat Genetics: From Genome to Field. Proceedings of the 12th International Wheat Genetics Symposium* (eds. Y. Ogihara, S. Takumi and H. Handa), 167–173. Springer.

IWGSC (2018). Shifting the limits in wheat research and breeding using a fully annotated reference genome. *Science* 361 (6403): eaar7191. https://doi.org/10.1126/science.aar7191.

Jannink, J.-L., Lorenz, A.J., and Iwata, H. (2010). Genomic selection in plant breeding: from theory to practice. *Briefings in Functional Genomics* 9: 166–177. https://doi.org/10.1093/bfgp/elq001.

Johnson, A.A.T., Kyriacou, B., Callahan, D.L. et al. (2011). Constitutive overexpression of the *OsNAS* gene family reveals single-gene strategies for effective iron- and zinc-biofortification of rice endosperm. *PLoS One* 6: e24476. https://doi.org/10.1371/journal.pone.0024476.

Kalunke, R.M., Tundo, S., Sestili, F. et al. (2020). Reduction of allergenic potential in bread wheat RNAi transgenic lines silenced for CM3, CM16 and 0.28 ATI genes. *International Journal of Molecular Sciences* 21: 5817. https://doi.org/10.3390/ijms21165817.

Kasarda, D.D. (2013). Can an increase in celiac disease be attributed to an increase in the gluten content of wheat as a consequence of wheat breeding? *Journal Agricultural Food Chemistry* 61: 1155–1159. https://doi.org/10.1021/jf305122s.

Khan, S., Anwar, S., Yu, S. et al. (2019). Development of drought-tolerant transgenic wheat: achievements and limitations. *International Journal of Molecular Sciences* 20: 3350. https://doi.org/10.3390/ijms20133350.

Kidane, Y.G., Gesesse, C.A., Hailemariam, B.N. et al. (2018). A large nested association mapping population for breeding and quantitative trait locus mapping in Ethiopian durum wheat. *Plant Biotechnology Journal* 17: 1380–1393. https://doi.org/10.1111/pbi.13062.

Kindred, D.R. and Gooding, M.J. (2005). Heterosis for yield and its physiological determinants in wheat. *Euphytica* 142 (1): 149–159. https://doi.org/10.1007/s10681-005-1250-y.

Kishii, M. (2019). An update on recent uses of *Aegilops* species in wheat breeding. *Frontiers in Plant Science 10*: 585. https://doi.org/10.3389/fpls.2019.00585.

Klumper, W. and Qaim, M. (2014). A meta-analysis of the impacts of genetically modified crops. *PLoS ONE* (11): e111629. https://doi.org/10.1371/2Fjournal.pone.0111629.

Korte, A. and Farlow, A. (2013). The advantages and limitations of trait analysis with GWAS: a review. *Plant Methods* 9: 29. https://doi.org/10.1186/1746-4811-9-29.

Krasileva, K.V., Vasquez-Gross, H.A., Howell, T. et al. (2017). Uncovering hidden variation in polyploid wheat. *Proceedings of the National Academy of Science, USA* 114 (6): E913–E921. https://doi.org/10.1073/pnas.1619268114.

Ladejobi, O., Elderfield, J., Gardner, K.A. et al. (2016). Maximising the potential of multi-parental populations. *Applied and Translational Genomics* 11: 9–17. https://doi.org/10.1016/2Fj.atg.2016.10.002.

Laidig, F., Piepo, H.-P., Rentel, D. et al. (2017). Breeding progress, environmental variation and correlation of winter wheat yield and quality traits in German official variety trials and on-farm during 1983-2014. *Theoretical and Applied Genetics* 130: 223–245. https://doi.org/10.1007/s00122-016-2810-3.

Lamacchia, C., Shewry, P.R., Di Fonzo, N. et al. (2001). Endosperm-specific activity of a storage protein gene promoter in transgenic wheat seed. *Journal of Experimental Botany* 52: 243–250. https://doi.org/10.1093/jexbot/52.355.243.

Lee, S., Jeon, U.S., Kim, Y.-K. et al. (2009). Iron fortification of rice seeds through activation of the *nicotianamine synthase* gene. *Proceedings of the National Academy of Sciences, USA* 106: 22014–22019. https://doi.org/10.1073/pnas.0910950106.

Li, A., Liu, D., Yang, W. et al. (2018). Synthetic hexaploid wheat: yesterday, today and tomorrow. *Engineering* 4: 552–558. https://doi.org/10.1016/j.eng.2018.07.001.

Li, J., Li, Y., and Ma, L. (2021). Recent advances in CRISPR/Cas9 and applications for wheat functional genomics and breeding. *aBIOTECH* https://doi.org/10.1007/s42994-021-00042-5.

Lovegrove, A., Pellny, T.K., Hassall, K. et al. (2020). Historical changes in the contents and compositions of fibre components and polar metabolites in white wheat flour. *Scientific Reports* 10: 5920. https://doi.org/10.1038/s41598-020-62777-3.

Ma, F., Li, M., Yu, L. et al. (2013). Transformation of common wheat (*Triticum aestivum* L.) with *avenin-like b* gene improves flour mixing properties. *Molecular Breeding* 32: 853–865. https://doi.org/10.1007/s11032-013-9913-1.

Mackay, I., Horwell, J., Garner, J. et al. (2011). Reanalyses of the historical series of UK variety trials to quantify the contributions of genetic and environmental factors to trends and variability in yield over time. *Theoretical and Applied Genetics* 122: 225–238. https://doi.org/10.1007/s00122-010-1438-y.

Malik, K.A. and Maqbool, A. (2020). Transgenic crops for biofortification. *Frontiers in Sustainable Food Systems* 4: 571403. https://doi.org/10.3389/fsufs.2020.571402.

Martignago, D., Rico-Medina, A., Blasco-Escamez, D. et al. (2020). Drought resistance by engineering plant tissue-specific responses. *Frontiers in Plant Science 10*: 1676. https://doi.org/10.3389/fpls.2019.01676.

Martin, A., Alvarez, A., Martin, L.M. et al. (1999). The development of Tritordeum: a novel cereal for food processing. *Journal of Cereal Science* 30: 85–95. https://doi.org/10.1006/jcrs.1998.0235.

Mather, K. and Jinks, J.L. (1982). *Biometrical Genetics. The Study of Continuous Variation*, 3e. Chapman and Hall.

McAllister, C.H., Beatty, P.H., and Good, A.G. (2012). Engineering nitrogen use efficient crop plants: the current status. *Plant Biotechnology Journal* 10: 1011–1025. https://doi.org/10.1111/j.1467-7652.2012.00700.x.

McCallum, C.M., Comai, L., Greene, E.A. et al. (2000). Targeting induced local lesions in genomes (TILLING) for plant functional genomics. *Plant Physiology* 123: 439–442. https://doi.org/10.1104/pp.123.2.439.

Menguer, P.K., Vincent, T., Miller, A.J. et al. (2018). Improving zinc accumulation in cereal endosperm using HvMTP1, a transition metal transporter. *Plant Biotechnology Journal* 16: 63–71. https://doi.org/10.1111/pbi.12749.

Michalcová, V., Dušinský, R., Sabo, M. et al. (2014). Taxonomical classification and origin of Kamut® wheat. *Plant Systematics and Evolution* 300 (7): 1749–1757. https://doi.org/10.1007/s00606-014-1001-4.

Morris, C.E. and Sands, D.C. (2006). The breeder's dilemma – yield or nutrition? *Nature Biotechnology* 24: 1078–1080. https://doi.org/10.1038/nbt0906-1078.

Niu, Z., Jiang, A., Haddad, W.U. et al. (2014). Review of doubled haploid production in durum and common wheat x maize hybridization. *Plant Breeding* 133: 313–320. https://doi.org/10.1111/pbr.12162.

Oladosu, Y., Rafii, M.Y., Abdullah, N. et al. (2016). Principle and application of plant mutagenesis in crop improvement: a review. *Biotechnology and Biotechnological Equipment* 30: 1–16. https://doi.org/10.1080/13102818.2015.1087333.

Ort, D.R., Merchant, S.S., Alric, J. et al. (2015). Redesigning photosynthesis to sustainably meet global food and bioenergy demand. *Proceedings of the National Academy of Sciences, USA* 112: 8529–8536. https://doi.org/10.1073/pnas.1424031112.

Pellegrino, E., Bedini, S., Nuti, M. et al. (2018). Impact of genetically engineered maize on agronomic, environmental and toxicological traits: a meta-analysis of 21 years of field data. *Scientific Reports* 8: 3113. https://doi.org/10.1038/s41598-018-21284-2.

Pellny, T.K., Patil, A., Wood, A.J. et al. (2020). Loss of TaIRX9b gene function in wheat decreases chain length and amount of arabinoxylan in grain but increases cross-linking. *Plant Biotechnology Journal* 18: 2316–2327. https://doi.org/10.1111/pbi.13393.

Pont, C., Leroy, T., Seidel, M. et al. (2019). Tracing the ancestry of modern bread wheats. *Nature Genetics* 51: 905–911. https://doi.org/10.1038/s41588-019-0393-z.

Popineau, Y., Deshayes, G., Lefebvre, J. et al. (2001). Prolamin aggregation, gluten viscoelasticity, and mixing properties of transgenic wheat lines expressing 1Ax and 1Dx high molecular weight glutenin subunit transgenes. *Journal of Agricultural and Food Chemistry* 49: 395–401. https://doi.org/10.1021/jf001015j.

Prandi, B., Tedeschi, T., Folloni, S. et al. (2017). Peptides from gluten digestion: a comparison between old and modern wheat cultivars. *Food Research International* 91: 92–102. https://doi.org/10.1016/j.foodres.2016.11.034.

Pronin, D., Borner, A., Weber, H. et al. (2020). Wheat (*Triticum aestivum* L.) breeding from 1891 to 2010 contributed to increasing yield and glutenin contents but decreasing gliadin contents. *Journal of Agricultural and Food Chemistry* 68: 13247–13256. https://doi.org/10.1021/acs.jafc.0c02815.

Pronin, D., Borner, A., and Scherf, K.A. (2021). Old and modern wheat (*Triticum aestivum* L.) cultivars and their potential to elicit celiac disease. *Food Chemistry* 339: 127952. https://doi.org/10.1016/j.foodchem.2020.127952.

Przewieslik-Allen, A.M., Wilkinson, P., Burridge, A. et al. (2021). The role of gene flow and chromosomal instability in shaping the bread wheat genome. *Nature Plants* 7: 172–183. https://doi.org/10.1038/s41477-020-00845-2.

Raffan, S. and Halford, N.G. (2019). Acrylamide in food: progress in and prospects for genetic and agronomic solutions. *Annals of Applied Biology* 173: 259–281. https://doi.org/10.1111/aab.12536.

Regina, A., Bird, A., Topping, D. et al. (2006). High-amylose wheat generated by RNA interference improves indices of large-bowel health in rats. *Proceedings of the National Academy of Sciences, USA* 103: 3546–3551. https://doi.org/10.1073/pnas.0510737103.

Ribeiro, M., Martinez-Quijano, M., Nunes, F.M. et al. (2016). New insights into wheat toxicity: breeding did not seem to contribute to a prevalence of potential coeliac disease's immunostimulatory epitopes. *Food Chemistry* 213: 8–18. https://doi.org/10.1016/j.foodchem.2016.06.043.

Rogniaux, H., Pavlovic, M., Lupi, R. et al. (2015). Allergen relative abundance in several wheat cultivars as revealed via a targeted quantitative approach using MS. *Proteomics* 15: 1736–1745. https://doi:10.1002/pmic.201400416.

Sanchez-Leon, S., Gil-Humanes, J., Ozuna, C.V. et al. (2018). Low-gluten, nontransgenic wheat engineered with CRISPR/Cas9. *Plant Biotechnology Journal* 16: 902–910. https://doi.org/10.1111/pbi.12837.

Sears, E.R. (1954). *The Aneuploids of Common Wheat. Research Bulletin 572.* Missouri Agricultural Experiment Station and U.S. Department of Agriculture. https://mospace.umsystem.edu/xmlui/bitstream/handle/10355/57878/AESResearchBulletin.pdf?sequence=1

Sears, E.R. (1975). The wheats and their relatives. In: *Handbook of Genetics*, vol. 2 (ed. R.C. King), 59–91. Plenum Press.

Selitrennikoff, C.P. (2001). Antifungal proteins. *Applied and Environmental Microbiology* 67: 2883–2894. https://doi.org/10.1128/AEM.67.7.2883-2894.2001.

Shaw, L.M., Turner, A.S., Herry, L. et al. (2013). Mutant alleles of *Photoperiod-1* in wheat (*Triticum aestivum* L.) that confer a late flowering phenotype in long days. *PLoS One* 8 (11): e79459. https://doi.org/10.1371/journal.pone.0079459.

Shewry, P.R. (2018). Do ancient types of wheat have health benefits compared with modern bread wheat. *Journal of Cereal Science* 79: 469–476. https://doi.org/10.1016/j.jcs.2017.11.010.

Shewry, P.R. and Hey, S. (2015). Do "ancient" wheat species differ from modern bread wheat in their contents of bioactive components? *Journal of Cereal Science* 65: 236–243. https://doi.org/10.1016/j.jcs.2015.07.014.

Shewry, P.R. and Jones, H.D. (2020). Improving wheat quality for breadmaking. In: *Breadmaking: Improving Quality* (ed. S. Cauvain), 237–258. Woodhead Publishing Ltd.

Shewry, P.R. and Lucas, J.A. (1997). Plant proteins that confer resistance to pests and pathogens. In: *Advances in Botanical Research*, vol. 26 (ed. J. Callow), 135–192. Academic Press https://doi.org/10.1016/S0065-2296(08)60120-2.

Shewry, P.R., Baudo, M., Lovegrove, A. et al. (2007). Are GM and conventionally bred cereals really different? *Trends in Food Science and Technology* 18: 201–209. https://doi.org/10.1016/2Fj.tifs.2006.12.010.

Shewry, P.R., Gebruers, K., Andersson, A.A.M. et al. (2011). Relationship between the contents of bioactive components in grain and the release dates of wheat lines in the HEALTHGRAIN diversity screen. *Journal of Agricultural and Food Chemistry* 59: 928–933. https://doi.org/10.1021/2Fjf103860x.

Shewry, P.R., Pellny, T.K., and Lovegrove, A. (2016). Is modern wheat bad for health? *Nature Plants* 2: 1–3. https://doi.org/10.1038/nplants.2016.97.

Shewry, P.R., Corol, D., Jones, H.D. et al. (2017). Defining genetic and chemical diversity in wheat grain by ^{1}H-NMR spectroscopy of polar metabolites. *Molecular Nutrition & Food Research* 61: 1600807. https://doi.org/10.1002/mnfr.201600807.

Simmonds, J.R., Fish, L.J., Leverington-Waite, M.A. et al. (2008). Mapping of a gene (Vir) for a non-glaucous, viridescent phenotype in bread wheat derived from Triticum dicoccoides, and its association with yield variation. *Euphytica* 159 (3): 333–341. https://doi.org/10.1007/s10681-007-9514-3.

Singh, S.P., Keller, B., Gruissem, W. et al. (2017). Rice *NICOTIANAMINE SYNTHASE 2* expression improves dietary iron and zinc levels in wheat. *Theoretical Applied Genetics* 130: 283–292. https://doi.org/10.1007/s00122-016-2808-x.

Slade, A.J., Fuerstenberg, S.I., Loeffler, D. et al. (2005). A reverse genetic, nontransgenic approach to wheat crop improvement by TILLING. *Nature Biotechnology* 23: 75–81. https://doi.org/10.1038/nbt1043.

Smith, G.P. and Gooding, M.J. (1999). Models of wheat grain quality considering climate, cultivar and nitrogen effects. *Agricultural and Forest Meteorology* 94 (3-4): 159–170. https://doi.org/10.1016/S0168-1923(99)00020-9.

Smyth, S.J. (2020). The human health benefits from GM crops. *Plant Biotechnology Journal* 18: 887–888. https://doi.org/10.1111/pbi.13261.

Sollid, L.M., Tye-Din, J.A., Qiao, S.-W. et al. (2020). Update 2020: nomenclature and listing of celiac-disease relevant gluten epitopes recognised by $CD4^+$ T cells. *Immunogenetics* 72: 85–88. https://doi.org/10.1007/s00251-019-01141-w.

Spisni, E., Imbesi, V., Giovanardi, E. et al. (2018). Different physiological responses elicited by ancient and heritage wheat cultivars compared to modern one. *Nutrients* 11: 2879. https://doi.org/10.3390/nu11122879.

Tabbita, F., Pierce, S., and Barneix, A.J. (2017). Breeding for increased grain protein and micronutrient content in wheat: ten years of the *GPC-B1* gene. *Journal of Cereal Science* 73: 183–191. https://doi.org/10.1016/j.jcs.2017.01.003.

Thomas, H., Thomas, H.M., and Ougham, H. (2000). Annuality, perenniality and cell death. *Journal of Experimental Botany* 51: 1781–1788. https://doi.org/10.1093/jexbot/51.352.1781.

Tiong, J., Sharma, N., Sampath, R. et al. (2021). Improving nitrogen use efficiency through overexpression of alanine aminotransferase in rice, wheat and barley. *Frontiers in Plant Science 12*: 628521. https://doi.org/10.3389/fpls.2021.628521.

Tosi, P., Masci, S., Giovangrossi, A. et al. (2005). Modification of the low molecular weight (LMW) glutenin composition of transgenic durum wheat: effects on glutenin polymer size and gluten functionality. *Molecular Breeding* 16: 113–126. https://doi.org/10.1007/s11032-005-5912-1.

Tucker, E.J., Baumann, U., Kouidri, A. et al. (2017). Molecular identification of the wheat male fertility gene *Ms1* and its prospects for hybrid breeding. *Nature Communications* 8: 869. https://doi.org/10.1038/s41467-017-00945-2.

Uauy, C., Distenfeld, A., Fahima, T. et al. (2006). A NAC gene regulating senescence improves grain protein, zinc, and iron content in wheat. *Science* 314: 1298–1301. https://doi.org/10.1126/science.1133649.

Vaquero, L., Comino, I., Vivas, S. et al. (2018). Tritordeum: a novel cereal for food processing with good acceptability and significant reduction in gluten immunogenic peptides in comparison with wheat. *Journal of the Science of Food and Agriculture* 98: 2201–2209. https://doi.org/10.1002/jsfa.8705.

Vasil, V., Castillo, A.M., Fromm, M.E. et al. (1992). Herbicide resistant fertile transgenic wheat plants obtained by microprojectile bombardment of regenerable embryogenic callus. *Bio-Technology* 10: 667–674. https://doi.org/10.1038/nbt0692-667.

Walkowiak, S., Gao, L., Monat, C. et al. (2020). Multiple wheat genomes reveal global variation in modern breeding. *Nature* 558: 277–283. https://doi.org/10.1038/s41586-020-2961-x.

Wiley, P., Tosi, P., Evrard, A. et al. (2007). Promoter analysis and immunolocalization show that puroindoline genes are exclusively expressed in starchy endosperm cells of wheat grain. *Plant Molecular Biology* 64: 125–136. https://doi.org/10.1007/s11103-007-9139-x.

Winfield, M., Allen, A., Burridge, A. et al. (2016). High density SNP genotyping array for hexaploidy wheat and its secondary and tertiary gene pool. *Plant Biotechnology Journal* 14: 1195–1206. https://doi.org/10.1111/2Fpbi.12485.

Wingen, L.U., Orford, S., Goram, R. et al. (2014). Establishing the A.E. Watkins landrace cultivar collection as a resource for systematic gene discovery in bread wheat. *Theoretical and Applied Genetics* 127: 1831–1842. https://doi.org/10.1007/s00122-014-2344-5.

Wingen, L.U., West, C., Leverington-Waite, M. et al. (2017). Wheat landrace genome diversity. *Genetic* 205: 1657–1676. https://doi.org/10.1534/genetics.116.194688.

von Wiren, N., Klair, S., Bansal, S. et al. (1999). Nicotianamine chelates both Fe^{III} and Fe^{II}. Implications for metal transport in plants. *Plant Physiology* 119: 1107–1114. https://doi.org/10.1104/pp.119.3.1107.

Wrigley, C.W. and Shepherd, K.W. (1973). Electrofocusing of grain proteins from wheat genotypes. *Annals of the New York Academy of Sciences* 209: 154–162. https://doi.org/10.1111/j.1749-6632.1973.tb47526.x.

Zhang, Y., Liang, Z., Zong, Y. et al. (2016). Efficient and transgene-free genome editing in wheat through transient expression of CRISP/Cas9 DNA or RNA. *Nature Communications* 7: 12617. https://doi.org/10.1038/ncomms12617.

Zheng, L., Cheng, Z., Ai, C. et al. (2010). Nicotianamine, a novel enhancer of rice iron bioavailability to humans. *PLoS One* 5: e10190. https://doi.org/10.1371/journal.pone.0010190.

Zheng, W., Ma, Z., Zhao, M. et al. (2020). Research and development strategies for hybrid *japonica* rice. *Rice* 13: 36. https://doi.org/10.1186/s12284-020-00398-0.

11

Epilogue: Wheat in Conflict and in Peace

At the beginning of this book, we briefly mentioned some of the conflicts to which wheat has contributed during its rise: the battle between the grasses and the trees, the change from mobility to sedentism, and the rise of states and state control. There are clearly many other frictions that were not mentioned but have become apparent in subsequent chapters.

We have spent a significant amount of time examining the environmental conflict. By definition, the development of agriculture has required us to change the environment to support new agroecosystems to provide us with food, fuel, and fibre. Furthermore, these changes have been rapid compared to the processes that have delivered the resources and environment that our agroecosystems rely on and we cannot assume that they will result in a stable outcome. In fact, experience from the past and present offers ample evidence of the unsustainability of many current practices and it is clear that these will continue to change, presenting new challenges.

Nonetheless, we have emphasised the need to support intensive production (i.e. land-use efficiency) in areas that are suited to such an approach in order to meet global demands for food. In such areas, technology can be used to improve the precision of farming, to ensure that resources are delivered in the right form, at the right time, in the right quantity, and in the right place. Similarly, it is crucial that modern crop improvement technologies are not unduly constrained as plant breeders attempt to meet the challenges of ever-evolving biotic and abiotic stresses, increasing yield potential, using resources more efficiently, and improving crop quality. These technologies include gene editing, which is becoming widely used as a research tool in wheat.

Significant advances can also be achieved in the recycling of nutrients and reducing the reliance on fossil fuels. Although there is a role for less-intensive systems, their widespread adoption would mean that a larger area would need to be farmed to realise future requirements for food, placing more pressure on fragile systems with associated damage to the environment. However, increasing the production on some areas should facilitate withdrawal from agriculture in others (for, e.g. rewilding).

The question of demand brings us to a further conflict: that of too little versus too much. Demand for grain has increased with population growth, economic prosperity, consumption of products from animals fed with grain, and in the production of biofuels. In Chapter 1,

Wheat: Environment, Food and Health, First Edition. Mike J. Gooding and Peter R. Shewry.

we reflected that increases in wheat yields since 1960 have exceeded population growth. So far, wheat production has been able to keep pace with, or even exceed, demand (as shown by the very low wheat prices of the 2000s). This remarkable achievement is, in part, testament to scientific advances and the efforts of plant breeders, equipment manufacturers, agronomists and farmers.

However, the question arises as to whether we are demanding too much. An analysis of the politics and socioeconomics of food supply and demand is beyond the scope of this book, but some observations can be made. In the early 1960s the global per capita production of wheat for all uses (e.g. including for animal feed) was around 75 kg/annum. By 2020 this had reached 100 kg/annum. Put another way, if the per capita consumption had remained at the level in the 1960s we would already be producing enough wheat for over 10 billion people. Similar conclusions can be reached for total production of calories from all cereal crops. This increased production has not, however, solved world hunger. In fact, at time of writing the Food and Agriculture Organization of the United Nations estimate that 700 million people (9% of the global population) are undernourished. However, the same number are obese, and around two billion are overweight. This is because many of us have access to more food than we require, leading to overconsumption and waste. Furthermore, this food is often available at a cost below the true cost when the more diffuse and longer-term environmental and health costs are included. It is clear that changing diets can have a significant effect on demand and thus also on the environmental cost and sustainability of agricultural practices.

The issue of diet raises the potential conflict between wheat and health. Although we have been aware of adverse effects of wheat consumption on health for many years, the prevalence was low, at similar levels to those of other food-related disorders, and it had little impact on the ability of wheat to satisfy the nutritional requirements of a substantial proportion of the global population. The current concern about increases in adverse impacts of wheat on heath, which is already affecting wheat consumption in some countries, therefore needs to be taken seriously. Furthermore, the fact that much of the concern has been propagated via social media and popular publications rather than conventional scientific channels means that we also need to adopt new strategies for communication. This is currently the most active topic of research in wheat, and it is therefore likely that we will develop a more complete understanding of the various interactions of wheat with health in the next few years. The challenge is then to communicate this to a public who are increasingly sceptical of conventional science and mistrusting of health professionals.

Finally, we should reflect that we have been living in times of unprecedented change for food production. Figure 11.1 increases the timescale for population growth to the last 10 000 years, i.e. broadly the agricultural era, and replots the wheat production data of Figure 1.4 on the same scale. This again demonstrates that the rapid and large increases in wheat production and yield have more than kept pace with the increasing number of humans on the planet. In effect, wheat production has been highly responsive and flexible. It is therefore not surprising that practices that have been developed to achieve such rapid change are not sustainable, particularly over the timescale of a century or more. The need to increase yield further, with greater levels of sustainability, with greater regional and

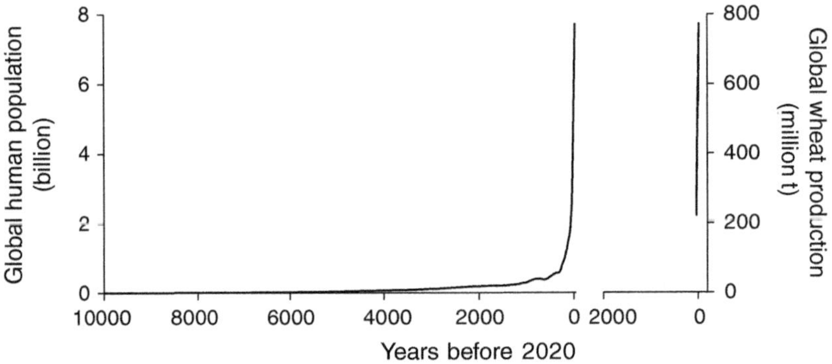

Figure 11.1 Global human population and wheat production.

social equality, and in the face of climate change therefore remains one of humanity's greatest challenges (Evans 1998). However, the future may not be as gloomy as the current situation suggests as the rate of population growth is already declining, and the global population may well reach a plateau of around 11 billion by 2100. A more stable demand for food may therefore allow new and more considered approaches as to how future generations are fed.

Reference

Evans, L.T. (1998). *Feeding the Ten Billion*. Cambridge University Press.

Index

Page numbers referring to figures are in italics and those referring to tables are in bold.

Wheat: Environment, Food and Health, First Edition. Mike J. Gooding and Peter R. Shewry.

t